Processes in Continental Lithospheric Deformation

Edited by

Sydney P. Clark, Jr.
Department of Geology and Geophysics
Yale University
P.O. Box 6666
New Haven, Connecticut 06511

B. Clark Burchfiel
Department of Earth, Atmospheric, and Planetary Sciences
Massachusetts Institute of Technology
Cambridge, Massachusetts 02139

and

John Suppe
Department of Geological and Geophysical Sciences
Princeton University
Princeton, New Jersey 08544

THE GEOLOGICAL SOCIETY OF AMERICA · 1888 ·

SPECIAL PAPER

218

Published by The Geological Society of America, Inc.
3300 Penrose Place, P.O. Box 9140, Boulder, Colorado 80301

GSA Books Science Editor Campbell Craddock

Printed in U.S.A.

Library of Congress Cataloging-in-Publication Data

Processes in continental lithospheric deformation / Sydney P. Clark, Jr., B. Clark Burchfiel, and John Suppe, editors.
p. cm. -- (Special paper / Geological Society of America ; 218)
Papers from a symposium held at Yale University Feb. 21-22, 1985.
Includes bibliographies.
ISBN 0-8137-2218-7
1. Orogeny--Congresses. 2. Earth--Crust--Congresses. I. Clark, Sydney P. (Sydney Procter), 1929- . II. Burchfiel, B. C., 1934- III. Suppe, John. IV. Series: Special papers (Geological Society of America) ; 218.
QE621.P73 1988
551.8--dc19 87-37344
CIP

Cover photo: A portion of John Rodgers' favorite mountain range: the folded Appalachians, mostly in central Pennsylvania. Photo courtesy of Earth Satellite Corp., Bethesda, Maryland, and Nicholas M. Short, NASA.

Contents

Preface

This volume is composed of eight papers presented at a symposium on "Processes of Continental Lithospheric Deformation" held at Yale University on 21–22 February 1985 in honor of John Rodgers on his retirement as Professor. Each paper displays a broad vision of some significant aspect of the Earth's crust, yet the collection is diverse, reflecting the full range of modern approaches to tectonics.

The first five papers are on regional tectonics and cover such significant segments of the Earth's crust as the Mediterranean, the central Andes, or the Canadian Cordillera. The last three papers focus on key processes that shape continental crust, or attempt to interpret mountain belts in terms of simple mechanical models that incorporate the dominant phenomena, such as rock strength, detachment zones, and erosion. Each contribution is introduced below.

Axial projections and models of crustal thickening, eastern Wopmay orogen, northeast Canadian shield by Paul F. Hoffman, Rein Tirrul, Janet E. King, Marc R. St-Onge, and Stephen B. Lucas of the Geological Survey of Canada paints a picture of Proterozoic mountain building in the Canadian Shield that is amazingly like Phanerozoic collisional belts around the world. Furthermore they show that many of the concepts and techniques developed in the analysis of younger belts—such as balanced cross sections and palinspastic reconstructions of overthrust belts—can be fruitfully applied in the Precambrian.

Tectonics of the central Andes, 30° to 33°S latitude by Victor A. Ramos of the Geological Survey of Argentina introduces the entire sweep of central Andean tectonic history from the early Paleozoic accretionary events on the western margin of Gondwanaland, through the principal orogenic events from late Paleozoic to Recent. The Gondwanides, Patagonides, and Andean orogenies are all presented in the context of plate motions.

Mesozoic and early Cenozoic magmatic evolution of the Canadian Cordillera by Richard Lee Armstrong of the University of British Columbia provides a comprehensive summary of magmatic activity in the various terranes of the Canadian Cordillera based on analysis of about 3,000 isotopic ages. A key feature of the record is episodic activity with distinct changes in geographic distribution from episode to episode. Major episodes are shown to correspond to times of increased relative or absolute plate convergence.

A possible Jurassic-Cretaceous transform system in the Alps and the Carpathians by Rudolf Trümpy of the ETH in Zurich applies a comprehensive understanding of regional stratigraphic relations in the Alpine–Carpathian chain, together with concepts of plate motions, to develop a rational model of Mesozoic paleogeography and tectonics. The Penninic ocean basins—with intervening continental slices—are considered to be a complex transform zone between the Piemont–Ligurian ocean basin on the west and the Vardar ocean basin on the east.

Plate kinematics and tectonics leading to the Alpine belt formation; A new analysis by Xavier Le Pichon, Francoise Bergerat, and Marie-José Roulet of Paris presents a kinematic analysis of the closure of the Tethyan belt that allows quantitative estimates of the shortening due to subduction and collision. This analysis implies that about half the total continental crust involved in the collision has been subducted. This paper also presents the evolution of stress trajectories in the west European platform since the Eocene.

Detachment faulting in continental extension; Perspectives from the Southwestern U.S. Cordillera by Gregory A. Davis and G. S. Lister of the University of Southern California and the Australian BMR presents a comprehensive summary of detachment faulting that sorts out and explains many apparently conflicting field observations as natural results of the evolution of crustal extension. Furthermore we are given a good indication of the absolute rates at which these systems evolve.

Mechanics, growth, and erosion of mountain belts by F. A. Dahlen and John Suppe of Princeton University presents a summary of the role of critical wedge taper in the mechanics of fold-and-thrust belts and accretionary wedges. Critical taper is then combined with mass-balance considerations to develop a simple kinematic model of compressive mountain belts. It is concluded that erosion rate exerts a first-order control over the evolution of mountain belts.

Some simple aspects of the support, structure, and evolution of mountain belts by Peter Molnar and Hélène Lyon-Caen of MIT explores the mechanical roles of strength and gravity as resisting forces in the development of mountain belts. In particular, it is concluded that the mean elevations of high plateaus should serve as approximate gauges of the average compressive stress pushing material into the plateaus at their margins.

Sydney P. Clark, Jr.
B. Clark Burchfiel
John Suppe

Dedication

John A. Rodgers

The symposium entitled "Processes in Continental Lithospheric Deformation" was held at Yale University on February 21 and 22, 1985, honoring John Rodgers, Silliman Professor of Geology. The symposium took place during the year in which John retired, and it might be considered that the symposium was held to honor that occasion. However, those who know him cannot really imagine John Rodgers in retirement. Rather it might better be stated that the symposium was to honor a man who had had, up to this point, one of the most extraordinary careers in the earth sciences. His achievements and influence are well known to all earth scientists. Formal retirement for John should only be considered a milestone in a career that we fully expect to continue for many more years. John has recently begun to learn Chinese and has just completed an extensive trip in China. He is still on the road, adding a new language to his repertoire and several more mountain ranges to his collection.

John is well known for his contributions to the understanding of mountain belts. The underlying theme of his work has always been that field relationships must be understood in order to set the boundary conditions for understanding the processes that operate within the Earth. Some of John's graduate students will remember that the text for his course in advanced structural geology in the late 1950s was Jean Goguel's book *Introduction à l'étude mécanique des déformations de l'écorce terrestre.* This book was little known in the U.S. at that time, but was one of the early works stressing the mechanics and processes of deformation in the Earth's crust. John's interest in understanding how the Earth works made "Processes in Continental Lithospheric Deformation" an appropriate theme for the symposium.

Twelve eminent earth scientists were invited to present their views on continental lithospheric deformation. Most of the papers presented at the symposium are included in this volume. The organizers were pleased to see the extraordinary attendance at the symposium; about 400 people filled the large hall to overflowing. They were not disappointed. The quality of the presentations and their scientific content were superb, befitting the man they were honoring. The fact that the majority of the audience consisted of young scientists and students was especially appropriate. One of the hallmarks of John's career has been the intellectual stimulation he has provided to his colleaues and especially to students.

Perhaps the highlight of the two-day symposium was a piano recital by John in the Great Hall of the Peabody Museum. The program contained five sonatas by Franz Joseph Haydn. The piano was set amongst the exhibits and the audience was seated throughout the hall, adjacent to the great Brontosaurus and beneath the world-famous mural by Rudolph Zallinger. The musical score and the backdrop made for a truly memorable evening.

John Rodgers has throughout his career provided intellectual stimulation and friendship to countless colleagues and students throughout the world. This symposium and this volume are a small expression of appreciation for the contribution he has made to all of us.

The National Science Foundation, ARCO, Sohio, Texaco, and Exxon provided financial support for the symposium. The Department of Geology and Geophysics of Yale University and the organizers of the symposium are grateful for their support.

Geological Society of America
Special Paper 218
1988

Axial projections and modes of crustal thickening, eastern Wopmay orogen, northwest Canadian shield

Paul F. Hoffman, Rein Tirrul*, Janet E. King, Marc R. St-Onge, and Stephen B. Lucas
Lithosphere and Canadian Shield Division, Geological Survey of Canada, Ottawa, Ontario K1A 0E4, Canada

ABSTRACT

Eastern Wopmay orogen comprises an external thin-skinned thrust-fold belt and a contiguous metamorphic-plutonic internal zone produced by the closing of a short-lived (1.90–1.88 Ga) ensialic(?) back-arc(?) basin on the west margin of Slave craton, an Archean granite-greenstone terrane in the northwest Canadian shield. Sets of high-amplitude, basement-involved folds subsequently developed parallel (D2) and oblique (D3) to the strike of the thin-skinned (D1) thrust-fold belt. The D3 folds permit the construction, by means of axial projections, of composite cross sections of the D1 and D2 structures through a depth interval of 30 km. The cross sections show that the sole thrust of the externides and the underlying autochthonous cover continue beneath the metamorphic-plutonic internal zone, which was thrust as a hot allochthon onto the relatively cold Archean craton.

In the externides, thrusting produced a minimum of 40 percent east-west shortening and thickening of the allochthonous cover. This is manifested by about 25 mappable imbrications of a lower cratonic shelf sequence (stratigraphic thickness of 1.5 to 2.5 km), and upright chevron-type folds in overlying foredeep flysch and molasse (minimum, 5 km of stratigraphic thickness). Thrust propagation was eastward (toward the craton), and an active accretionary wedge of relatively low taper is indicated by high basal "step-up" angles, common back-thrusts, and lack of overall change in exposed stratigraphic level across the belt. The sole thrust is consistently positioned 100 to 300 m above the unfaulted basement surface, independent of footwall lithology, for at least 100 km west of the frontal thrusts.

In the internides, a basal allochthonous assemblage of rift-facies clastics and bimodal volcanics is characterized by recumbent isoclinal folds, most of which are east-vergent. Allochthonous syn-rift deposits occur in both the eastern and western internides, but the overlying slope-rise facies of the cratonic margin sequence and succeeding synorogenic foredeep deposits are limited to the east. The slope-rise and foredeep strata are deformed by east-vergent thrusts and folds with steeply dipping axial surfaces. The medial and western internides are occupied by a salic-through-mafic plutonic suite (Hepburn Intrusive Suite), emplaced as the basin closed. North-trending folds and thrusts both predate and postdate prograde metamorphic isograds that envelope the composite Hepburn batholith of the medial internides. Eastward translation of the hot batholith on the sole thrust caused a minimum 250°C inversion of the contemporaneous geothermal gradient beneath the allochthon. This is evident from inverted metamorphic isograds that transect the sole thrust and the underlying, upward-facing, autochthonous cover.

Following thin-skinned deformation, the allochthon and autochthon were congruently shortened to form large-scale north-trending folds (D2) having a ca. 35 km wave-

*Deceased, May 1987

length and an apparent structural relief of 13 and 6 km on the basement surface in the internides and externides, respectively. Thrusting of the basement-cover contact is insignificant, and the folds, which are completely exposed in oblique cross section, are clearly not the result of basement involvement in D1 thrusting. However, the existence of a blind thrust at depth is not ruled out.

The cross folds (D3) are similar in style to the D2 folds but trend northeast and have first-order wavelengths of 80 to 140 km and amplitudes at the basement-cover interface of up to 15 km. Locally, D3 produced belts of well-developed higher order folds, typically having cuspate-lobate basement-cover profiles, and inhomogeneous but locally strong northeast-striking foliations. In common with D2 folds, the D3 folds have no significant associated thrusts, and are well developed in areas where the contemporary basement temperature exceeded a threshold of ca. 350°C. Zones of strong D3 deformation occur east of the frontal D1 thrusts, far beyond the limit of tectonically transported heat.

The observed evolution in modes of crustal thickening—early thin-skinned thrusting followed by coaxial, and later transverse, thick-skinned folding—appears to have occurred in several other well-exposed, broadly coeval, orogenic belts in the Canadian shield. Detailed mapping of the steeply plunging segments of these orogens will allow their three-dimensional structural and metamorphic configurations to be elucidated by the use of axial projections, as successfully employed in Wopmay orogen.

INTRODUCTION

It is commonly assumed that mountainous topography, deep drilling, and/or seismic reflection profiling are required to resolve the depth dimension of orogenic belts. Conversely, many orogenic belts have been folded across strike, providing structural relief from which composite cross sections can be constructed by means of axial projection (Stockwell, 1950). For example, cross folds in eastern Wopman orogen (1.9 Ga) expose an external thrust-fold belt and a contiguous metamorphic-plutonic internal zone through a depth range of ca. 30 km. Geologic maps therefore constitute oblique cross sections, linking shallow crustal levels comparable to those of young mountain belts, with deeper crustal levels.

Mapping of a 300-km strike length of the orogen in the last decade is sufficient to validate the critical prerequisite for axial projection—that the structures are approximately cylindrical over the distances of the projections. Superb bedrock exposure allows the structure to be resolved in far greater detail than with seismic or drilling methods, at a fraction of the cost, and the gentle glaciated topography allows every outcrop to be examined and sampled with ease.

The purpose of this chapter is to discuss the modes of crustal thickening documented by integrated structural and metamorphic studies carried out during coordinated systematic mapping of both the externides and internides of the orogen. We document an evolution from thin- to thick-skinned modes of crustal thickening: from early imbrication and folding of sedimentary and volcanic rocks above a flat sole thrust, to late basement-involved folding of regional scale without significant contemporaneous thrusting. The geometry of metamorphic isograds relative to the sole thrust demonstrates that, during the thin-skinned deformation, a hot metamorphic-plutonic allochthon was translated onto a cold cratonic autochthon. Subsequently, high-relief, essentially unbroken basement folds were produced, oriented parallel and transverse to the earlier thin-skinned thrust-fold belt. The importance of basement folding without thrusting contrasts with interpretations of many younger, less deeply eroded orogenic belts, in which basement folds are commonly attributed to the development of thrust duplexes in the basement.

Although representing the results of 75 person-months of field work in eastern Wopmay orogen, this is an interim report, subject to revision as field work continues to the south. Individually, our contributions to this chapter are: RT and JEK, structure of the externides and internides, respectively; MRS and SBL, metamorphism of the internides and externides, respectively; PFH, chief instigator and author.

REGIONAL SETTING AND STRATIGRAPHY

Eastern Wopmay orogen (Figs. 1, 2) lies between basement rocks of Slave craton (pre-2.5 Ga) to the east and postorogenic plutonic, volcanic and derived sedimentary rocks of Great Bear magmatic zone (1.88 to 1.84 Ga) to the west. The orogen incorporates a depositional prism, the Coronation Supergroup, that was deformed and metamorphosed during the Calderian orogeny, during which time it was intruded by the 1.89- to 1.88-Ga Hepburn plutonic suite (Hoffman and Bowring, 1984).

The Coronation Supergroup (Figs. 3, 4) comprises three stratigraphically distinct sequences (Hoffman and Bowring, 1984): (1) a western basal rift-facies assemblage (Akaitcho Group) composed of immature subaqueous clastic sediments and

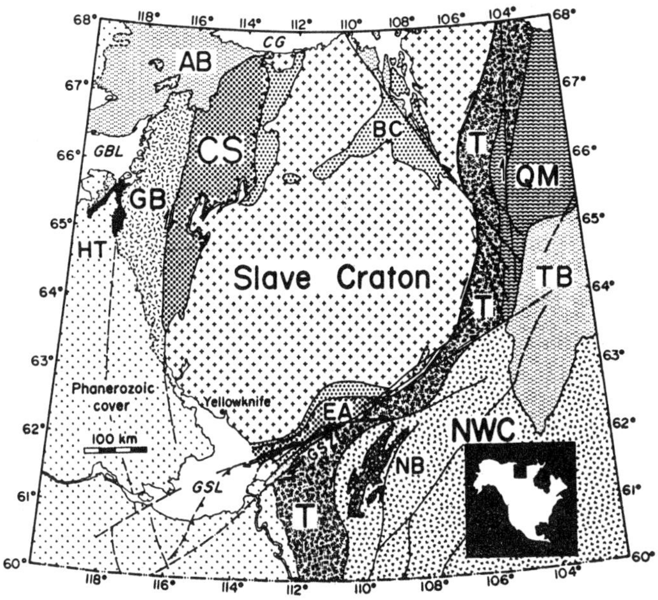

Figure 1. Regional setting of Archean Slave craton and its bounding early Proterozoic orogens. Coronation Supergroup (CS) of eastern Wopmay orogen was deformed during the Calderian orogeny and is flanked by two calc-alkaline magmatic arcs, the pre-Calderian Hottah terrane (HT) and the post-Calderian Great Bear magmatic zone (GB). East margin of Slave craton is bounded by Thelon-Taltson magmatic zone (T), developed on the leading edge of reactivated Archean crust of northwest Churchill province (NWC). East Arm (EA) and Bear Creek (BC) foreland basins and associated thrust-fold belts are related to Slave-Churchill collision, as are Queen Maud uplift (QM), Great Slave lake shear zone (GSZ), and the Nonacho intermontane basin (NB) of the Churchill hinterland (see Hoffman, 1987c and references therein). Amundsen (AB) and Thelon (TB) basins contain postorogenic middle Proterozoic cover. Major water bodies are Great Slave Lake (*GSL*), Great Bear Lake (*GBL*), and Coronation Gulf (*CG*). Most of the region is accessible only by light aircraft operating from the logistical center at Yellowknife.

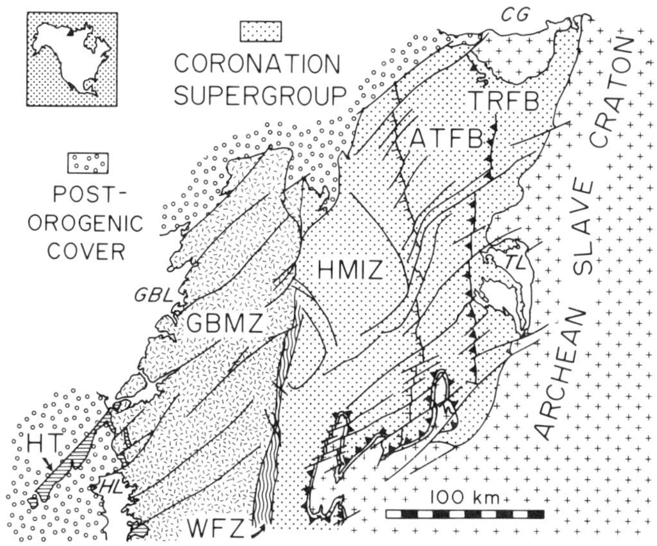

Figure 2. Tectonic zonation of northern Wopmay orogen (compare with Fig. 3): HT, Hottah terrane; GBMZ, Great Bear magmatic zone; WFZ, Wopmay fault zone; HMIZ, Hepburn metamorphic-plutonic internal zone; ATFB, Asiak thrust-fold belt; TRFB, Tree River fold belt. Water bodies: Coronation Gulf (*CG*), Great Bear Lake (*GBL*), Hottah Lake (*HL*) and Takijuq Lake (*TL*).

succeeding bimodal volcanics (Easton, 1981); (2) a passive, west-facing, shelf-rise sequence (Epworth Group) composed of storm-dominated shelf-facies siliciclastics (Odjick Formation) and cyclic dolomite-argillite (Rocknest Formation) in the externides, and contiguous slope-rise facies in the internides (Grotzinger, 1986a,b); and (3) a synorogenic, eastward-younging, foredeep sequence (Recluse Group), invaded by syndepositional gabbro sills (Hoffman, 1987a), comprising a basal ferruginous siltite (Tree River Formation), euxinic pelite (Fontano Formation), feldspathic wacke turbidites (Asiak Formation) intertongued with concretionary and laminated calcareous argillite (Kikerk and Cowles formations, respectively), and fluvial molasse (Takiyuak Formation). U-Pb zircon chronology (Hoffman and Bowring, 1984; S. A. Bowring and W. R. Van Schmus, in preparation) shows that the tripartite succession was deposited in about 15 m.y. The syn-rift volcanics have ages clustered about 1,900 Ma and the Hepburn plutonic suite (Fig. 3), which intrudes in the internal foredeep turbidites, have ages clustered about 1885 Ma. Therefore, the Calderian orogeny involved the closing of a basin that was still underlain by thin lithosphere, not thermally recovered from the rifting event that initiated the basin (St-Onge and King, 1987a,b). Furthermore, recent evidence suggests that the rifting took place in a back-arc setting (Reichenbach, 1986; Hildebrand and others, 1986, 1987), implying an elevated thermal regime for eastern Wopmay orogen throughout the period in question. As we will show, this is reflected in the metamorphic and structural style of the orogen.

DEFORMATIONAL HISTORY

Four periods of compressional deformation (D1 through D4) are evident throughout most of eastern Wopmay orogen. D1

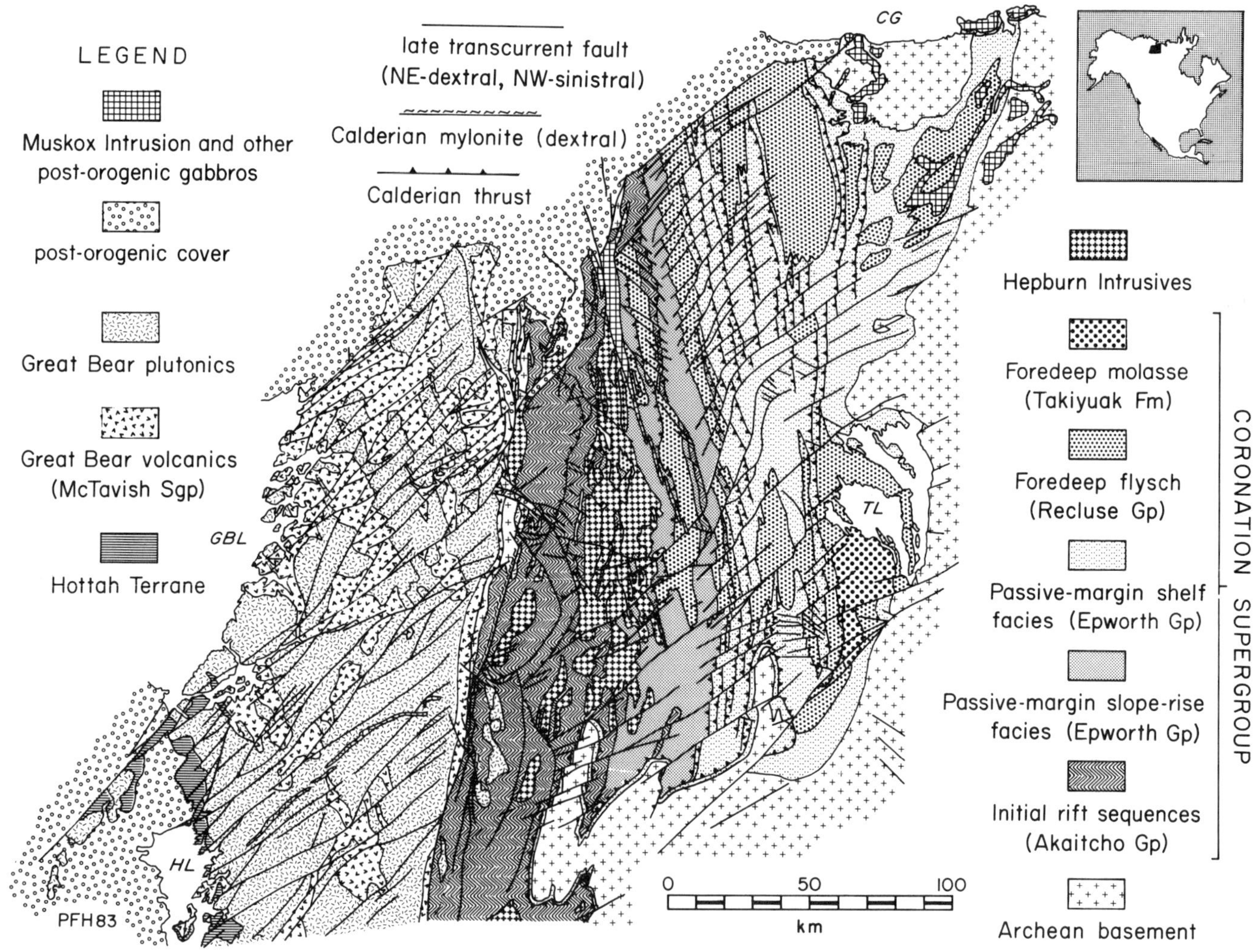

Figure 3. Simplified geologic map of northern Wopmay orogen.

and D2 are thin- and thick-skinned deformations, respectively, resulting from east-west shortening during the Calderian Orogeny. D3 is manifested by the regional cross-folds, and D4 by a system of conjugate transcurrent faults that postdate the Great Bear magmatic zone (Fig. 3). An additional deformation of post-D3, pre-D4 age produced major northwest-trending folds in the Great Bear magmatic zone (Hoffman, 1984) but is only locally evident in eastern Wopmay orogen and is not discussed in this chapter.

D1 produced north-trending imbricate thrusts and folds of the Coronation Supergroup above a sole thrust that broke along a "flat," located 100 to 300 m stratigraphically above the unbroken (nonstretched) Archean basement surface, extending for almost 100 km transverse to the strike of the belt (Figs. 5, 6). In the south half of the belt, the front of the allochthon is defined by a major thrust ramp that steps up through 6 km of footwall strata. The western half of the allochthon (internides) carries the syn-rift Akaitcho Group and slivers of Archean basement. These provide indirect evidence of a major ramp in the sole thrust west of Exmouth antiform (Fig. 5), because they originate stratigraphically below the exposed "flat." This ramp may be located at the transition from stretched to nonstretched basement in the autochthon, approximately coincident with the shelf-slope break in the Epworth Group prior to thrusting (Fig. 4). Based on shortening estimated from the cross sections of the externides, the palinspastic location of the shelf-slope break would be on strike with the west flank of Exmouth antiform.

D2 produced north-trending, symmetrical to weakly east-vergent folds, as much as 13 km in structural relief, involving the basement, its autochthonous cover, and the overlying D1 allochthon. The thick-skinned nature of this deformation is seen in the system of basement folds, including the Exmouth, Scotstoun, and Carousel antiforms (Fig. 5), exposed in oblique cross section on the flanks of Redrock culmination. The only significant fault

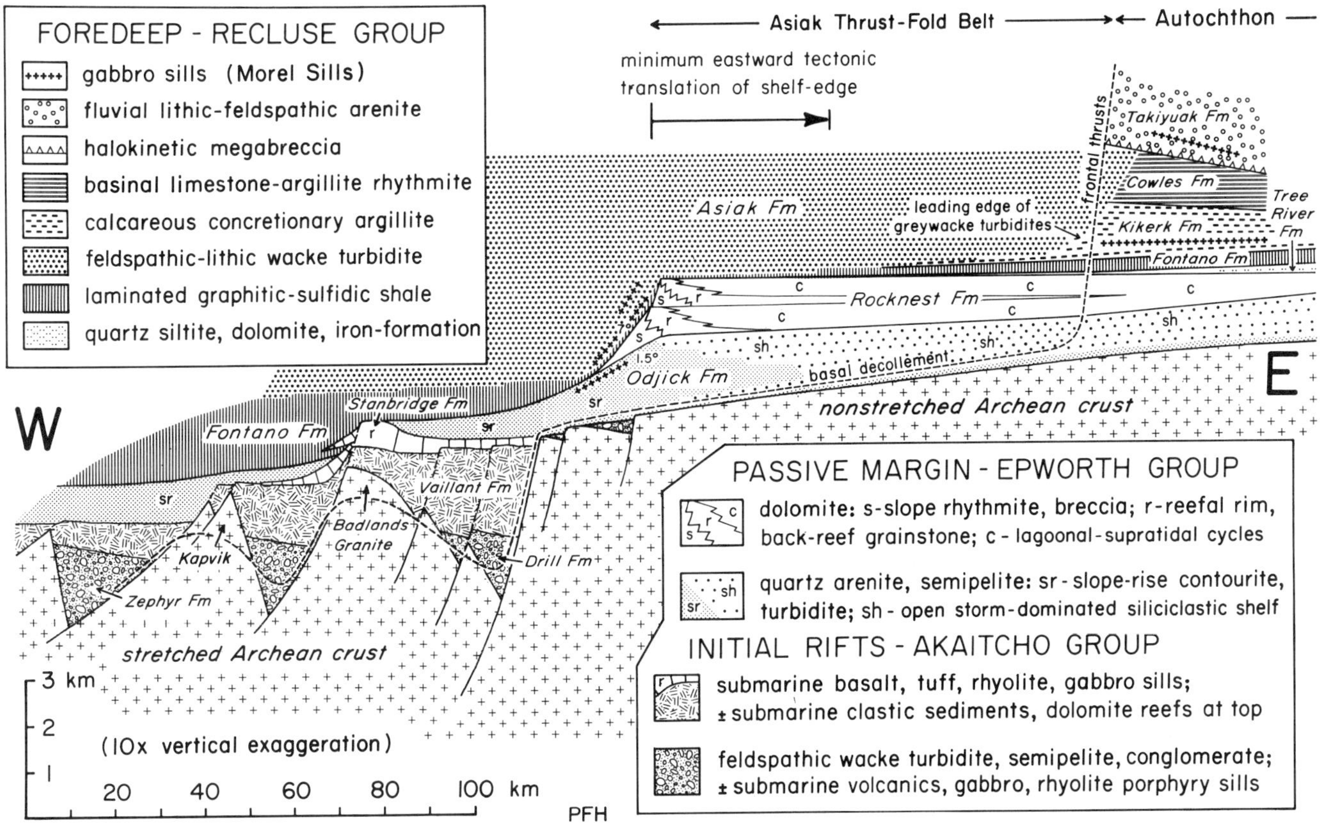

Figure 4. Stratigraphic reconstruction of the Coronation Supergroup, showing inferred trajectory of sole thrust (basal décollement). Note vertical exaggeration.

related to D2 is a high-angle thrust, having about 3 km of slip, located on the east limb of the Scotstoun antiform (Fig. 6). Based on metamorphic evidence discussed later, the evolution from thin- to thick-skinned deformation (D1 to D2) during the Calderian orogeny was probably continuous.

D3 produced east-northeast–trending cross-folds and related fabrics superimposed on the Calderian structures. First-order cross-folds are responsible for the regional-scale structural culminations and saddles (Fig. 5), and typically have limb dips of 10° to 15°, locally of 30° (King, 1986). The D3 deformation is strongest in the Tree River fold belt adjacent to Eokuk uplift (Fig. 5), where D3 folds clearly postdate the youngest D1 thrusts. The D3 folding produced as much as 30 km of structural relief but no significant associated thrusting. D3 is thought to be genetically unrelated to the Calderian orogeny because it extends at least 300 km east of the limit of Calderian thrusting and because

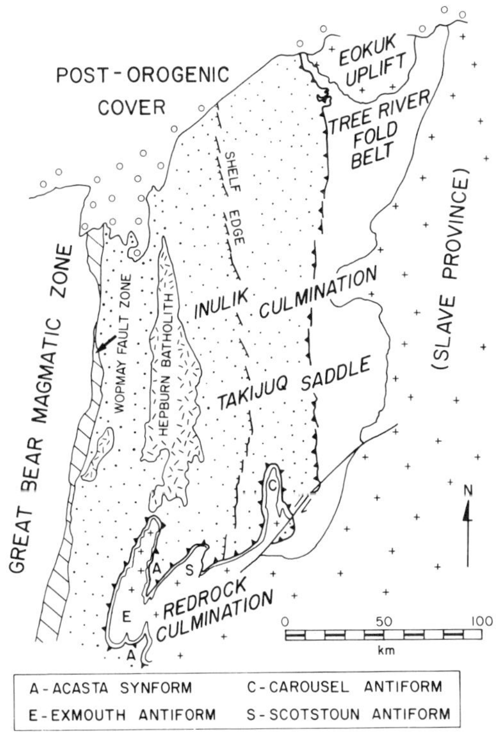

Figure 5. ENE–WSW trending D3 cross-folds and their spatial relationship to D1 Calderian allochthon (stippled) and D2 Exmouth (E), Scotstoun (S), and Carousel (C) basement antiforms.

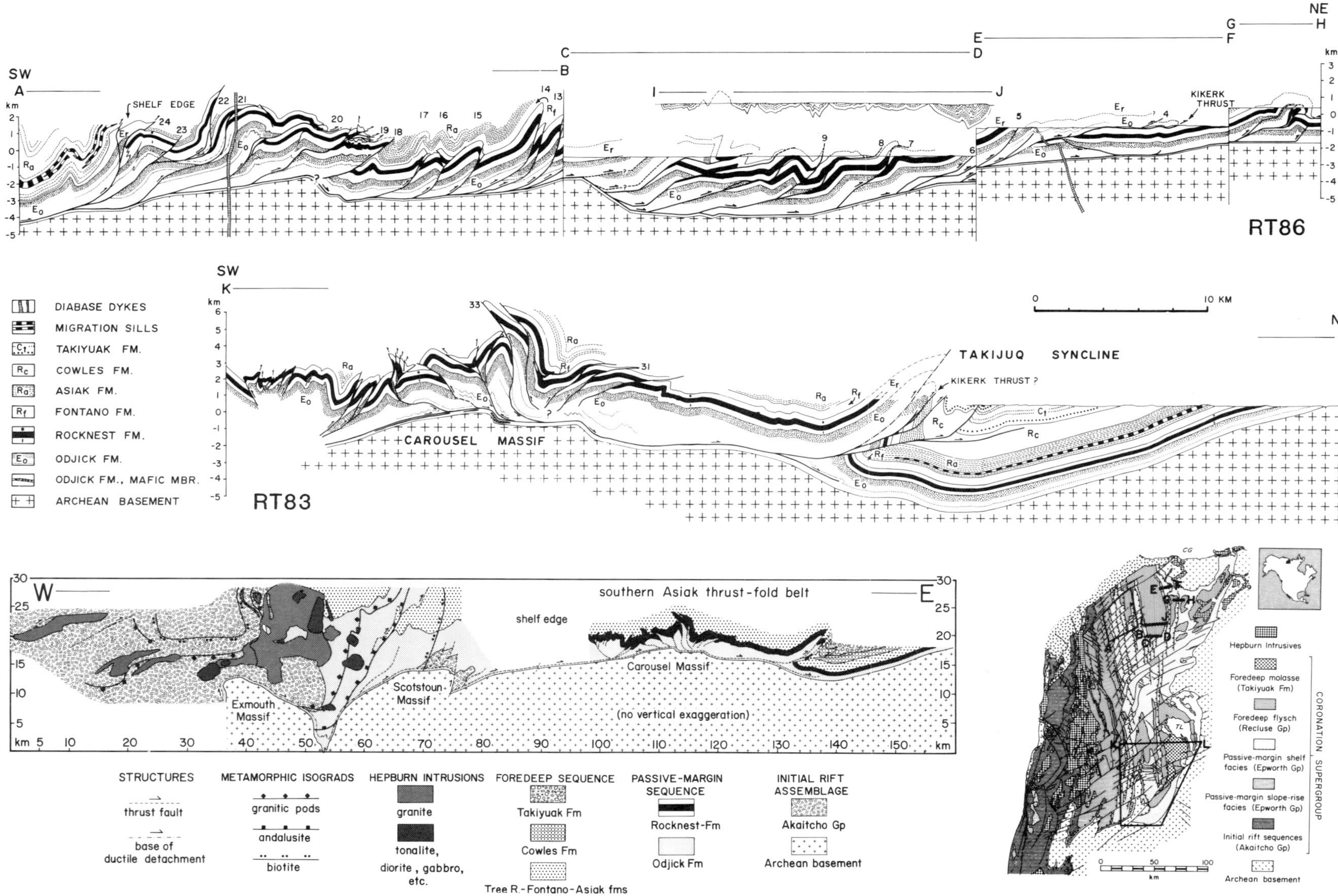

Figure 6. Composite structural cross sections (no vertical exaggeration) of Asiak thrust-fold belt constructed by means of axial projections on limbs of oblique D3 cross-folds. Sections: W–E, full width of eastern Wopmay orogen based on projections between Redrock culmination and Takijuq saddle (see Fig. 5); A–B, plunge projection at level of Rocknest Formation, extrapolated to basement; C-H, balanced structure cross sections; I–J, profile of base of foredeep turbidites (Asiak Fm) based on surface dips along section line; K–L, plunge projection down to basement surface. Inset map shows locations.

the inferred shortening direction (NNW–SSE) is incompatible with evidence that the Calderian orogeny is fundamentally a manifestation of dextral transpression of the west margin of Slave craton, and therefore that the direction of maximum regional Calderian shortening should be oriented ENE–WSW.

D4 involved east-west shortening and north-south extension in a dominantly pure-shear regime approximating horizontal plane strain. This was accomplished by block rotations about vertical axes in a system of conjugate transcurrent faults (Fig. 3). Relative to domains lacking transcurrent faults, Calderian structures are rotated clockwise in domains of northwest-trending sinistral faults and counter-clockwise in domains of northeast-trending dextral faults. In the externides, transcurrent slip is proportional to rotation of fault planes toward the extension direction, and regional bulk shortening exceeds 20 percent, locally exceeding 30 percent (Tirrul, 1987a). Only minor changes in crustal thickness occur in localized domains of normal or reverse faulting. The D4 deformation postdates the 1.88 to 1.84-Ga arc magmatism (Hoffman and Bowring, 1984; Hildebrand and others, 1987; S. A. Bowring and W. R. Van Schmus, in preparation) in the Great Bear magmatic zone (Figs. 2–3), which in turn appears to postdate the D3 cross-folding event. As the D4 deformation does not involve crustal thickening, it is not discussed further in this chapter, but its effects have been corrected for in our synthesis of the earlier deformations.

ASIAK THRUST-FOLD BELT

Asiak thrust-fold belt is the most external zone of Wopmay orogen, being located between the autochthonous cover of Slave craton and the Hepburn metamorphic-plutonic internal zone (Fig. 2). The external-internal transition is placed at the western limit of thick shelf-facies dolomite in the Rocknest Formation, which is marked by a narrow shelf-edge reef complex (Grotzinger, 1986a) that passes abruptly westward into a much thinner argillaceous slope facies (Fig. 4). In the underlying Odjick Formation, the same transition is expressed as a gradational change from a more arenaceous outer shelf facies to a more argillaceous upper slope facies. The observed absence of normal faults at the basal unconformity or significant fault-related sedimentary facies in the autochthonous cover indicates that the shelf sequence was deposited on nonstretched Archean basement (Fig. 4).

The structure of the externides is best seen in composite cross sections representing the exposed north and south ends of the belt (Fig. 6). The cross sections were constructed from axial projections of the limbs of D3 cross-folds (Tirrul, 1983). The southern cross section is obliquely exposed in its entirety on the north flank of Redrock culmination (Fig. 5). The complex imbricate structure of the allochthon, exposed structurally above (i.e., north of) the autochthonous basement massif of Carousel antiform, illustrates how surface mapping (Fig. 7) provides the basis for a well-constrained structural cross section (Fig. 8). In constructing the cross sections, projected outcrop control was supplemented by bed-length balancing requirements, and inferences concerning fold and fault geometries derived from surface observations (Figs. 9 through 12).

The northern cross section is derived from the flanks of Inulik and Eokuk culminations (Fig. 5), and the depths to the sole thrust and underlying basement are inferred rather than projected from outcrop as in the southern cross section. As drawn, the northern cross-section is balanced and the estimated minimum 41 percent shortening compares well with the minimum 43 percent shortening estimated for the southern cross section (Tirrul, 1983).

Regional variation in metamorphic grade in the externides and adjacent autochthonous cover is shown in Figure 13 (Lucas, 1984). It reflects the combined effects of variation in exposed structural depth and lateral variation in metamorphic grade at a given structural level. A close association of structural and metamorphic "highs" is observed for both D2 (Carousel antiform) and D3 (Eokuk uplift) structures.

Thin-skinned (D1) Deformation in the Externides

Structural geometry is best known at the level of the Rocknest Formation (Tirrul, 1982, 1983) because of greatly superior outcrop and internal stratigraphic control relative to other formations. Ten members are recognized in the 1-km-thick cyclic shelf facies, which have been mapped throughout Asiak thrust-fold belt (Grotzinger, 1986c). Approximately 150 regressive 1- to 5-m-thick cycles can be distinguished, and many can be correlated for 210 km parallel and 120 km perpendicular to strike (Grotzinger, 1986b,c).

The structure of the Rocknest Formation is dominated by closely spaced, steeply ramped imbricate thrusts and related folds of overall eastward vergence (Fig. 6). The thrusts typically have straight map traces, excluding D3 cross-folding and D4 transcurrent faulting, and terminate along strike in folds. Back thrusts are relatively common, especially in the south half of the belt, but no consistent cross-cutting relationship between fore and back thrusts is established. Thrust sequence, as determined by progressive folding of older thrust planes above younger ramps, is mainly prograde (progressing from west to east, or from top to bottom in the imbricate stack) but retrograde, or "out-of-sequence" thrusts also occur.

In areas of anchizonal metamorphism (Fig. 13), the Rocknest Formation, which underwent complete early diagenetic marine dolomitization (J. P. Grotzinger, personal communication, 1986), displays low penetrative strain and curved parallel fold geometry (Fig. 9B). In areas of greenschist facies, dolomite has been inhomogeneously strained such that early thrust-related folds were tightened, and thrust imbricates steepened (Fig. 8).

Many of the thrusts appear to die out upward in the foredeep flysch (Fig. 6). Some disappear in the incompetent shale of the Fontano Formation, but others climb well up into the Asiak Formation turbidites (Fig. 11B). The frontal thrust system west of Takiyuak syncline penetrates the entire preserved stratigraphy (Fig. 6). Tight upright chevron-style folds are characteristic of the turbidites, and the shortening estimated from average limb dip

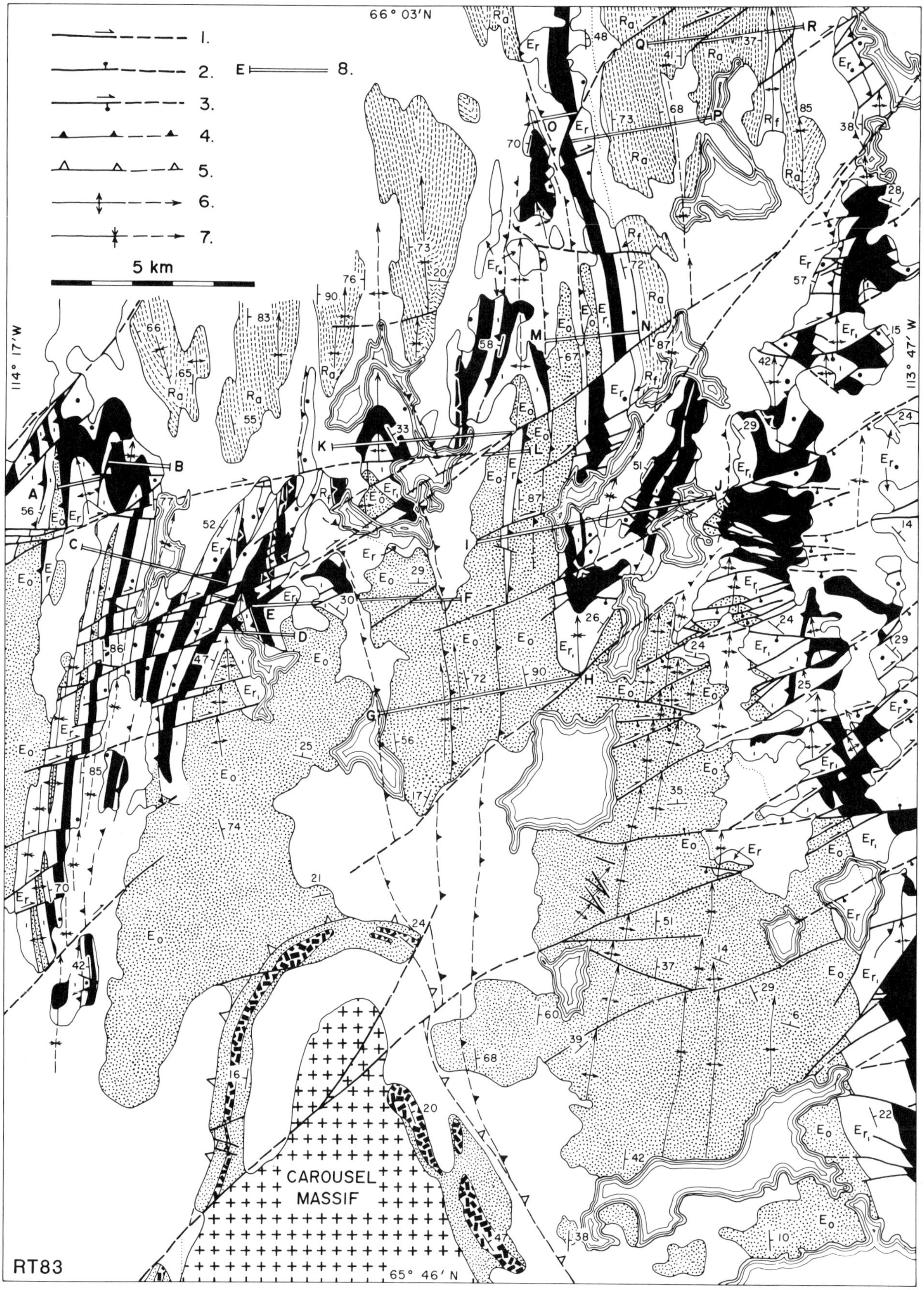
66° 03′N
1.
2.
3.
4.
5.
6.
7.
8.
5 km
114° 17′W
113° 47′W
CAROUSEL
MASSIF
RT83
65° 46′ N

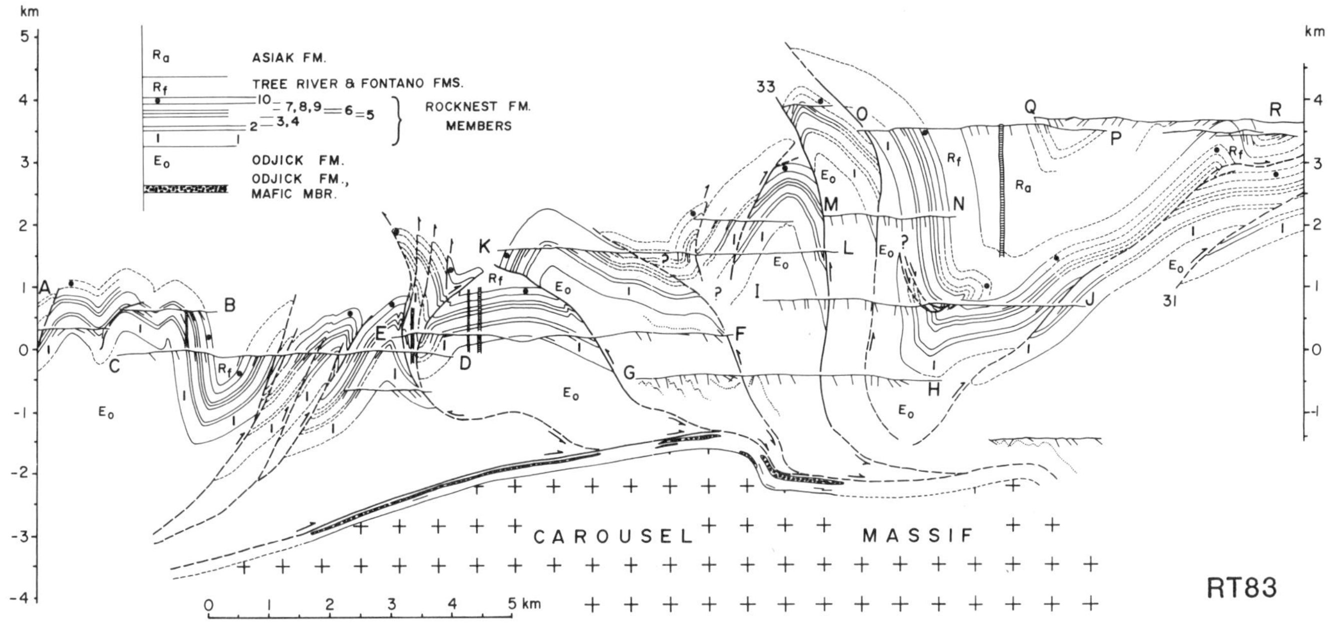

Figure 8. Composite axial (plunge) projection of structure at north end of Carousel basement antiform. Levels of control (A–B, etc.) are keyed to Figure 7; contacts are dashed where no down-plunge information is available. Rocknest members 2 through 5 are combined to make the middle bold unit shown in Figure 7. Stratigraphic units (Eo, etc.) are explained in caption to Figure 7.

Figure 7. Simplified geologic map of plunging north end of Carousel basement antiform showing location of selected transects projected to control structural cross section in Figure 8. Map units: Eo, Odjick Fm; Er, Rocknest Fm; Rf, foredeep euxinic shale (Fontano Fm); Ra, foredeep turbidites (Asiak Fm). Structural symbols: 1, transcurrent fault; 2, normal fault; 3, oblique-slip fault; 4, thrust fault; 5, sole thrust; 6, plunging anticline; 7, plunging syncline; 8, projected structural level.

Figure 9. Character of D1 folds (looking north) in Asiak thrust-fold belt. A, Angular and curved nonparallel folds in quartzitic semipelite (outer shelf facies) at lower greenschist metamorphic grade in allochthonous Odjick Formation. B, Curved parallel fold in cyclic peritidal dolomite at anchizonal metamorphic grade in allochthonous Rocknest Formation. Width of fold shown is ca. 100 m.

Figure 10. Outcrop appearance of D1 thrusts in Asiak thrust-fold belt. A, Minor east-vergent thrust (arrows) in Rocknest dolomite. Vertical scale in right center is 1.5 m. B, Exhumed surface of Kikerk Thrust near line of cross section E–F in Figure 6. Footwall is Top Member dolomite of Rocknest Formation and hanging wall (upper left) is ferruginous quartzite of lower Odjick Formation. Thrust has a stratigraphic displacement of 2 km and a minimum horizontal translation of 8 km. Persons on skyline provide scale; boulders on thrust plane are glacial erratics.

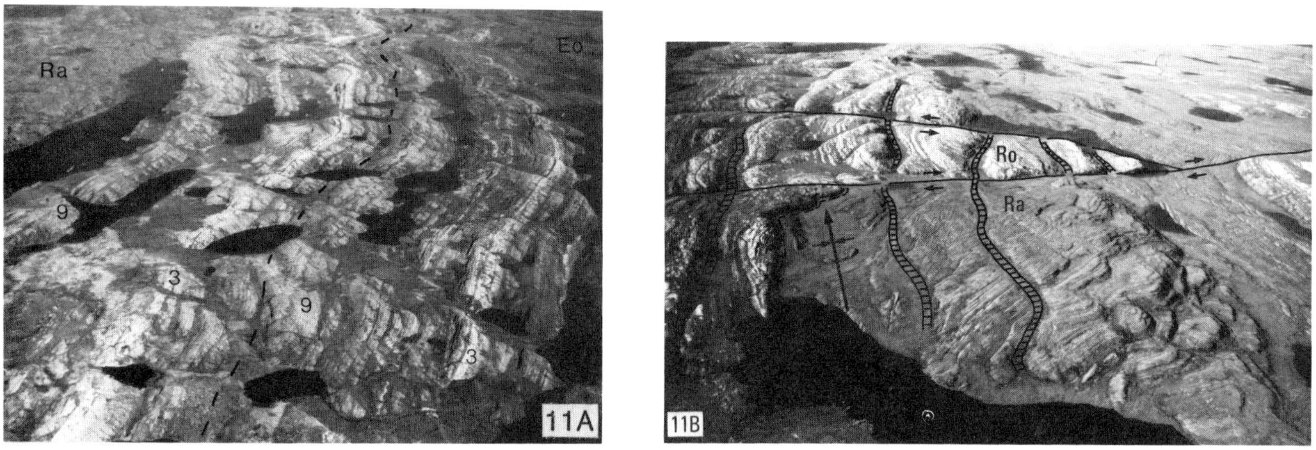

Figure 11. Aerial views (looking north) of D1 thrusts in Asiak thrust-fold belt. A, West-dipping thrust 6 (section C–D, Fig. 6) duplicates 0.7 km of Rocknest dolomite. Dashed line indicates thrust trace; numbers refer to members of Rocknest Formation. Note D3 cross-folds in far distance. Width of view is 1.5 km in foreground. B, West-dipping thrust 23 (section A–B, Fig. 6) at head of lake in left foreground places light-colored Rocknest dolomite (Er) over subthrust syncline of dark-colored foredeep turbidites (Ra). Note D4 right-lateral and left-lateral transcurrent faults (arrows) that offset thrust-trace in background. Field of view is 2 km wide in foreground. Ladder symbols mark traces of post-tectonic diabase dikes.

approaches that achieved by thrusting and folding in the Rocknest Formation (Tirrul, 1983, 1987b). Thus, there is no incompatibility in shortening between the upper domain (Recluse Group) dominated by folds and the lower domain (Epworth Group) dominated by thrusts.

Fold geometries and stratigraphic repetitions in the Odjick Formation suggest, given the limitations of outcrop, that most thrusts ultimately splay from a sole thrust consistently located 100 to 300 m above the basement surface (Fig. 6). However, some thrusts appear to die out downward, and some of the shortening and thickening achieved by thrusting of the Rocknest Formation is balanced by relatively small-scale folds and steeply dipping cleavage in the Odjick Formation. Although semipelite above the sole thrust is strongly folded on the outcrop scale (Fig. 9A), the

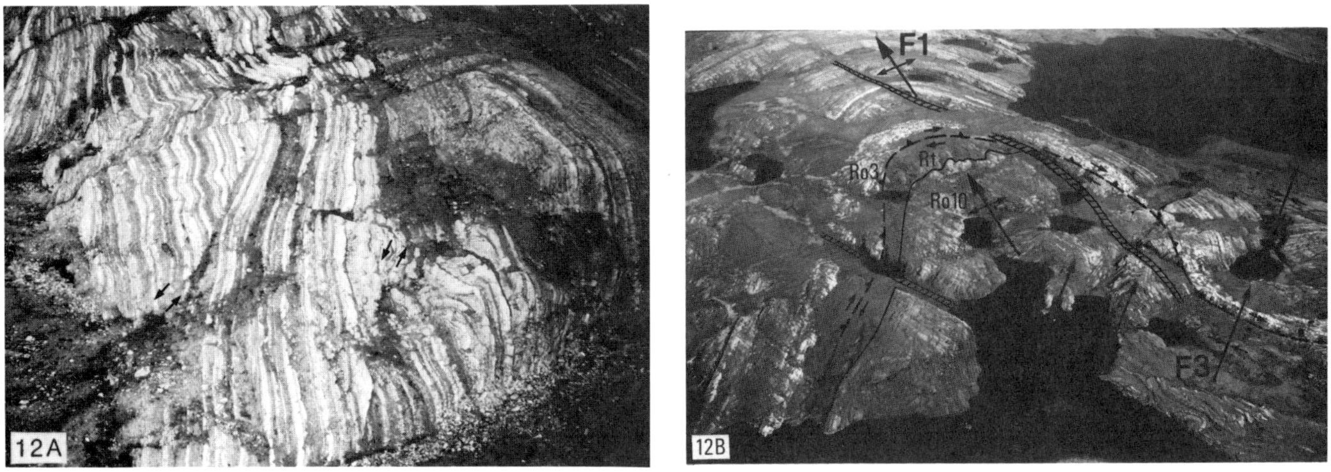

Figure 12. Aerial views (looking north) of D1 thrusts in Asiak thrust-fold belt. A, Pair of well-defined lateral ramps (arrows) in subvertical second-order imbricates of cyclic Rocknest dolomite, just east of section A–B, Figure 6. Stratigraphic top is to left; width of view in foreground is 170 m. B, Approximate down-plunge view of anticlinorium west of thrust 9 (section C–D, Fig. 6). Folded thrust (dashed line) is a hangingwall flat that juxtaposes 0.7 km of Rocknest dolomite over recessive basal foredeep pelite (Tree River Formation) along western limb but directly onto the top of the Rocknest Formation on the eastern limb, indicating that the thrust cuts down-section (10's of m only) in the inferred transport direction. Higher order folds in footwall (in right-center) are D3 folds unrelated to thrusting. Width of view in foreground is 1.3 km.

autochthonous pelite sequence is but very weakly cleaved, bedding-cleavage relations being compatible with a component of bedding-parallel, eastward-vergent simple shear (King, 1986). The vergence is consistent on both limbs of Carousel antiform, indicating that the cleavage is not due to flexure or to diapiric rise of the basement, but is compatible with a steep shear strain gradient downward from the active sole thrust. The sole thrust itself occupies a narrow drift-covered valley, and drilling would be required to elucidate its nature and evolution at the outcrop and grain scale. Where exposed, the frontal ramps are discrete; in places knife-sharp surfaces separate different types of fault rock. Quartzite and dolomite grade irregularly into fine cataclasites, characterized by multiple phases of brecciation and cementation, usually less than 10 cm thick, adjacent to the fault. Pelite, particularly the lower Odjick Formation, grades into a tectonic melange, up to 10 m thick, composed of phacoidally cleaved phyllite with lenses of indigenous dolomite or quartzite.

D2 Deformation in the Externides

Carousel, Scotstoun, and Exmouth antiforms are broad open structures, coaxial with D1 folds but congruently involving the basement, its autochthonous cover, the Calderian sole thrust, and overlying allochthon (Fig. 6). The antiforms have an east-vergent crestal asymmetry, and the overall structural relief of the basement surface ranges from an estimated 6 km in the south to 2.2 km in the north. The greater relief in the south, and, in particular, the prominence of Carousel antiform, may be related to a node of relatively high syn-D1 metamorphic grade in that area (Fig. 13). This suggests a close temporal association of D1 and D2.

The basement folds in the north are inferred rather than directly observed as in the south. If basement folds do not occur and the observed structural highs are assumed to be solely the result of cover imbrication, then the amount of shortening required would greatly exceed that observed in the south, and the northern cross section could not be balanced given the constraints from axial projections.

The D2 deformation produced no visible penetrative fabric in the externides, reflecting the small amount of shortening involved. The weak cleavage in the autochthonous cover, described earlier, is related to D1. Gneissic foliation in the basement granitoids (tonalite, granodiorite, monzogranite) of Carousel antiform strikes northeast and is erosionally truncated at the basement-cover contact.

Excluding D4 transcurrent faults, the continuity of the basement-cover contact in the externides is broken by only one high-angle reverse fault, having a throw of less than 200 m. At variance with observations and inferences in many other thrust-fold belts, the basement-cored D2 antiforms ("external massifs") of Wopmay orogen do not result from basement involvement in thrusting (e.g., Hsü, 1979; Boyer and Elliott, 1982; Platt and others, 1986).

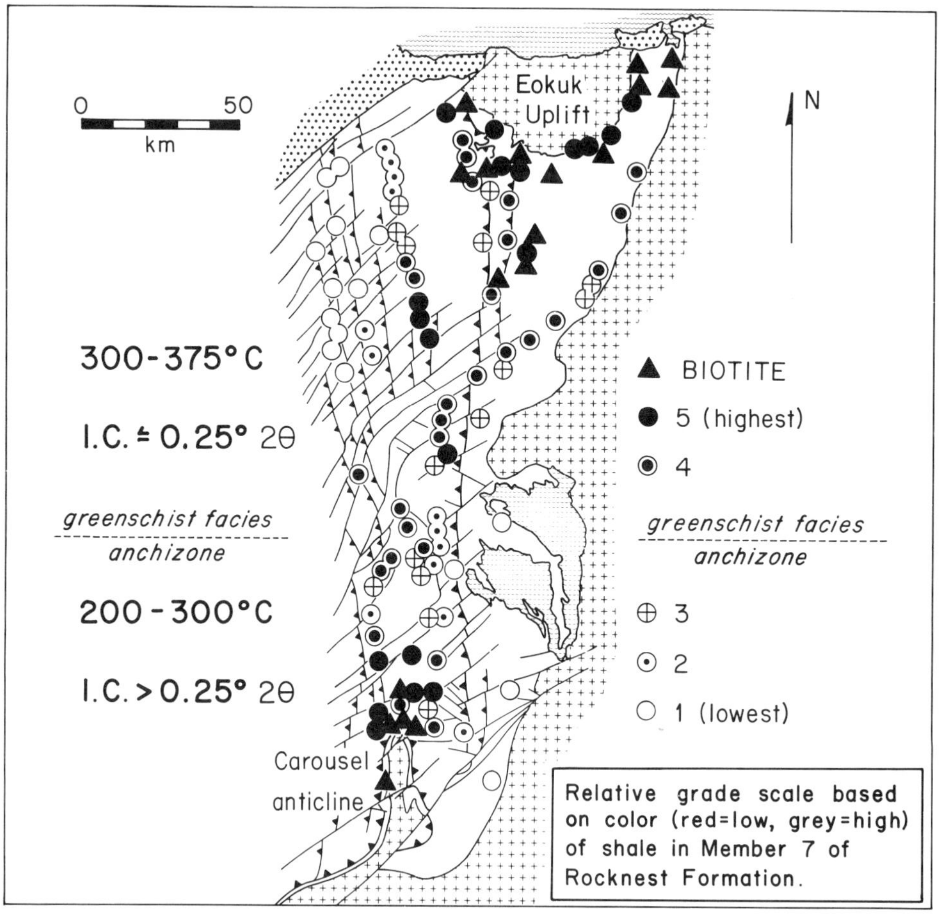

Figure 13. Variation in metamorphic grade in externides (Asiak thrust-fold belt and autochthon) according to relative grade scale (right), based on color and fabric development in Red Shale Member of Rocknest Formation, and correlative illite crystallinity values (left). Solid triangles indicate occurrence of metamorphic biotite in Rocknest or Odjick Formation. Lateral variation predominates over variation with depth relative to basement surface. Note spatial association of D2 (Carousel antiform) and D3 (Eokuk uplift) basement structural highs with thermal metamorphic highs, and anticorrelation of metamorphic grade with structural depth between Carousel antiform and autochthon (compare with Fig. 6).

D3 Deformation in the Externides

Regional scale cross-folding of the externides is evident in the areal distribution of stratigraphic units (compare Figs. 3 through 5). Cross-folding is also documented by variation in plunge of D1 and D2 structures, which may exceed 20°. First-order cross-folds define three structural culminations and two intervening saddles (Fig. 5). Structural relief of the southern (Takijuq) saddle is greater than 6 km; for the northern (Tree River) saddle it is at least 2 km. The wavelength of these structures is approximately 120 km. A cross section of the northern saddle is shown in Figure 14D.

Higher order D3 folds are best developed south of Eokuk uplift (Fig. 14A), where a sinuous belt of discontinuous northeast-trending folds that refold the D1 structures to the west is spectacularly exposed (Fig. 15). The folds are typically open, upright to steeply inclined, and polyharmonic, with characteristic wavelengths of 600 to 700 m and 60 to 100 m. Thrusting is strictly limited to small, rootless, "out-of-syncline" thrusts.

The Archean basement bounding the Tree River fold belt is folded but not involved in thrusting. Most of the basement-cover contact has been carefully mapped, and, with minor exceptions, is an intact depositional contact. The basement has been deformed into structures that resemble large-scale mullions, with convex-upward lobes ranging from a few meters to hundreds of meters wide, separated by pinched cusps. The margins of the lobes are commonly overturned, as shown in Figure 14C. Basement deformation is associated with inhomogeneous retrograde foliations that are particularly well developed in some cusp zones. Basement deformation also occurred by means of discontinuous

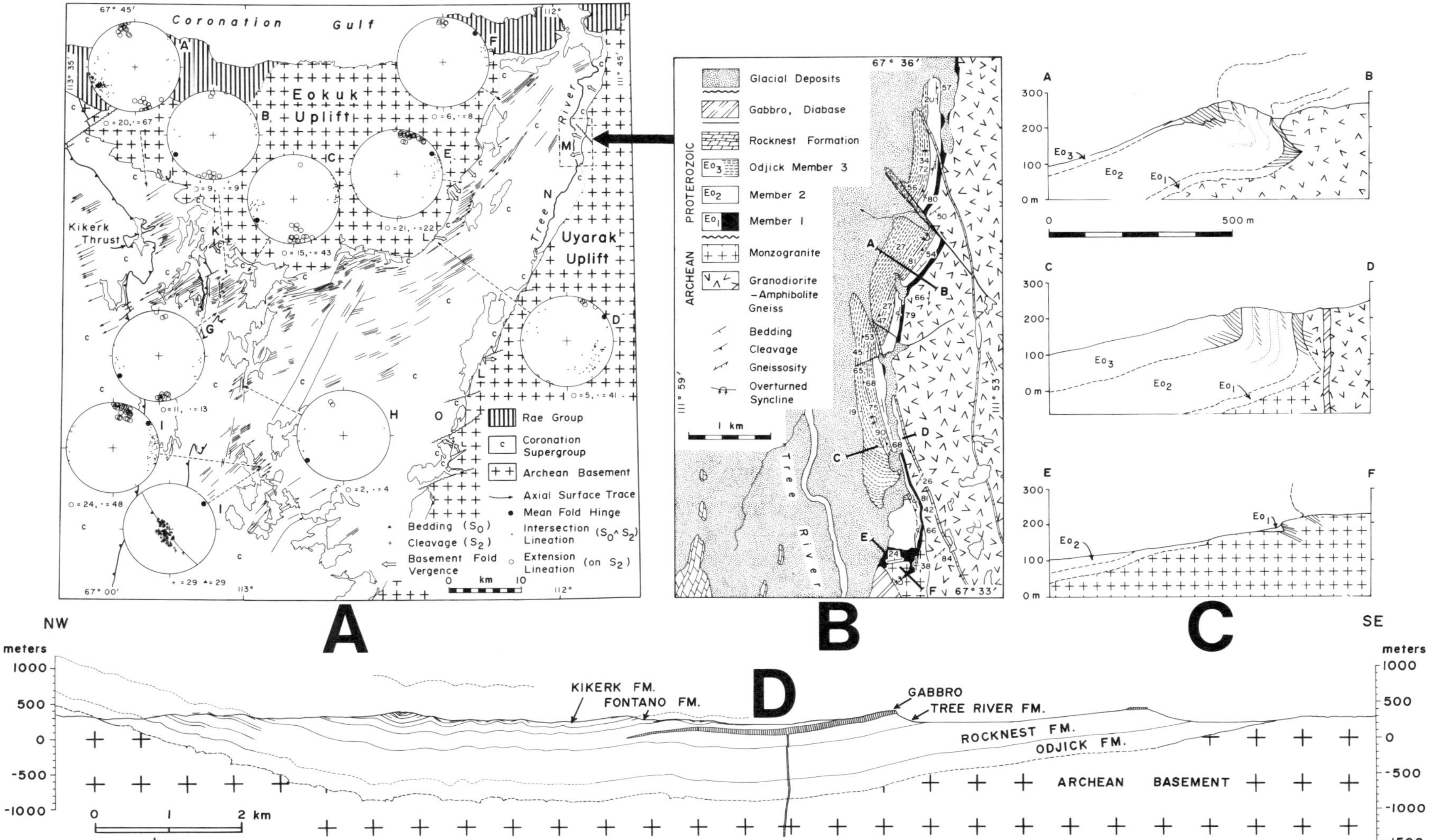

Figure 14. Tree River fold belt. A, D3 fold trajectory map of Tree River fold belt showing basement-cover contact and selected structural data. Note the consistent trend of fold hinges and extension lineations on the southeast and southwest margins of Eokuk uplift, showing that the fold belt manifests regional northwest-southeast shortening, and therefore, that basement uplift is a consequence, not a cause, of folding. As discussed in text, the oblique angular relationship between fold hinges and extension lineation on S2 suggests a pre-D3 southward-vergent deformation event. B, C, Geologic map and cross sections A–F showing lobate-cuspate geometry typical of basement-cover contact at location M. Note that bedding in overturned limbs has a steeper dip than associated cleavage in basement and cover, indicating that the limbs were not passively rotated by ductile basement shear zones (cf. Ramsay, 1980). Topographic relief is adequate for excellent control. D, Composite structural cross section of D3 Tree River synclinorium between Eokuk and Uyarak uplifts, constructed from surface information on line of section and axial projections along northwest margin. Line of section is close to dashed arrow from stereonet "D."

Figure 15. Aerial views of D3 cross-folds in Tree River fold belt. A, Approximately axial view (looking west-southwest) of typical discontinuous, disharmonic, curved parallel folds in Rocknest dolomite on footwall of Kikerk thrust (section E–F, Fig. 6). Fold wavelengths are 0.4 km in foreground and 0.9 km in background. B, View looking north of steeply plunging cross-folds in Rocknest dolomite of thrust imbricate 5 (section E–F, Fig. 6) in Asiak thrust-fold belt. Fold wavelength is about 0.8 km.

small-scale conjugate shear zones that accommodated northwest-southeast shortening and subvertical extension.

Compared with the fold pattern observed in the Rocknest Formation and younger units, the basement surface geometry in the Tree River fold belt is complex. Although the dominant trend is northeast, the lobes and cusps show a wide variation in orientation and vergence. For example, folds at location J (Fig. 14A) are overturned to the northwest; at K, to the southeast; at L, to the east; and at M, to the west. Each is associated with one to three generations of cozonal cleavages in the adjacent cover, different in orientation from neighboring cusps of different trend. From field relations alone, the relative ages of these structures are unclear, but their development may have been broadly coeval.

Penetrative fabrics resulting from NNW–SSE shortening of highly variable intensity occur over a wide area (Fig. 16). In the externides, north-trending Calderian (D1) structures are transected by northeast-trending D3 cleavage. In the Tree River fold belt, bedding-cleavage relations indicate that cleavage developed both before and during D3 folding. Although cleavage is commonly axial planar to folds, or oriented consistently with flexural shear during folding, the dominant cleavage close to Eokuk uplift (S2 in Fig. 14A) is developed at a low angle (about 20°) to bedding in pelitic interbeds between quartzite or dolomite. This cleavage is obliquely folded by the dominant Tree River (D3) folds, and, when bedding is unfolded, the cleavage dips exclusively to the north. Its development involved a large component of bedding-parallel, top-directed-southward, simple shear. This is indicated by strong cleavage refraction and poor development or absence of cleavage in competent layers, by restored north-plunging stretching lineations (Fig. 14A) and by folded extension veins that were rotated through vertical to a northerly inclination prior to D3 folding.

The overall shortening of both cover and basement rocks is interpreted to be caused by NNW–SSE compression throughout the thickness of the crust. The reason for the wide variation in orientation of basement structures may be due to the inability of an infinite half-space (analogous to the basement-cover contact) to buckle. Instead, relief is produced by penetrative strain and shape modification of initial perturbations into lobate-cuspate form (Smith, 1979; Ramberg, 1981, p. 151–156). Presumably, variably oriented weak zones both enhance amplification and distort regional strain trajectories.

A tectonic interpretation of the north-dipping folded cleavage (S2) is difficult. The angular difference in implied principal strains, as well as the difference in age, makes it uncertain as to whether the cleavage is broadly cogenetic with D3 or is due to a distinct earlier event. Cogenesis is favored by the fact that both the cleavage and Tree River folds are well developed close to Eokuk uplift and poorly developed away from it. The cleavage was not produced by gravitational sliding off the flanks of a domal uplift since the cleavage and related lineations are not concentrically and radially disposed, respectively, about the uplift (Fig. 14A).

There is a clear spatial and temporal relationship between the intense D3 deformation and relatively high metamorphic grade (Fig. 13) associated with the Tree River fold belt. Peak metamorphic recrystallization was synchronous with the development of D3 fabrics. The situation is analogous to, but not correlative with, the Calderian metamorphic "high" associated with Carousel antiform. The ultimate cause of the Tree River thermal anomaly is unknown, as is the cause of the D3 deformation itself. However, the deformation is consistent in age and orientation with folding of the western Bear Creek foreland basin (Fig. 1), thrusting in the East Arm (Fig. 1) thrust-fold belt (Hoff-

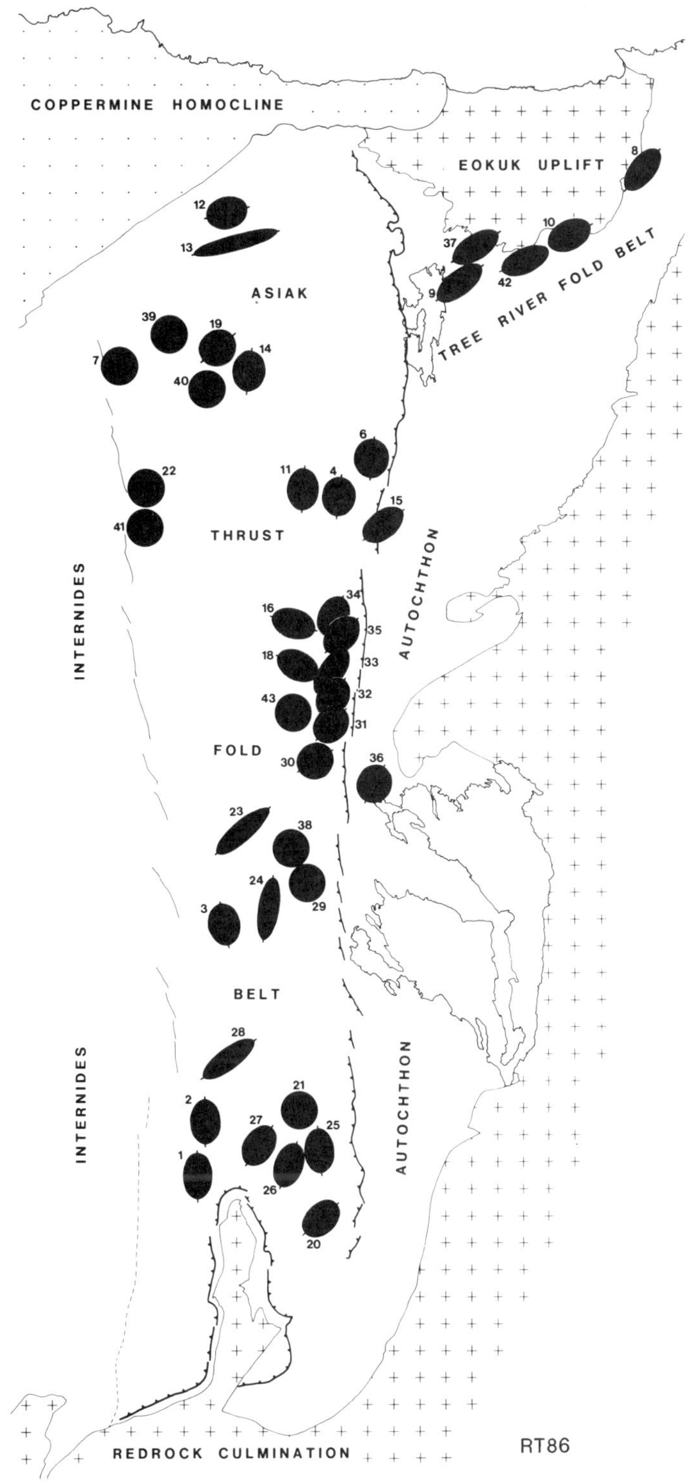

Figure 16. Summary of finite strain determinations from Rocknest dolomite in Asiak thrust-fold belt and autochthon. Locations 1 through 7 are based on variations in stratigraphic thickness; strain ellipses are fold profile sections plotted along axial surface traces. Locations 11 through 24 are total strain determinations on ooid packstones; XZ section is plotted along trend of the YZ plane, regardless of the plunge of Z (>80°). Most of the remaining determinations are based on deformed *Conophyton* and originally hemispheric stromatolite domes, measured within gently dipping bedding planes. Locations 16 and 18 are within blocks bounded by left-slip D4 faults that have probably rotated 30° clockwise. Most other samples have been variably rotated counterclockwise due to D4 transcurrent faulting.

man, 1981, 1987b; Bowring and others, 1984), and collision in the Trans-Hudson orogen (Van Schmus and others, 1987), 1300 km to the southeast.

HEPBURN METAMORPHIC-PLUTONIC INTERNAL ZONE

Structurally, the internides are continuous with the externides; the distinction is based on primary lithology and metamorphic grade. The internides expose mainly subaqueous rift deposits of the Akaitcho Group, off-shelf facies of the cratonic margin sequence (Odjick Formation), relatively fine-grained foredeep deposits (Recluse Group), and granitic through gabbroic intrusions of the Hepburn Intrusive Suite (Figs. 3, 4). Deposition is presumed to have been initiated by stretching of the Archean basement west of Exmouth antiform (Fig. 5). The Hepburn plutons were intruded as the basin closed. Metamorphic grade, recorded in pelite, ranges from phyllite containing chlorite + muscovite to migmatite containing sillimanite + K-feldspar (Fig. 17). The lateral transition from externides to internides does not correspond to a significant change in structural depth as defined by exposed stratigraphic level (Fig. 3).

Structures produced by each of the three deformations are continuous between the two zones, although they change somewhat in character and magnitude, reflecting differences in lithology and metamorphic conditions. The D1 sole thrust continues 100 to 300 m above the basement surface from Carousel antiform west to Exmouth antiform (Fig. 17), separating drastically different structural and stratigraphic domains. The easternmost rift-facies allochthon (Cloos nappe of Hoffman and Pelletier, 1982) is exposed 10 to 20 km west of the Rocknest shelf edge (Fig. 3). The fact that no normal faults cut basement-cover contact as far west as Exmouth antiform implies a minimum of 50 km of eastward translation of Cloos nappe. This agrees with the minimum 40 km of eastward translation of the Rocknest shelf edge estimated from the cross sections of the externides (Tirrul, 1983).

Like Carousel antiform, the north-trending Exmouth and Scotstoun antiforms are D2 structures that congruently fold both the autochthon and allochthon (King, 1987). The configuration of thick-skinned D3 cross-folds, as documented by plunge variation of Calderian folds and linear fabrics (Fig. 18) and by metamorphic pressure data (St-Onge, 1984a,b, 1985, 1987), controls the regional variation in level of exposure with respect to Calderian structures. The total structural relief provided by cross-folding apparently exceeds 30 km (King, 1986).

As in the external zone, the deepest levels are exposed around Redrock culmination; the shallowest levels occur in the twin saddles to the north (compare Figs. 5, 18). Plutons of Hepburn batholith (Hoffman and others, 1980; Lalonde, 1986) are limited to shallower levels, and most are clustered along a north-trending zone in the Calderian allochthon that separates a western domain of rift deposits from an eastern domain underlain by passive-margin and foredeep strata (Fig. 3).

D1 Deformation in the Internides

Composite cross sections of the Calderian structure and metamorphic zones of the internides (Fig. 19) were constructed by stacking axially projected data from detailed transects at different structural levels (Fig. 17). King (1986) gives a fuller description of the methodology and its rationale. Obviously, most of the structures projected are only approximately cylindrical on the regional scale; however, we believe that the essential geometric relations between stratigraphic contacts, structural features, and metamorphic zones are reliably portrayed. This is supported by the continuity along strike of the geology mapped at shallow levels north of Redrock culmination, the essential similarity of deep-level cross sections derived from the north and south limbs of Redrock culmination, the vertical coherence of the cross sections, and the qualitatively corroborative metamorphic pressure data (St-Onge, 1987; St-Onge and King, 1987a,b).

Certain of the Hepburn plutons, although clearly not cylindrical, were also projected onto the cross section to outline the general form and position of the batholith with respect to metamorphic isograds and structures. The inferred rootless form of the batholith is consistent with the observed absence of Hepburn plutons in the autochthonous basement and cover of Redrock culmination (Fig. 17).

D1 structures within the Calderian allochthon vary according to both stratigraphic and structural level. At intermediate and high structural levels east of the batholith, upright open folds in passive-margin and foredeep semipelites have axial surfaces that dip vertically or steeply to the west (Fig. 19). East-vergent thrusts are folded coaxially with thrust-ramp folds; a regional cleavage, defined by alignment of metamorphic phyllosilicates, is axial planar to the folded thrusts. In contrast, the rift deposits are characterized by recumbent asymmetric folds. Recumbent folds of bedding at relatively low metamorphic grades have strong axial planar foliations; relatively high metamorphic grades are characterized by isoclinal folds of foliations axial planar to earlier coaxial folds (Fig. 20A). Sheath folds have not been observed, and fold axes mostly strike north-south. The folds (Fig. 20A) are mostly eastward vergent, except for an anomalous area at an intermediate structural level about 20 km west of Hepburn batholith (Fig. 19), where, for unknown reasons, the recumbent folds are westward vergent.

Stretching lineations (such as oriented mineral grains, deformed amygdules, spherulites, lapilli, and pillows are only well developed in certain of the rift deposits. In Cloos nappe, stretching lineations plunge southwest to west, compatible with dextral oblique convergence of the allochthon relative to Slave craton. In the western internides, stretching lineations are mostly oriented north-south, and recumbent folds typically have very straight hinges (Fig. 20A) compatible with axis-parallel extension (cf. Watkinson, 1975). The longitudinal stretching fabrics must be the result of simple shear; otherwise, there is a strain incompatibility between the western internides and the externides, where no longitudinal stretching is observed. The longitudinal stretching

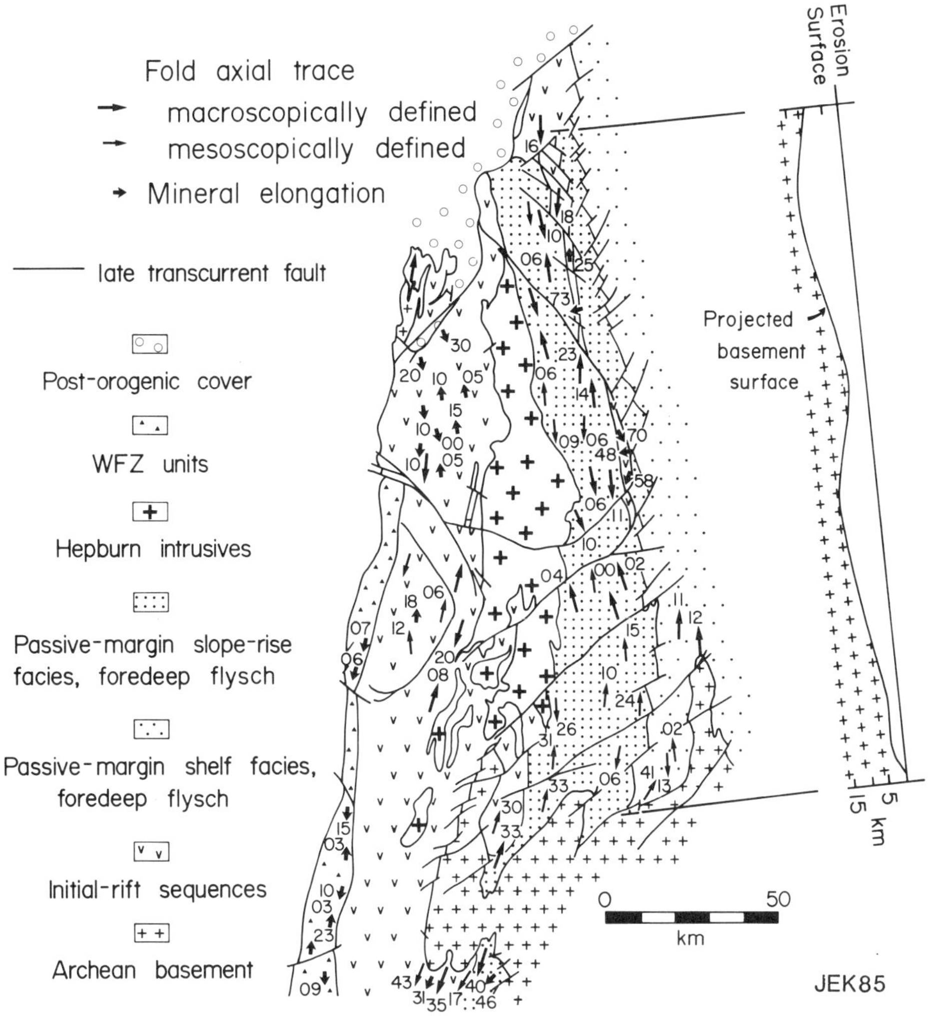

Figure 18. Compilation of representative plunges of Calderian (D1 to D2) linear fabric elements in internides. On the right is a length section of eastern internides constructed from undulating Calderian plunges. Note the overall similarity in form of basement profile to that in the externides, which is evident from distribution of foredeep, passive-margin and basement rocks (Fig. 3). Where plunge value is not indicated, direction of plunge was derived from stratigraphic relationships, but structural data are insufficient to derive actual plunge. WFZ: Wopmay fault zone.

fabrics may record the strike-shear component of oblique convergence in a ductile equivalent of the vectorial partitioning model of Fitch (1972), as has been proposed in a geologically similar situation in the southern Canadian Cordillera (Van Den Driessche and Maluski, 1986; Ellis, 1986). Strike-shear in the internides may also be due to lateral extension and/or tectonic "escape" of the hinterland as proposed for the northern Himalaya and Tibet (Tapponnier and others, 1986). Very high shear strain is indicated by the fact that stretching lineations are subparallel, rather than oblique, to the strike of the Wopmay internides.

Although the trace of the sole thrust is usually drift covered, its existence is inferred from an abrupt change structurally downward from the allochthonous domain of tight polyphase recumbent folds to autochthonous strata lacking D1 folds (Fig. 21). The low state of strain in the autochthonous strata is remarkable considering the minimum 30 km of burial implied by the cross section (Fig. 19) and by metamorphic pressure data (St-Onge, 1987), and the high grade of (static) metamorphism observed around Exmouth antiform (Fig. 19). Stromatolites in forsterite marble are virtually unstrained (Fig. 20B), and even incompetent lithologies have only spaced zones of foliation subparallel to bedding. Aside from the single high-angle D2(?) thrust on the east limb of Scotstoun Antiform (Figs. 17, 19), there is no Calderian structural repetition of the basement-cover contact.

Figure 17. Simplified geologic map of Hepburn metamorphic-plutonic zone (internides) of eastern Wopmay orogen. Transects (A–A′, etc.) were used in constructing composite cross section, as indicated in upper panel of Figure 19. Structures: BF, Belleau fault; CA, Carousel antiform; CF, Calder fault; CN, Cloos nappe; EA, Exmouth antiform; MT, Marceau thrust; OF, Okrark fault; SA, Scotstoun antiform; SN, Sitiyok nappe; WFZ, Wopmay fault zone. Water bodies: BL, Bent Lake; BiL, Bishop Lake; CL, Calypso Lake; CR, Coppermine River; RL, Redrock Lake.

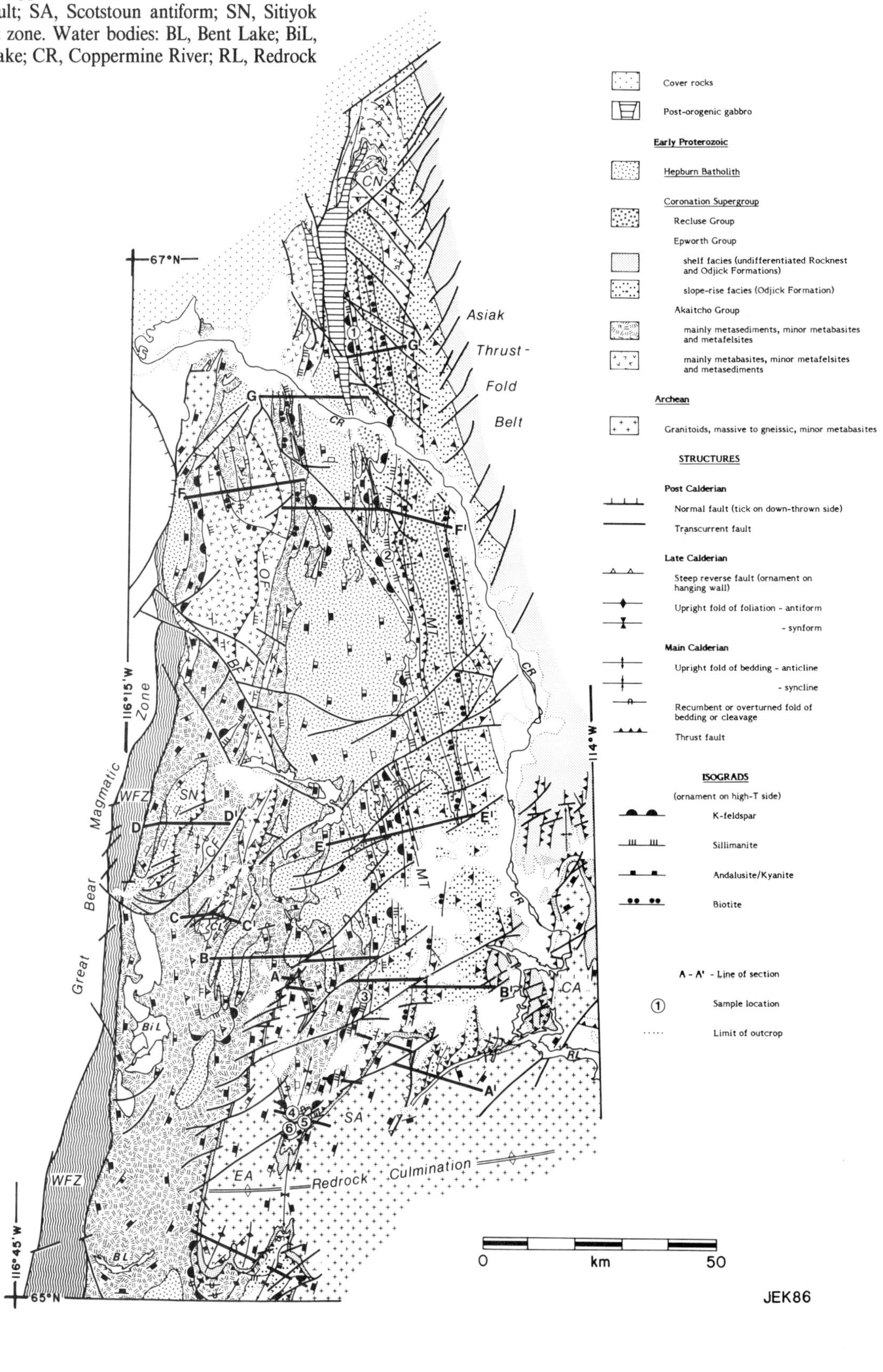

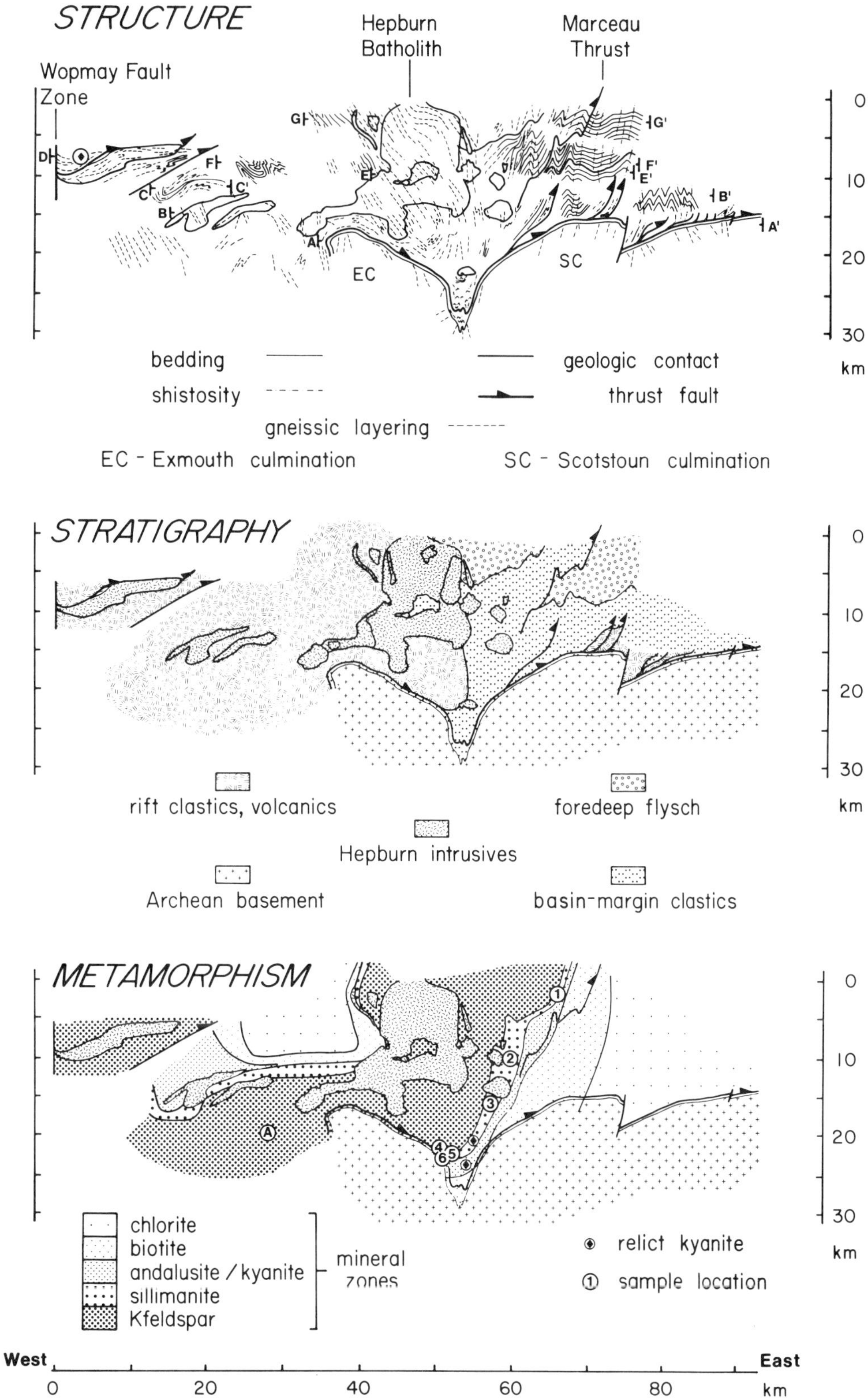

Figure 19. Composite cross sections of internal metamorphic-plutonic zone constructed from axial projections. See Figure 17 for locations of transects (A–A′, etc., in upper panel) used to control cross section. Circled numbers in lower panel are locations of samples for which *P-T* paths have been determined (see Fig. 22); circled A indicates area of K-feldspar + sillimanite paragneiss containing relict andalusite. For relation to externides, see section W–E in Figure 6.

Figure 20. Contrasting deformation states across sole thrust in the internides. A, View looking south at outcrop-scale, recumbent, angular nonparallel D1 folds in thin-bedded meta-arkose of allochtonous rift facies in transect C–C′ of Figures 17 and 19. Note straightness of the fold hinges, consistent with strain fabrics in this domain indicative of axis-parallel stretching. Fold style and intensity is typical of Akaitcho Group. B, Stromatolitic marble (assemblage forsterite-dolomite-quartz) from autochthonous cover in south-plunging Acasta synform between Exmouth and Scotstoun antiforms (see Fig. 21). Pen is 15 cm long. Stromatolite is virtually unstrained despite more than 30 km of tectonic burial (see Fig. 19).

D1 Metamorphism in the Internides

The configuration of prograde metamorphic zones (Fig. 22) as projected onto the cross section (Fig. 19) is very revealing. At intermediate and high structural levels, the Hepburn batholith is asymmetrically contained within a pronounced thermal ridge, bulbous in cross section. The inversion (inward dip) of isograds on the flanks of the thermal ridge was previously demonstrated (St-Onge, 1981, 1983) from geometric relations independent of axial projection. It can also be demonstrated that thrusting occurred both before and after prograde metamorphic equilibration. In the far western internides (Fig. 19), a set of west-dipping postmetamorphic thrusts superposes pelitic assemblages containing kyanite + K-feldspar above and andalusite + cordierite below, implying a difference in metamorphic pressure of at least 3 kbar. In contrast, the andalusite and staurolite isograds east of Hepburn batholith postdate the D1 Marceau thrust and associated (D1?) folds, but the adjacent sillimanite + K-feldspar isograd is deformed by D2(?) folds.

At low structural levels, the set of isograds east of Hepburn batholith curves downward to the west, passing beneath the leading edge of the batholith, transecting the sole thrust and autochthonous cover, and finally intersecting basement in the crest of Exmouth antiform (Figs. 17, 19). Accordingly, the entire prograde sequence of isograds is inverted in the synform to the east (Fig. 19). The fact that isograds in the statically metamorphosed autochthonous cover are inverted but the stratigraphy is upright means that the inversion cannot be the result of structural rotation. A simple explanation is that heat was conducted downward from the Calderian allochthon into a relatively cold autochthon (colder by at least 250°C, based on the estimated temperature range of the inferred mineral reactions). If so, thrusting must have occurred while the allochthon was near peak metamorphic temperature, given that isograds cross the sole thrust without offset. Furthermore, because the isograds appear to postdate the tight D2 basement synform between Exmouth and Scotstoun antiforms (Fig. 19), it follows that the D2 deformation must have occurred very soon after D1 thrusting.

The Hepburn batholith is spatially, and by inference, causally related to the allochthonous thermal ridge (Fig. 19). Therefore, the batholith is probably also allochthonous. This is consistent with the notable absence of Hepburn intrusions in the autochthon (Fig. 17). The batholith is intrusive into stratigraphic units as young as the foredeep turbidites, implying that emplacement occurred while tectonic shortening was in progress. The intrusive suite has an age range of about 1895 to 1878 Ma (S. A. Bowring and W. R. Van Schmus, in preparation) and comprises more than 200 discrete plutons ranging widely in size, composition, tectonic fabric, and level of emplacement as implied by contact metamorphic assemblages (Hoffman and others, 1980;

Figure 21. Simplified structural map and vertical air photo (boxed area on map) of southwest part of Redrock culmination, showing simple, open D2 folds of basement and autochthonous cover, structurally overlain by allochthonous domain of east-vergent recumbent folds. Heavy dashed line on map is trace of sole thrust. →

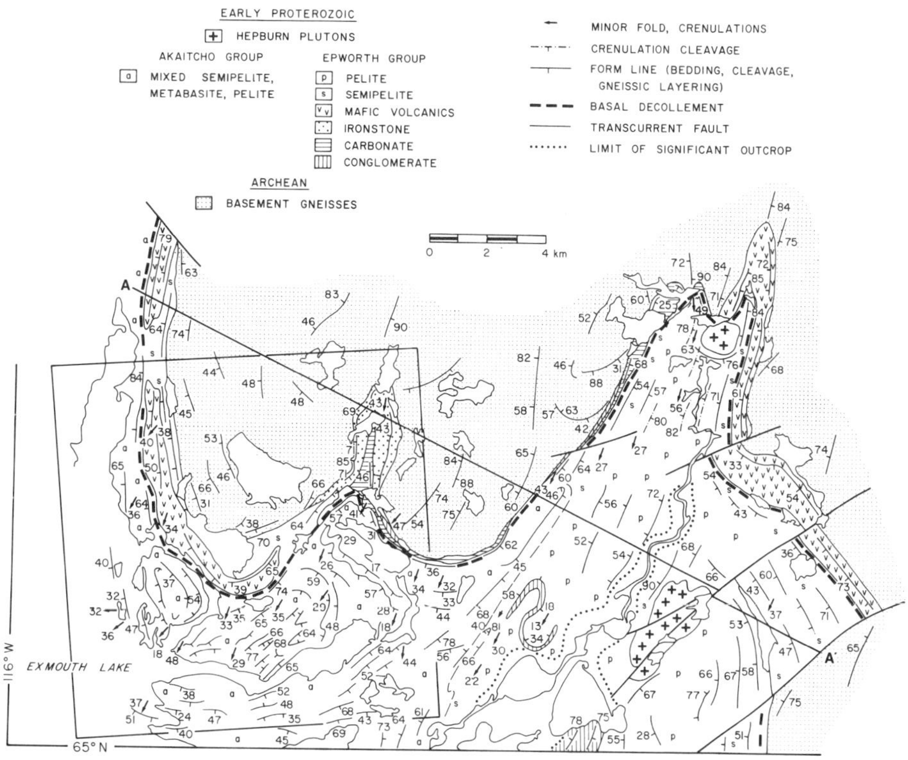
EARLY PROTEROZOIC
HEPBURN PLUTONS
AKAITCHO GROUP
MIXED SEMIPELITE, METABASITE, PELITE
EPWORTH GROUP
PELITE
SEMIPELITE
MAFIC VOLCANICS
IRONSTONE
CARBONATE
CONGLOMERATE
ARCHEAN
BASEMENT GNEISSES
MINOR FOLD, CRENULATIONS
CRENULATION CLEAVAGE
FORM LINE (BEDDING, CLEAVAGE, GNEISSIC LAYERING)
BASAL DECOLLEMENT
TRANSCURRENT FAULT
LIMIT OF SIGNIFICANT OUTCROP
0 2 4 km
A
A'
EXMOUTH LAKE
116°W
65°N

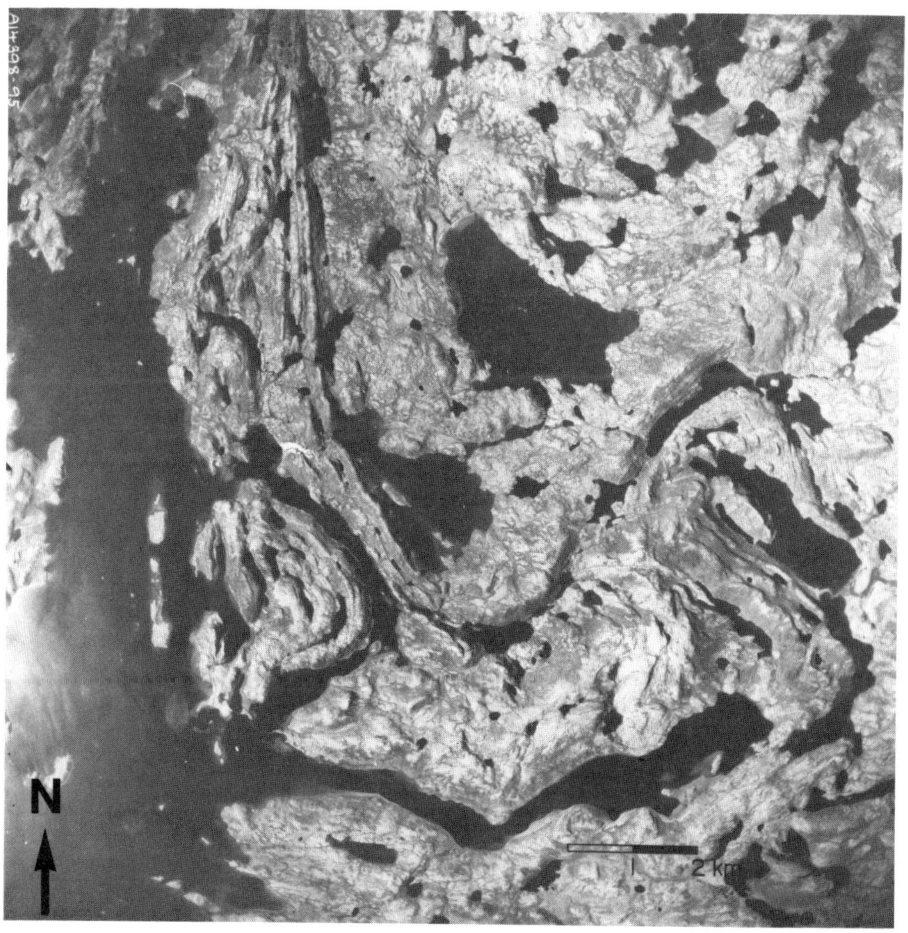
N
2 km

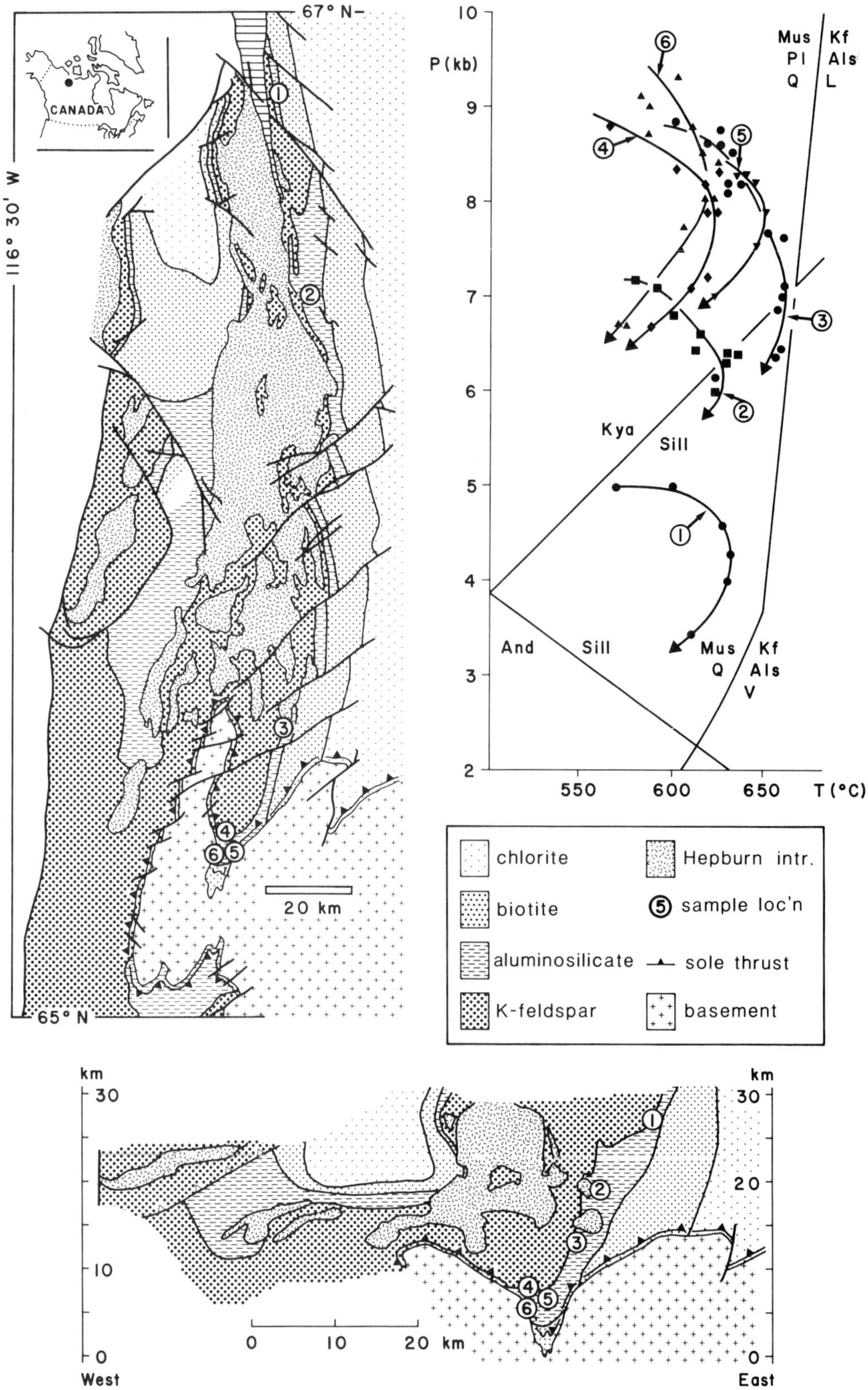

Figure 22. Distribution of metamorphic zones in map view (upper left) and cross section (below) of the internides, showing locations of samples used for *P-T* path determinations (upper right), calculated from core-to-rim microprobe analyses of zoned poikiloblastic garnets (St-Onge, 1987) in muscovite + aluminosilicate schist of Odjick Formation. Samples 1 through 5 are from increasing structural depth in allochthon; sample 6 is from autochthonous cover.

St-Onge and others, 1982, 1983, 1984; Lalonde, 1986). The largest and most abundant intrusions are variably foliated, megacrystic, biotite monzogranites, typically containing metasedimentary xenoliths, accessory garnet, and sillimanite. The granites are peraluminous, have low magnetic susceptibility, and moderately heavy (δO^{18} = 8 to 13) whole-rock oxygen isotope ratios (Lalonde, 1986). These characteristics, which imply a strong metasedimentary contribution by anatexis and/or assimilation, are counterbalanced by the presence of many sizeable bodies of tonalite, quartz diorite, granodiorite, and smaller bodies of diorite, norite, anorthosite, and various ultramafic rock types. Petrochemical data do not uniquely define the tectonic setting, but the suite is evidently the product of mixed crust- and mantle-derived magmas that were emplaced into a prism of young wet sediment undergoing tectonic shortening, causing partial assimilation of sediment and quenching of the magmas with little or no surface volcanism. Such a scenario is compatible with the petrography of the foredeep turbidites (Jeletzky, 1974), which contain abundant plutonic and metamorphic detritus but limited evidence of contemporaneous intermediate and felsic volcanism.

The metamorphic evolution of the internides requires a compound causal development (St-Onge and King, 1987a,b). The metamorphic evolution in *P-T* space for each part of the internal zone is constrained by overprinting mineral assemblages and multipoint *P-T* paths derived from core-to-rim microprobe analyses of single, zoned, poikiloblastic garnets containing inclusions of biotite, plagioclase, quartz ± aluminosilicate (St-Onge, 1987). An early stage of high-*T* low-*P* metamorphism is indicated by relict andalusite porphyroblasts in Akaitcho Group sillimanite + K-feldspar paragneiss in the region west of Exmouth antiform (Fig. 17 and lower panel of Fig. 19). As metamorphic grade in this region increases systematically with stratigraphic depth and as the region is devoid of Hepburn plutons to provide local heat sources, it is postulated that the steep geothermal gradient (ca. 70°C/km) required is a direct consequence of lithospheric stretching (St-Onge and King, 1987a,b), possibly in a back-arc setting (Hildebrand and others, 1986).

The pronounced thermal ridge enclosing Hepburn batholith is attributed to advection of heat by the rising batholith (St-Onge and King, 1987b). The asymmetric disposition of the ridge with respect to the batholith (Fig. 19) may be due to eastward convection of heat by fluid flow within the allochthon. The form and preferred stratigraphic position of the batholith near the top of the rift sequence prompts the suggestion that the batholith has been rotated clockwise (looking north) in the plane of the cross section. Such a rotation must have preceded peak metamorphic equilibration because there is no systematic east-to-west increase in metamorphic pressure across the batholith (St-Onge, 1981).

Six calculated *P-T* paths (Fig. 22) for aluminosilicate + muscovite schist at different structural levels on the east flank of the thermal ridge document poikiloblastic garnet growth during decompression (uplift). The initial heating and subsequent cooling during decompression, recorded by all samples, is consistent with the England and Richardson (1977) model of tectonically thickened crust undergoing simultaneous thermal relaxation and uplift. For the autochthon (Sample 6), heating would include contributions from above and below. Because garnets near the base of the allochthon (Samples 4 and 5) do not show initial cooling related to underthrusting by the relatively cold autochthon (cf. Davy and Gillet, 1986), thrusting must have occurred prior to the uplift recorded in the zoned garnets. The similar form of *P-T* paths derived for samples from both the autochthon and allochthon suggests that the documented uplift of the allochthon occurred after it was thrust onto the craton, essentially in the tectonic framework and structural position depicted in Figure 19. This is corroborated by the qualitative agreement between the structural levels of the six samples inferred by the axial projection and the maximum pressures documented by paths 1 through 3 and inferred by paths 4 through 6 (Fig. 22).

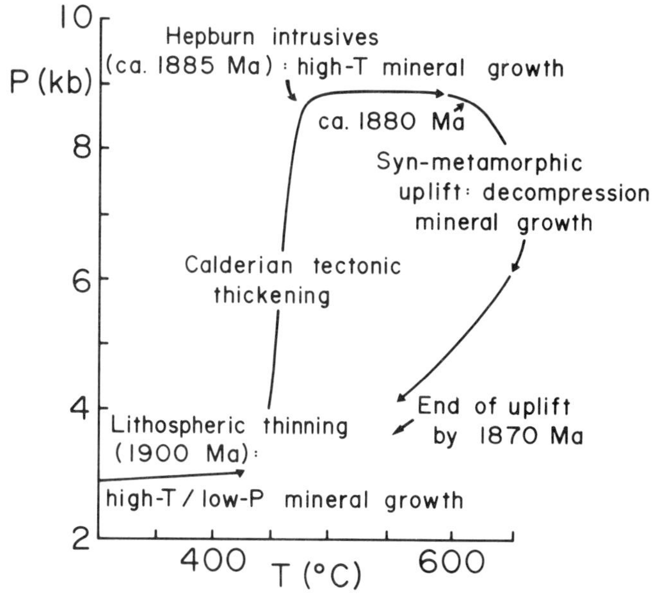

Figure 23. Synopsis of metamorphic history of central internides as constrained by overprinted mineral assemblages (see text and St-Onge and King, 1987a,b), garnet *P-T-t* paths (see Fig. 22 and St-Onge, 1987) and geochronologic (U-Pb zircon) data (S. A. Bowring and W. R. Van Schmus, in preparation).

In short, regional metamorphism combines three overlapping, thermally additive tectonic regimes (Fig. 23): initial increase in heat flux resulting from lithospheric stretching, advection of heat during Hepburn plutonism, and thermal relaxation concurrent with uplift following crustal thickening. The rapid evolution of the basin (ca. 20 Ma) minimizes cooling, which is reflected in the high metamorphic temperatures ultimately attained.

D2 and D3 Deformations in the Internides

The thick-skinned D2 and D3 folds are similar in style but higher in amplitude compared with those of the externides (King,

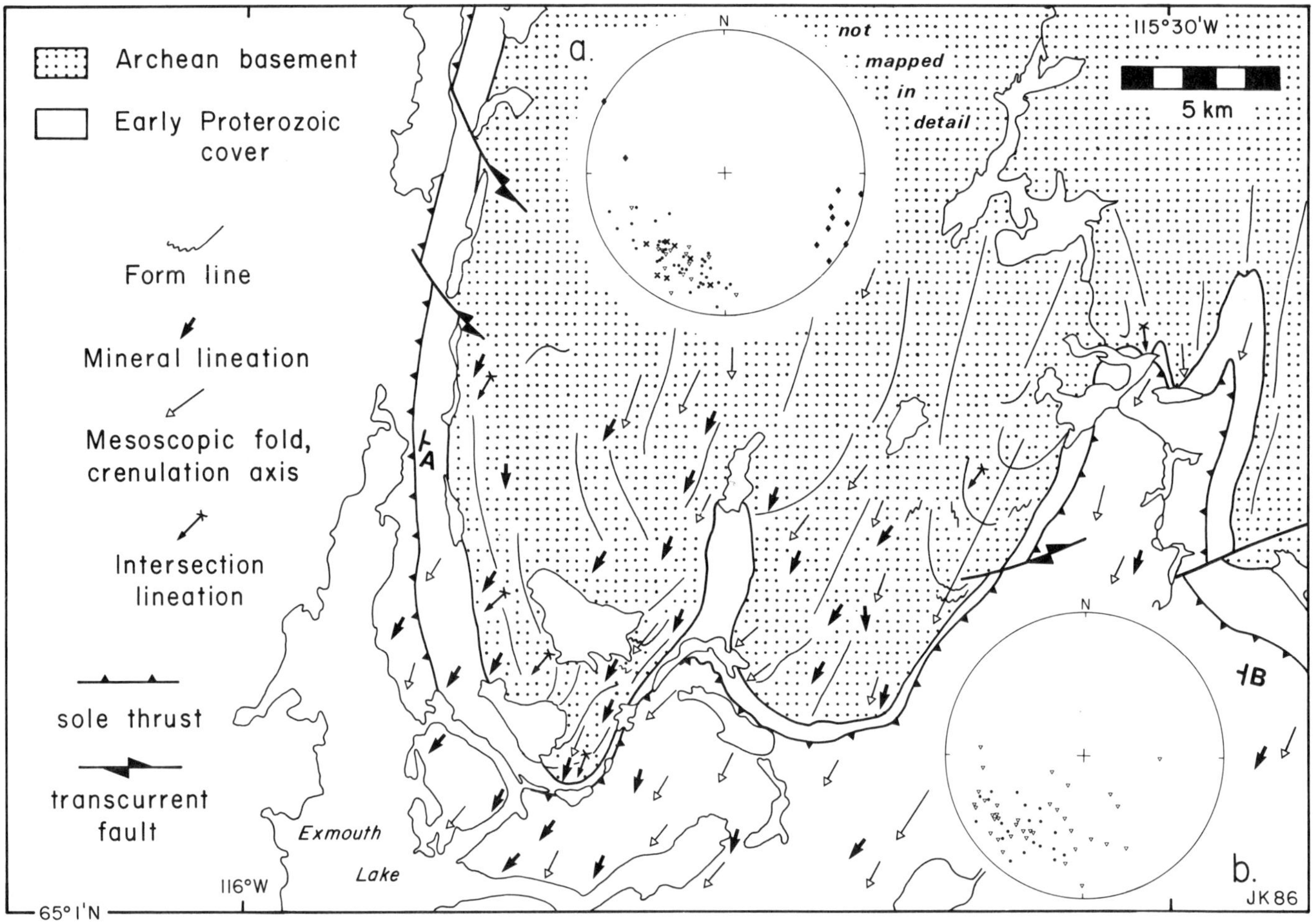

Figure 24. Simplified geologic map of south end of Exmouth antiform, emphasizing structural elements in basement. For form lines in the cover, see Figure 21. Only D2 folds are indicated in the cover. Equal area stereonets show linear elements in (a) basement and (b) cover. Symbols: dots, mineral lineations and streaking; x's intersections between Archean gneissic layering and Calderian cleavage; triangles, fold axes; diamonds, poles of Calderian axial planar cleavage in basement. A–B is line of section for Figure 25.

1987). The estimated structural relief on the basement surface between Exmouth and Scotstoun (D2) antiforms is about 13 km (Fig. 19). The greater amplitude is probably related to higher mean basement temperature in the internal zone. The only significant D2 basement fault is located on the east limb of Scotstoun antiform and has an estimated 3.5 km of high-angle reverse slip (Fig. 19).

Tight outcrop- and map-scale D2 folds with steep axial surfaces congruently deform the autochthon and allochthon in the tightly pinched basement synform between the two antiforms. Elsewhere in the Akaitcho Group, D1 foliations are crenulated about steep north-trending D2 axial surfaces. In the Epworth and Recluse groups at intermediate and high structural levels, retrograde D2 foliations are subparallel to steeply dipping synmetamorphic D1 foliations.

The manifestations of D2 deformation in basement rock have been documented in the southern part of Exmouth antiform (King and others, 1987). The Archean gneissic layering within the core of the antiform is deformed by tight, upright, centimeter- to kilometer-scale folds that also plunge moderately to the southwest (Fig. 24). In the hinge zones of the folds, a steeply dipping, axial planar cleavage defined by the planar alignment of quartz and feldspar can be recognized; on the limbs of the folds, similar planar alignment of minerals occurs but cannot be unambiguously distinguished from the earlier gneissic foliation. A pronounced lineation, defined by streaks of recrystallized quartz and feldspar, and occasionally by alignment of hornblende, coaxial with the axes of Calderian folds (including the main Exmouth antiform) and the intersection between Archean gneissic layering and the Calderian cleavage, occurs in the southern and western part of the basement core (Fig. 24). The pre-D3 strain pattern of upright folds, steep cleavage, and coaxial lineations throughout the basement core is compatible with a history of east-west shortening and layer-parallel (hinge-line) extension (cf. Watkinson, 1975) during Calderian deformation. The folds and lineations are coaxial with late metamorphic (D2) folds and lineations in the cover strata (Fig. 24).

The angular relationships between gneissic layering in the Archean gneisses and the folded basement-cover interface show that the prefolding attitude of the gneissic layering was inclined to the interface. On the east side of both lobes of the antiform, gneissic layering near the interface was drawn into macroscopic antiforms that are not expressed in the trace of the folded interface (Fig. 24). Elsewhere, the layering is roughly parallel to the limbs and axial plane of Exmouth antiform. This pattern suggests that the gneissic layering originally dipped westward from the basement-cover interface (Fig. 25). Where the angular discordance was drawn into the east limbs of the fold lobes, it is modified but is preserved; curvature of the layering records the folding of the originally obliquely inclined surface in this part of the fold. In the hinge zone, the layering remained at an acute angle to the basement-cover interface but is not itself strongly curved, although it is probably modified. On the west limbs, the original angular discordance is sufficiently modified to appear nearly parallel to the limb and axial plane of the antiform.

Diapiric amplification of the basement antiforms is possible given the greater density of the metasedimentary cover relative to the basement (M. D. Thomas, personal communication). The amount of D2 shortening is therefore difficult to estimate because constant bed length cannot be assumed if stretching due to diapiric amplification has occurred on the steeply dipping fold limbs.

The map-scale D3 cross-folding is well documented by D1 and D2 plunge reversals (Fig. 18), the configuration in map view of the sole thrust, and regional variation in metamorphic pressure inferred from mineral assemblages and microprobe data (St-Onge, 1984a, 1987). Outcrop-scale D3 deformation has not been systematically investigated in the internides, but kink-bands, compatible with a north-south component of shortening, are ubiquitous in phyllites and schists.

The garnet core-to-rim *P-T* paths (Fig. 22) are not "nested," contrary to the pattern expected for the uplift of different structural levels at a common uplift rate (cf., England and Thompson, 1984). Differential uplift rates, ranging from about 1.5 to 2.7 mm/yr given the isotopic age constraints (St-Onge, 1987), may be the causal factor for the varying *P-T* paths in Wopmay orogen. The differential rates are spatially related to regional cross folding (St-Onge and King, 1987b; St-Onge, 1987), thus indicating that the D3 deformation occurred as garnet growth continued. This implies a very rapid evolution from D1 through D3.

SUMMARY AND DISCUSSION

Eastern Wopmay orogen experienced three stages and two distinct modes of crustal thickening. Early thin-skinned thrusting and related folding (D1) evolved into coaxial thick-skinned folding (D2) of 35-km wavelength, producing an estimated 6 to 13 km of structural relief on the autochthonous basement surface. Subsequent thick-skinned cross-folding (D3) of 80- to 140-km wavelength produced 10 and 30 km of structural relief in the externides and internides, respectively, allowing composite cross sections of the earlier structures to be constructed by means of axial projections. Neither of the thick-skinned deformations has significant associated thrusting. Apparently they represent buckling instabilities developed in a lithosphere undergoing horizontal compression (cf. Ricard and Froidevaux, 1986).

D1 thin-skinned deformation involved east-west shortening and eastward translation above a subhorizontal sole thrust located within pelitic strata 100 to 300 m above the Archean basement surface. A minimum of 40 percent tectonic shortening and thickening above the sole thrust was achieved in overlapping domains of contrasting structural style, corresponding to different

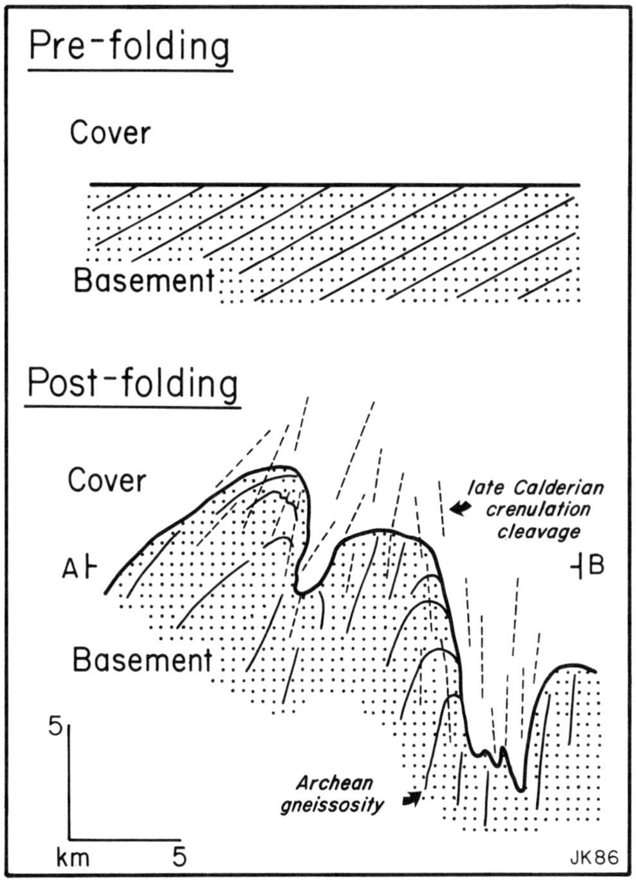

Figure 25. Cross sections showing modification of attitude of Archean gneissic layering in southern part of Exmouth antiform by Calderian folding. Form of basement-cover interface is based on axial projection modified from King (1986). Dashed lines indicate D2 cleavage. Line of section A–B is shown in Figure 24.

stratigraphic-structural levels. Tight recumbent folds characterize the western basal syn-rift deposits, imbricate thrust sheets and related folds typify the overlying west-facing shelf-rise prism, and tight upright chevron-type folds predominate in the synorogenic foredeep turbidites. The D1 structural wedge, prior to thick-skinned folding, had an overall geometry and internal structure essentially like that seen in multichannel seismic reflection profiles of modern oceanic accretionary wedges (e.g., Westbrook and others, 1982; Aoki and others, 1983). A particular point of similarity is the presence of a thin autochthonous sedimentary cover, "glued" to the basement, which extends beneath the allochthon for at least 100 km behind the frontal thrust. A significant difference compared with many oceanic accretionary wedges and foreland thrust-fold belts is the absence of contemporaneous normal faults developed at the top of the subducting plate (cf. Lillie, 1984).

According to theoretical mechanical models (Chapple, 1978; Davis and others, 1983; Stockmal, 1983; Dahlen and others, 1984; Dahlen, 1984; Zhao and others, 1986; Dahlen and Suppe, this volume), accretionary wedges can be characterized according to their "critical taper" (the shape for which the wedge is on the verge of failure under horizontal compression everywhere, including the sole thrust). The taper angle is a function of the resistance to frictional sliding along the sole thrust relative to the internal yield strength of the allochthonous wedge. The following geometric evidence suggests that the active D1 Calderian wedge had a relatively low taper (Tirrul, 1983) east of Hepburn batholith: (1) numerous reversals of local structural vergence (back-thrusts); (2) large "step-up" angles between the sole thrust and the imbricate thrusts; and (3) lack of progressive east-west change in average erosion level relative to basement, as reflected by the stratigraphic levels exposed in any east-west transect (Fig. 3). The mechanical implication is that the sole thrust was weak, and/or that the wedge was internally strong, relative to many other foreland thrust-fold belts. As the stratigraphic position of the sole thrust does not appear to be related to any particular lithological layer, variation in pore-fluid pressure may have been the critical factor in localizing the sole thrust (cf. Westbrook and Smith, 1983).

Thrusting of hot allochthons over relatively cold autochthons is now recognized in many Phanerozoic orogenic belts (e.g., Thompson and others, 1968; Andreasson and Lagerblad, 1980; Bordet and others, 1981; Hodges and others, 1982; Burg and others, 1984; Watkins, 1985). Evidence that Hepburn batholith and its associated high-grade thermal metamorphic ridge are rootless, having been translated onto an autochthon at least 250°C colder, includes the configuration of inverted metamorphic isograds beneath the batholith (i.e., static inverted metamorphism of stratigraphically upward-facing autochthonous cover), and the absence of Hepburn intrusives in the autochthon beneath the batholith. The batholith apparently was intruded through the zone of stretched basement west of Exmouth antiform, from which it was structurally detached and translated, while still hot, onto the cratonic margin. The high metamorphic temperatures attained in the allochthon are due to rapid evolution of the basin from initial stretching in a back-arc setting to final shortening and uplift supplemented by intrabasinal plutonic heat advection.

The two thick-skinned folding events represent unconventional modes of crustal thickening in response to horizontal shortening. The observed coherence of complete large- and small-scale basement folds, lacking structural repetitions of the basement-cover contact, rules out basement-involved thrusting, duplexing, or wedging at the levels of observation. In the externides, there is a clear spatial association of thick-skinned deformation and a contemporary metamorphic threshold of ca. 350°C, which apparently allows the basement to deform by generating systems of ductile shear zones. At the level of the basement-cover interface, D2 and D3 folds typically have cuspate-lobate profiles controlled by ductility contrast (Ramsay, 1982). The ca. 35-km wavelength of first-order D2 folds, if treated as a buckling phenomenon, suggests the possibility of a blind décollement in the lower crust (cf. Ricard and Froidevaux, 1986). Conversely, the 80- to 140-km wavelength of first-order D3 folds would be compatible with elastic-plastic buckling of a relatively thin lithosphere, compared with the 130- to 250-km wavelength of folds observed in 40- to 70-Ma oceanic lithosphere of the central Indian Ocean (Weissel and others, 1980; McAdoo and Sandwell, 1985; Levchenko and others, 1985; Zuber, 1987). A relatively thin effective lithosphere is neither surprising in view of an ensialic back-arc setting, nor in view of the thermal implications of continued garnet growth in the internides and local biotite-grade syn-D3 metamorphism in the externides. Analysis of the gravity field may clarify the degree of lithospheric mantle participation in the thick-skinned folding. A quantitative analysis is required because the density of the cover may exceed that of the basement, meaning that the cover and the mantle have counteractive effects on the gravity anomalies produced by folding.

It has been shown that the D3 cross-folds provide the basis for axial projections, from which composite cross sections of the D1 and D2 structures may be constructed. Although limited to structures that are approximately cylindrical on the regional scale, geologists have long urged the use of axial projections in elucidating the depth dimension of orogenic belts (e.g., Dale, 1894; Argand, 1911; Mackin, 1950; Stockwell, 1950). In Wopmay orogen, the composite cross sections so constructed have greater resolution than seismic reflection profiles, and the structural relief exposed in this essentially flat terrain greatly exceeds the topographic relief of the highest mountains on earth.

Several other early Proterozoic orogenic belts in the Canadian shield appear to have evolved structurally from early thin-skinned thrusting through coaxial and transverse thick-skinned folding. Examples include the Cape Smith belt (Hoffman, 1985; St-Onge and others, 1986, 1987; St-Onge and Lucas, 1987) and Labrador trough (Hoffman, 1985; N. Goulet, personal communication, 1986) of northern Quebec, the Bear Creek Hills (Tirrul, 1985; R. Tirrul and J. P. Grotzinger, in preparation) and East Arm of Great Slave Lake (Hoffman, 1987b) thrust-fold belts on the east margin of Slave craton (Fig. 1), and the western Foxe fold belt (Henderson, 1984) and Amer thrust-fold belt (Tippett and Heywood, 1978; Patterson, 1986) in the northwestern Churchill Province. This means that the three-dimensional structure and metamorphic zonal configuration of these belts may also be elucidated by axial projections. The essential prerequisites are detailed structural maps of key orogenic segments of sufficient strike length to link the shallowest and deepest structural levels. When this is accomplished, some basis may exist for drawing inferences concerning long-term secular change in tectonic processes (conditioned by the earth's declining thermal regime), a subject of much current speculation.

The common occurrence of long-wavelength basement folds in the ca. 1.9-Ga orogenic belts of the Canadian shield may record a fundamental characteristic of continental lithospheric deformation at that time, or it may be specifically related to the rapid succession (documented geochronologically) of microcontinental collisions that occurred during the primal assembly of the North American craton (Hoffman, 1988). Lithospheric buckling, as opposed to failure by thrusting, would be favored by a relatively thin but stiff rheologically dominant layer. Whatever their cause, the presence of high-amplitude thick-skinned folds permits the linkage of shallow and deep crustal levels through the ease of surface mapping. There is no reason for shield geologists to suffer "mountain envy."

ACKNOWLEDGMENTS

We gratefully acknowledge our principal co-workers in Wopmay Orogen—Sam Bowring, Mike Easton, John Grotzinger, Robert Hildebrand, André Lalonde and Ingrid Reichenbach—for original information, new ideas, and critical discussions. We also thank Jean-Pierre Burg, Darrel Cowan, Simon Hanmer, and Tony Morse for critically reading earlier drafts of the manuscript and making helpful comments. Pianistic pleasures were provided by Sonny Clark, Tommy Flanagan, Elmo Hope, and Hank Jones. The manuscript is Geological Survey of Canada Contribution 28386.

REFERENCES CITED

Andreasson, P. G., and Lagerblad, B., 1980, Occurrence and significance of inverted metamorphic gradients in the western Scandinavian Caledonides: Journal of the Geological Society of London, v. 137, p. 219–230.

Aoki, Y., Tamano, T., and Kato, S., 1983, Detailed structure of the Nankai Trough from migrated seismic sections, *in* Watkins, J. S., and Drake, C. L., eds., Studies in continental margin geology: American Association of Petroleum Geologists Memoir 34, p. 309–324.

Argand, E., 1911, Les nappes de recouvrement des Alpes Pennines et leur prolongements structuraux: Matiere carte geologique de Suisse, N.S. 31, Berne, 26 p.

Bordet, P., Colchen, M., Le Fort, P., and Pecher, A., 1981, The geodynamic evolution of the Himalaya; Ten years of research in central Nepal Himalaya and some other regions, *in* Gupta, H. K., and Delaney, F. M., eds., Zagros, Hindu Kush, Himalaya; Geodynamic evolution: American Geophysical Union, Geodynamics Series v. 3, p. 149–168.

Bowring, S. A., Van Schmus, W. R., and Hoffman, P. F., 1984, U-Pb zircon ages from Athapuscow Aulacogen, East Arm of Great Slave Lake, N.W.T., Canada: Canadian Journal of Earth Sciences, v. 21, p. 1315–1324.

Boyer, S. E., and Elliott, D., 1982, Thrust systems: American Association of Petroleum Geologists Bulletin, v. 66, p. 1196–1230.

Burg, J. P., Layreloup, A., Marchand, J., and Matte, Ph., 1984, Inverted metamorphic zonation and large scale thrusting in the Variscan Belt: An example in the French Massif Central, *in* Hutton, D.H.W., and Sandserson, D. J., eds., Variscan tectonics of the North Atlantic region: Geological Society of London Special Publication 14, p. 47–61.

Chapple, W. M., 1978, Mechanics of thin-skinned fold-and-thrust belts: Geological Society of America Bulletin, v. 89, p. 1189–1198.

Dahlen, F. A., 1984, Noncohesive critical Coulomb wedges; An exact solution: Journal of Geophysical Research, v. 89, p. 10125–10133.

Dahlen, A., Suppe, J., and Davis, D., 1984, Mechanics of fold-and-thrust belts and accretionary wedges; Cohesive Coulomb theory: Journal of Geophysical Research, v. 89, p. 10087–10101.

Dale, T. N., 1894, Mount Greylock; Its areal and structural geology, *in* Pumpelly, R., Wolff, J. E., and Dale, T. N., eds., Geology of the Green Mountains in Massachusetts: U.S. Geological Survey Monograph 23, p. 125–203.

Davis, D., Suppe, J., and Dahlen, F. A., 1983, Mechanics of fold-and-thrust belts and accretionary wedges: Journal of Geophysical Research, v. 88, p. 1153–1172.

Davy, P., and Gillet, P., 1986, The stacking of thrust slices in collision zones and its thermal consequences: Tectonics, v. 5, p. 913–929.

Easton, R. M., 1981, Stratigraphy of the Akaitcho Group and the development of an early Proterozoic continental margin, Wopmay Orogen, N.W.T., *in* Campbell, F.H.A., ed., Proterozoic basins of Canada: Geological Survey of Canada Paper 81-10, p. 79–95.

Ellis, M. A., 1986, Structural morphology and associated strain in the central Cordillera (British Columbia and Washington); Evidence of oblique tectonics: Geology, v. 14, p. 647–650.

England, P. C., and Richardson, S. H., 1977, The influence of erosion upon the mineral facies of rocks from different metamorphic environments: Journal of the Geological Society of London, v. 134, p. 201–213.

England, P. C., and Thompson, A. B., 1984, Pressure-temperature-time paths of regional metamorphism; I. Heat transfer during the evolution of regions of thickened continental crust: Journal of Petrology, v. 25, p. 894–928.

Fitch, T. J., 1972, Plate convergence, transcurrent faults, and internal deformation adjacent to southeast Asia and the western Pacific: Journal of Geophysical Research, v. 77, p. 4432–4460.

Grotzinger, J. P., 1986a, Evolution of early Proterozoic passive-margin carbonate platform, Rocknest Formation, Wopmay Orogen, Northwest Territories, Canada: Journal of Sedimentary Petrology, v. 56, p. 831–847.

——, 1986b, Cyclicity and paleoenvironmental dynamics, Rocknest platform, northwest Canada: Geological Society of America Bulletin, v. 97, p. 1208–1231.

——, 1986c, Measured stratigraphic sections and palinspastic cross-sections, Rocknest Formation (1.9 Ga), Wopmay Orogen, Northwest Territories: Geological Survey of Canada Open File 1278.

Henderson, J. R., 1984, Description of a virgation in the Foxe Fold Belt, Melville Peninsula, Canada, *in* Kröner, A. and Greiling, R., eds., Precambrian tectonics illustrated: Stuttgart, Schweizerbart'sche, p. 251–262.

Hildebrand, R. S., Hoffman, P. F., and Bowring, S. A., 1986, Wopmay orogen as an evolving continental arc; Magmatic response to contrasting stress regimes, *in* Geological Association of Canada Program with Abstracts, v. 11, p. 81.

Hildebrand, R. S., Hoffman, P. F., and Bowring, S. A., 1987, Tectonomagmatic evolution of the 1.9 Ga Great Bear magmatic zone, Wopmay orogen, northwestern Canada: Journal of Volcanology and Geothermal Research, v. 32, p. 99–118.

Hodges, K. V., Bartley, J. M. and Burchfiel, B. C., 1982, Structural evolution of an A-type subduction zone: Lofoten-Rombak area, northern Scandinavian Caledonides: Tectonics, v. 1, p. 441–462.

Hoffman, P. F., 1981, Autopsy of Athapuscow Aulacogen; A failed arm affected by three collisions, *in* Campbell, F.H.A., ed., Proterozoic basins of Canada: Geological Survey of Canada Paper 81-10, p. 97–102.

——, 1984a, Geology, northern internides of Wopmay Orogen, District of Mackenzie, Northwest Territories: Geological Survey of Canada Map 1576A, scale 1:250,000.

——, 1985, Is the Cape Smith Belt (northern Quebec) a klippe?: Canadian Journal of Earth Sciences, v. 22, p. 1361–1369.

——, 1987a, Proterozoic foredeeps, foredeep magmatism and Superior-type iron-formations in the Canadian Shield, *in* Kröner, A., Proterozoic lithospheric evolution: Washington, D.C., American Geophysical Union Geodynamics Series, v. 17, p. 85–98.

——, 1987b, Geology and tectonics, east arm of Great Slave Lake, District of Mackenzie, Northwest Territories: Geological Survey of Canada Map 1628A, scale 1:250,000.

——, 1987c, Continental transform tectonics: Great Slave Lake shear zone (ca. 1.9 Ga), northwest Canada: Geology, v. 15, p. 785–788.

——, 1988, United plates of America, the birth of a craton: Early Proterozoic assembly and growth of Laurentia: Annual Review of Earth and Planetary Sciences, v. 16 (in press).

Hoffman, P. F., and Bowring, S. A., 1984, Short-lived 1.9 Ga continental margin and its destruction, Wopmay Orogen, northwest Canada: Geology, v. 12, p. 68–72.

Hoffman, P. F., and Pelletier, K. S., 1982, Cloos Nappe in Wopmay Orogen: significance for stratigraphy and structure of the Akaitcho Group, and implications for opening and closing of an early Proterozoic continental margin, *in* Current research, Pt. A: Geological Survey of Canada Paper 82-1A, p. 109–115.

Hoffman, P. F., St-Onge, M. R., Easton, R. M., Grotzinger, J. P., and Schulze, D. E., 1980, Syntectonic plutonism in north-central Wopmay Orogen (early Proterozoic), Hepburn Lake map area, District of Mackenzie, *in* Current research, Pt. A: Geological Survey of Canada Paper 80-1A, p. 171–177.

Hsü, K. J., 1979, Thin-skinned plate tectonics during neo-Alpine orogenesis: American Journal of Science, v. 279, p. 353–366.

Jeletzky, O. L., 1974, Unroofing of the Hepburn batholith; A petrographic study of the Aphebian Recluse Formation, Epworth Group, Northwest Territories [B.Sc. thesis]: Ottawa, Carleton University, 73 p.

King, J. E., 1986, The metamorphic internal zone of Wopmay Orogen (Early Proterozoic), Canada: 30 km of structural relief in a composite section based on plunge projection: Tectonics, v. 5, p. 973–994.

——, 1987, Style and processes of autochthonous basement involvement in Wopmay Orogen, Canada: Journal of Structural Geology (in press).

King, J. E., Barrett, P. D., and Relf, C. D., 1987, Contrasting styles of basement deformation and longitudinal extension in the metamorphic-internal zone of Wopmay orogen, Northwest Territories, *in* Current research, Pt. A: Geological Survey of Canada Paper 87-1A, p. 515–531.

Lalonde, A. E., 1986, The intrusive rocks of the Hepburn metamorphic-plutonic zone of the central Wopmay orogen, N.W.T. [Ph.D. thesis]: Montreal, McGill University, 258 p.

Levchenko, O. V., Merklin, L. R., and Neprochnov, Yu.P., 1985, Folded structures in the central trough of the Indian Ocean: Geotectonics, v. 19, p. 10–15.

Lillie, R. J., 1984, Tectonic implications of subthrust structures revealed by seismic profiling of Appalachian-Ouachita orogenic belt: Tectonics, v. 3, p. 619–646.

Lucas, S. B., 1984, Low-grade metamorphism in the externides of Wopmay Orogen, N.W.T. [B.Sc. thesis]: Kingston, Queen's University, 163 p.

Mackin, J. H., 1950, The down-structure method of viewing geologic maps: Journal of Geology, v. 58, p. 55–72.

McAdoo, D. C., and Sandwell, D. T., 1985, Folding of oceanic lithosphere: Journal of Geophysical Research, v. 90, p. 8563–8569.

Patterson, J. G., 1986, The Amer belt; Remnant of an Aphebian foreland fold and thrust belt: Canadian Journal of Earth Sciences, v. 23, p. 2012–2023.

Platt, J. P., Coward, M. P., Deramond, J., and Hossack, J., 1986, Thrusting and deformation: Journal of Structural Geology, Special Issue, v. 8, p. 215–492.

Ramberg, H., 1981, Gravity, deformation, and the Earth's crust: London, Academic Press, 452 p.

Ramsay, J. G., 1980, Shear zone geometry: A review: Journal of Structural Geology, v. 2, p. 83–99.

——, 1982, Rock ductility and its influence on the development of tectonic structures in mountain belts, *in* Hsü, K. J., ed., Mountain building processes: London, Academic Press, p. 111–128.

Reichenbach, I., 1986, The Bell Island Group: Evidence for an ensialic marginal basin in the early Proterozoic Wopmay Orogen, Northwest Territories [M.Sc. thesis]: Ottawa, Carleton University, 120 p.

Ricard, Y., and Froidevaux, C., 1986, Stretching instabilities and lithospheric boudinage: Journal of Geophysical Research, v. 91, p. 8314–8324.

Smith, R. B., 1979, The folding of a strongly non-Newtonian layer: American Journal of Science, v. 279, p. 272–287.

Stockmal, G. S., 1983, Modeling of large-scale accretionary wedge deformation: Journal of Geophysical Research, v. 88, p. 8271–8287.

Stockwell, C. H., 1950, The use of plunge in the construction of cross-sections of folds: Proceedings of the Geological Association of Canada, v. 3, p. 97–121.

St-Onge, M. R., 1981, "Normal" and "inverted" metamorphic isograds and their relation to syntectonic Proterozoic batholiths in the Wopmay Orogen, Northwest Territories, Canada: Tectonophysics, v. 76, p. 295–316.

——, 1983, "Normal" and "inverted" metamorphic isograds and their relation to syntectonic Proterozoic batholiths in the Wopmay Orogen, Northwest Territories, Canada; Reply: Tectonophysics, v. 91, p. 180–182.

——, 1984a, Geothermometry and geobarometry in pelitic rocks of north-central Wopmay Orogen (early Proterozoic), Northwest Territories, Canada: Geological Society of America Bulletin, v. 95, p. 196–208.

——, 1984b, Evolution of the early Proterozoic Calderian thermal culmination; Mobility in pressure, temperature, and physical space, Wopmay Orogen, N.W.T.: Geological Association of Canada Program with abstracts, v. 9, p. 108.

——, 1985, The muscovite-melt bathograd and low-P isograd suites in north-central Wopmay Orogen, Northwest Territories, Canada: Journal of Metamorphic Petrology, v. 2, p. 315–326.

——, 1987, Zoned poikiloblastic garnets; Documentation of *P-T* paths and syn-metamorphic uplift through thirty kilometers of structural depth, Wopmay Orogen, Canada: Journal of Petrology, v. 28, p. 1–28.

St-Onge, M. R., and King, J. E., 1987a, Thermo-tectonic evolution of a metamorphic internal zone documented by axial projections and petrological *P-T* paths, Wopmay orogen, northwest Canada: Geology, v. 15, p. 155–158.

——, 1987b, Evolution of regional metamorphism during back-arc stretching and subsequent crustal shortening in the 1.9 Ga Wopmay orogen, Canada: Philosophical Transactions of the Royal Society of London Ser. A, v. 321, p. 199–218.

St-Onge, M. R., and Lucas, S. B., 1987, Structural and thermal history of the Cape Smith thrust-fold belt, northern Quebec, *in* Lewry, J. F., and Stauffer, M. R., eds., The early Proterozoic Trans-Hudson orogen: Lithotectonic correlations and evolution: Geological Association of Canada Special Paper (in press).

St-Onge, M. R., King, J. E., and Lalonde, A. E., 1982, Geology of the central Wopmay Orogen (early Proterozoic), Bear Province, District of Mackenzie; Redrock Lake and the eastern portion of Calder River map areas, *in* Current research, Pt A: Geological Survey of Canada Paper 82-1A, p. 99–108.

St-Onge, M. R., Lalonde, A. E., and King, J. E., 1983, Geology, Redrock Lake and eastern Calder River map areas, District of Mackenzie; The central Wopmay Orogen (early Proterozoic), Bear Province, and the western Archean Slave Province, *in* Current research, Pt. A: Geological Survey of

Canada Paper 83-1A, p. 147–152.

St-Onge, M. R., King, J. E., and Lalonde, A. E., 1984, Deformation of the Coronation Supergroup and its basement in the Hepburn metamorphic-plutonic zone of Wopmay Orogen; Redrock Lake and the eastern portion of Calder River map areas, District of Mackenzie, *in* Current research, Pt. A: Geological Survey of Canada Paper 84-1A, p. 171–180.

St-Onge, M. R., Lucas, S. B., Scott, D. J., and Bégin, N. J., 1986, Eastern Cape Smith Belt; An early Proterozoic thrust-fold belt and basal shear zone exposed in oblique section, Wakeham Bay and Cratere du Nouveau Québec map areas, northern Quebec, *in* Current research, Pt. A: Geological Survey of Canada Paper 86-1A, p. 1–14.

St-Onge, M. R., Lucas, S. B., Scott, D. J., and Bégin, N. J., 1987, Tectonostratigraphy and structure of the Lac Watts–Lac Cross–Riviere Déception area, central Cape Smith belt, northern Quebec, *in* Current research, Pt. A: Geological Survey of Canada Paper 87-1A, p. 619–632.

Tapponnier, P., Peltzer, G., and Armijo, R., 1986, On the mechanics of the collision between India and Asia, *in* Coward, M. P., and Ries, A. C., eds., Collision tectonics: Oxford, Blackwell Scientific Publications, p. 115–157.

Thompson, J. B., Robinson, P., Clifford, T. N., and Trask, N. J., 1968, Nappes and gneiss domes in west-central New England, *in* Zen, E., White, W. S., Hadley, J. B., and Thompson, J. B., Studies in Appalachian geology; Northern and maritime: New York, John Wiley Interscience Publishers, p. 203–218.

Tippett, C. R. and Heywood, W. W., 1978, Stratigraphy and structure of the northern Amer Group (Aphebian), Churchill Structural Province, District of Keewatin, *in* Current research, Part B: Geological Survey of Canada Paper 78-1B, p. 7–11.

Tirrul, R., 1982, Frontal thrust zone of Wopmay Orogen, Takijuq Lake map area, District of Mackenzie, *in* Current research, Pt. A: Geological Survey of Canada Paper 82-1A, p. 119–122.

——, 1983, Structure cross-sections across Asiak foreland thrust and fold belt, Wopmay Orogen, District of Mackenzie, *in* Current research, Pt. B: Geological Survey of Canada Paper 83-1B, p. 253–260.

——, 1985, Nappes in the Kilohigok Basin, and their relation to the Thelon Tectonic Zone, District of Mackenzie, *in* Current research, Pt. A: Geological Survey of Canada Paper 85-1A, p. 407–420.

——, 1987a, Geology, Calliope Lake area, Wopmay orogen, District of Mackenzie: Geological Survey of Canada Map 1654A, scale 1:50,000 (in press).

——, 1987b, Geology, Asiak River area, northern Asiak thrust and fold belt, Wopmay orogen, District of Mackenzie: Geological Survey of Canada Map 1653A, scale 1:50,000 (in press).

Van Den Driessche, J., and Maluski, H., 1986, Mise en évidence d'un cisaillement ductile dextra d'age cétacé moyen dans la région de Tete Jaune Cache (nord-est du complexe métamorphique Shuswap, Columbie Britannique): Canadian Journal of Earth Sciences, v. 23, p. 1331–1342.

Van Schmus, W. R., Bickford, M. E., Lewry, J. F., and Macdonald, R., 1987, U-Pb geochronology in the Trans-Hudson Orogen, northern Saskatchewan, Canada: Canadian Journal of Earth Sciences, v. 24, p. 407–424.

Watkins, K. P., 1985, Geothermometry and geobarometry of inverted metamorphic zones in the west-central Scottish Dalradian: Journal of the Geological Society of London, v. 142, p. 157–165.

Watkinson, A. J., 1975, Multilayer folds initiated in bulk plane strain, with the axis of no change perpendicular to the layering: Tectonophysics, v. 28, p. T7–T11.

Weissel, J. K., Anderson, R. N., and Geller, C. A., 1980, Deformation of the Indo-Australian plate: Nature, v. 287, p. 284–291.

Westbrook, G. K., and Smith, M. J., 1983, Long décollements and mud volcanoes: evidence from the Barbados Ridge complex for the role of high pore-fluid pressure in the development of an accretionary complex: Geology, v. 11, p. 279–283.

Westbrook, G. K., Smith, M. J., Peacock, J. H., and Poutler, M. J., 1982, Extensive underthrusting of undeformed sediment beneath the accretionary complex of the Lesser Antilles subduction zone: Nature, v. 300, p. 625–628.

Zhao, W.-L., Davis, D. M., Dahlen, F. A., and Suppe, J., 1986, Origin of convex accretionary wedges; Evidence from Barbados: Journal of Geophysical Research, v. 91, p. 10246–10258.

Zuber, M. T., 1987, Compression of oceanic lithosphere; An analysis of intraplate deformation in the Central Indian Basin: Journal of Geophysical Research, v. 92, p. 4817–4825.

MANUSCRIPT ACCEPTED BY THE SOCIETY JUNE 1, 1987
GSC CONTRIBUTION 28386

Printed in U.S.A.

Geological Society of America
Special Paper 218
1988

The tectonics of the Central Andes; 30° to 33° S latitude

Victor A. Ramos
Departamento Ciencias Geológicas, Universidad de Buenos Aires, Pabellón 2–Ciudad Universitacia (1428) Buenos Aires, Argentina

ABSTRACT

The tectonic evolution of the central Andes is depicted through analysis of the 30° to 33°S segment, which encompasses the highest part of the Andean Cordillera. A period of rifting, starting as early as 600 Ma, was a milestone in the evolution of the different tectonic regimes that are responsible for the present geological composition and structure of the Andes.

The early Paleozoic was an important period of continental accretion when allochthonous terranes, such as Chilenia, were incorporated onto the western margin of Gondwanaland. The subduction zone was located about 300 km east of the present trench. An early Paleozoic magmatic arc was developed in the western Sierras Pampeanas. Sedimentary facies of that age record a continental margin between Chilenia and the magmatic arc, which is associated with a disrupted ophiolite sequence. The Famatinides orogeny produced the first deformation of the Andes and the uplift of the Protoprecordillera during middle to late Devonian times.

The Gondwanides orogeny is characterized by subduction of an oceanic plate beneath a continental margin, with the accretion of minor exotic terranes in the southern part of the Andes. Magmatism, eastward migration of the volcanic front, sedimentation pattern, and deformation defined an evolving orogenic sequence. This is correlated to a varying convergence history linked to variations in the apparent polar wandering path of western Gondwanaland during late Paleozoic–early Mesozoic times.

The Patagonides orogeny was also governed by changes in relative plate motions during middle to late Mesozoic times. The paleotectonic history suggests that two orogenic styles were produced: a stage with little compression and back-arc volcanism, and a stage with high compression without volcanism but with important deformation and emplacement of postorogenic granitoids.

The Andean orogenic cycle is distinguished by conspicuous segments of the orogen that are controlled by the segmentation of the subducted oceanic Nazca plate related to the subduction of aseismic ridges. But the compressive phases and mountain building are more closely related to changes of plate motion, which affected thousands of kilometers of the continental margin, exceeding the length of any individual segment. The age of the subducted oceanic slab is an important factor that controls the magmatic activity and the presence or absence of retroarc magmatism.

The relative influence of each proposed mechanisms varies substantially among the different segments.

INTRODUCTION

Several recent syntheses have shown that a simplistic model is far from a realistic approach to the tectonic evolution of the Andes. A variety of complex processes have been proposed for the mountain building of the Andean Cordillera: variations in plate motions, oblique and normal subduction, docking of exotic or allochthonous terranes, collision of oceanic ridges, or the buoyancy effects of a young oceanic lithosphere being subducted. Most of the proposals, however, have insufficient field constraints due to the scanty basic information in some cases, as well as difficult access to the vast Spanish geological literature.

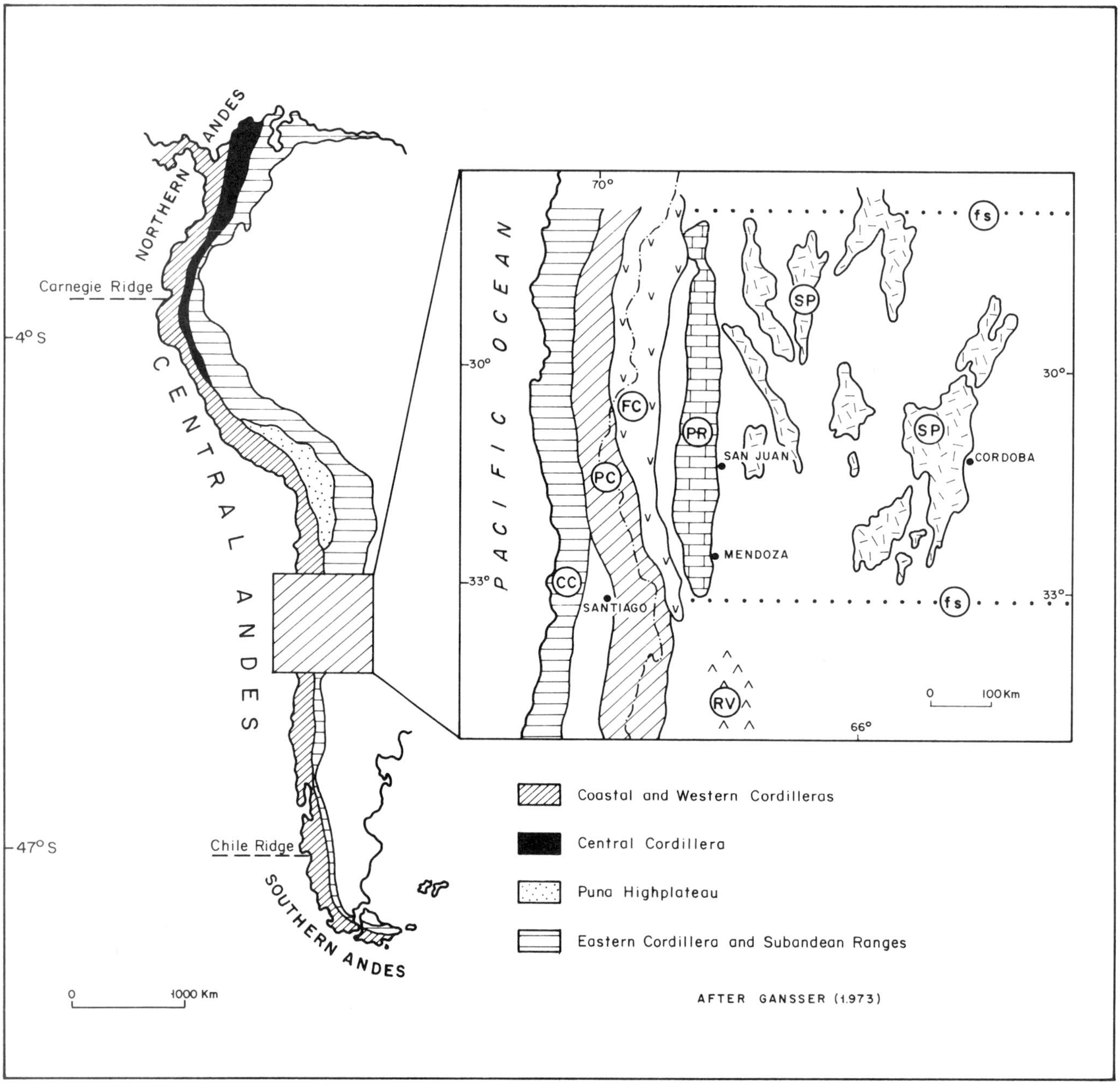

Figure 1. Main subdivisions of the Andean Cordillera (after Gansser, 1973), with location of the analyzed area. The insert shows geological provinces of central Andes between 28° and 35° S lat. CO = Coastal Cordillera; PC = Principal Cordillera; FC = Frontal Cordillera; PR = Precordillera; SP = Sierras Pampeanas; RV = retroarc volcanic province. (fs) indicates the approximate limit of the flat-slab subduction segment (after Ramos and others, 1984; Isacks and others, 1982). Dotted lines are the outlines of San Juan and Mendoza provinces.

The objective of this chapter is to summarize the geological framework of the different tectonic cycles that have been recognized, their main diastrophic phases, and the possible tectonic mechanisms that controlled the evolution of this part of the Andes. The Andes consists of many segments with quite variable geologic and diastrophic histories: the present discussion deals only with the highest segments of the Andes (30° to 33° S lat.; see Fig. 1). Its complex evolution is somewhat representative of the type of processes and the timing of the Phanerozoic structural evolution of the cordillera. Some of the hypotheses discussed here are based on sound geological bases, but others are still working hypotheses deserving of more geological and geophysical studies.

TABLE 1. CENTRAL ANDES GEOLOGIC PROVINCES AND OROGENIC CYCLES*

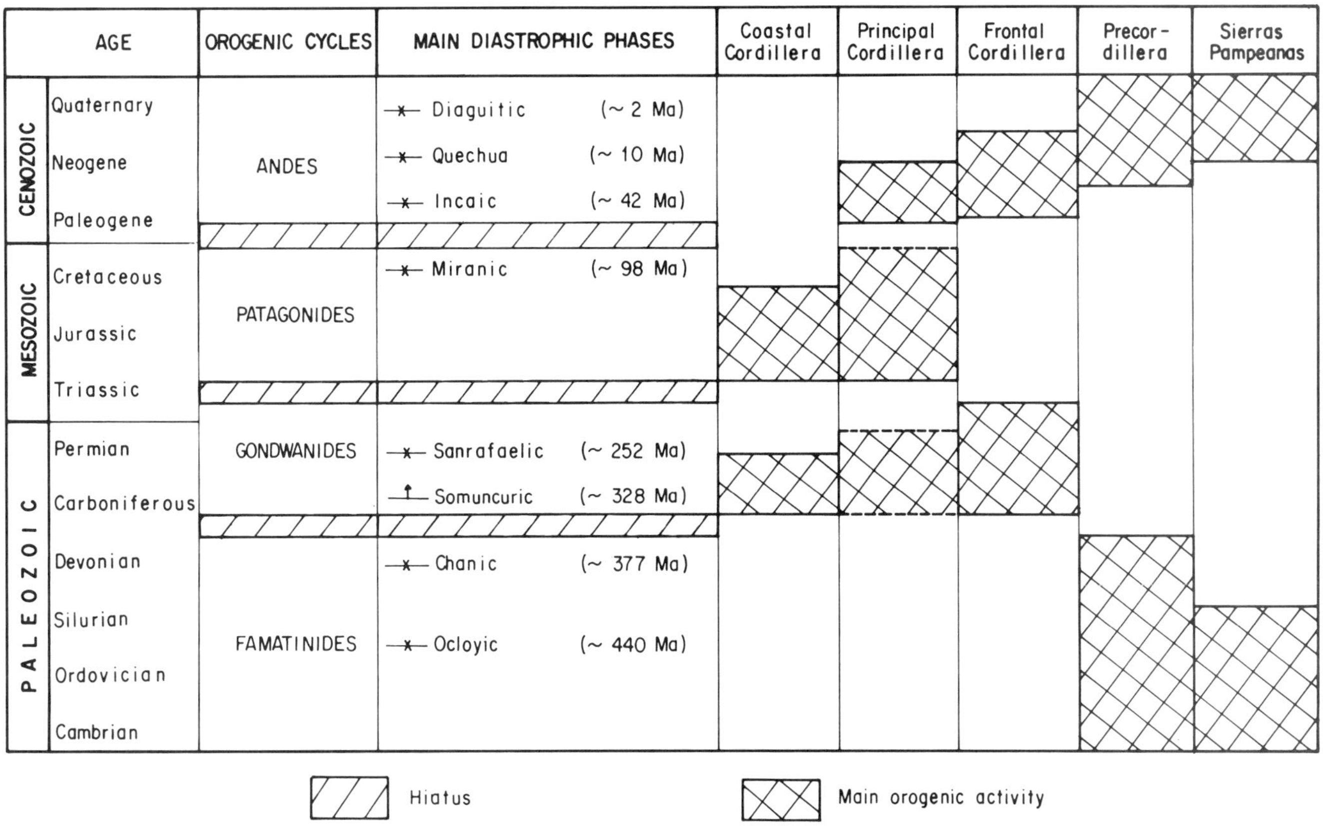

*Based on Groeber (1918, 1946, 1953), Keidel (1921), Yrigoyen (1976, 1979), Aceñolaza and Toselli (1973, 1976), and Ramos and Ramos (1979).

OROGENIC CYCLES OF THE CENTRAL ANDES

The superposition of different tectonic cycles has formed a series of geologic provinces with different geologic histories in this segment of the Andes (Fig. 1). In general, the successive orogens migrated westward by a combination of collision and subduction tectonics.

The Precordillera (PR) is an Andean thrust-and-fold belt that has been developed on early to late Paleozoic sedimentary rocks. The western edge of the Precordillera was strongly deformed during the Famatinian orogeny by the collision and accretion of an allochthonous terrane in late Devonian times (Ramos and others, 1984, 1986; Kay and others, 1984; Hervé and others, 1987).

The Frontal Cordillera (FC) and the Coastal Cordillera (CC) were built during the Gondwanides orogeny in late Paleozoic–early Mesozoic times. They are the results of an Andean-type of subduction orogenesis with subordinate accretion of minor exotic terranes (Helwig, 1972; Hervé and others, 1981; Mpodozis and Forsythe, 1983; Ramos and others, 1984).

The Principal Cordillera (PC), or Main Andes, was the locus of the Patagonides and Andean orogenies during Mesozoic and Cenozoic times. The Andean mountain building also influenced the Sierras Pampeanas (Pampean Ranges). These are Precambrian–early Paleozoic basement rocks, uplifted during Tertiary times, resulting in structures closely resembling those of the Laramide region of the United States (Jordan and others, 1983).

Table 1 summarizes the timing of the different orogenies, their main diastrophic phases, and the locus of the deformation at each time.

Strong angular unconformities marking the main orogenies are recognized between early to middle Devonian and Carboniferous rocks (Chanic movements), early and late Permian rocks (Sanrafaelic movements), late Eocene and early Oligocene rocks (Incaic movements), middle and late Miocene deposits (Quechua movements), and between Pliocene and early Pleistocene strata (Diaguitic movements) (Turner and Méndez, 1975; Stipanicic and Rodrigo, 1970a,b; Yrigoyen, 1976, 1979; Ramos and Ramos, 1979). Also, strong angularities may be seen locally between Early and Late Cretaceous deposits (Patagonides movements of Keidel, 1921).

FAMATINIAN OROGENIC CYCLE

An ancient passive continental margin was developed in the westernmost border of Gondwanaland: it is represented by the Early Cambrian to Ordovician carbonate miogeoclinal facies that crops out in the Precordillera (Baldis and others, 1982). Tectonic subsidence analysis of this sequence by Bond and others (1984) indicates that the main period of rifting that produced the passive margin apparently started about 600 Ma. The timing for the breakup of the supercontinent fits well with regional geology; it occurred immediately after the important late Proterozoic collisional episodes of the Brasilian orogenic cycle (Almeida and Hasui, 1983). This cycle resulted in the accretion of the different cratons of South America.

During early Paleozoic time, the western Gondwana continental edge was located east of the Andes at the western border of the Precordillera (Fig. 2). An extensive carbonate platform, with faunas closely linked with the Appalachian or the North America Cordilleran ones (Baldis and others, 1982), is identified in the central and eastern Precordillera. This platform was bounded by slope facies to the west, associated with mafic to ultramafic rocks (Borrello, 1969). The geochemical and petrographic analyses of these rocks indicate oceanic affinities (Kay and others, 1984); they have been interpreted as a disrupted ophiolite sequence (Haller and Ramos, 1984). The mafic rocks were obducted during the Ocloyic and Chanic movements and were incorporated into the Ordovician to early Devonian subduction complexes of the western Precordillera (Ramos and others, 1984, 1986).

A volcano-plutonic arc, known as the Faja Eruptiva de la Puna (Méndez and others, 1972), formed to the east of the present continental edge from 22° to 28° S lat. at the same time (Ramos and Ramos, 1979; Coira and others, 1982). This magmatic belt may be traced southward through early Paleozoic deeply eroded plutons in the western Sierras Pampeanas (Fig. 2) (González and others, 1985) as far south as 38° S (Llambías and Rapela, 1985). The Cambrian to Late Ordovician igneous rocks of the Sierras Pampeanas are interpreted as a magmatic arc related to the subduction zone interpreted to lie along the western margin of the Precordillera. This is corroborated by the geochemical and petrological characteristics of Sierras Pampeanas granitoids (Rapela and others, 1982; Rapela, 1982; Hervé and others, 1987). These igneous rocks belong to a calc-alkalic series with low $^{87}Sr/^{86}Sr$ ratios that originated in the upper mantle and have only minor contamination by the sialic crust (Rapela and others, 1982). The magmatism was most active during the latest Ordovician, which is the age of the Ocloyic movements (ca. 440 Ma), the Faja Eruptiva de la Puna, as well as the western Sierras Pampeanas. The magmatic activity is associated with an intense regional synkinematic metamorphism in the Sierras Pampeanas (Dalla Salda and Varela, 1982). The Famatinian paleogeography is depicted in Figure 2.

The schematic sections of Figure 3 summarize the evolution of the Famatinian orogenic cycle. The closure of the oceanic basin, lying between the allochthonous terrane of Chilenia and the western margin of Gondwanaland, produced intense Chanic deformation, accretion of the metamorphic basement of the Frontal Cordillera, cessation of magmatic activity in the Sierras Pampeanas, and migration of the subduction zone to the west. As a result of the collision, a new trench was formed more than 250 km westward along the eastern margin of Panthalassa (or the ancestral Pacific Ocean).

Time constraints set by the main unconformities in the stratigraphic sequences of the miogeocline (Baldis and others, 1982), magmatic climaxes in the Sierras Pampeanas (Ramos and Ramos, 1979), and metamorphism and deformation in the foreland (Dalla Salda and Varela, 1982) show two definite phases of tectonism: the Ocloyic and Chanic movements (Table 1). These were represented by a sequence of successive deformational events, with latitudinal variations in intensity and distribution. Tectonic activity reached a maximum at the end of Ordovician time (Ocloyic movements) and in the middle of Devonian time (Chanic movements).

The Famatinian orogenic cycle, in the study segment (30° to 33° S lat.), consists of a series of interrelated climaxes that built the Famatinides at the end of the early Paleozoic, somewhat similar to the Taconic orogeny in the Appalachians, as proposed by Rodgers (1971). As a result of the Famatinian orogeny, the submarine Tontal structural high developed in a forearc setting and was uplifted to form the Protoprecordillera as a distinctive subaerial morphostructural unit at the end of Devonian time (Rolleri and Baldis, 1967).

GONDWANIC OROGENIC CYCLE

The late Paleozoic tectonic cycle began with the inception of subduction along the present continental margin. Simple subduction of an oceanic plate beneath the continental margin was responsible for sedimentary and tectonic accretion on the westward facing continental edge.

A subduction complex developed along the western slope of the Coastal Cordillera from Late Carboniferous to Late Triassic time (Thiele and Hervé, 1984). Paired metamorphic belts of blueschists and high to medium P/T metamorphic facies crop out along the Pacific margin, associated with metamorphosed pillow basalts and metacherts that occur in discontinuous boudinaged bands and rootless intrafolial folds (Hervé and others, 1984).

Farther south along the present coast, minor exotic terranes have been identified in the Madre de Dios region (see Fig. 4) by Mpodozis and Forsythe (1983), who confirmed the previous hypotheses of Helwig (1972). The accreted rocks there are composed of mafic and ultramafic oceanic rocks, and fusulinid-bearing platform carbonates associated with flysch and pelagic chert facies, which were docked into the Gondwanic subduction complex (see Fig. 4). These terranes have lithologies and geological histories in common with many similar exotic terranes identified along the circum-Pacific margin of the United States and Canada (Monger, 1984) and the Soviet Union (Musatov and others, 1984; Yanshin and others, 1984).

These terranes indicate the widespread dispersal of a Pacific subequatorial landmass capped by fusulinid-bearing carbonates, which could represent the proposed lost Pacifica continent of Nur and Ben-Avraham (1977).

North of 32° S lat., Mesozoic and younger continental ero-

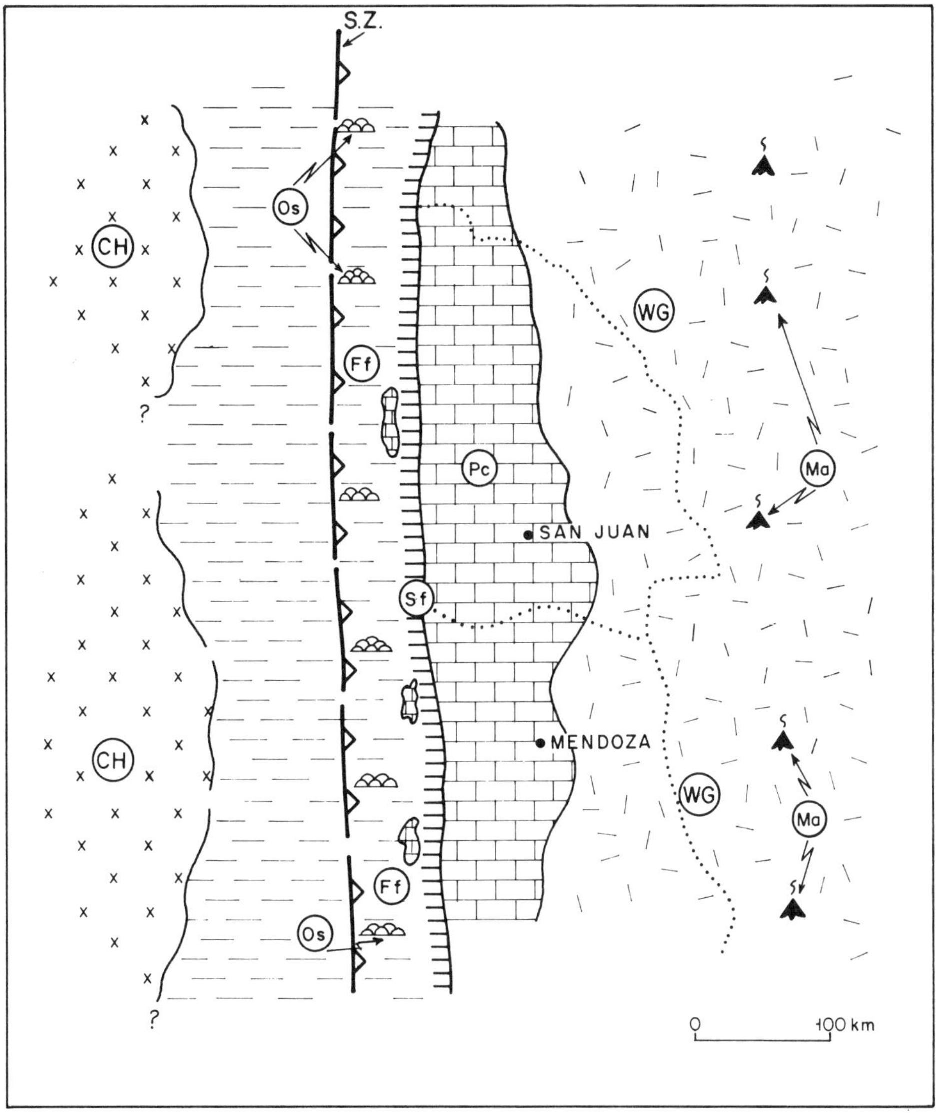

Figure 2. Early Paleozoic elements during the Famatinian orogenic cycle prior to the accretion of Chilenia (after Ramos and others, 1984). CH = Chilenia terrane; WG = West Gondwanaland; Os = ophiolite sequence; Ff = flysch and pelagic facies; Sf = slope facies with olistoliths; Pc = carbonate platform; Ma = magmatic arc; SZ = inferred subduction zone. Dotted lines indicate the outlines of San Juan and Mendoza provinces.

sion by subduction has eliminated great portions of the Gondwanic subduction complexes, as suggested by Rutland (1971) and Schweller and Kulm (1978); however, a reconstruction of the Gondwanic orogen transverse to the main structural trend is still possible (Figs. 5, 6). Most of the Gondwanic sedimentary and metamorphic rocks indicate an important lateral accretion of the continental margin, probably from discontinuous and isolated pre-Carboniferous basement blocks that constitute the substratum of the Frontal Cordillera. These blocks are composed of medium-grade metamorphic rocks that constitute the Chilenia terrane. A few Rb/Sr isotopic data available for these basement rocks point to a late Precambrian–Early Cambrian age (Caminos and others, 1979).

The Coastal Cordillera is interpreted as an accretionary wedge developed over an oceanic basement, which was highly deformed and metamorphosed during the Gondwanic cycle (Hervé and others, 1982). The existence in this area of minor blocks of sialic pre-Gondwanic basement cannot be ruled out on the basis of the present state of knowledge of coastal basement.

The Gondwanic cycle is distinguished by two important

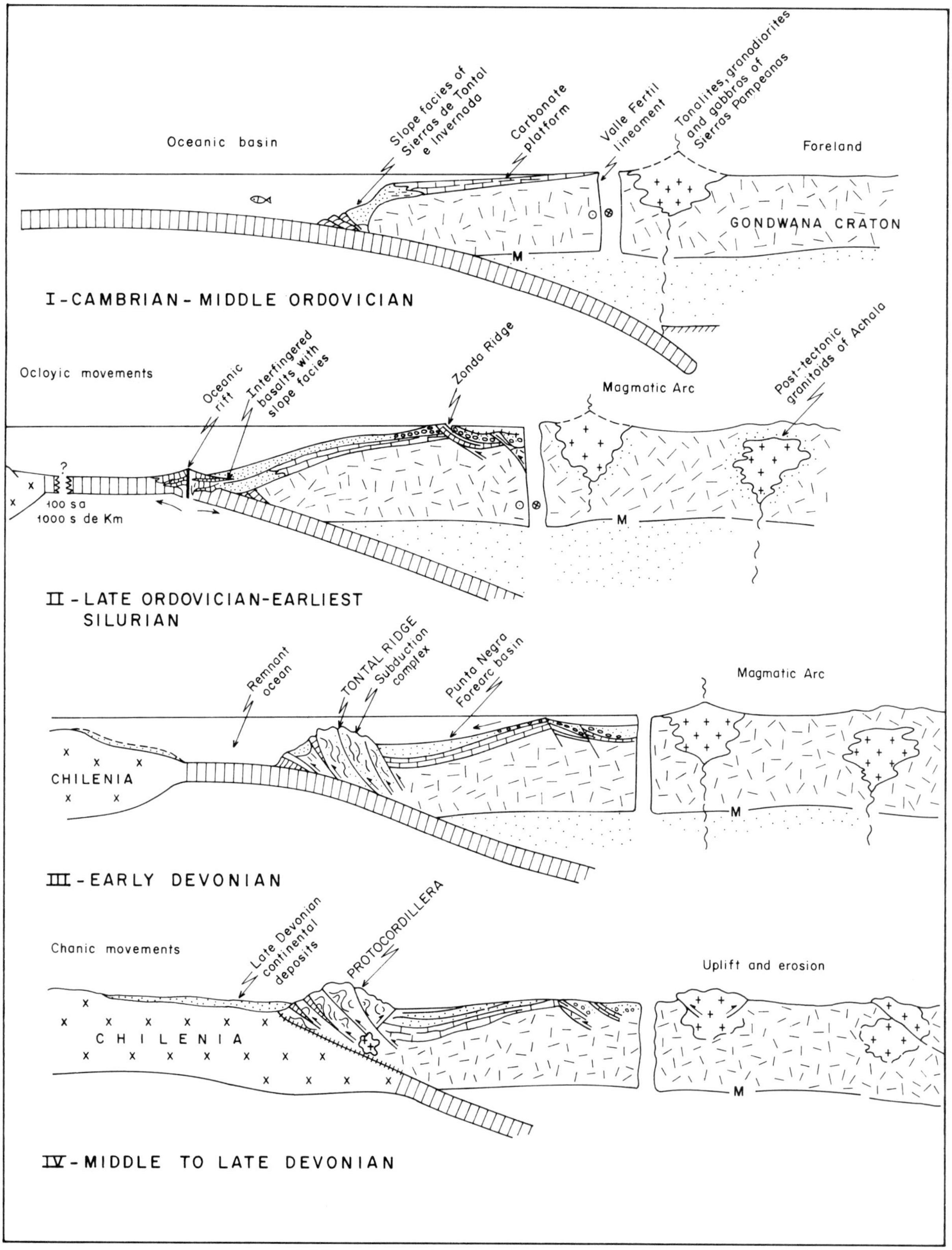

Figure 3. Schematic sections showing the Famatinian orogenic cycle of the central Andes as a consequence of the collision of Chilenia (after Ramos and others, 1985).

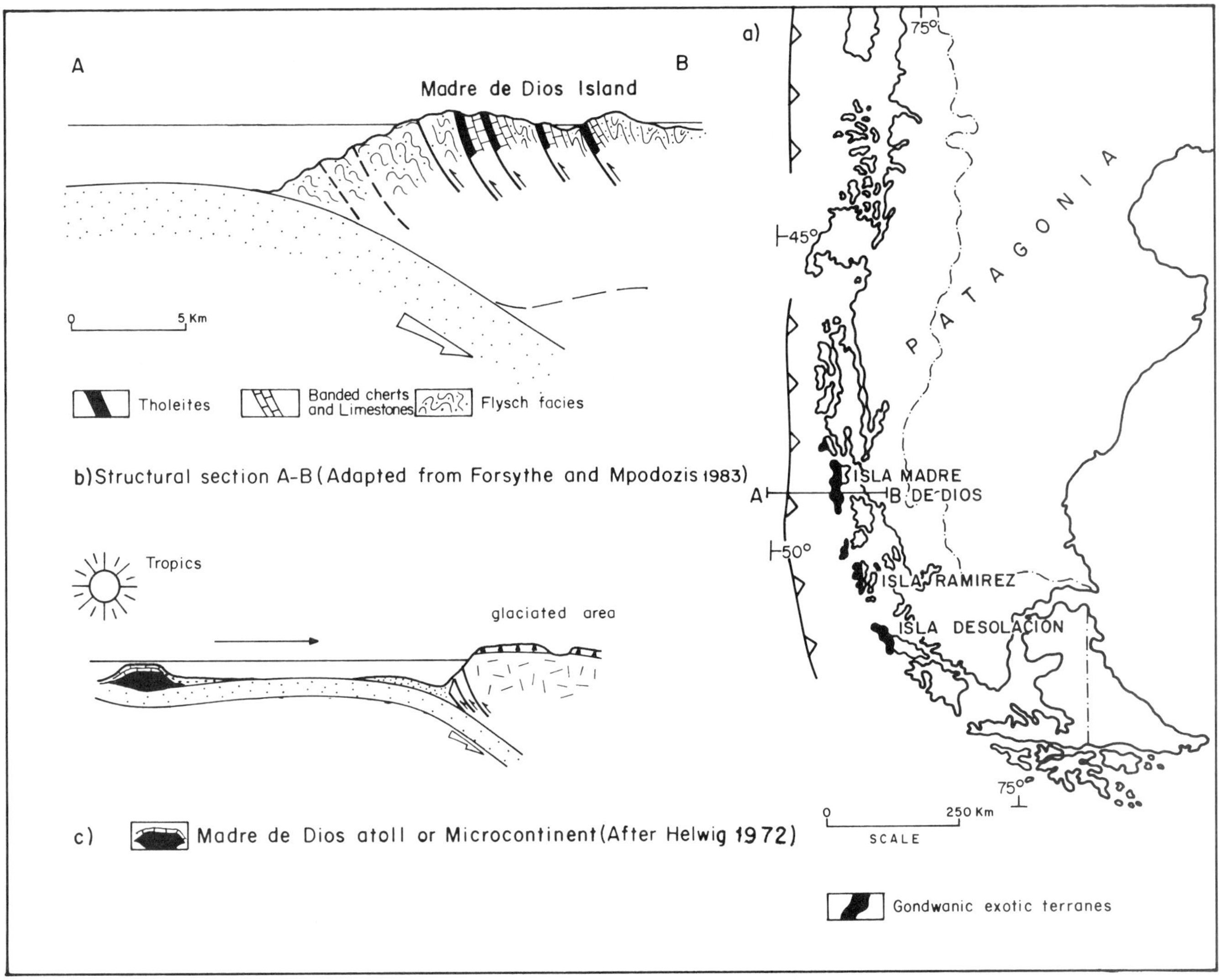

Figure 4. Exotic terranes of Madre de Dios, southern Chile, accreted to the continental edge during late Paleozoic–early Mesozoic time. a and b, Distribution and schematic section (adapted from Forsythe and Mpodozis, 1983). c, Proposed mechanism of accretion (after Helwig, 1972).

diastrophic phases: the Somuncuric and the Sanrafaelic (Fig. 6). The first is associated with an Early Carboniferous metamorphic event; it is also related to important magmatic activity that can be traced from 26° to 41° S lat. (Hervé and others, 1982; Ramos and Cortés, 1984; Llambías and Rapela, 1985). The Sanrafaelic orogenic phase produced the intense folding and thrusting responsible for the strong angular unconformity between the Permo-Triassic volcanics and the Late Carboniferous–Early Permian flysch facies that crops out in the Principal Cordillera. This phase of deformation was followed by an important eastward migration of the magmatic arc.

The magmatic arc is represented by a composite calc-alkalic batholith with synorogenic tonalites and granodiorites bearing hornblende and biotite as the main mafic minerals (Mpodozis and others, 1976). The postorogenic magmatism is characterized by acidic granitoids and widespread volcanics (Hervé and others, 1987). The locus of all this magmatic activity is concentrated in the Coastal and Frontal Cordilleras, although in the Sierras Pampeanas some minor stocks of Carboniferous granitoids are also present. When the geochemical composition of the Sierras Pampeanas granitoids is compared with the precollisional granitoids of the Famatinian cycle, striking differences are apparent. Gondwanic granitoids of the Sierras Pampeanas show a progressive enrichment of incompatible elements with decreasing age (Rapela, 1982), and have a high initial $^{87}Sr/^{86}Sr$ ratio (0.710). This implies that they were derived by the partial melting of older

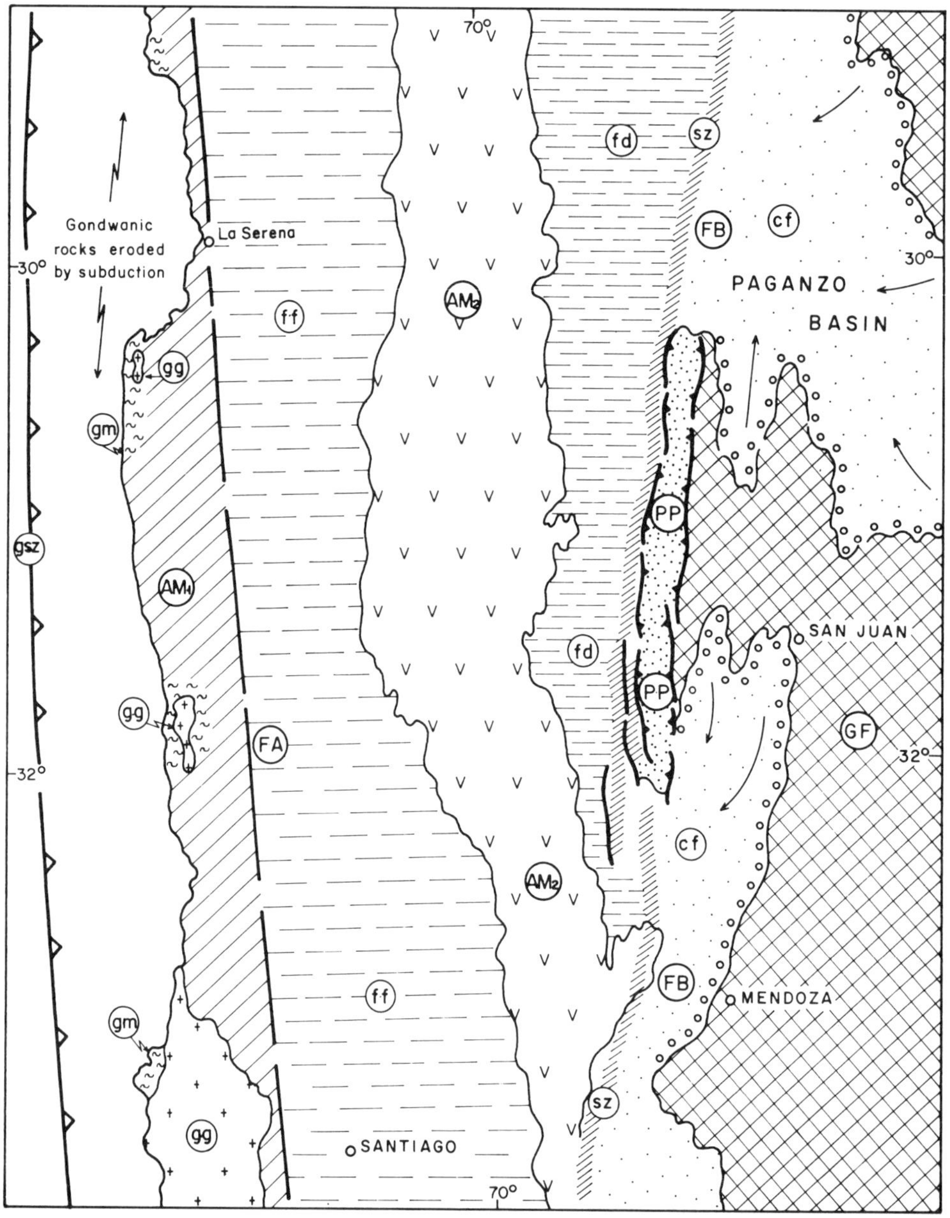

Figure 5. Gondwanic paleogeography and main tectonic elements during late Paleozoic times (after Ramos and others, 1985). GF = Gondwana foreland; FB = foreland basins; PP = Protoprecordillera; AM = arc massif. (1) pre-Sanrafaelic phase, and (2) post-Sanrafaelic phase. FA = forearc accretionary prism; sz = suture zone between Chilenia and Gondwanaland; gsz = Gondwanic subduction zone; cf = continental facies; fd = fluvio-deltaic and estuarine facies; ff = flysch facies; gg = possible Carboniferous granites; gm = metamorphic remnants of the subduction complex.

crustal rocks (Rapela and others, 1982) after the important deformation that resulted from the accretion of Chilenia.

There is no evidence of collision at these latitudes during late Paleozoic time. However, western Gondwanaland was highly mobile from late Carboniferous to late Permian time, as inferred from the South American paleomagnetic studies (Oviedo and Vilas, 1984). The quasi-static period in the apparent polar wandering path (APWP) from latest Permian to Jurassic time is coincident with a pronounced hook in the APWP (Valencio and others, 1983), which indicates significant changes in direction and velocity of drifting. Such a change might have produced important variations in plate motion at the subduction zone, and its beginning coincides in time with the Sanrafaelic phase.

Important transcurrent faulting at the end of the Gondwanic

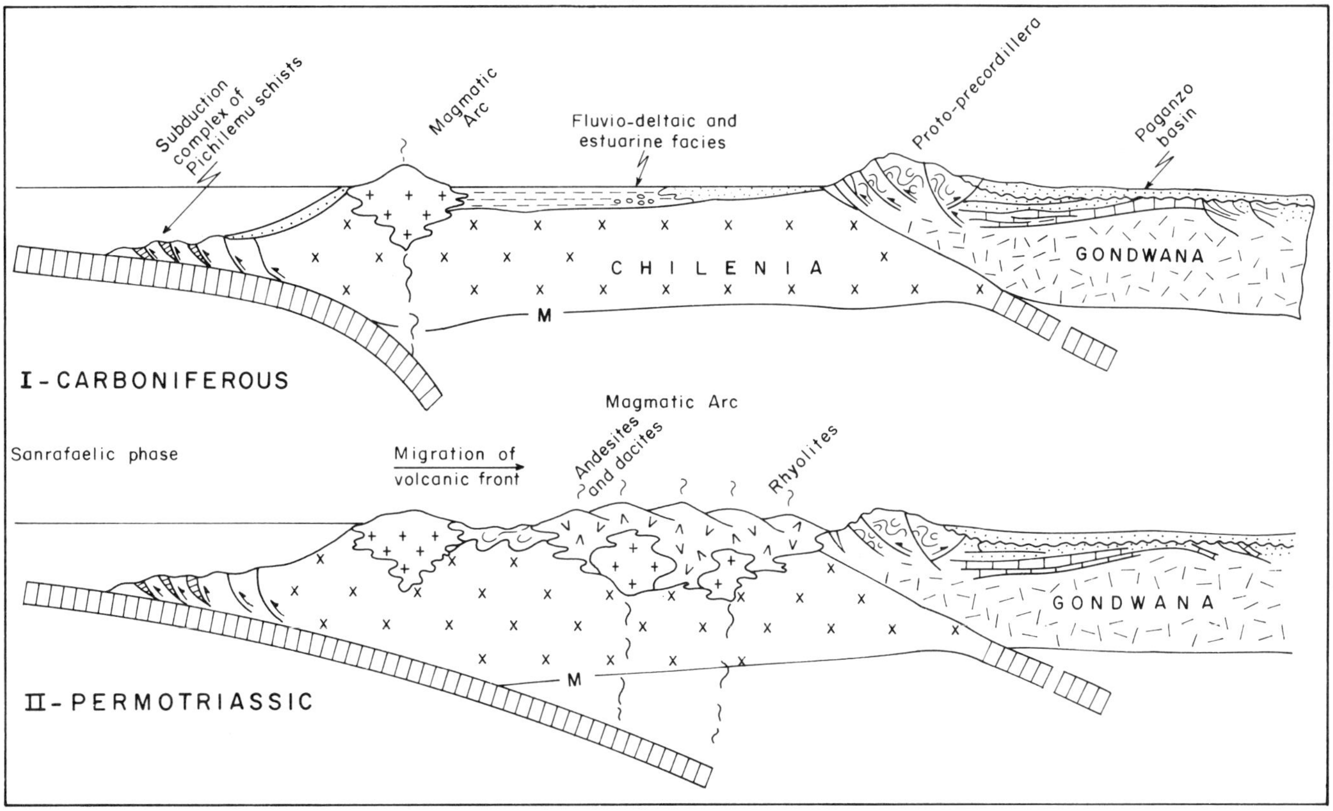

Figure 6. Schematic sections showing the accretionary prism and the subduction complex of the Coastal Cordillera, the Gondwanic magmatic arc, and the migration of the volcanic front toward the foreland after the Sanrafaelic phase (after Ramos and others, 1984).

orogenic cycle is inferred on the basis of conspicuous strike-slip movements along the Tupungato, Polvaredas, and Cruz de Caña faults (Fuentes and others, 1986). These may represent either a component of oblique subduction or lateral displacements of the Chilenia terrane, mainly in the basement of the Frontal Cordillera, similar to the displaced terranes observed in North America by Monger (1984) and Frei and others (1984). Some paleomagnetic data of the Chilenia terrane obtained on Late Paleozoic rocks indicate a plausible important anticlockwise rotation of the Sierra del Tigre block in the Frontal Cordillera (Vilas and Valencio, 1982). This deformation would have been the final attachment of the Chilenia terrane to Gondwanaland, acting from that time on as a unified basement for Andean evolution. However, the heterogeneous nature of this basement strongly constrained the style of deformation and magmatism of the Mesozoic and Cenozoic sequences.

PATAGONIDES OROGENIC CYCLE

The development of this orogenic cycle along the western active continental margin of South America during Mesozoic time was first proposed by Keidel (1921). It encompasses a series of deformational phases starting in Jurassic and ending in Late Cretaceous time. The principal compressive episode is assigned to the main Miranic phase (see Table 1) (Stipanicic and Rodrigo, 1970a,b; Malumián and others, 1983).

The Patagonides cycle is closely linked in time with the pre-rifting, rifting, and drifting stages of the eastern margin of the continent produced by the opening of the South Atlantic (Malumián and others, 1983; Malumián and Ramos, 1984). Although the main transgressions were from the Pacific side in the orogenic area, they coincide with Atlantic transgressions that were confined to the present continental platform.

The Patagonides as a morphostructural unit are clearly preserved south of 39° S lat. where the products of the mountain-building cycle have been little modified by the Andean orogeny. North of this latitude, in the segment analyzed, the stronger superimposed Andean deformation overprinted the structure of the previous cycle. Nonetheless, several authors have recognized an important compressive phase during the Cretaceous that was responsible for the first deformation of the Mesozoic sequences at these latitudes (Vicente and others, 1973; Auboin and others, 1973), whereas others are inclined to credit the Incaic phase (late Eocene) as the first conspicuous compressive deformation of this segment of the Cordillera (Digregorio and Uliana, 1980).

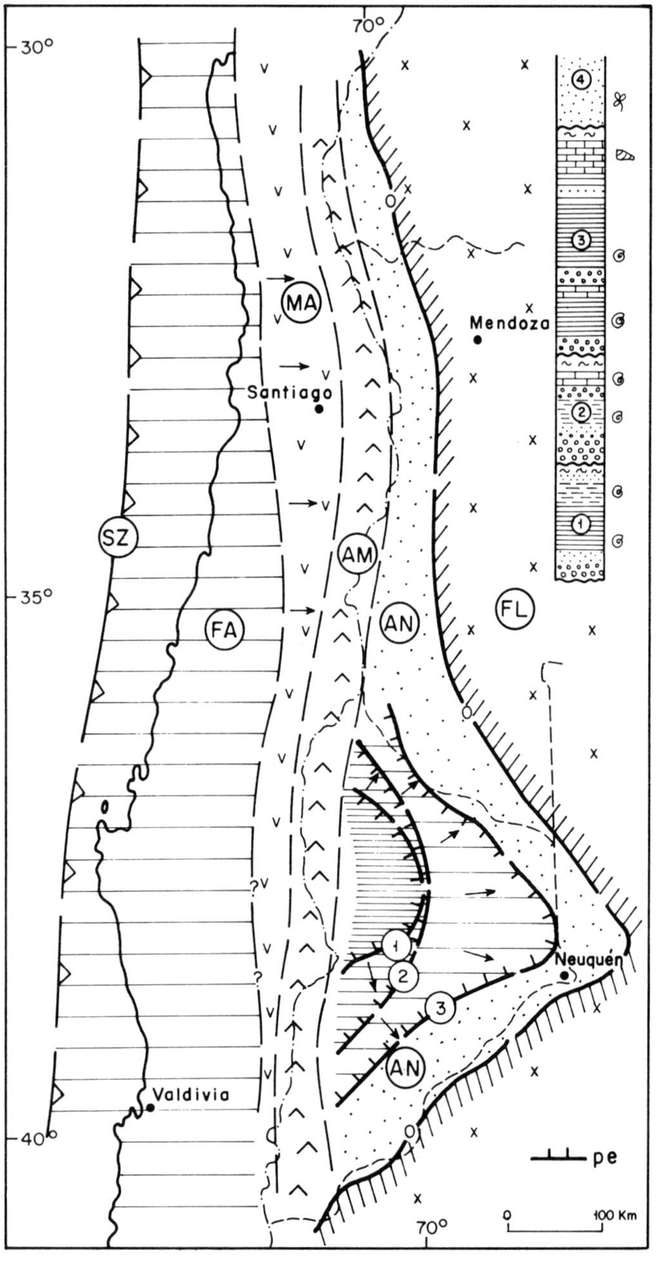

Figure 7. Basin development during the Patagonides orogenic cycle and generalized stratigraphic section (adapted from Digregorio and others, 1984; Gulisano and others, 1984a,b). FL = Foreland; AN = Aconcaguan-Neuquenian basin; AM = arc massif; MA = migrated arc; FA = forearc region; SZ = subduction zone. The platform edge (pe) defined by facies analyses was shifted toward the foreland during the *Cuyano* (1) Early Jurassic), *Chacayano-Loteniano* (2) (Middle to Late Jurassic), and *Mendociano* (3) (Tithono-neocomian), which indicates trend of generalized subsidence without structuring. (4) Continental deposits of the *Rayosiano* (Middle Cretaceous).

One of the outstanding features of this cycle is the development of many Mesozoic marine sequences controlled by Pacific transgressions and regressions. Those sequences are grouped into four sedimentary cycles: the Cuyano, Chacayano-Loteniano, Mendociano, and Rayosiano, which are separated by regional first-order unconformities (Digregorio and Uliana, 1980; Gulisano and others, 1984a,b; Mitchum and Uliana, 1985). These sequences are clearly depicted in the Neuquén Basin, a Liassic to Neocomian marine foreland basin developed behind the magmatic arc in the eastern foothills of the Cordillera. Figure 7 shows the extension of the basin and a generalized stratigraphic column with the main cycles and unconformities. The Neuquén basin is linked to the north with the Aconcaguan basin, which has a different paleogeography with a larger participation of volcanic rocks, but similar stratigraphic cycles in the middle Jurassic to Cretaceous evolution.

No deformation occurred until the Miranic phase (early Cenomanian, according to Stipanicic and Rodrigo, 1970b), when a foredeep filled with continental deposits was developed to the east of the fold and thrust belt (Ramos, 1978). Conspicuous lateral adjustment and flower structures were developed along northwest-trending features, such as the Huincul fault that offsets the southern part of the Neuquén basin. Seismic profiling shows deformation and truncations of early Cretaceous deposits at this time (Orchuela and others, 1981).

There are interesting parallels between the evolution of the basin and the magmatic arc history. The main periods of regional unconformities in the foreland (Charrier, 1981; Malumián and Ramos, 1984) are coincident with the times at which the magmatic arc migrated eastward (Rivano, 1984) (see Fig. 8). The intermittent nature of the magmatic activity, as well as the spatial variation of the volcanic front in the Andes, is closely related to changes in plate motion controlled by variation in the spreading velocities of the Pacific and Atlantic oceanic ridges (Charrier, 1973; Frutos, 1981; Ramos, 1985b). Continental erosion of the margin by subduction cannot by itself explain the magnitude of the shifting of the different volcanic fronts.

Most of the Chilean and the westernmost Argentine Mesozoic basins of the Principal Cordillera and the Patagonian Andes are intra-arc basins controlled by the development of two distinctive arcs (Nasi and Thiele, 1982; Ramos and Palma, 1983; Charrier, 1984): an inner, more active arc concentrated the main andesitic activity, and an intermittent, outer arc produced rock suites that are mainly of bimodal composition (Ramos and others, 1982). This tectonic regime can be recognized from Early Jurassic until Neocomian time, from 47° S lat. to north of 33° S lat. (Ramos and others, 1982; Nasi and Thiele, 1982).

Several authors have proposed an extensional regime in the arc massif region mainly during Early Cretaceous time, which was responsible for the intra-arc basin development. This process has been envisaged as an intra-continental spreading that has controlled, during a period of low convergence rate at the trench (ca. 5 cm/yr; Larson and Pitman, 1972), the rapid subsidence of the volcanic pile where burial metamorphism closely followed

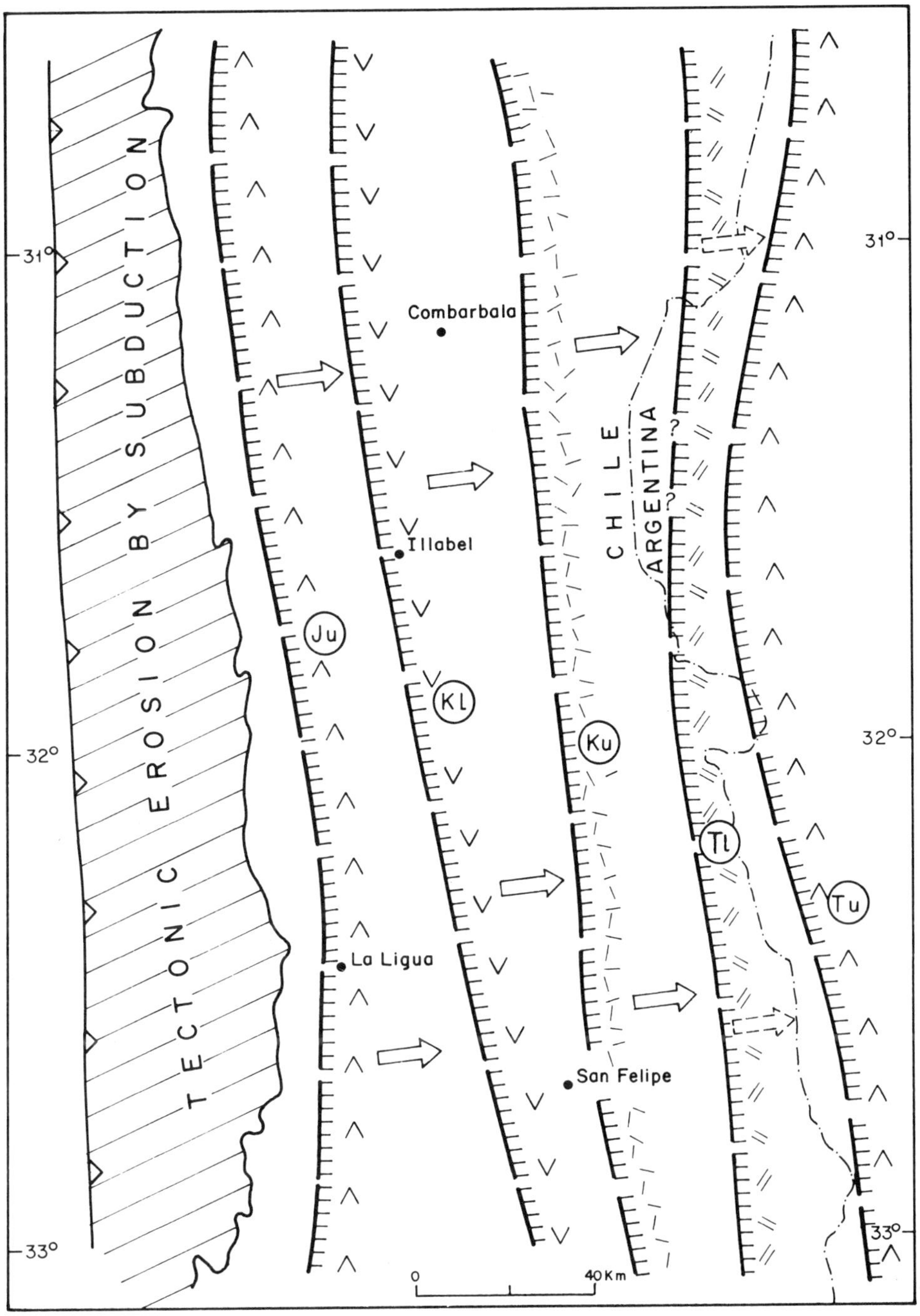

Figure 8. Migration of the magmatic arc during the Patagonides and Andean orogenic cycles (after Rivano, 1984). Ju = Jurassic; Kl = Early Cretaceous; Ku = Late Cretaceous; Tl = pre-incaic phase; Tu = pre-Quechua phase volcanic fronts.

extrusion (Levi and Aguirre, 1981; Levi and Nystrom, 1982; Aberg and others, 1984).

The intermittent activity of the outer arc is closely related to periods of low stands of sea level in both the intra-arc and foreland basins. These local sea-level changes have been partially correlated with the worldwide eustatic onlap cycles by Mitchum and Uliana (1985). Figure 9 illustrates the relationships between development of an outer arc and the low-stand intervals.

The intra-arc basins were active until Early Barremian time, when an important eastward migration of the main magmatic arc occurred (Fig. 8, position K1). It coincided with a low-stand period in the retroarc and intraarc basins and the development of a single and expanded central arc. The retroarc easternmost basin was exclusively continental from this time on, and the Pacific seas no longer reached the eastern side of the cordillera. This important paleogeographic change is closely linked with the beginning of one of the periods of higher spreading rates within the South Atlantic (5 cm/yr) and the Pacific-Phoenix (ca. 18 cm/yr) ocean floor (Larson and Pitman, 1972; Piraces, 1979).

Although the absolute motion of South America relative to the subducting ocean floor is unknown, it is evident that a faster convergence rate increased horizontal compression and inhibited intra-arc basin formation. Thus, the variation observed in the intra-arc basin development (Fig. 9), the intermittent nature of the outer arc, and the eastward migration of the volcanism may be connected events that are related to changes in plate motion rates. Unfortunately, there is no detailed knowledge of the Mesozoic plate motions at these latitudes. Even so, three different stages in the tectonic evolution of the cordillera can be objectively described (Fig. 10). The first one coincided with a regime of low compressive stress and slow plate motion, resulting in intra-arc basin development. With an increase in plate motion, and therefore of the compressive stress, the intra-arc basins disappeared, and the arc moved toward the foreland. This second step is recognized at about 110 Ma when the Rayosiano sedimentary cycle started. The eastward migration of the magmatic arc could be related to a decrease in the subducted slab dip, coeval with a higher convergence rate, because the large amount of eastward shifting of the volcanic arc cannot be explained by tectonic erosion at the subduction zone, although it could be enhanced by this process.

The maximum sea-floor spreading at the Pacific and the South Atlantic spreading centers, reached during Late Cretaceous time (ca. 80 to 110 Ma; Larson and Pitman, 1972), may be responsible for the third stage in the evolution of the Patagonides cycle. At this final stage, a mountain chain was built on the Argentine side. This deformation of the strata on the eastern flank of the orogen as a fold-and-thrust belt caused formation of a foreland basin at the leading edge of the deformation, possibly related to the tectonic loading of the adjacent thrust belt (Jordan, 1981). At the final stage of compression, several granitoid stocks were emplaced in the arc massif (Munizaga and Vicente, 1982). These three different stages are summarized in Figure 10.

As a concluding remark on the Patagonides orogenic cycle, it is necessary to emphasize that the different paleogeographic settings, the major depositional sequences and the successive tectonic regimes seem to be controlled by the subduction geometry, as inferred from the migration of the volcanic front. The changes in the subduction geometries were probably controlled by plate motion rates and orientations. No evidence of collisional tectonics is known yet during the Mesozoic; furthermore, the synchroneity of the paleogeographical variations through space and the deformation with compressive regimes from 22° to 52° S lat. rules out major collisions as an effective cause of the variable tectonic settings described (Coira and others, 1982; Ramos and others, 1982; Dalziel, 1985).

Another important relation is the connection between periods of slow spreading rate and transgressions (high stands) in contrast to general regressions (low stands) and faster rates of convergence in the orogenic area. This relationship of regression with higher convergence in orogenic areas was already noticed in the central Andes (Piraces, 1979; Malumián and Ramos, 1984; Ramos, 1985b), although some other authors have found evidence supporting the opposite links (Noble and others, 1974; Francis, 1984).

ANDEAN OROGENIC CYCLE

This tectonic cycle, responsible for the present configuration of the Andes, began in early Tertiary time and is still developing. One of the outstanding features of its evolution is the segmented nature of the orogen (Isacks and Barazangi, 1977; Jordan and others, 1983b), which produced conspicuous longitudinal changes along discrete segments of the Andes. This segmentation is also known in previous orogenic cycles (Ramos, 1983; Jordan and others, 1984), although the Cenozoic provides the best-documented evidence. The modern configuration of segments can probably be extended back in time only to the late Miocene (Jordan and others, 1983b).

Structural setting

The analyzed area overlies one of the presently flat segments of the Benioff zone between the Nazca and the South American plates. The development of a high compressive regime, a fact that controlled the formation of thin-skinned tectonics on the previously mentioned Mesozoic sequences (Figs. 11, 12), seems to be linked with the beginning of the period of flat subduction.

The structural style obtained from detailed mapping of the area is significantly different from that previously proposed for this section of the Andes. Former workers favored high-angle reverse faults (upthrusting) as the dominant fault geometry (for example, Gonzalez Bonorino, 1950; Zeil, 1979), ignoring both the existence of low-angle thrusts in the study area and also the evidence for thrusting and overthrusting described elsewhere in the cordillera (Schiller, 1912; Polanski, 1964, 1972; Vicente, 1972; Vicente and others, 1979).

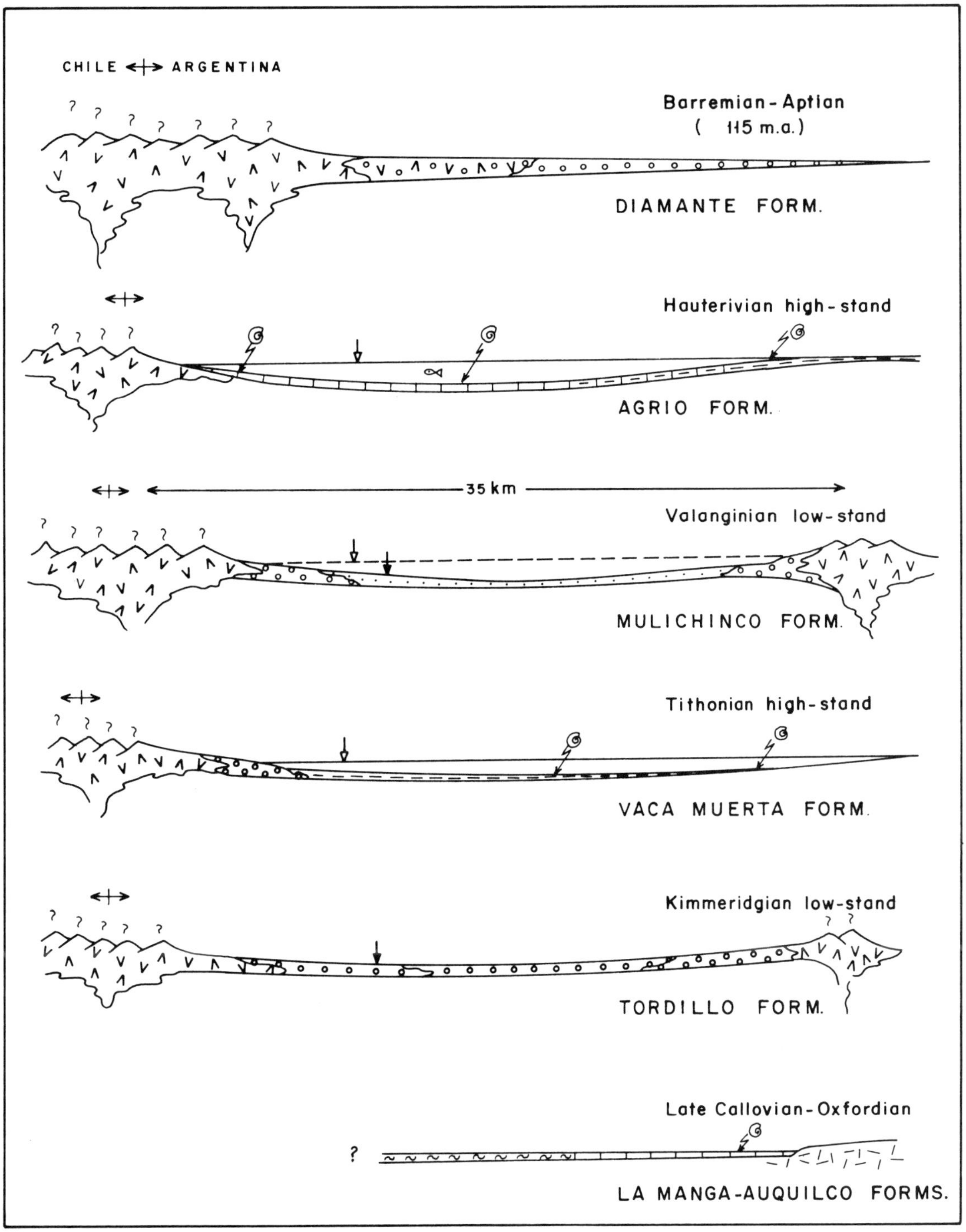

Figure 9. Evolution of the inner and outer arcs during the Patagonides orogenic cycle. Note that maximum eastward shifting of volcanism is coincident with low stands in basin (after Ramos, 1985b).

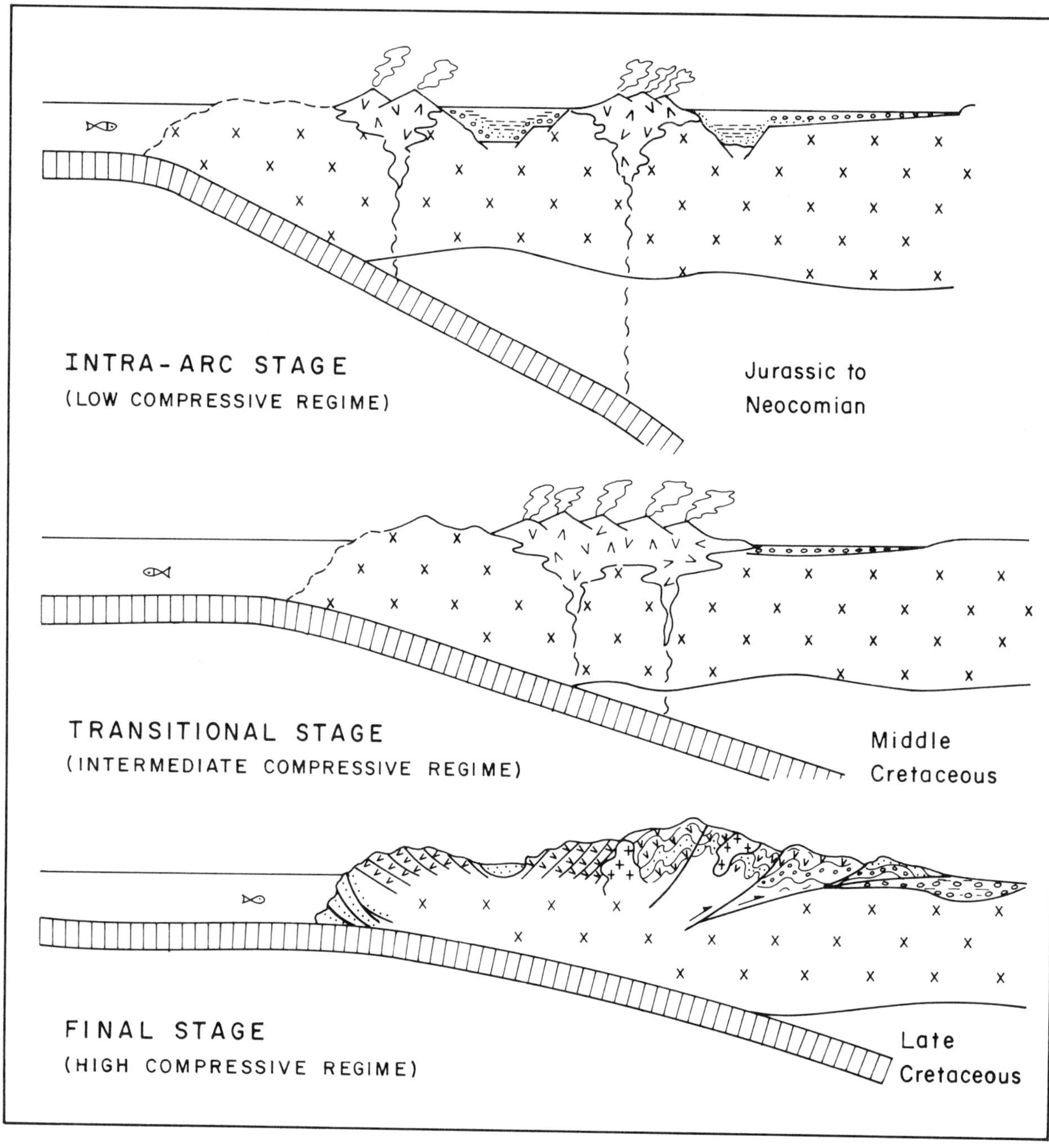

Figure 10. Development of the different tectonic stages of the Patagonides cycle from Jurassic to Cretaceous times (adapted from Vicente and others, 1973; Frutos, 1981; Charrier, 1981, 1984; Ramos, 1985b).

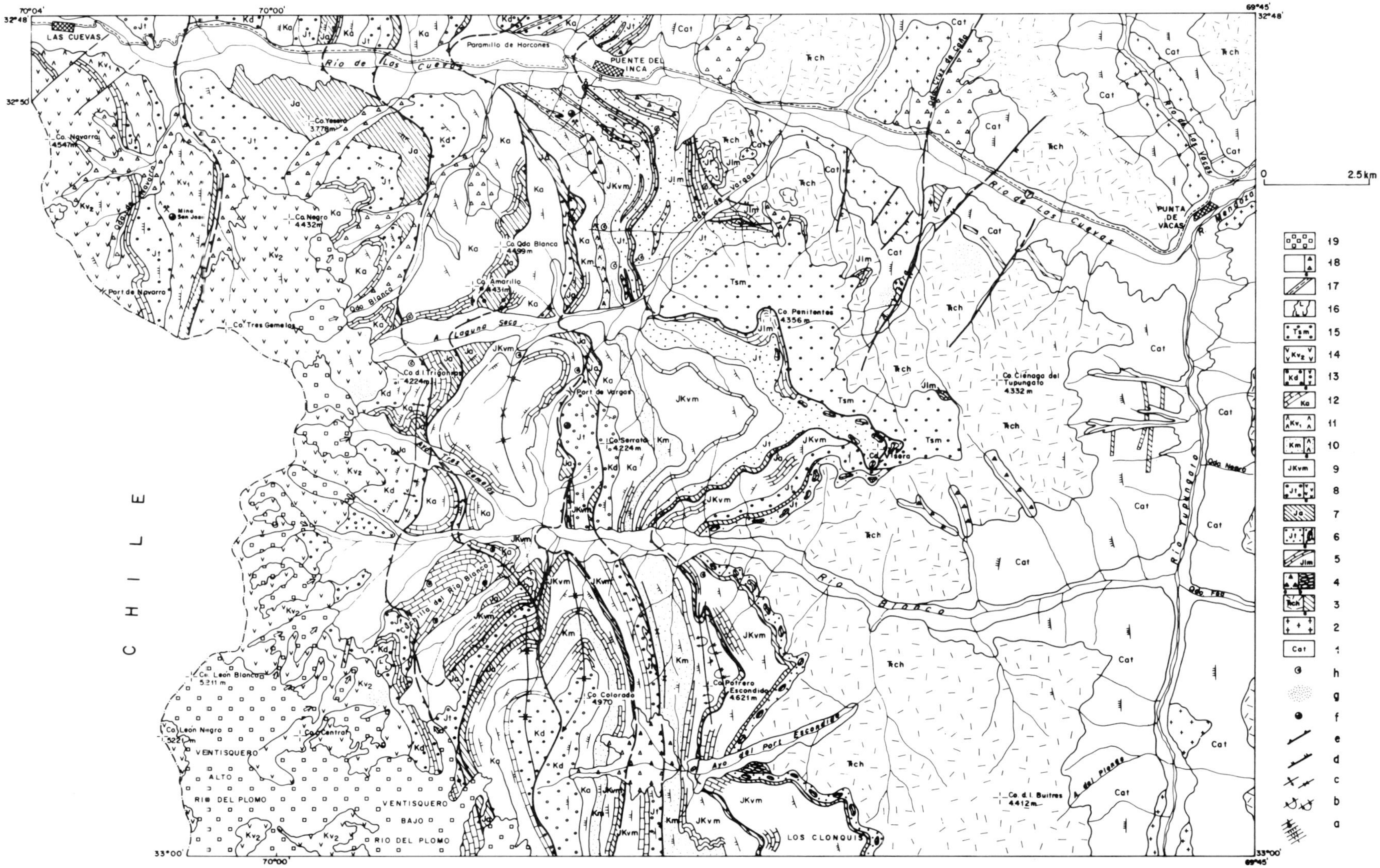

Figure 11. Geological map of the High Cordillera of Mendoza Province depicting the thin-skinned thrust-and-fold belt developed during the Andean cycle (after Ramos, 1985a). Key: 1 = Carboniferous deposits; 2 = Permian granitoids; 3 = Permotriassic volcanics of Choiyoi Formation; 4 = Liassic? volcanics; 5 = La Manga Limestones; 6 = Late Jurassic red beds; 7 = Auquilco Gypsum; 8 = Tordillo Formation; 9 = Vaca Muerta Formation; 10 = Mulichinco Formation, a = volcanics; 11 = Early Cretaceous volcanics; 12 = Agrio Formation; 13 = Diamante Formation, a = volcanics; 14 = Middle to Late Cretaceous volcanics; 15 = Tertiary Santa María Conglomerate; 16 = Farellones Volcanics; 17 = Trachytic dykes; 18 = Quaternary deposits, a = glacial deposits; 19 = ice. a = bed dips, b = overturned fold axis, c = anticline, syncline, d = fault, e = thrust, f = mineral occurrence, g = hydrothermal alteration zone, h = fossil locality.

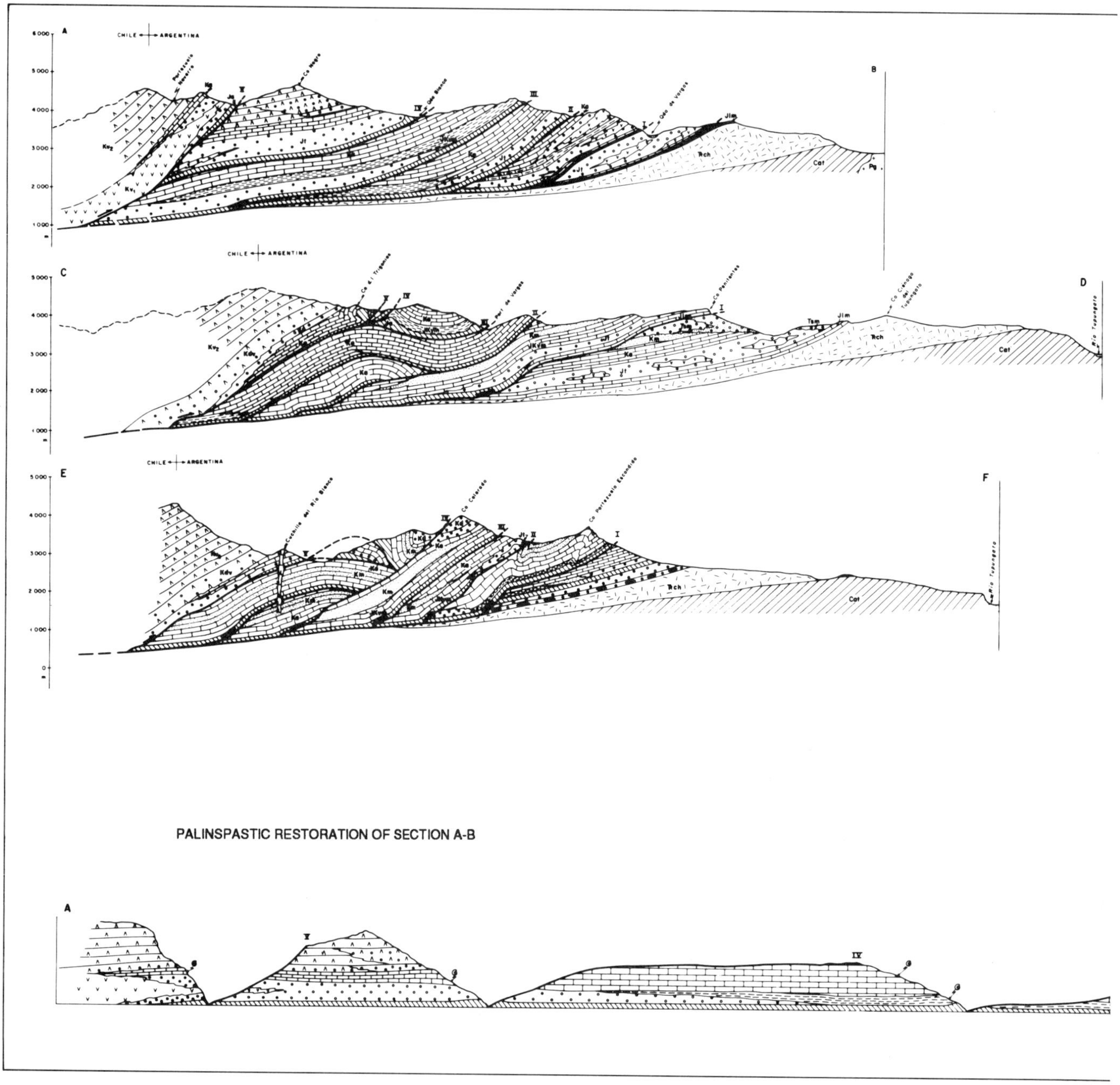

Figure 12 (continued on following page). Balanced cross sections of the Principal Cordillera of Mendoza with a palinspastic restoration of the section A-B. Note the reconstructed intra-arc basin and the Mesozoic sequences. Location of the structural sections on Figure 13. No vertical exaggeration (after Ramos, 1985b).

The leading thrust of the Principal Cordillera at these latitudes crops out at Cerro Penitentes (section C-D, Fig. 12). The fault plane is clearly exposed with more than 1 km of vertical relief, and a conspicuous ramp structure can be observed. The ramp structure dips 30° to 35° to the west, but the leading edge of the thrust above the ramp dips less than 10° to 15°, producing a very irregular surface trace (Fig. 11). Some of the westernmost thrusts have steeper dips, but this could be a result of subsequent rotation during an eastward migration of the thrust front.

Another outstanding feature is the diapiric emplacement of gypsum domes observed close to the termination surface of the thrust sheets. The hanging walls of the domes are extremely deformed and most commonly overturned. The diapir is composed of gypsum with a sheath of shales interbedded with some thin

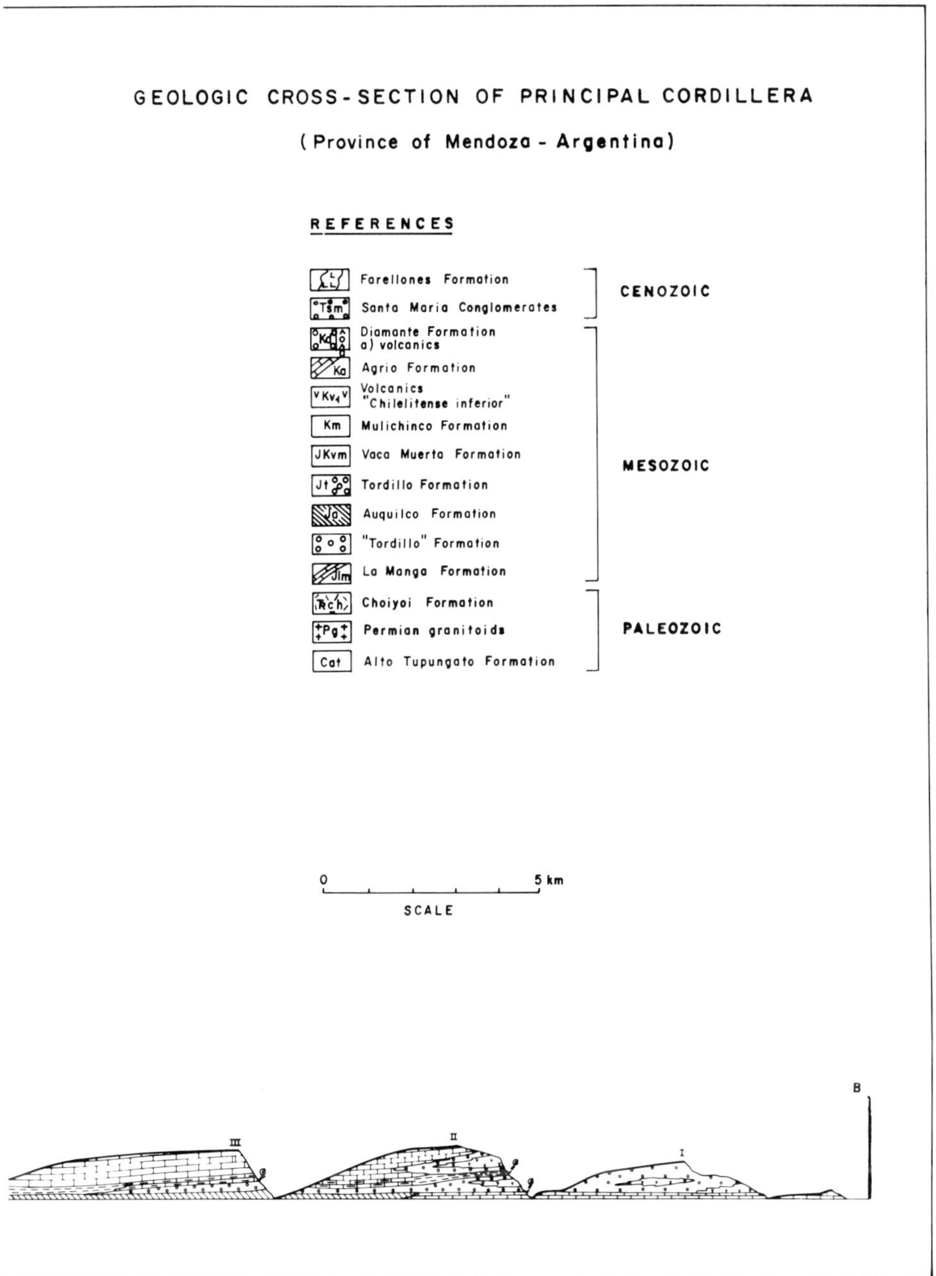

layers of carbonates. They range in diameter from a few hundred meters to more than 2 km. The gypsum beds also control the inception of the detachment surfaces. Antiformal stack duplexes are developed in the southernmost section (E-F, Fig. 12), similar to the ones described by Boyer and Elliot (1982).

Palinspastic restoration of the sections, using limited thickness control and assuming negligible volume changes due to plastic flow of gypsum, indicates minimum shortening, which ranges from 65 to 68 percent. These values are within the range observed by Vásquez and Gorroño (1980) for the sub-Andean fold-thrust belt of Argentina, which varies from 50 to 72 percent. The shortening obtained, although representative of the Principal Cordillera in the flat-slab zone (30° to 33° S lat.) is not characteristic of the Andes farther south. The tectonic style of the Principal Cordillera between 34° and 38° S lat. is dominated by folding, associated with subordinate thrusts, with considerably less shortening (Burckhardt, 1900; Kozlowski, 1984). The main thrusting period of the Principal Cordillera is accepted as being produced by the Quechua phase (mid- to late Miocene) (Charrier and Vicente, 1972; Charrier, 1981), which coincides with the time at which the subducted Nazca plate acquired its present segmented form (Jordan and othes, 1983b). The segmentation seems to control a change in the compressive regime along the strike, with lower compressive stress where the Benioff zone has a 30° dip south of 34° S lat. The higher compressional regime may be enhanced by the greater age of the subducted slab in the flat segment. The net result is greater deformation in the flat slab segment of the Principal Cordillera. These differences in shortening are in agreement with the variations in compressive regime of normal and flat-slab segments measured from earthquake activity (Isacks and others, 1982).

Structural evolution

The Cenozoic sedimentary history records the eastward migration of the orogenic front. Thick sequences of continental deposits (Santa María Conglomerates of Schiller, 1912) unconformably overlie the Mesozoic rocks in the Principal Cordillera. The angular unconformity is clearly seen east of Cerro Aconcagua and west of Cerro Penitentes. Those conglomeratic deposits are interpreted as alluvial-fan sediments interfingered with the lower section of the volcanics of the Farellones Formation (25 to 10 Ma; Munizaga and Vicente, 1982). Based on fossil vertebrate records, a late Eocene–Oligocene age was assigned to the continental deposits (Gorroño and others, 1979; Ramos, 1985a). This well-developed regional unconformity is thereby assigned to the Incaic phase (late Eocene–earliest Oligocene) according to Charrier and Vicente (1972) and Ramos (1985b).

The Tertiary deposits farther east of the High Cordillera are represented by distal fluvial facies partially synchronous with the Santa María Conglomerate and the volcanism of the Farellones Formation. The Tertiary extra-Andean sequences at these latitudes (30° to 33° S lat.) range from Eocene to middle Miocene (Rolleri and Fernandez Garrasino, 1979), and contain several tuff layers that attest the cordilleran volcanic activity at that time.

An unconformity separates the Chasiquense beds of La Pilona (Pascual and Odreman Rivas, 1973), presently interpreted as late Miocene in age, from the older Tertiary beds in the Uspallata valley and the Cacheuta area. This unconformity was produced by the Quechua phase, which folded and thrust the Santa María Conglomerate and the Farellones volcanics about 10 Ma (Charrier and Vicente, 1972; Munizaga and Vicente, 1982; Malumián and Ramos, 1984). It is also responsible for the deformation of the Frontal Cordillera, uplifted during this phase in middle to late Miocene time (Polanski, 1964).

The eastern border of the Frontal Cordillera was uplifted during the Diaguitic phase, which in the latest Pliocene gave rise to the present configuration of the Principal and Frontal Cordilleras (Polanski, 1964; Yrigoyen, 1979). The late Miocene and Pliocene deposits of the Uspallata and Cacheuta regions were folded and thrust at that time with the subsequent deposition of the alluvial fan deposits of the Mogotes Formation during the Pleistocene (Fig. 13).

Therefore, the Tertiary sedimentary facies shows a migration of the coarse alluvial fan facies from: (1) the inner area of Cerro Penitentes in the High Andes, postdating the Incaic phase; to (2) the Uspallata valley and Cacheuta in the Quechua phase; to (3) the outer foothills of Mendoza City during the Diaguitic phase. Even the Pleistocene fanglomerates of the Mogotes Formation cropping out west of Mendoza City and other younger alluvial fans have been deformed by neotectonic activity, such as the one described by Polanski (1962) in the extra-Andean area of the southern Mendoza region.

The seismic sections of the plains located eastward of the Precordillera clearly show that the present orogenic front is composed of a set of imbricated overthrusts (Bettini, 1981; Bettini and Turic, 1981). This structural style was corroborated by drilling, and it has somewhat similar characteristics to the Incaic or Quechua orogenic fronts (see Fig. 14).

The Diaguitic deformation front on the eastern side of the Precordillera is still active. Intense compressive deformation, as inferred from earthquake focal mechanisms and displacements on the alluvial fans, is continuing today (Bastías and others, 1984).

Tectonic interpretation

The Andean structural evolution of the Central Andes is the result of a combination of several tectonic mechanisms. Their relative importance varies among the different segments recognized along the cordillera.

The Andean orogenic cycle started at these latitudes with an interval of normal subduction and volcanic activity. This was interrupted by the Incaic phase (Table 1) of deformation. The timing of the interruption of the volcanism is poorly constrained at these latitudes. Based on the fossil record of the alluvial fan sediments deposited after the Incaic deformation phase, the deformation might be dated as pre-Late Eocene. If regional data are considered, the Incaic phase may be about 42 Ma (Malumián and Ramos, 1984), an age that clearly matches the maximum of normal convergence rate between 42 and 40 Ma postulated by Pilger (1984). There is a striking coincidence between the increase of plate motion rates, the cessation of magmatism, and the compressive deformation at the Incaic orogenic front.

Most of the Oligocene was quiescent, with low rates of convergence (Pilger, 1981, 1984) coincident with the lack of volcanic activity in the Cordillera. This quiescence ended at 25 Ma when the volcanic activity of the Farellones Formation began (Munizaga and Vicente, 1982). Two interrelated tectonic features demonstrate that change in the geometry of the Benioff zone followed: (1) eastward migration of the subduction-related magmatic foci from a position about 180 km from the trench at about 25 Ma to 600–700 km from the trench at 6 Ma (Leveratto, 1976; Munizaga and Vicente, 1982; Kay, in preparation); and (2) eastward shifting of the orogenic front since the Incaic phase (275 km from the trench at 42 Ma) to the Quechua phase (325 km at 10 Ma) and Diaguitic phase (350–365 km at 2 Ma)—this implies an average shifting rate of 2.5 mm/yr of the orogenic front since early Tertiary time, although the shifting was probably episodic.

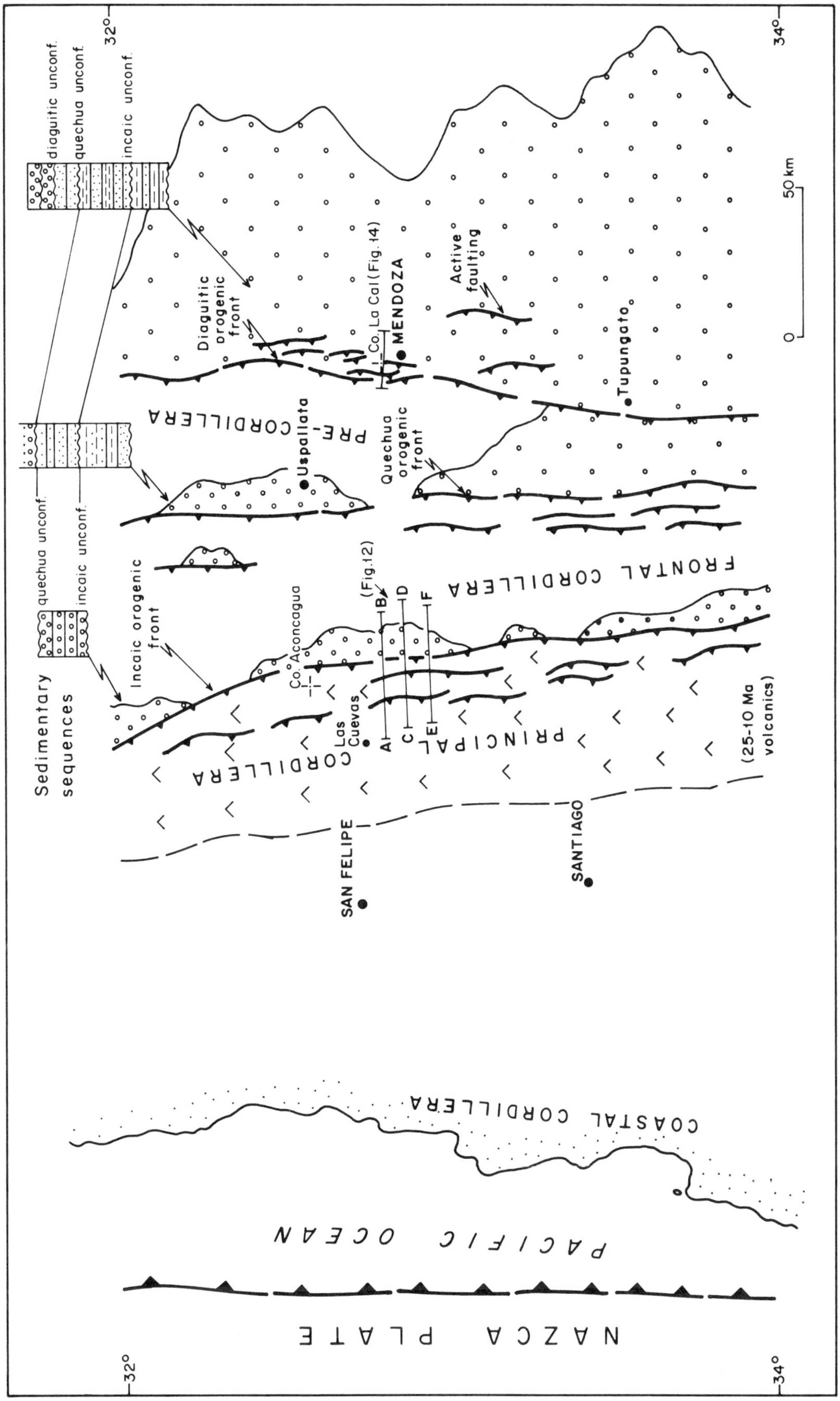

Figure 13. Migration of the orogenic front from the early Tertiary up to the present in the central Andes at about 32° to 34°S lat.

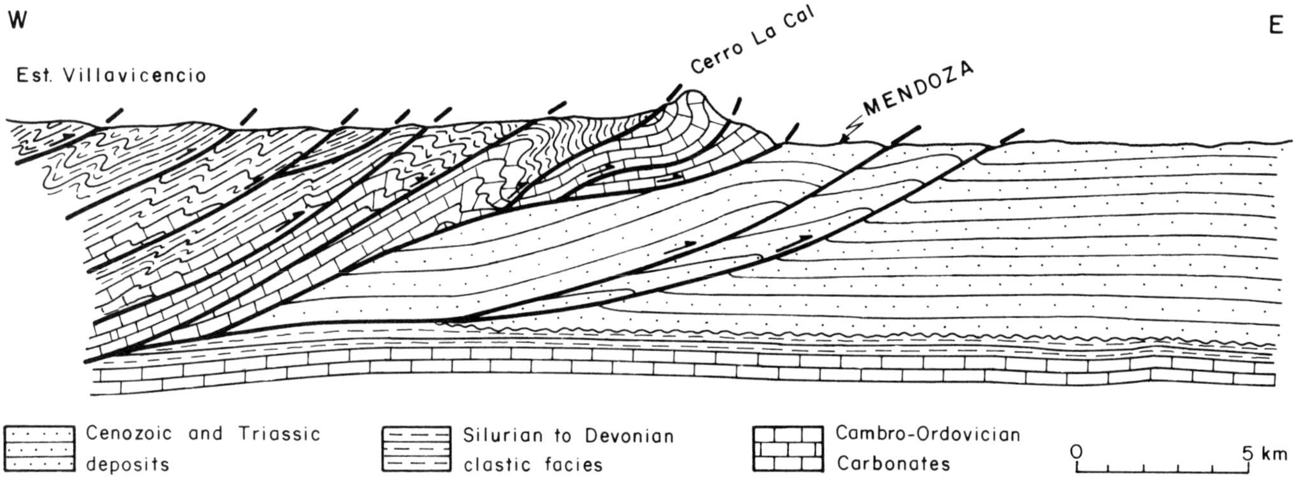

Figure 14. Structural section of the Diaguitic orogenic front at the eastern foothills of the Precordillera in Cerro La Cal and adjacent plains around Mendoza City (after Bettini and Turic, 1981). Location as in Figure 13.

The causes of the change in the Benioff zone geometry and segmentation are probably complex and multifaceted. The following factors seem to be important:

1. The breakup of the Farellones plate into the Cocos and Nazca plates, which occurred at 25 Ma, seems to mark the beginning of a period of higher convergence rates (Handschumacher, 1976). This age coincides with the initiation of Farellones magmatism and is a milestone in the geodynamic evolution of the area.

2. The increase in plate convergence from 25 to 26 Ma up to 10 Ma (defined by Pilger, 1984) was interrupted by the Quechua phase, when the present deceleration began.

3. An age decrease of the subducted oceanic plate would favor the resorption of the oceanic slab and increase volcanic activity, as noticed by Wortel (1984). Conversely, according to Wortel the subduction of younger oceanic lithosphere will increase buoyancy effects and thereby reduce the dip of the Benioff zone. This proposal is not in agreement with the geometry of the Benioff zones north and south of 34°S. To the north, where older oceanic crust is being subducted, the Benioff zone is flatter than to the south, where younger oceanic crust has a steeper dip (Isacks and Barazangi, 1977). Wortel (1984) has proposed that a transitional stage, which involved the detachment of the lower part of the slab, producing deep earthquakes, has taken place in the last 15 m.y. The timing of the detachment, although poorly constrained and documented, may coincide with the Quechua phase. The age-dependent effects of the subducted slab are better evidence in the southern segment (34°–37°S lat.). There, the younger lithosphere being consumed controlled the location of very active retroarc volcanism (Fig. 1), which may coincide with the complete resorption of the oceanic slab (Jordan and others, 1983a, b). This retroarc volcanic activity is absent farther north where oceanic lithosphere is older.

4. Several authors have explained the present segmentation of the subducted Nazca plate as controlled by the collision of aseismic ridges (Pilger, 1978, 1981; Nur and Ben-Avraham, 1981). In Pilger's interpretation, the buoyancy effects produced by subduction of those ridges, along with the younger age of the subducted slab, will combine to diminish the angle of the Benioff zone. The effects of the subduction of the Juan Fernandez hotspot trace began about 15 Ma, according to Pilger (1984). There is a 5-m.y. gap between the phase with the subsequent uplift and faulting of the Sierras Pampeanas (Jordan and others, 1983b). This gap may be due to the lack of time constraints in the interaction between the Juan Fernandez ridge and the trench, or a normal delay time between the initial encounter at the trench and the deformation phase.

The flat slab segment is unique in comparison to the tectonic evolution of the other segments. South of the region of flat subduction there is an age decrease in the subducted slab, and therefore a higher thermal gradient than in the flat slab segment. Even so, the present angle of subduction is about 30° (Isacks and others, 1982). Although an increase in convergence rate diminishes the subduction angle (Yokokura, 1981), increased convergence has affected all segments uniformly, and thus must be disregarded as an important mechanism contributing to flat subduction of individual segments.

Therefore, collision of the Juan Fernandez aseismic ridge still seems to be the most important mechanism that controlled the flat subduction segment. However, the age dependence of the subducted slab and the convergence rate strongly influenced the development of the magmatic activity. Lack of definition on the available convergence rates for the last 10 m.y. hinders correlation with the Diaguitic and subsequent deformation episodes.

CONCLUSIONS

The central Andes records a complex history of interaction between oceanic and continental plates since as early as 600 Ma when apparently the southwestern margin of Gondwanaland was broken up. The orogenic cycles recognized suggest diverse mechanisms that could be summarized as follows:

1. The Famatinian orogeny was a period of important allochthonous terrane accretion on the southwestern margin of Gondwanaland. The timing of collisions controlled the different diastrophic phases and mountain-building processes.

2. The Gondwanides orogeny, although registering the docking of some minor exotic terranes, was dominated by tectonics with normal subduction without terrane accretions. Correlation between periods of major drifting of west Gondwanaland and the magmatic activity, as well as periods of quiescence in the apparent polar wandering path with those of compressive tectonics, support this assessment.

3. The Patagonides orogeny is entirely dominated by subduction tectonics, as far as the correlation of presently available oceanic data and continental geology permits. An outstanding inference is that intra-arc basins developed during periods of low compressive stress at slow convergence rates. Another important inference is that faster periods of plate motions correlate with local sea-level low stands and high volcanic activity in the orogenic areas.

4. The Andean orogeny shows a complex interaction between the subduction of aseismic ridges with the segmentation of the cordillera and the development of longitudinal changes in the geometry of the Benioff zones. Nevertheless, diastrophic phases identified along many thousand of kilometers of continental margin extend beyond the individual collided segments. Therefore, the Incaic, Quechua, and Diaguitic compressive phases may be better explained by changes in plate motions.

Finally, it must be emphasized that the precise timing of the orogenic phases and the mountain building associated with them may vary slightly from one segment to another. These spatial variations are the result of the interaction of the different tectonic mechanisms.

ACKNOWLEDGMENTS

I am indebted to Yale University for inviting me to participate in the symposium honoring John Rodgers, and to Clark Burchfiel and Sydney Clark for encouraging the writing of this study, as well as to Alberto Riccardi, Richard Allmendinger, Teresa Jordan, and Sydney Clark for their helpful comments on the manuscript. Gratitude is also expressed to the Servicio Geológico Nacional of Argentina for many years of logistical support for field studies on the Andes, and to many colleagues of the Servicio for their cooperation, advice, and comments.

REFERENCES CITED

Aberg, G., Aguirre, L., Levi, B. and Nystrom, J. O., 1984, Spreading subsidence and generation of ensialic marginal basin; An example from the Early Cretaceous of Central Chile, *in* Kokelaar, B. P., and Howells, M. F. eds., Marginal basin geology: Geological Society of London Special Publication 16, p. 185–193.

Aceñolaza, F. G., and Toselli, A., 1973, 1976, Consideraciones estratigráficas sobre el Paleozoico inferior del Noroeste Argentino: II° Congreso Latinamericano de Geología, Memorias, 2, p. 755–763.

Auboin, J., Borrello, A. V., Cecioni, G., Charrier, R., Chotin, P., Frutos, J., Thiele, R., and Vicente, J.C., 1973, Esquisse Paléogéographique et structurale des Andes Meridionales: Revue de Geographie Physique et de Geologie Dynamique, v. 15, no. 1–2, p. 11–72.

Almeida, F. F. de, and Hasui, Y., 1983, Precambriano de Brasil, p. 1–378: Sao Paulo, Editora E. Blucher.

Baldis, B., Beresi, M. S., Bordonaro, O., and Vaca, A., 1982, Síntesis evolutiva de la Precordillera Argentina: V° Congreso Latinoamericano de Geología, Actas 4, p. 399–445.

Bastías, H. E., Weidmann, N.E., and Perez, A. M., 1984, Dos zonas de fallamiento plio-cuaternario en la Precordillera de San Juan: IX° Congreso Geológico Argentino, Actas 2, p. 329–341.

Bettini, F. H., 1981, Nuevos conceptos tectónicos del centro y borde occidental de la Cuenca Cuyana: Asociación Geológica Argentina, Revista 35 no. 4, p. 579–580.

Bettini, F. H., and Turic, M. A., 1981, Sobrecorrimientos en el borde occidental de la Cuenca Cuyana: XLIII Reunión a nivel de expertos, ARPEL, p. 1–16.

Bond, G. C., Nickelson, P. A., and Kominz, M. A., 1984, Breakup of a supercontinent between 625 Ma and 55 Ma; New evidence and implications for continental histories: Earth and Planetary Science Letters, v. 70, p. 325–345.

Borrello, A. V., 1969, Los geosinclinales de la Argentina: Dirección Nacional de Geología y Minería, Anales 14, p. 1–136.

Boyer, S. E., and Elliott, D., 1982, Thrust system: American Association of Petroleum Geologists Bulletin, v. 69, no. 9, p. 1196–1230.

Burckhardt, C., 1900, Coupe Geologique de la Cordillere entre las Lajas et Caracautin: Museo de La Plata, Anales 2, p. 1–136.

Caminos, R., Cordani, U. G., and Linares, E., 1979, Geología y geocronología de las rocas metamórficas y eruptivas de la Precordillera y Cordillera Frontal de Mendoza, República Argentina: II° Congreso Geológico Chileno, Actas 1(F), p. 45–61.

Charrier, R., 1973, Interruptions of spreading and the compressive tectonic phases of Meridional Andes: Earth and Planetary Science Letters, v. 20, p. 212–249.

——, 1981, Mesozoic and Cenozoic stratigraphy of the Central Argentinian–Chilean Andes (32°–35°S) and chronology of their tectonic evolution: Zentralblatt fur Geologie und Palaontologie, Teil 1, p. 344–355.

——, 1984, Areas subsidentes en el borde occidental de la cuenca de tras-arco jurásico-cretácica, Cordillera Principal Chilena entre 34° y 34°30′, S: IX° Congreso Geológico Argentino, Actas 2, p. 107–124.

Charrier, R., and Vicente, J. C., 1972, Liminary and geosyncline Andes; Major orogenic phases and synchronical evolution of the central and austral sectors of the southern Andes: Conferencia sobre Problemas de la Tierra Sólida, Buenos Aires, 2, p. 451–470.

Coira, B. L., Davidson, J. D., Mpodozis, C., and Ramos, V. A., 1982, Tectonic and magmatic evolution of the Andes of northern Argentina and Chile: Earth Science Review, v. 18, no. 3–4, p. 303–332.

Dalla Salda, L. and Varela, R., 1982, La estructura del basamento del tercio sur de la Sierra de Pie de Palo, Provincia de San Juan, Argentina: V° Congreso

Latinamericano de Geología, Actas 1, p. 451–468.

Dalziel, I.W.D., 1985, Collison and Cordilleran Orogenesis; An Andean perspective, *in* Collision tectonics: London Geological Society Special Publication, v. II, p. 389–404.

Digregorio, J. H., and Uliana, M., 1980, Cuenca Neuquina, *in* Segundo Simposio de Geología Regional Argentina: Academia Nacional de Ciencias, v. 2, p. 985–1032.

Digregorio, R. E., Gulisano, C. A., Gutierrez Pleimling, A. R., and Minitti, S. A., 1984, Esquema de la evolución geodinámica de la Cuenca Neuquina y sus implicancias paleogeográficas: IX° Congreso Geológico Argentino, Actas 2, p. 147–162.

Forsythe, R. and Mpodozis, C., 1983, Geología del basamento pre-Jurásico superior en el Archipiélago Madre de Dios, Magallanes, Chile: Servicio Nacional Geología y Minería Boletín 39, 63 p.

Francis, E. H., 1984, Magma and sediment; II. Problems of interpreting paleovolcanics buried in the stratigraphic column: Journal of the Geological Society of London, v. 140, p. 165–183.

Frei, L. S., Magill, J. R., and Cox, A., 1984, Paleomagnetic results from the central Sierra Nevada: Constraints on reconstructions of the western United States: Tectonics, v. 3, no. 2, p. 157–177.

Frutos, J., 1981, Andean tectonics as a consequence of sea-floor spreading: Tectonophysics, v. 72, p. T21–T32.

Fuentes, J., Ramos, V. A., and Velo, R., 1986, La falla Río Tupungato; Una fractura de cizalla gondwánica, Mendoza, Argentina: Comunicaciones, Una Revista de Geología andina, Santiago de Chile, v. 37, p. 1–15.

Gansser, A., 1973, Facts and theories on the Andes: Journal of the Geological Society of London, v. 129, p. 93–131.

González, R., Cabrera, M. A., Castellote, P., Omil, M., Bortolloti, P., Moyano, R., and Ojeda, J., 1985, Esquematización de la ubicacion espacial y temporal de la eruptividad en Sierras Pampeanas Noroccidentales, República Argentina: IV° Congreso Geológico Chileno, Actas III, no. 4, p. 138–150.

Gonzalez Bonorino, F., 1950, Geologic cross-section of the Cordilera de Los Andes at about parallel 33°S L. (Argentina–Chile): Geological Society of America, Bulletin, v. 61, p. 17–26.

Gorroño, R., Pascual, R., and Pombo, R., 1979, Hallazgo de mamíferos eógenos en el sur de Mendoza; Su implicación en la datación de los "rodados lustrosos" y del primer episodio orogénico del Terciario en esa región: VII° Congreso Geológico Argentino, Actas 2, p. 475–488.

Groeber, P. 1918, Edad y extensión de las estructuras de la Cordillera entre San Juan y Nahuel Huapi: Physis, Revista de la Sociedad Argentina de Ciencias Naturales, v. 4, p. 208–240.

——, 1946, Observaciones geológicas a lo largo del meridiano 70°; 1. Hoja Chos Malal: Sociedad Geológica Argentina, Revista 1, no. 3, p. 177–208.

——, 1953, Mesozoico, *in* Geografía de la República Argentina: Sociedad Argentina de Estudios Geográficos GAEA, v. 2, p. 1–541.

Gulisano, C. A., Gutierrez Pleimling, A. R., and Digregorio, R. E., 1948a, Esquema estratigráfico de la secuencia jurásica al oeste de la provincia del Neuquén: IX° Congreso Geológico Argentino, Actas 1, p. 236–259.

——, 1984b, Análisis estratigráfico del intervalo Tithoniano–Valanginiano (Formaciones Vaca Muerta, Quintuco y Mulichinco) en el suroeste de la provincia del Neuquén: IX° Congreso Geológico Argentino, Actas 1, p. 221–235.

Haller, M., and Ramos, V. A., 1984, Las ofiolitas famatinianas (Eopaleozoico) de las provincias de San Juan y Mendoza: IX° Congreso Geológico Argentino, Actas 2, p. 66–83.

Handschumacher, D. W., 1976, Post-Eocene plate tectonics of the Eastern Pacific, *in* Sutton, G. H., Manghnani, M. H., and Moberly, R., eds., The geophysics of the Pacific Ocean and its margins: Washington, D.C., American Geophysical Union, p. 117–202.

Helwig, J., 1972, Stratigraphy, sedimentation, paleogeography, and paleoclimates of Carboniferous ("Gondwana") and Permian of Bolivia: American Association of Petroleum Geologists Bulletin, v. 56, p. 108–1033.

Hervé, F., Davidson, J., Godoy, E., Mpodozis, C., and Covacevich, V., 1981, The Late Paleozoic in Chile; Stratigraphy, structure, and possible tectonic framework: Anais da Academia Brasileira de Ciencias, v. 53, p. 361–373.

Hervé, F., Kawasita, K., Munizaga, F., and Bassei, M., 1982, Edades Rb-Sr de los cinturones pareados de Chile Central: III° Congreso Geológico Chileano, Actas 2 (D), p. 116–135.

——, 1984, Rb-Sr isotopic ages from Late Paleozoic metamorphic rocks of central Chile: Journal of the Geological Society of London, v. 141, p. 877–884.

Hervé, F., Godoy, E., Parada, M., Ramos, V. A., Rapela, C., Mpodozis, C., and Davidson, J., 1987, A general view of the Chilean-Argentine Andes, with emphasis on their early history, *in* Monger, J., and Francheteau, J., eds., Circumpacific orogenic belts and evolution of the Pacific Ocean basin, Geodynamic series, v. 18, p. 97–113.

Isacks, B., and Barazangi, M., 1977, Geometry of Benioff zones; Lateral segmentation and downwards bending of the subducted lithosphere, *in* Talwani, M., and Pitman, W., eds., Island arcs, deep sea trenches and back arc basins: Washington, D.C., American Geophysical Union, Ewing Series 1, p. 99–114.

Isacks, B., Jordan, T., Allmendinger, R., and Ramos, V. A., 1982, La segmentación tectónica de Los Andes Centrales y su relación con la geometría de la placa de Nazca subductada: V° Congreso Latinoamericano de Geología, Actas 3, p. 587–606.

Jordan, T., 1981, Thrust loads and foreland basin evolution; Cretaceous, western United States: American Association of Petroleum Geologists Bulletin, v. 65, no. 12, p. 2504–2520.

Jordan, T., Isacks, B., Allmendinger, R., Brewer, J., Ramos, V. A., and Ando, C., 1983, Andean tectonics related to geometry of subducted plates: Geological Society of America Bulletin, v. 94, no. 3, p. 341–361.

Jordan, T., Ramos, V. A., Allmendinger, R., and Isacks, B., 1984, Andean tectonics related to geometry of subducted Nazca plate; Discussion and reply: Geological Society of America Bulletin, v. 95, no. 7, p. 877–880.

Kay, S. M., Ramos, V. A., and Kay, R., 1984, Elementos mayoritarios y trazas de las vulcanitas ordovícicas en la Precordillera Occidental; Basaltos de rift oceánicos tempranos (?) próximos al margen continental: IX° Congreso Geológico Argentino, Actas 2, p. 48–65.

Keidel, J., 1921, Sobre la distribución de los depósitos glaciares del Pérmico conocidos en la Argentina y su significación para la estratigrafía de la serie del Gondwana y la paleogeografía del Hemisferio Austral: Academia Nacional de Ciencias Boletín, v. 25, p. 239–368.

Kozlowski, E., 1984, Interpretación estructural de la Cuchilla de La Tristeza, provincia de Mendoza: IX° Congreso Geológico Argentino, Actas 2, p. 381–395.

Larson, R. L., and Pitman, W. C., 1972, World-wide correlation of Mesozoic magnetic anomalies, and its implications: Geological Society of America Bulletin, v. 83, no. 12, p. 3645–3661.

Leveratto, M. A., 1976, Edad de intrusivos cenozoicos en la Precordillera de San Juan y su implicancia estratigráfica: Asociación Geológica Argentina, Revista 31, no. 1, p. 53–58.

Levi, B., and Aguirre, L., 1981, Ensialic spreading-subsidence in the Mesozoic and Paleogene Andes of central Chile: Journal of the Geological Society of London, v. 138, p. 75–81.

Levi, B., and Nystrom, J. O., 1982, Spreading subsidence and subduction in central Chile; A preliminary geochemical test in Mesozoic-Paleogene rocks: III° Congreso Geológico Chileno, Actas 1 (B), p. 28–36.

Llambías, E., and Rapela, C. W., 1985, Geología de los complejos eruptivos de la Esperanza, provincia de Río Negro: Asociación Geológica Argentina, Revista 39 no. 3–4, p. 220–243.

Malumián, N., and Ramos, V. A., 1984, Magmatic intervals, transgression-regression cycles and oceanic events in the Cretaceous and Tertiary of southern South America: Earth and Planetary Science Letters, v. 67, p. 228–237.

Malumián, N., Nullo, F. E., and Ramos, V. A., 1983, The Cretaceous of Argentina, Chile, Paraguay, and Uruguay, *in* Moullade, M., and Nairn, A.E.M., eds., The Phanerozoic geology of the world; II, The Mesozoic B: New York, Elsevier, p. 265–304.

Méndez, V., Navarini, A., Plaza, D., and Viero, O., 1972, Faja Eruptiva de la Puna oriental: V° Congreso Geológico Argentino, Actas 4, p. 89–100.

Mitchum, R. M., and Uliana, M. A., 1985, Seismic stratigraphy of carbonate depositional sequences, Upper Jurassic–Lower Cretaceous, Neuquén Basin, Argentina: American Association of Petroleum Geologists Memoir (in press).

Monger, J.W.H., 1984, Cordilleran tectonics; A Canadian perspective: Societé Geológique de France Bulletin, v. 7-26, no. 2, p. 255–278.

Mpodozis, C., and Forsythe, R., 1983, Stratigraphy and geochemistry of accreted fragments of the ancestral Pacific floor in southern South America: Palaeogeography, Palaeoclimatology, and Palaeoecology, v. 41, p. 103–124.

Mpodozis, C., Parada, M. A., Rivano, S., and Vicente, J. C., 1976, Acerca del plutonismo tardi-hercínico de la Cordillera Frontal entre los 30° y 33° sur (Provincias de Mendoza y San Juan, Argentina; Coquimbo, Chile): VI° Congreso Geológico Argentino, Actas I, p. 143–172.

Munizaga, F., and Vicente, J. C., 1982, Acerca de la zonación plutónica y del vulcanismo miocénico en los Andes de Aconcagua (lat 32°–33°S); Datos radiométricos K/Ar: Revista Geològica de Chile, v. 16, p. 3–21.

Musatov, D. I., Fedorovsky, V. S., Afanasyev, T., Zonenshain, L. P., Kleschchev, K. A., Kovalev, A. A., Mezhelovsky, N. V., and Sorokhtin, O. G., 1984, Some aspects of geology and geological history of the USSR territory according to new geotectonic concepts, *in* Geology of the U.S.S.R.: Colloquium 01, v. 1, 27° International Geological Congress, p. 129–144.

Nasi, C., and Thiele, R., 1982, Estratigrafía del Jurásico y Cretácico de la Cordillera de la Costa al sur del Río Maipo, entre Melipilli y Laguna de Aculeo (Chile Central): Revista Geológica de Chile, v. 16, p. 81–99.

Noble, D., McKee, E. H., Farrar, E., and Petersen, U., 1974, Episodic cenozoic volcanism and tectonism in the Andes of Perú: Earth and Planetary Science Letters, v. 21, p. 213–220.

Nur, A., and Ben-Avraham, Z., 1977, Lost Pacifica continent: Nature, v. 270, p. 4143.

—— , 1981, Volcanic gaps and the consumption of aseismic ridges in South America: Geological Society of America Memoir 154, p. 729–740.

Orchuela, I. A., Ploszkiewicz, J. V., and Viñes, R., 1981, Reinterpretación estructural de la denominada dorsal neuquina: VIII° Congreso Geológico Argentino, Actas 3, p. 281–293.

Oviedo, E. S., and Vilas, J.F.A., 1984, Movimientos recurrentes en el permotriásico entre el Gondwana occidental y el oriental: IX° Congreso Geològico Argentino, Actas 3, p. 97–114.

Pascual, R., and Odreman Rivas, O., 1973, Las unidades estratigráficas del Terciario portadoras de mamíferos; Su distribución y sus relaciones con los acontecimientos diastróficos: V° Congreso Geológico Argentino, Actas 3, p. 293–338.

Pilger, R. H., 1978, A method for finite plate reconstructions with applications to Pacific-Nazca plate evolution: Geophysical Research Letters, v. 5, p. 469–472.

—— , 1981, Plate reconstructions, aseismic ridges, and low-angle subduction beneath the Andes: Geological Society of America Bulletin, v. 92, p. 448–456.

—— , 1984, Cenozoic plate kinematics, subduction, and magmatism; South American Andes: Journal of the Geological Society of London, v. 141, p. 793–802.

Piraces, R., 1979, Velocidades de depositación de las secuencias mesozoicas en los Andes a la latitud de Chile Central; Su relación con los eventos magmáticos y tectónicos: II° Congreso Geológico Chileno, Actas 1 (F), p. 83–106.

Polanski, J. , 1962, Estratigrafía, Neotectónica y geomorfología del Pleistoceno pedemontano entre los Ríos Diamante y Mendoza: Asociación Geológica Argentina, Revista, v. 17, no. 3-4, p. 127–349.

—— , 1964, Descripción geológica de la Hoja 25 a Volcán San José, Provincia de Mendoza: Dirección Nacional de Geología y Minería Boletín 98, p. 1–94.

—— , 1972, Descripción geológica de la Hoja 24 ab Cerro Tupungato, Provincia de Mendoza: Dirección Nacional de Geología y Minería Boletín 128, p. 1–98.

Ramos, E. D., and Ramos, V. A., 1979, Los ciclos magmáticos de la República argentina: VII° Congreso Geológico Argentino, Actas 1, p. 771–786.

Ramos, V. A., 1978, Estructura, *in* Geología y Recursos Naturales del Neuquén: VII° Congreso Geológico Argentino, Relatorio, p. 99–118.

—— , 1983, Evolución tectónica y metalogénesis de la Cordillera Patagónica: II° Congreso Nacional Geología Económica, Actas 1, p. 108–124.

—— , 1985a, El Mesozoico de la Alta Cordillera de Mendoza; Facies y desarrollo estratigráfico, Argentina: IV° Congreso Geológico Chileno, Actas 1 (1), p. 492–513.

—— , 1985b, El Mesozoico de la Alta Cordillera de Mendoza; Reconstrucción tectónica de sus facies, Argentina: IV° Congreso Geológico Chileno, Actas 1 (2), p. 104–118.

Ramos, V. A., and Cortés, J. M., 1984, Estructura e interpretación tectónica, *in* Geología y Recursos Naturales de la Provincia de Río Negro: IX° Congreso Geológico Argentino, Relatorio, p. 317–346.

Ramos, V. A., and Palma, M. A., 1983, Las lutitas pizarreñas del cerro Dedo, Lago La Plata, provincia del Chubut: Asociación Geológica Argentina, Revista, v. 38, no. 2, p. 148–160.

Ramos, V. A., Niemeyer, H., Skarmeta, J., and Muñoz, J., 1982, The magmatic evolution of the Austral Patagonian Andes: Earth Science Reviews, v. 18, no. 3-4, p. 411–443.

Ramos, V. A., Jordan, T. E., Allmendinger, R. W., Kay, S. M., Cortés, J. M., and Palma, M. A., 1984, Chilenia; Un terreno alóctono en la evolución paleozoica de los Andes Centrales: IX° Congreso Geológico Argentino, Actas 2, p. 84–106.

Ramos, V. A., Jordan, T. E., Allmendinger, R. W., Mpodozis, C., Kay, S. M., Cortés, J. M., and Palma, M. A., 1986, Paleozoic terranes of the central Argentine-Chilean Andes: Tectonics, v. 6, no. 6, p. 855–880.

Rapela, C. W., 1982, Aspectos geoquímicos y petrológicos del Batolito de Achala, provincia de Córdoba: Asociación Geológica Argentina, Revista, v. 37, no. 3, p. 313–330.

Rapela, C. W., Heaman, L. M., and McNutt, R. J., 1982, Rb/Sr geochronology of granitoid rocks from the Pampean Ranges, Argentina: Journal of Geology, v. 909, p. 574–582.

Rivano, S., 1984, Geología del Meso-Cenozoico entre los 31 y 33° de latitud sur, *in* Seminario; Actualización de Geología de Chile. Apuntes: Servicio Nacional de Geología y Minería, p. K1–K17.

Rolleri, E., and Baldis, B., 1967, Paleogeography and distribution of Carboniferous deposits in the Argentine Precordillera: Coloquio de la IUGS: La Estratigrafía del Gondwana, Ciencias de La Tierra, v. 2, p. 1005–1024.

Rolleri, E., and Fernandez Garrasino, C. A., 1979, Comarca septentrional de Mendoza, *in* Segundo Simposio de Geología Regional Argentina: Academia Nacional de Ciencias de Córdoba, v. 1, p. 771–810.

Rodgers, J., 1971, The Taconic Orogeny: Geological Society of America Bulletin, v. 82, p. 1141–1178.

Rutland, R.W.R., 1971, Andean orogeny and sea floor spreading: Nature, v. 233, p. 252–255.

Schiller, W., 1912, La Alta Cordillera de San Juan y Mendoza y parte de la provincia de San Juan: Anales del Ministerio de Agricultura, Sección Geología, Mineralogía y Minería, v. 7, no. 5, p. 1–68.

Schweller, W. J., and Kulm, L. D., 1978, Extensional rupture of oceanic crust in the Chile trench: Marine Geology, v. 28, p. 271–291.

Stipanicic, P. N., and Rodrigo, F., 1970a, El diastrofismo Jurásico en Argentina y Chile: IV° Jornadas Geológicas Argentinas, Actas 2, p. 353–368.

—— , 1970b, El diastrofismo eo y mesocretácico en Argentina y Chile, con referencias a los movimientos jurásicos de la Patagonia: IV° Jornadas Geológicas Argentinas, Actas 2, p. 334–352.

Thiele, R., and Hervé, F., 1984, Sedimentación y tectónica de antearco en los terrenos preandinos del Norte Chico, Chile: Revista Geológica de Chile, v. 22, p. 61–75.

Turner, J.C.M., and Méndez, V., 1975, Geología del sector oriental de los Departamentos de Santa Victoria e Iruya, provincia de Salta, Repùblica Argentina: Academia Nacional de Ciencias Boletín, v. 51, no. 1-2, p. 11–24.

Valencio, D. A., Vilas, J. F., and Pacca, I. G., 1983, The significance of the paleomagnetism of a sequence of red beds of the middle and upper sections of Paganzo Group (Argentina) and the correlation of Upper Paleozoic–Lower Mesozoic rocks: Geophysical Journal of the Royal Astronomical

Society, v. 51, p. 59–74.
Vásquez, J. and Gorroño, R. A., 1980, Límite de la Faja Plegada de la República Argentina: Asociación geológica Argentina, Revista, v. 35, no. 4, p. 582–585.
Vicente, J. C., 1972, Apercu sur l'organization et l'evolution des Andes argentino-chiliennes centrales au paralléle de Aconcagua: XXIV° International Geological Congress, section 3, p. 42–436.
Vicente, J. C., Charrier, R., Davidson, J., Mpodozis, C., and Rivano, S., 1973, La orogénesis subhercínica; Fase mayor de la evolución paleogeográfica y estructural de los Andes Argentino-Chilenos Centrales: V° Congreso Geològico Argentino, Actas 5, p. 81–98.
Vicente, J. C., Sequeiros, F., Valdivia, M. A., and Zavala, J., 1979, El sobre-escurrimiento de Cuicha-Lluta; Elemento del accidente mayor andino al noroeste de Arequipa: Sociedad Geológica del Perú, Boletín 61, p. 67–100.
Vilas, J. F. and Valencio, D., 1982, Implicaciones geodinámicas de los resultados paleomagnéticos de formaciones asignadas al Paleozoico tardío–Mesozoico temprano del centro-oeste argentino: V° Congreso Latinoamericano de Geología, Actas 3, p. 743–758.
Wortel, M.J.R., 1984, Spatial and temporal variations in the Andean subduction zone: Journal of the Geological Society of London, v. 141, p. 783–791.
Yanshin, A. L., Khain, V. E., and Gatinsky, Y. G., 1984, The principal problems in the tectonics of Asia, *in* Tectonics of Asia: 27th International Geological Congress, Colloquium 05, v. 5, p. 3–12.
Yokokura, T., 1981, On subduction dip angles: Tectonophysics, v. 77, p. 66–77.
Yrigoyen, M., 1976, Observaciones geológicas alrededor del Aconcagua: I° Congreso Geológico Chileno, Actas 1 (A), p. 169–190.
——, 1979, Cordillera Principal, *in* Segundo simposio de Geología Regional Argentina: Academia Nacional de Ciencias de Córdoba, v. 1, p. 651–694.
Zeil, V., 1979, The Andes; A geological review: Gebrüder Borntaeger, p. 1–260.

MANUSCRIPT ACCEPTED BY THE SOCIETY JUNE 1, 1987

Printed in U.S.A.

Geological Society of America
Special Paper 218
1988

Mesozoic and early Cenozoic magmatic evolution of the Canadian Cordillera

Richard Lee Armstrong
Department of Geological Sciences, University of British Columbia, Vancouver, British Columbia V6T 2B4, Canada

One of the diseases of this age is the multiplicity of books: they doth so overcharge the world that it is not able to digest the abundance of idle matter that is everyday hatched and brought forth into the world.

Barnaby Rich, 1613

ABSTRACT

Approximately 3,000 Ar, Sr, and Pb isotopic age determinations for Canadian Cordilleran rocks have been cataloged, categorized as to reliability and significance, and plotted on histograms, distribution maps for different time intervals, and space-time plots to show the magmatic evolution in this 2,300-km portion of the Circum-Pacific Mobile Belt. The history revealed is episodic, with stable distribution patterns within episodes and distinct lulls and changes in distribution between the episodes.

From 230 to 214 Ma (during Late Triassic time), extensive mafic volcanism occurred in the Wrangell, Quesnel, and Stikine terranes. Volcanic-related ultramafic complexes are found scattered through the two latter terranes. Large calc-alkaline granitic plutons are known only in a belt crossing Stikinia in northern British Columbia. At the same time, blueschists formed in the Cache Creek accretion wedge.

From 214 to 200 Ma (end of Triassic and part of Early Jurassic time), Early to Middle Jurassic arc magmatism began in Wrangellia and in the northern Quesnel, Stikine, and Yukon terranes. A distinct magmatic event is recognizable only in southern Quesnellia. Magmatism was absent on the North American craton. The Cache Creek and Quesnel terranes were definitely linked, Stikine and Cache Creek terranes were probably linked, and a regional metamorphic episode was completed in the Yukon Terrane by this time.

From 200 to 155 Ma (late Early to early Late Jurassic time), magmatism was extensive in the Wrangell, Quesnel, Stikine, and Yukon terranes. Magmatism overlapped into North America *only* east of southern Quesnellia *after* about 180 Ma. By the middle of this time interval, the southern Quesnel–Slide Mountain–North America linkage was complete, and major deformation and metamorphism had affected the Omineca Belt in British Columbia. Early to Middle Jurassic magmatism in southern Wrangellia (Vancouver Island) is distinctly older than the Middle to Late Jurassic magmatism that occurred in central Wrangellia (Queen Charlotte Islands).

From 155 to 140 Ma (during Late Jurassic time), a few last-gasp plutons of the late Early to early Late Jurassic episode and other rocks with partially reset 155- to 145-Ma dates occur in the Wrangell, Quesnel, and Stikine terranes. Late Jurassic magmatism (160 to 140 Ma) occurred in the Alexander Terrane (Saint Elias region).

From 145 to 138 Ma (latest Jurassic and beginning of Early Cretaceous time), plutonism occurred in the Endako area of central British Columbia (Francois Lake suite) but is virtually unknown elsewhere.

From 135 to 125 Ma (during Early Cretaceous time), there was a magmatic lull of major significance present throughout western North America.

From 110 to 90 Ma (middle Cretaceous time), widespread plutonism occurred across all terranes. Dual culminations are evident: the Coast Plutonic and Ominica belts. *Before* this time all sutures except those outboard of Wrangellia had been closed.

From 80 to 70 Ma (during Late Cretaceous time), a narrow, sinuous belt of magmatism persisted, mostly in the southeastern Coast Plutonic Belt, southwestern Yukon Territory, and scattered across the Skeena and Stikine arches.

From 70 to 60 Ma (latest Cretaceous to Paleocene time), a distinct lull in magmatism occurred. Rare plutons of this time interval are known in the Coast Plutonic Belt, on the Skeena Arch, and in the southern Intermontane Belt.

From 55 to 45 Ma (latest Paleocene to Middle Eocene time), widespread and voluminous magmatism occurred in all terranes. The early Cenozoic volcanic front crossed the Coast Plutonic Complex from its east side in the south to its west side in the north. Associated thermal and tectonic effects were strong even into the Omineca Belt, producing large reset metamorphic areas in the Coast and Omineca belts. This was a short-lived event, synchronous from southern British Columbia through the Yukon Territory. West of the volcanic front, offshore of Wrangellia, Metchosin volcano growth was underway at this time. Late in this time interval, the 50?–45–36-Ma Catface–Leech River event(s) of southern Wrangellia occurred. There is also time overlap with a diffuse Massett magmatic event in the Queen Charlotte Islands, and with Baranoff Island and Yakutat–Saint Elias region magmatism.

Initial $^{87}Sr/^{86}Sr$ ratios and petrographic characteristics of Canadian Cordilleran igneous rocks are reviewed in the time frame just described. These reflect the nature of underlying crust, contemporaneous lithosphere thickness, and distance from the subduction zone.

Comparisons with other parts of the Circum-Pacific Magmatic Belt shows both out-of-phase magmatism (Japan and southwestern Alaska) and perfect matching of some episodes (Sierra Nevada). Major magmatic episodes correspond to times of increased westward motion of North America with respect to hot spots or to times of increased convergence between western North America and the Farallon Plate.

INTRODUCTION

The Canadian Cordillera has traditionally been subdivided into five tectonic belts (Fig. 1) and is more recently discussed in terms of numerous terranes (Fig. 2) that have internal stratigraphic coherence but are in various ways distinct from one another. These terranes are commonly juxtaposed across major faults, and in some cases they are suspected of having formed in widely separated places and later accreted into the Cordilleran collage we study today. Important syntheses of the geology and tectonic history have been provided by White (1959), McCrossan and Glaister (1964), Wheeler and Gabrielse (1972), Wheeler and others (1974), and Monger and Price (1979). Monger and others (1972), Monger and Price (1979), and Monger (1984) discussed the impact of plate tectonic concepts on the tectonic synthesis; the terrane subdivision has been developed at length by Jones and others (1977, 1983), Coney and others (1980), and Saleeby (1983).

White, in his 1959 review of Cordilleran orogenies, had to rely entirely on stratigraphic grounds for the ages of magmatic events. The age assignments available to him were not greatly refined from the interpretations of pioneers like Daly (1912). White sensed the need for isotopic geochronometry and was instrumental in establishing a K-Ar laboratory at the University of British Columbia (UBC) in the mid 1960s. A few years earlier, one of the world's first K-Ar labs had been established at the University of Alberta and had begun work on Cordilleran rocks. This chapter presents a review of a bit more than 20 years of geochronometric work toward the goal of knowing the ages of igneous rocks in the Canadian Cordillera. This research has been the major concern of the UBC lab, which now includes Rb-Sr and U-Pb dating in its arsenal of techniques. Other major contributors are the Geological Survey of Canada laboratory in Ottawa, Queens University, and the University of Alberta. I have been involved in this work since 1973.

Geochronometric knowledge has grown from modest beginnings. In 1961, Baadsgaard and others published 20 K-Ar dates, mostly on granitic rocks from British Columbia, and identified a number of magmatic episodes. In hindsight, their success was remarkable, documenting six periods of magmatic activity, including four of the seven episodes discussed in this chapter. In 1964, Gabrielse and Reesor discussed a protracted and complex

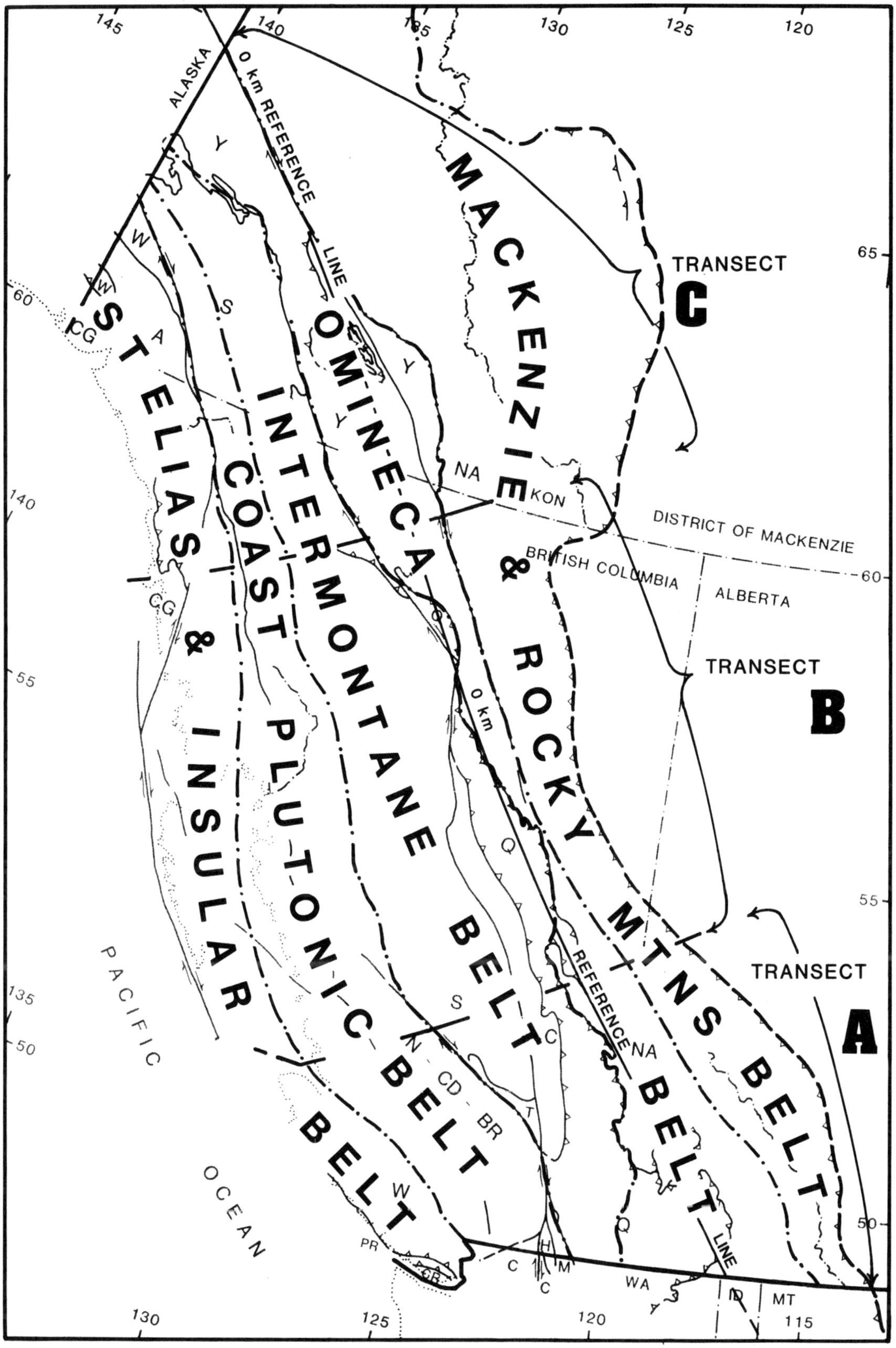

Figure 1. Tectonic belts of the Canadian Cordillera showing three transects for which geochronometry is compiled in subsequent figures.

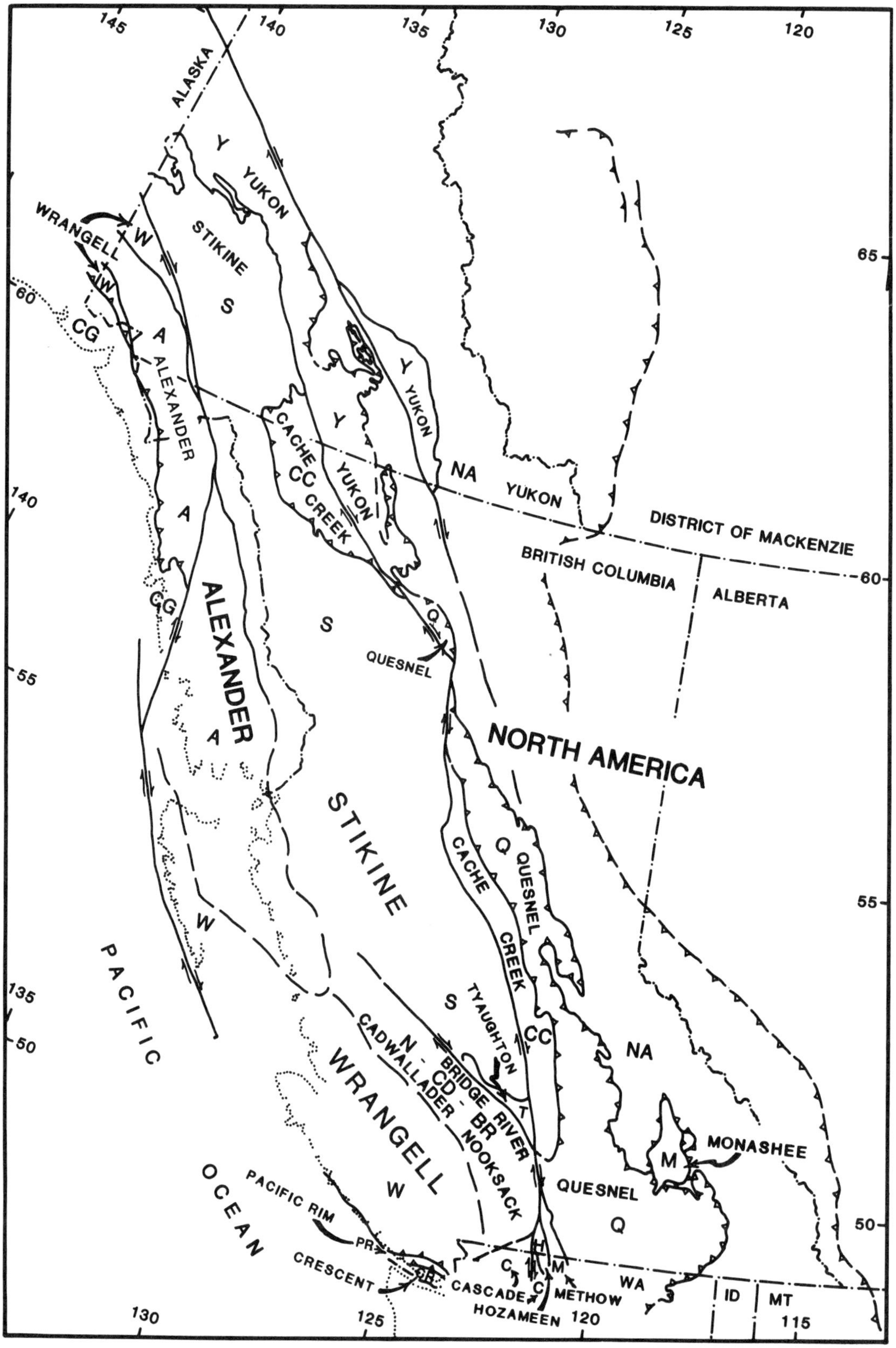

Figure 2. Simplified terrane map of the Canadian Cordillera (adapted from Monger and Price, 1979). The narrow Slide Mountain Terrane has been included with Quesnel and Yukon terranes that lie just to its west, and the Taku Terrane with the Alexander Terrane on its western side.

magmatic history based on 196 K-Ar dates, mostly done by the Geological Survey of Canada. Their perceptive analysis was confused on several issues by poor data quality, and it was even claimed that one could do as well with a random number generator (Ross, 1966), but again hindsight shows that they were more often correct than confused. This was because they relied on stratigraphic and geologic relationships as much as the new dates in creating their synthesis. In 1970 Wanless published his "Geochronometric Map of Canada" showing more than 300 Cordilleran localities with one or more dates. Petö (1974) plotted 394 K-Ar dates and cited 5 Rb-Sr studies in western Canada. That same year saw updated reviews of plutonism in the Coast Plutonic Complex (Roddick and Hutchison, 1974) and central and eastern Cordillera (Gabrielse and Reesor, 1974). Now, ten years later, we have nearly an order of magnitude increase in the available data! The present review is based on approximately 3,000 isotopic dates contained in the Canadian Cordilleran Geochron File (CCGF) at UBC as of September 1984.

The CCGF is a product of several years of effort to catalog all geochronometric data for the Canadian Cordillera. Dates from published sources, many theses, work done for industry and British Columbia Ministry of Energy Mines and Petroleum Resources geologists, and all unpublished UBC data for which suitable sample information exists have been recorded on standardized forms. All dates have been recalculated using the decay constants recommended by the IUGS Subcommission on Geochronology (Steiger and Jäger, 1977). Each date is categorized as to its reliability and probable significance—if analytically accurate, reasonably precise, and on suitable material, then whether it represents a volcanic or intrusive event, is from a metamorphic rock, or is partially or totally reset by a younger metamorphic, uplift and erosion, or alteration event. This is a somewhat subjective exercise but if there is any error, it is on the side of giving significant event status to partially reset dates. Isotopic dates that do not meet the numerous conditions required for giving the times of significant events will only blur any truly episodic history. The real events will be more sharply defined than any representation of them using CCGF data. From that total body of data a series of maps and other diagrams has been prepared that forms the basis for this chapter. The object has been to present a large amount of information in a condensed form so that the geologic significance is evident, and not to focus on individual results. This means that many contributors to the data set will go unheralded, but this is a necessary economy—the actual tables fill 20 loose-leaf notebooks, and the reference cards form a pile 20 cm high!

MAGMATIC EPISODES

The seven episodes delineated here are largely familiar to geologists working in the Canadian Cordillera. But the description that can now be provided is strengthened in detail, and some observations are new. Histograms of Cordilleran geochronometry, compiled separately for three transects, are presented in Figure 3. The southern transect (A) comprises data from southern Wrangellia (all of Vancouver Island) to Quesnellia. The middle transect (B) includes nearly all of Stikinia—from Skeena arch in the south to the Stikine arch in the north—and central Wrangellia (Queen Charlotte Islands). The northern transect (C) includes the Yukon Territory and northwesternmost British Columbia. Transect boundaries have been placed in areas where data are relatively sparse so that closely related isotopic ages are not split between diagrams. The compilation shown is a third-generation version, each an improved and more expressive representation of the data.

Distinction of episodes is based on peaks and low points of the histograms, on changes in geographic distribution of magmatism, and on the basis of compatibility with previous descriptions of geologic age relationships and natural groupings of rocks. Metamorphic and partially to completely reset dates on igneous rocks have been separated from dates close to the times of magmatic events on the histograms, and the two types of dates are plotted on separate maps. Regardless of whether the whole data set or just the part indicating igneous events is considered, the histograms are similar. The nonigneous dates smooth the apparent time distribution and shift the older peaks to somewhat younger values, as would be expected, but the important conclusion is that times of magmatism and metamorphic events are broadly synchronous. This is one reason that first impressions from reconnaissance geochronometry have been largely correct. The synchroneity is particularly striking in the case of the early Cenozoic episode. The linkage is tectonic as well as thermal—reset metamorphic areas are not simply contact metamorphic aureoles but more often are related to erosional or tectonic unroofing.

The vertical background lines on Figure 3 show the times chosen for separation of episodes: 40, 64, 84, 130, 155, 192, 214, and 235 Ma. These are the null points and times of change in geographic distribution. The nonrandom nature of the episodes is demonstrated by their independent appearance on more than one transect histogram. Each histogram is an entirely separate sample of Cordilleran rocks. Samples were collected and dated by different workers for a variety of motives, yet the same events are observed. Any sampling bias in a data set of this size is probably minor and mostly an inverse age effect, episodes becoming blurred and diminished as time passes. The effects of erosion would be to shift the ratio of volcanic to intrusive rocks representing each episode but would not create or erase episodes. Histogram peaks occur at 45–55, 70–80, 90–110, 138–145, 160–170, 180–195, 200–214, and 215–230 Ma. The first and third peaks are the most prominent. Resolution of Jurassic events is still poor. The problems include increasing analytical error with older dates, more frequent partial resetting of dates, and a genuinely more diverse history in terranes that may have been widely separated from one another.

A different representation of the data in the CCGF are the space-time plots of Figures 4, 5, and 6, one for each transect. In those plots all dates are projected onto imaginary planes perpen-

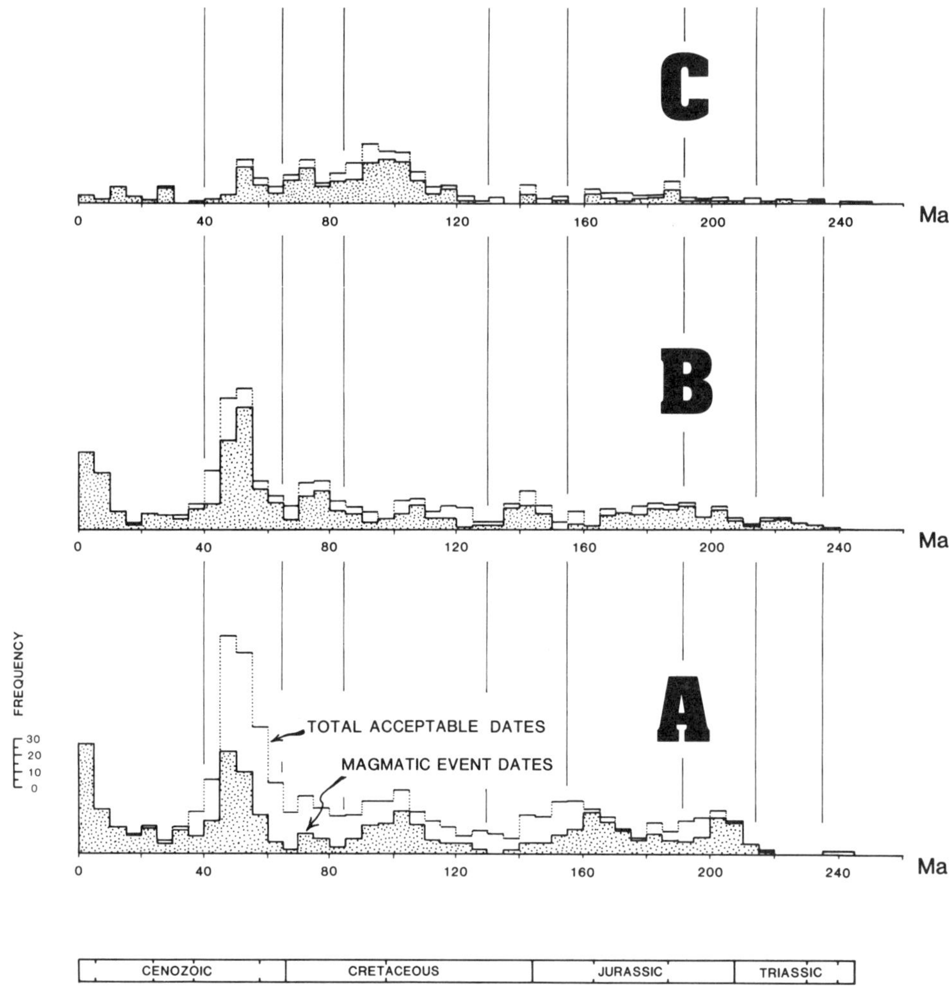

Figure 3. Histograms of age determinations in three transects of the Canadian Cordillera. Unreliable analyses and sedimentary samples have been excluded from this and all other plots. Dates inferred to be on or close to magmatic events are distinguished from the total collection, which includes dates on metamorphic rocks and variably reset dates on igneous rocks, but this does not greatly affect recognition of culminations of magmatic activity. A summation of all magmatic dates is shown in Figure 22. Vertical background lines show the natural divisions of magmatic history into time intervals centered on major magmatic episodes. Divisions are placed at times of reduced magmatism, which are also times of change in geographic distribution of magmatism. Each time interval bracketed by vertical lines is represented by a separate map (Figs. 9 to 15). The time scale is that proposed for DNAG (Palmer, 1983). Raw data are recorded in the Canadian Cordilleran Geochron File (CCGF), and are complete as of September 1984.

dicular to Cordilleran structural trends. The horizontal scale is distance east and west of an arbitrary reference line shown on Figure 1. The vertical scale is time so that each individual magmatic date is shown. Horizontal lines are the chosen time interval delimiters. On these figures the clustering into episodes is further evident. In addition, the changing position and band width of magmatism and time-transgressive trends are visible. In preparation for this chapter, two northwest-southeast–oriented projections were also made—one for Insular to Intermontane belt data, the other for Omineca Belt data—but these showed nothing more than the cross-strike transects and are not presented here.

The most striking time-transgressive feature, especially evident on transects A and B, is the steady regression of the western magmatic front toward cratonic North America. Through Jurassic and Cretaceous time the average regression rate has been about 1.8 km/Ma. This eastward migration was well known from previous studies in the Coast Plutonic Complex (Roddick and Hutchison, 1974). Its explanation may include widening of the

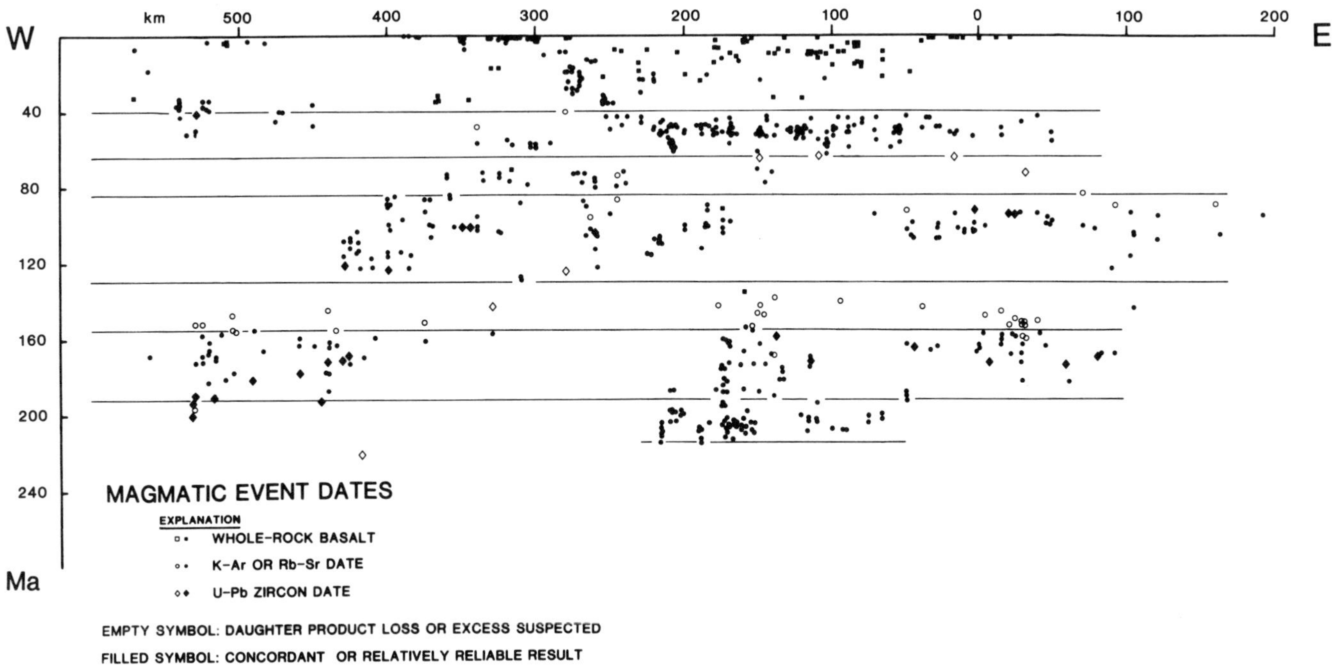

Figure 4. Space-time Profile for Transect A, showing isotopic dates for igneous events.

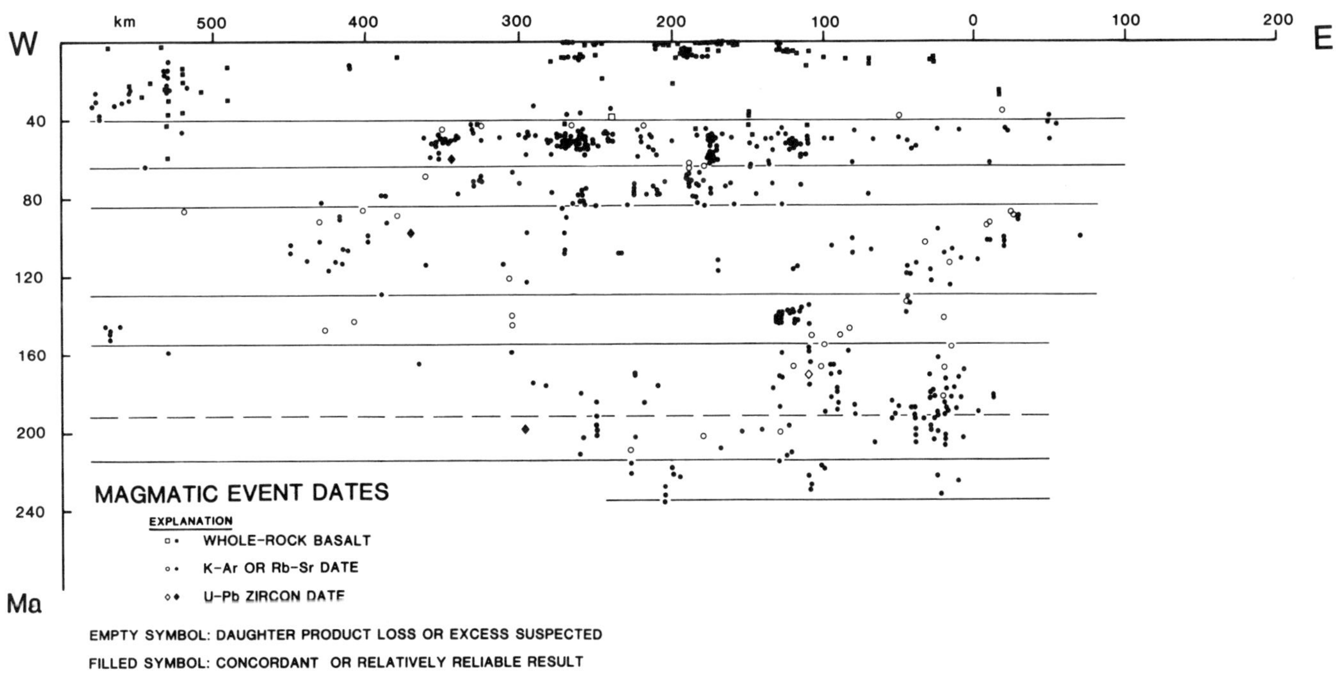

Figure 5. Space-time Profile for Transect B, showing isotopic dates for igneous events.

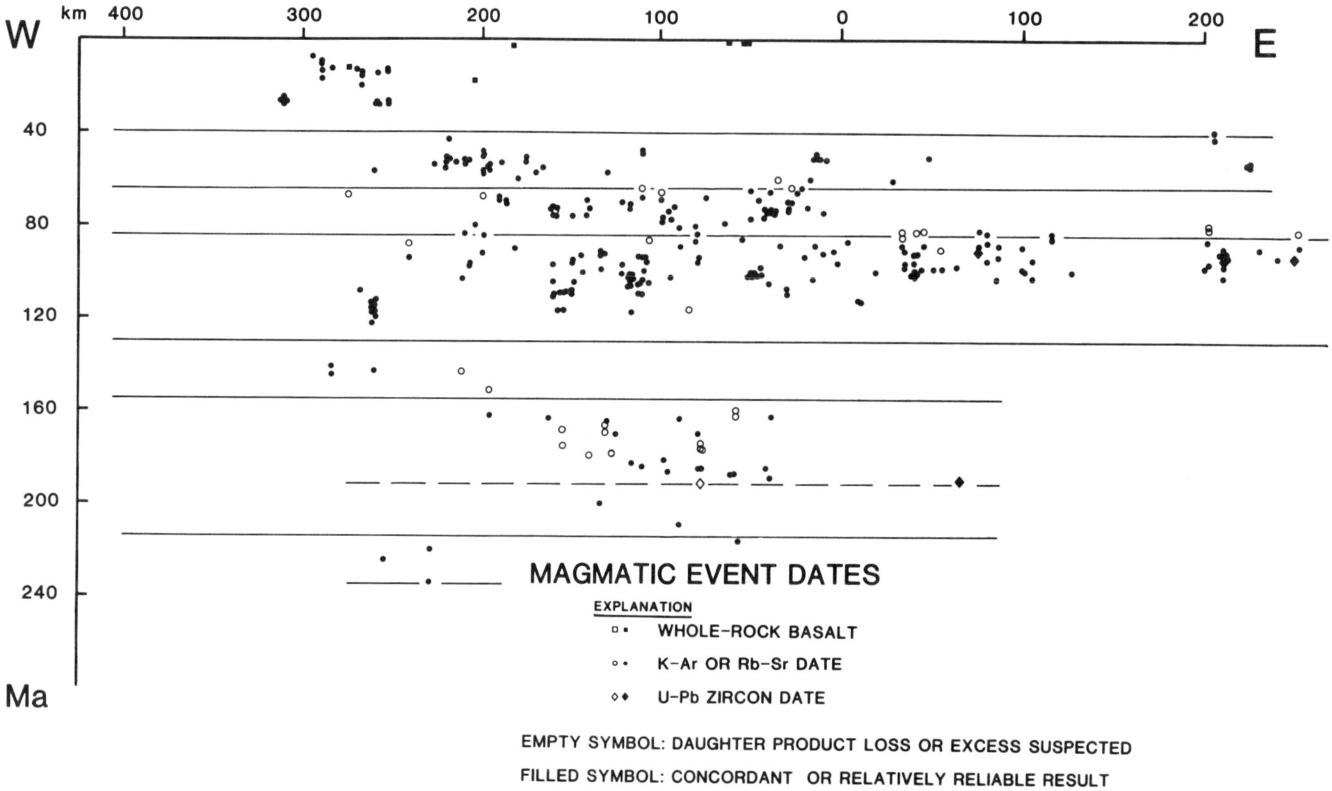

Figure 6. Space-time Profile for Transect C, showing isotopic dates for igneous events.

arc-trench gap with age (Dickinson, 1973) but must certainly include, and may be entirely due to, steady decretion of the continental margin by subduction erosion (Scholl and others, 1980; von Huene, 1984) or intermittent lateral removal of material into Alaska (in Mesozoic events analogous to the Cenozoic one discussed by Cowan, 1982).

Also of note is a slight asymmetry within individual episodes. Where the documentation is good, magmatism begins in inland areas later than areas near the magmatic front. This may be related to recession of the magmatic front, but the transgressive inception of magmatism appears to be an order of magnitude more rapid—a few centimeters per year.

Each magmatic episode is discussed in turn in the following sections of this chapter. On Figure 7, a schematic representation of the space-time profiles is used as background for names used in discussing volcanic rocks. The reviews of Cordilleran volcanism by Souther (1967, 1977) provide an explanation of most of these names. Figure 8 shows names of plutonic rocks against the same background. Together these figures provide a correlation guide to Cordilleran magmatic rocks. One point they illustrate well is the exact correspondence of volcanic and plutonic episodes—only mafic volcanic units may lack a corresponding granitic pluton suite. The only major plutonic suite lacking an equivalent volcanic unit is the Jurassic suite of southeastern British Columbia. This same area contains the broadest exposure of high-grade metamorphic rocks, so it is reasonable to infer that the absence of volcanic equivalents is due to their removal by erosion. Evidence supporting this view is the presence in southwestern Alberta of Jurassic volcanic and subvolcanic clasts in Cretaceous conglomerate (Norris and others, 1965), and Jurassic hornblende and feldspar in Cretaceous sands (Tater, 1965; Rahmani and Lerbekmo, 1975).

In organizing data for this chapter, a duality in the character of composite plutons was noted. Many batholiths are geochronometrically *singular*, that is, they yield tightly clustered dates. Their entire history occurred within one unresolvable instant of geologic time, and any single age determination is representative of the entire body. Examples of this behavior are the Guichon (White and others, 1967) and Francois Lake (White and others, 1970) batholiths and the Selwyn Plutonic Suite (Anderson, 1983b). Contrasting behavior is shown by the geochronometrically *protracted* composite plutons. These have histories spanning several magmatic episodes and are difficult to classify into episodes based on a few isotopic dates. Each requires a study of thesis magnitude, and it is these that slow progress, require application of multiple dating techniques, and contribute to confusion

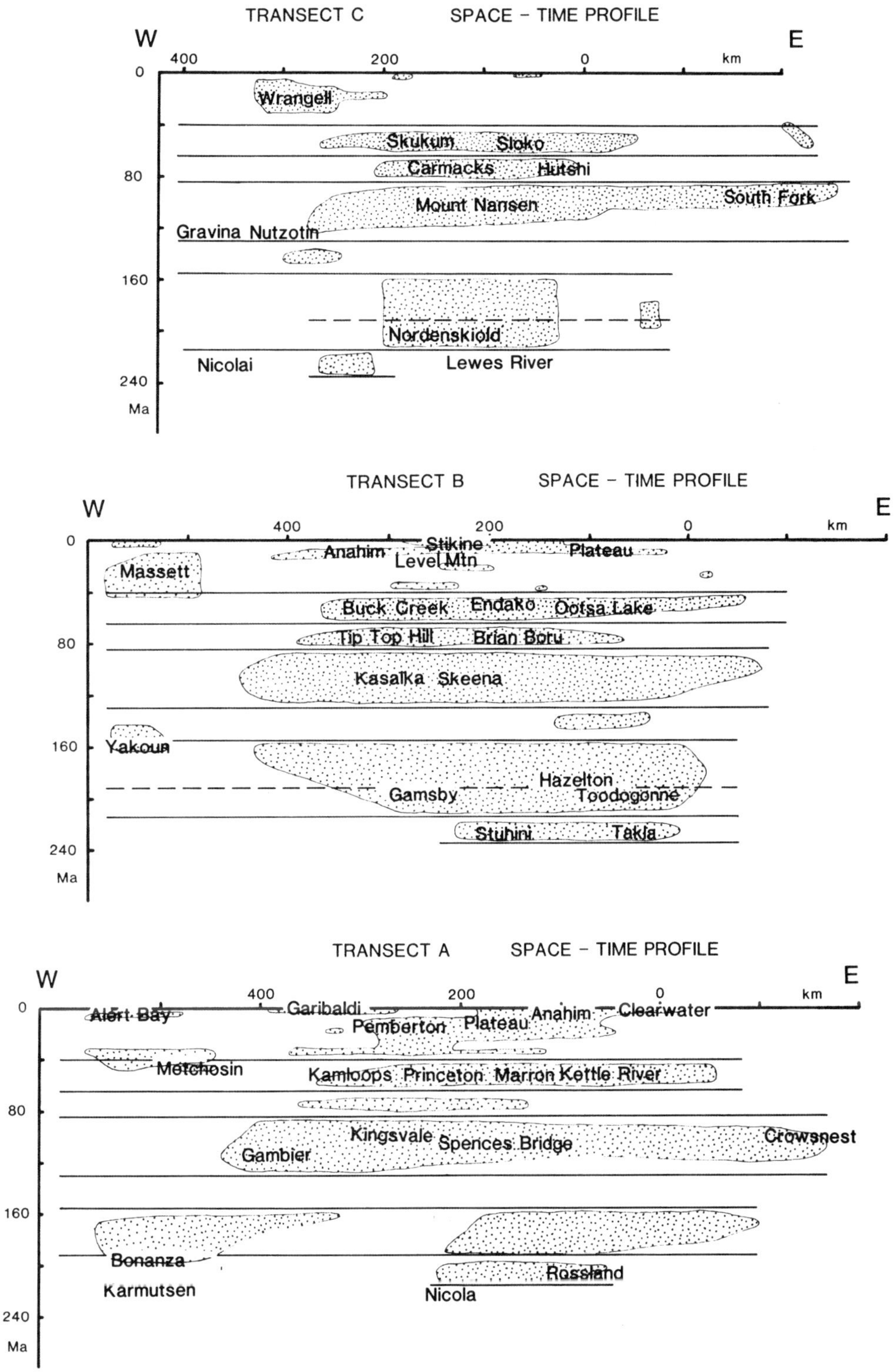

Figure 7. Space-time magmatic framework showing names of important and well-known volcanic units—formations, groups, or volcanic belts.

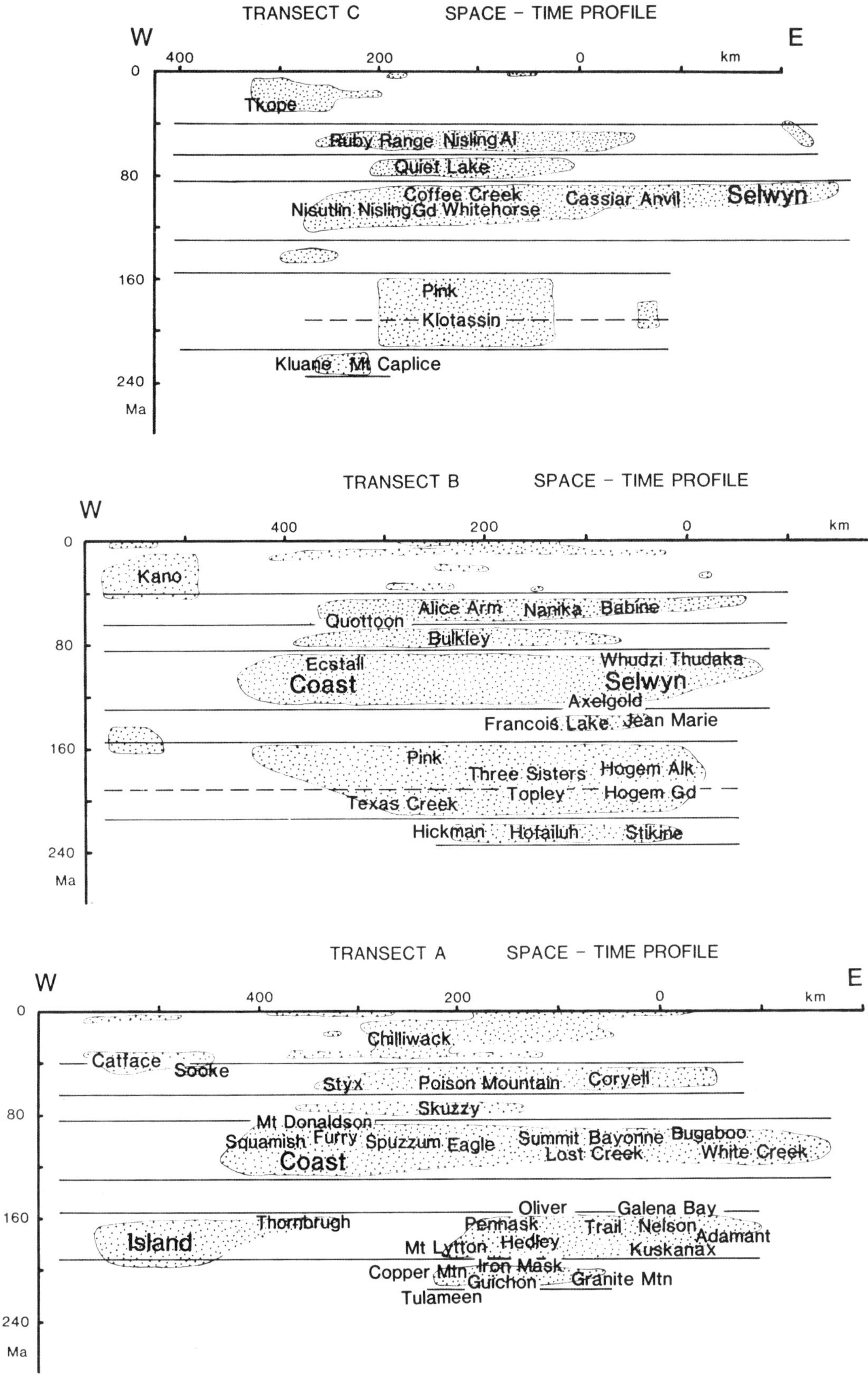

Figure 8. Space-time magmatic framework showing names of important and well-known plutonic units—intrusive bodies or plutonic suites.

and controversy. Examples include the Hotailuh (Anderson, 1983a) and Hogem (Eadie, 1976; Garnett, 1978) batholiths and the Coast Plutonic Complex (Roddick and Hutchison, 1974).

REVIEW OF MAGMATIC HISTORY

Figures 9 through 15 show the geographic distribution of magmatism within each of the time intervals chosen to encompass major magmatic episodes. For reference the sutures or cryptic boundaries between major terranes and the location of dates recorded in the CCGF are shown. Names of volcanic units provide linkage to Figure 7. Some geographic reference points are also labeled. Isotopic dates are translated into geologic time-scale units according to the DNAG scale (Palmer, 1983). The uncertainties in this correlation of relative and chronometric scales increase from 2 to 5 m.y. over the time interval of interest here, and are usually less than the analytical uncertainties of individual dates.

***From 235 to 214 Ma** (Late Triassic).* No igneous rocks of this age occur on North America proper, but mafic volcanic accumulations are common on the major terranes. In the terranes a distinct hiatus usually separates Upper Triassic strata from older rocks (Muller, 1977; Read and Okulitch, 1977). The record of Carboniferous to Mid-Triassic volcanism is obscure, and granitic plutons of that time are rare outside the Alexander Terrane in northwesternmost British Columbia. The Cassiar Orogeny at the end of the Permian (White, 1959) and the Tahitanian Orogeny separating Middle and Upper Triassic (Souther, 1972) are names assigned to this break, which preceded large-scale granitic magmatism in the Cordillera.

Figure 9 shows the few locations where mafic to ultramafic complexes related to Karmutsen, Nicolai, Nicola, Takla, and Stuhini volcanics have been dated. Fossil evidence shows the different volcanic units to be of slightly different age and time span, but this detail is not resolvable by existing geochronometry. Karmutsen is limited to part of the Carnian, Nicolai to the Norian; Nicola is mostly Norian and locally continues to Early Jurassic in Quesnellia. Takla, Stuhnini, and Lewes River groups include both Carnian and Norian strata.

The difference in stratigraphic age of volcanic rocks between Quesnel and Stikine terranes is small but significant. Even more striking is the contrast in plutonic rocks. No granitic plutons of mid-Late Triassic age have been discovered in Quesnellia, in spite of intensive study related to the Mesozoic porphyry copper deposits there. In Stikinia in northern British Columbia, there are several large calc-alkaline granitic batholiths of both singular and protracted character—Stikine, Hotailuh, and Hickman—and several smaller plutons, including the Mount Caplice stock in the Coast Mountains southwest of Atlin Lake, that are older than 215 Ma. Their ages overlap those inferred for the closely associated volcanic rocks (Anderson, 1983a, 1984; and unpublished UBC data). This granitic plutonic event is uniquely restricted to part of Stikinia, and likely occurred before Stikinia came to rest against other terranes. The orientation and location of this volcanic arc(?) in Late Triassic time is largely unresolved, but there is a west-to-east trend toward more alkaline and less sodic magmas. The Triassic volcanic rocks of southern Stikinia are either covered or unstudied.

Cache Creek rocks that are caught between Stikine and Quesnel terranes were part of an active accretion wedge during the Late Triassic (Monger, 1984). Blueschist blocks along the Pinchi Fault give Late Triassic dates (Patterson and Harakal, 1974), and Upper Triassic rocks of both oceanic and arc affinity are tectonically mixed into the Cache Creek (Travers, 1978; Orchard, 1981).

***From 214 to 192 Ma** (End of Triassic and Early Jurassic).* Figure 10 shows that magmatism during this time was still restricted to the terranes outboard of North America. Wrangellia was separated from Intermontane terranes by a Bridge River—Hozameen ocean. On Vancouver Island, Bonanza volcanism (Sinemurian and Pliensbachian in age, according to Tipper, 1984b; calc-alkaline, intermediate to silicic in character, according to Muller, 1977) began at this time but this is not demonstrated by isotopic data. Zircon Pb/Pb dates around 200 Ma for diorite to tonalite plutons of the Westcoast Complex of Vancouver Island confirm an Early Jurassic beginning of arc magmatism (Isachsen and others, 1985).

Southern Quesnellia stands out as the site where a latest Triassic–very early Jurassic magmatic episode is clearly demonstrable. In that terrane are numerous, mostly singular plutons with isotopic ages of 204 to 212 Ma. These include the porphyry-copper–related Guichon, Iron Mask, Copper Mountain, and Granite Mountain bodies (Christopher and Carter, 1976). The volcanic equivalent is late-Nicola and Rossland groups, which extend to Sinemurian and Toarcian time (Tipper, 1984a), respectively. The granitic bodies intrude the Cache Creek accretion wedge and thereby link it to Quesnellia by the end of Triassic time (Monger, 1984).

Central and northern Quesnellia and Stikinia are similar in granitic pluton geochronometry in that Early Jurassic magmatism cannot be distinguished from late Early to Middle Jurassic magmatism. On Stikinia a hiatus generally separates Upper Triassic (Takla Group, mafic) from Early to Middle Jurassic (Toodoggone and Hazelton groups, calc-alkaline, intermediate to silicic) volcanic rocks (Tipper and Richards, 1976; Monger and Church, 1977). On northern Quesnellia, Takla augite porphyry volcanism continues from Late Triassic to Sinemurian time (Tipper, 1984b). On Figure 10 a few dates from these igneous rocks are shown but most appear on Figure 11. Stikine and Cache Creek terranes were probably linked by this time (Coney and others, 1980; Monger and others, 1982; Monger, 1984; Tipper, 1984b).

In the Laberge Group of the Stikine Terrane in the Yukon, volcanic rocks such as the Nordenskold Dacite belong to this episode. The early Klotassin Batholith is included as well. Its age is indicated by stratigraphic relationships and one zircon U-Pb date (Templeman-Kluit and Wanless, 1980).

Metamorphic rocks at several localities in the Yukon Crystalline Terrane give K-Ar dates as old as, or nearly as old as, this

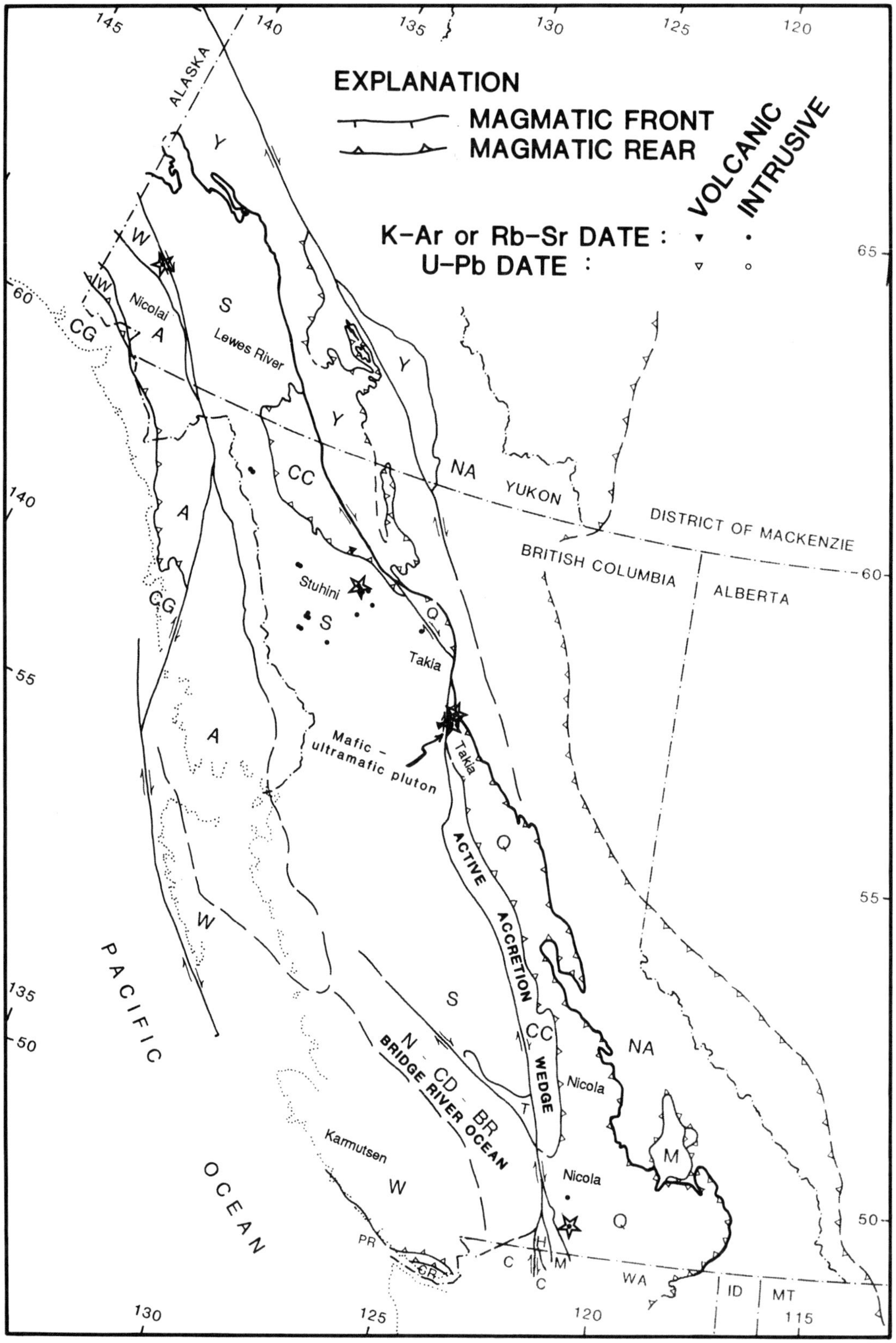

Figure 9. Map of the Canadian Cordillera showing distribution of dated magmatic rocks and outlines of magmatic belts for the time interval 235 to 214 Ma (Middle to Late Triassic) showing the Late Triassic magmatic culmination; granitic plutons of this age are restricted to Stikinia in northern British Columbia.

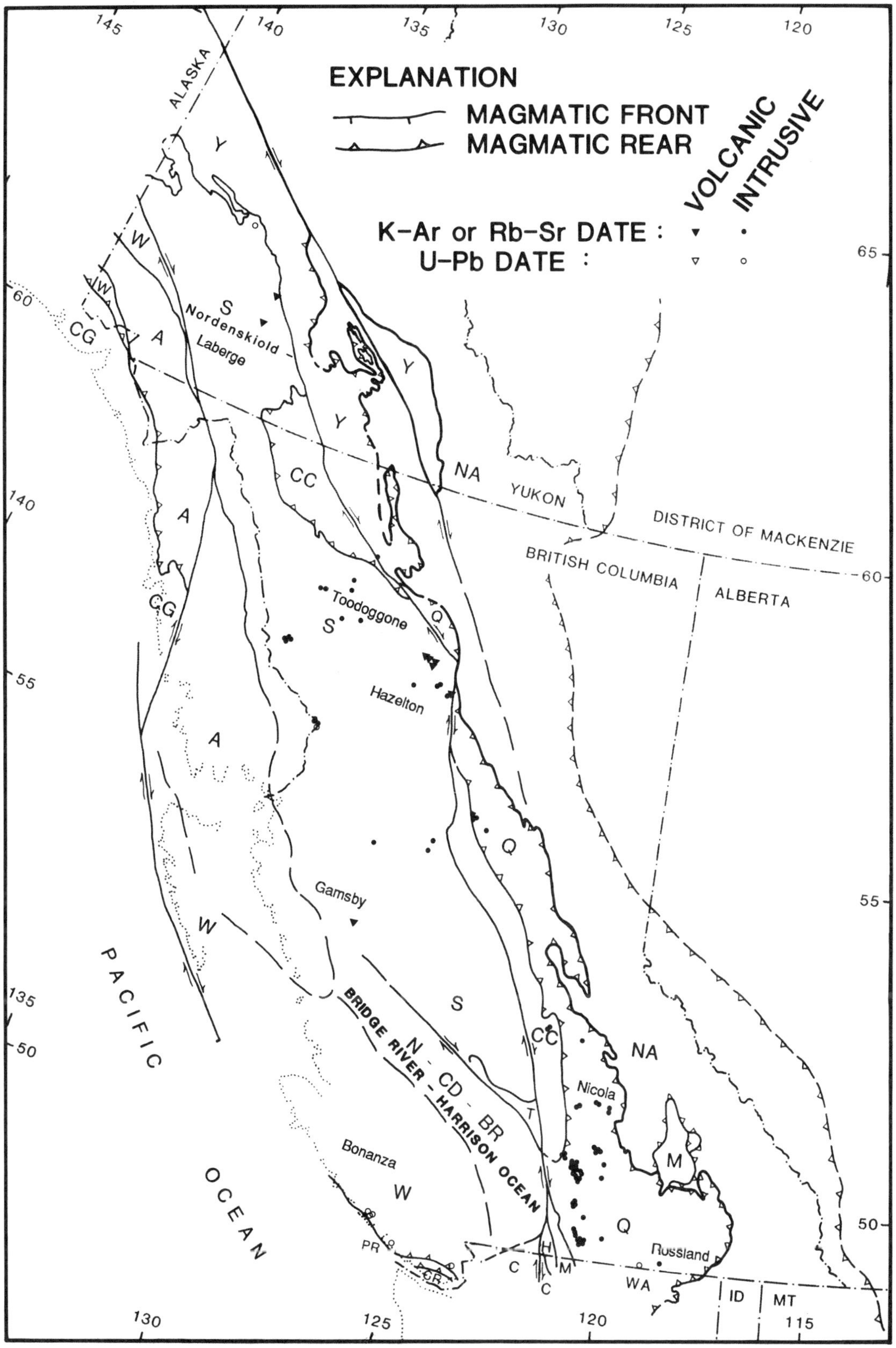

Figure 10. Map of the Canadian Cordillera showing distribution of dated magmatic rocks and outlines of magmatic belts for the time interval 214 to 192 Ma (end of Triassic and Early Jurassic) showing the Early Jurassic magmatic culmination, a feature distinctive of southern Quesnellia.

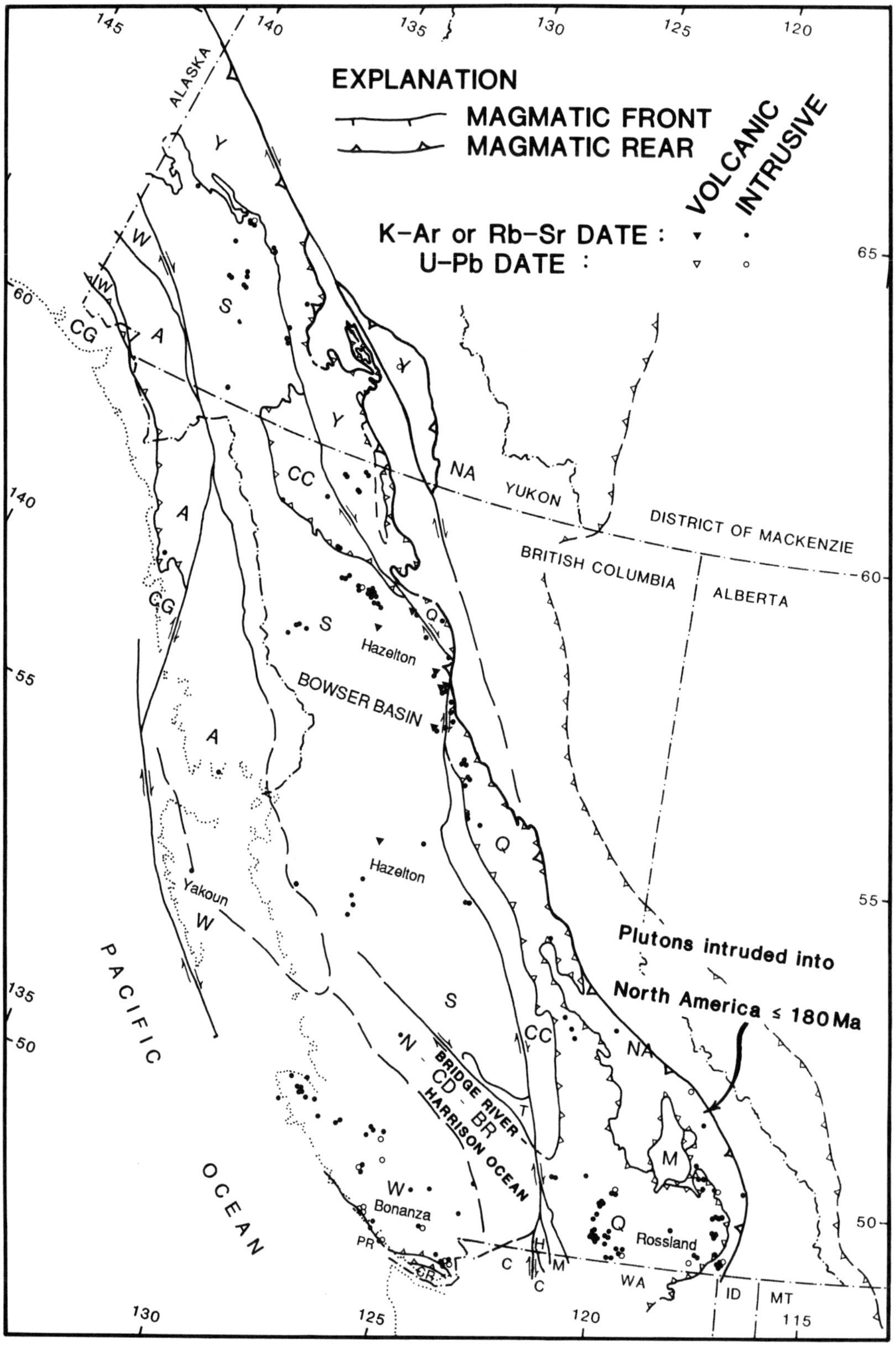

Figure 11. Map of the Canadian Cordillera showing distribution of dated magmatic rocks and outlines of magmatic belts for the time interval 192 to 155 Ma (late Early to middle Late Jurassic) showing extensive spread of Jurassic magmatism across all terranes. Data for Alaska and names of magmatic belts recognized in the Alaskan panhandle on this and subsequent maps are from Brew and Morrell (1983), Hudson (1983), Wilson and others (1979), and Dadisman (1980).

magmatic episode, and a Late Triassic–Early Jurassic regional metamorphism is recognized (Templeman-Kluit, 1976). In a few Yukon Terrane localities shown in Figure 16, metamorphic rock dates are even older, as old as 284 Ma (e.g., Htoon, 1977). These are not yet fully explained; most are probably relicts of Paleozoic metamorphic or igneous events. Only the Vedder Complex of the North Cascades in southernmost British Columbia (Armstrong and others, 1983) give similarly old dates (285 to 219 Ma).

***From 192 to 155 Ma** (Late Early to early Late Jurassic).* In Figure 11, magmatism can be seen to have spread over a larger region by this time. Most dates for Bonanza volcanic rocks and Island intrusions of southern Wrangellia appear on this map. Rb-Sr and U-Pb dates indicate synchronous volcanism and plutonism 185 to 190 Ma ago (Isachsen and others, 1985) and subsequent quiescence on Vancouver Island. The Yakoun Volcanics of the Queen Charlotte Islands are younger in stratigraphic age (Middle to early Late Jurassic) (Sutherland-Brown, 1968) than the chemically and petrographically similar Bonanza Volcanics of Vancouver Island (Early to early Middle Jurassic). Consistent with this, no Early or Middle Jurassic dates have been reported for plutonic rocks of the Queen Charlotte Islands; only a cluster of Late Jurassic hornblende K-Ar dates are available, and these appear mostly in Figure 12. A number of plutons of this time interval occur in the Alexander Terrane; its relationship to Wrangellia before Late Jurassic time is unclear. By Late Jurassic time, Wrangellia and the Alexander Terrane were joined (Coney and others, 1980), and there is increasing doubt that they were ever widely separated.

The youngest preserved strata of the Bridge River–Hozameen ocean between Wrangellia and the Intermontane terranes are Middle Jurassic (Monger and Price, 1979; Haugerud, 1984). The youngest oceanic strata in the Harrison Lake area are early Late Jurassic (Brookfield, 1973). Tipper (1984b) has suggested that Wrangell-Intermontane terrane amalgamation may have been as early as the end of Middle Jurassic time, but Cole and Tennyson (1977) and Monger and others (1982) prefer a mid-Cretaceous juncture. This issue is discussed below.

In southern Quesnellia, late Early Jurassic magmatism had shifted eastward, partially out of the "Nicola Belt" (Preto and others, 1979); in early Middle Jurassic time, magmatism encroached onto North America for the first time. Suturing of Quesnellia and the easterly adjacent Slide Mountain terrane to North America was apparently completed at about the time of demarcation between Early and Middle Jurassic magmatic episodes in southern British Columbia. The youngest displaced rocks (Rossland) are Toarcian (about 190 Ma). The oldest plutons cutting miogeosynclinal rocks of North America after suturing (e.g., the Kuskanax Batholith dated by Parrish and Wheeler, 1983) are 173 to 182 m.y. old. A somewhat earlier closure, or closure in progress, of the Anvil-Sylvester ocean (a northern Slide Mountain equivalent) is hinted at by clasts of metamorphic rocks in Upper Triassic clastic rocks deposited on North America in the Yukon (Templeman-Kluit, 1979a,b).

The Early to Mid-Jurassic plutons of southern Quesnellia may resolve into several groups, although the data could be argued to show a continuum of magmatic activity. The earliest group, 190 to 173 Ma, includes parts of the composite and protracted Mount Lytton Batholith, the Hedley Complex, and, on the east, the Kuskanax Batholith. The second group clusters around 167 to 162 Ma in age and includes the Pennask and Nelson batholiths. The final group is the muscovite-garnet–bearing Oliver and Galena Bay plutons of 157 to 153 Ma age. These plutons bracket phases of deformation in the Shuswap Metamorphic Complex and Kootenay Arc of the southern Omineca Belt. Plutons in the first group are deformed where inside the metamorphic complex; those in the second are post-tectonic relative to ductile structures in some areas, but deformed in others; and plutons in the third group postdate all structures in their wall rocks except late brittle faults (Armstrong, 1982; Okulitch, 1984).

In Stikinia and the Yukon Terrane, magmatism of the same age as the first subgroup of southern Quesnellia (190–173 Ma) was extensive and is not clearly resolved in time from that which occurred during the previous map time interval (214–192 Ma). Here, there is no overlap into parts of North America, only very scattered representation of a second Jurassic-age group (172–165 Ma), and no known example of a third (<165 Ma). Examples of the main group include the Three Sisters Pluton of the Hotailuh Batholith (184–170 Ma) (Anderson, 1983a) and the syenite phases of the Hogem Batholith (180–175 Ma). The second group is represented by Pink Quartz Monzonite (170–165 Ma) (Tempelman-Kluit and Wanless, 1975) and the late phase of the Hickman Batholith (172 Ma), which is also a distinctive pinkish quartz monzonite.

From Early Jurassic time onward, the central part of the area known as the Bowser Basin appears to have been largely devoid of magmatism (Tipper, 1984b). In Early Jurassic time, the basin was a site of shale deposition. In later Jurassic time it was a slowly subsiding basin receiving clastic debris from Stikine and Skeena arches and the Omineca Belt (Eisbacher, 1974). The subsiding and nonmagmatic character earn it the designation "cold spot." It stands empty on all the post-Triassic magmatic distribution maps.

***From 155 to 125 Ma** (Latest Jurassic and first half of Early Cretaceous).* This was a time of little magmatism over the entire Cordillera. Western Canada was no exception, as can be seen on the histograms of Figure 3. Likewise, the map (Fig. 12) is relatively empty even though its time interval is of comparable length to that for the other maps. Many of the dates shown could be reasonably considered to be partially reset from the previous episode. This is certainly true for all that appear in southern Wrangellia, Quesnellia, and most of Stikinia. The lull in dates for either plutonic or volcanic rocks is coincident with regional gaps in the stratigraphic record (Eisbacher, 1977) indicating a time of emergence, erosion, and external drainage.

The important exceptions to the lull following 155 Ma are the latest Jurassic–earliest Cretaceous Francois Lake Intrusions and Late Jurassic magmatism of the Wrangell-Alexander Ter-

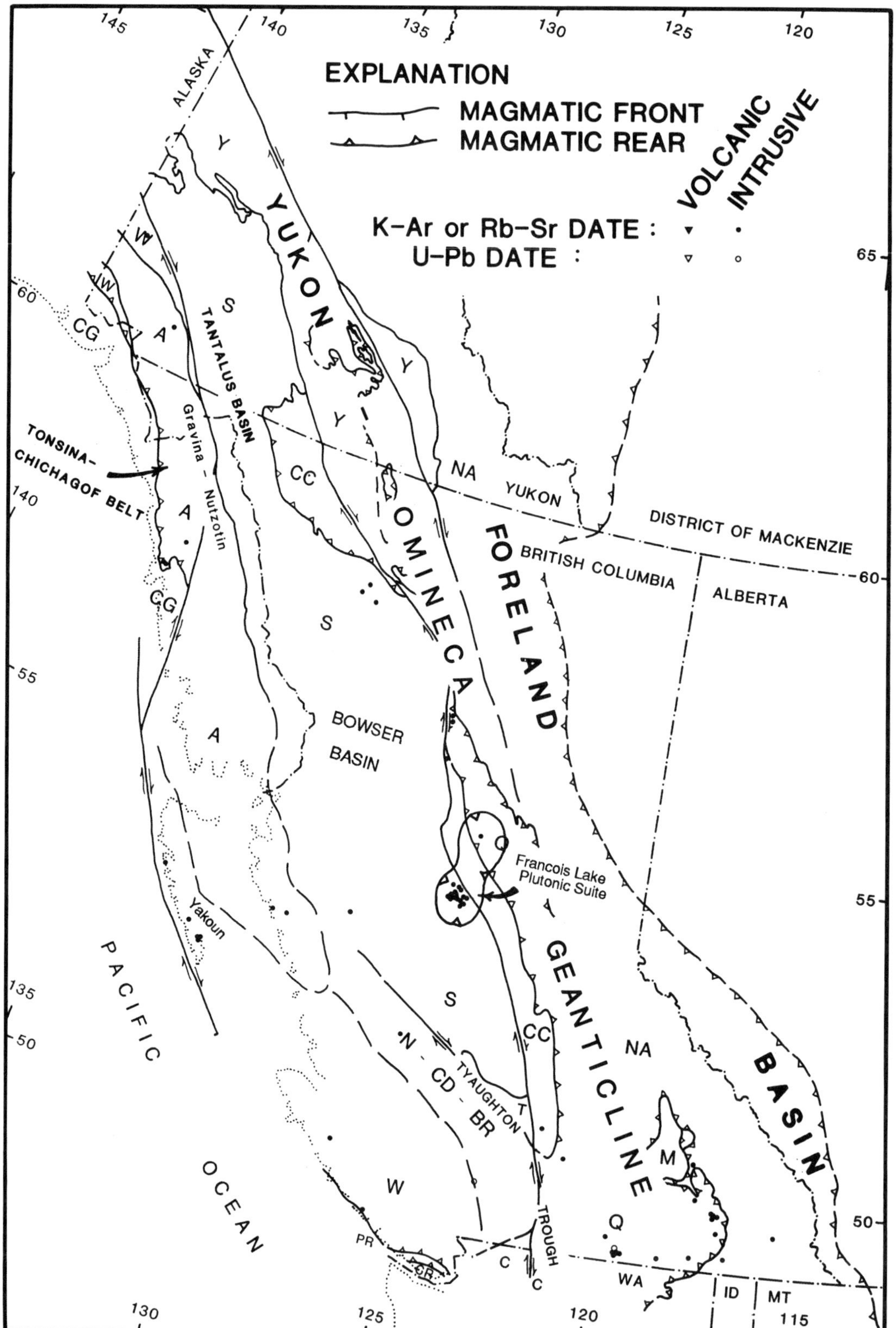

Figure 12. Map of the Canadian Cordillera showing distribution of dated magmatic rocks and outlines of magmatic belts for time interval 155 to 130 Ma (latest Jurassic to mid-Neocomian) showing the restricted extent of latest Jurassic magmatism. The beginning of the Early Cretaceous is a magmatic lull throughout the Cordillera.

rane. Nevertheless, between 135 and 125 Ma (centering on the time separating Figs. 12 and 13), absolute quiescence appears to have reigned. Dating of the Francois Lake Intrusions is of good quality. Numerous K-Ar dates (White and others, 1970) agree perfectly among themselves and with Rb-Sr whole-rock isochrons for several of the same bodies (J. McDonald and R. L. Armstrong, unpublished data). This suite seems to be quite unique, and singularly responsible for one histogram peak. A few similar dates are shown in the Wrangell-Alexander terrane, and many more in the 140- to 160-Ma range have been obtained for the Saint Elias project of the Geological Survey of Canada (C. Dodds, written communication, 1984), but these are still unpublished and thus do not appear on the figures in this paper. In Alaskan terms these plutons belong to the 164 to 136-Ma Tonsina-Chichagof Belt (Hudson, 1983). They may be related to contemporaneous volcanism of the Gravina-Nutzotin Belt (Berg and others, 1972), which links the Alexander and Wrangell terranes in Alaska in Late Jurassic time.

The linkage of the two major composite terranes (Insular and Intermontane) was complete before the beginning of the next magmatic episode. Granitic clasts derived from the Coast Plutonic Complex were being shed into the Tyaughton Trough, which lay just west of Quesnellia, by Late Jurassic time, according to Anderson (1976) and Tipper (1984b) or by Albian time, according to Tennyson and Cole (1978) and Monger and others (1982).

From 130 to 84 Ma (Middle Cretaceous). Figure 13 shows a largely assembled Cordilleran margin of North America on which an Andean magmatic arc has been established. The analogy of western North America and the Andes is an old one advocated by previous workers such as Hamilton (1969) and Dickinson (1981), and supported by recent studies of the Andes themselves (Thorpe and others, 1982; Jordan and others, 1983). Some later strike-slip disruptions of the mid-Cretaceous magmatic belt are suggested in the figure but their exact magnitudes are undefined because of the low angles of intersection between faults and magmatic belts. No major sutures are permissible east of the magmatic front of the Coast Plutonic Complex after middle Early Cretaceous time, and terrane distinctions are from this point onward relatively unimportant. It is at this time that the five *tectonic* belts become the established and controlling features of the Cordilleran landscape.

Mid-Cretaceous magmatism was concentrated in two geographic belts: the western, including the Coast Plutonic Belt (Alaska panhandle–southwestern Yukon–western British Columbia, and the eastern, including the Omineca Belt and western Rocky-Mackenzie Mountains. A few plutons and scattered accumulations of volcanic rocks of this age occur in the Intermontane belt (Kasalka volcanics and parts of the Skeena Group), but compared to previous times or to coeval flanking belts the activity there was subdued.

Plutons of the mid-Cretaceous episode make up the bulk of the southern Coast Plutonic Complex. The magmatic front is sharply defined and appeared suddenly after a long hiatus. Going northward, the overprint of younger Coast Plutonic episodes becomes stronger so that only a few plutons along the western margin of the Complex, such as the Ecstall (Woodsworth and others, 1983), can be surely included in the mid-Cretaceous. In Alaska the equivalent rocks are plutons of the Muir-Chichagoff and Admiralty-Revillagigedo belts and the belt of Alaskan-type ultramafic complexes (Brew and Morrell, 1983; Hudson, 1983; Gehrels and Berg, 1984; Brew and others, 1984). The Lower Cretaceous Gambier Group of the southern to central Coast Mountains is intruded by the mid-Cretaceous plutons, but this is probably a case of plutons invading their comagmatic volcanic cover. In the southwestern corner of the Intermontane Belt, both volcanic (Spences Bridge and Kingsvale Groups) and plutonic (e.g., Eagle and Verde Creek) rocks are representative of the mid-Cretaceous episode.

Mid-Cretaceous plutons of the eastern belt are distinctive in chemistry; many show some S-type characteristics and enrichment in Al and K. They also have unusual isotopic compositions, enriched in ^{87}Sr and ^{18}O (Dagenais, 1984). In this episode, magmatism extended the farthest into North America that it reached in Mesozoic time. During the entire Phanerozoic, this inward reach of magmatism was exceeded only by a few Eocene outliers. The magmatic *rear* parallels the Rocky-Mackenzie Mountain structural front (Fig. 13). This is significant in demonstrating that the major salients in the front are not just thin-skinned extensions of displaced cover rocks but involve deep crust, and presumably upper mantle, as well.

Mid-Cretaceous volcanic rocks occur in southwestern and eastern Yukon (Mount Nansen and South Fork volcanics, respectively), are missing from the Omineca Belt in British Columbia, and are represented only by the strongly alkaline Crowsnest Volcanics in the Rocky Mountains of southwestern Alberta (Folinsbee and others, 1961).

From 85 to 64 Ma (Late Cretaceous). In previous syntheses of Cordilleran magmatism, this episode was not clearly resolved from the Early Cenozoic episode (Gabrielse and Reesor, 1974), but the evidence for their separation is clear (Fig. 3), as has been recognized by Carter (1982) and by Woodsworth (1979). Compared to episodes before and afterward, the Late Cretaceous magmatic belt was narrow (Fig. 14). It bulged cratonward at the Skeena and Stikine arches and was very narrow at the latitude of the Bowser Basin "cold spot." In the Alaskan panhandle, coeval plutons are the foliated granitic sills of the Coast Plutonic Complex (Gehrels and others, 1984). The trace of the magmatic front meanders from the southeast corner of the Coast Plutonic Belt to the eastern edge of the Insular Belt in southeasternmost Alaska and back into the Intermontane Belt in the Yukon. Volcanic equivalents occur in scattered small patches in the Intermontane Belt. In the Yukon these volcanic rocks have been shown to become more potassic and alkaline toward the northeast (Grond and others, 1984). Most of the named plutons of this episode are small; several are mineralized; in central British Columbia they are referred to as the Bulkley Suite (Carter, 1982).

From 64 to 40 Ma (Paleocene to Middle Eocene). The final

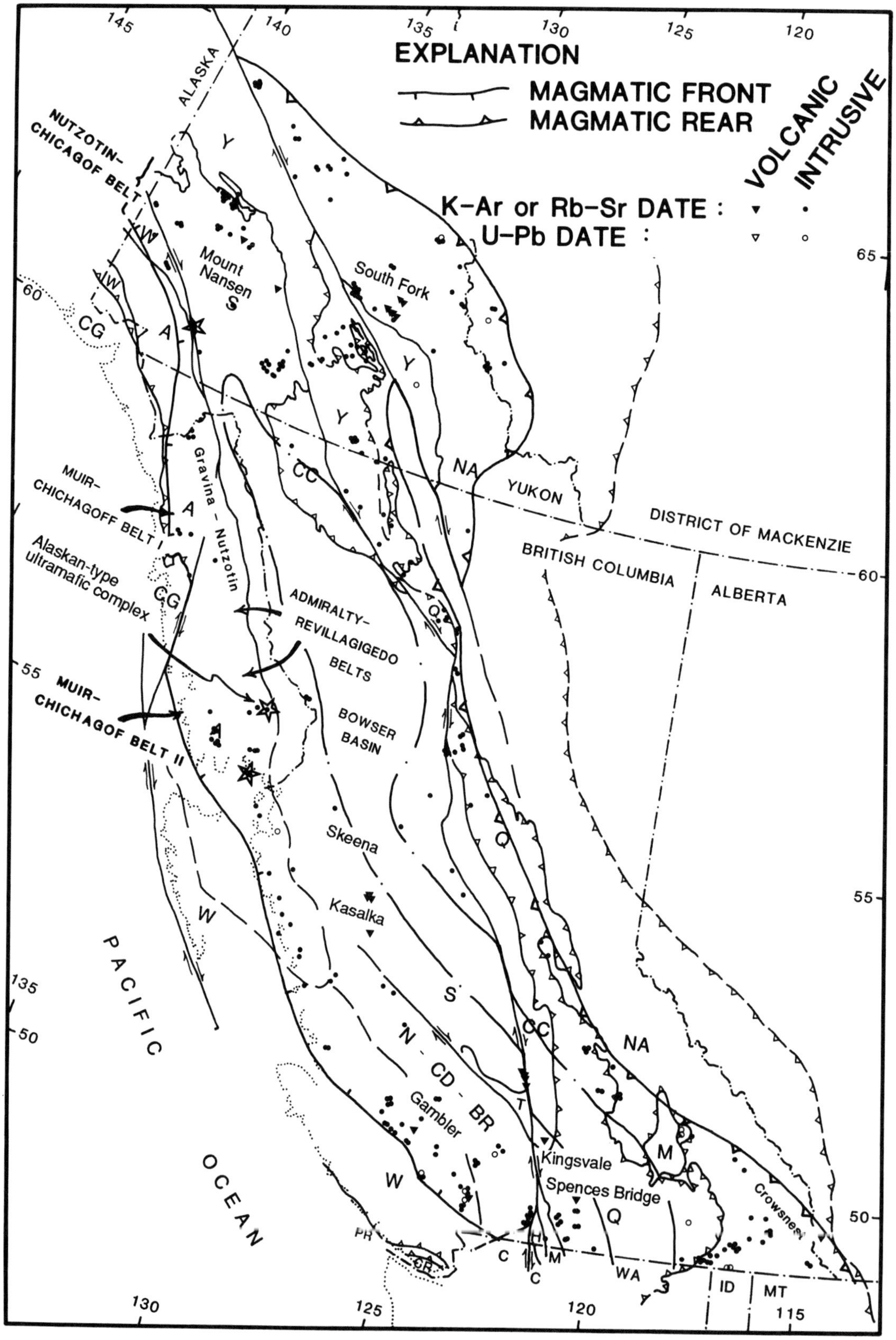

Figure 13. Map of the Canadian Cordillera showing distribution of dated magmatic rocks and outlines of magmatic belts for the time interval 130 to 84 Ma (middle Cretaceous = mid-Neocomian to mid-Late Cretaceous) showing the dual culmination Andean magmatic belt of middle Cretaceous time.

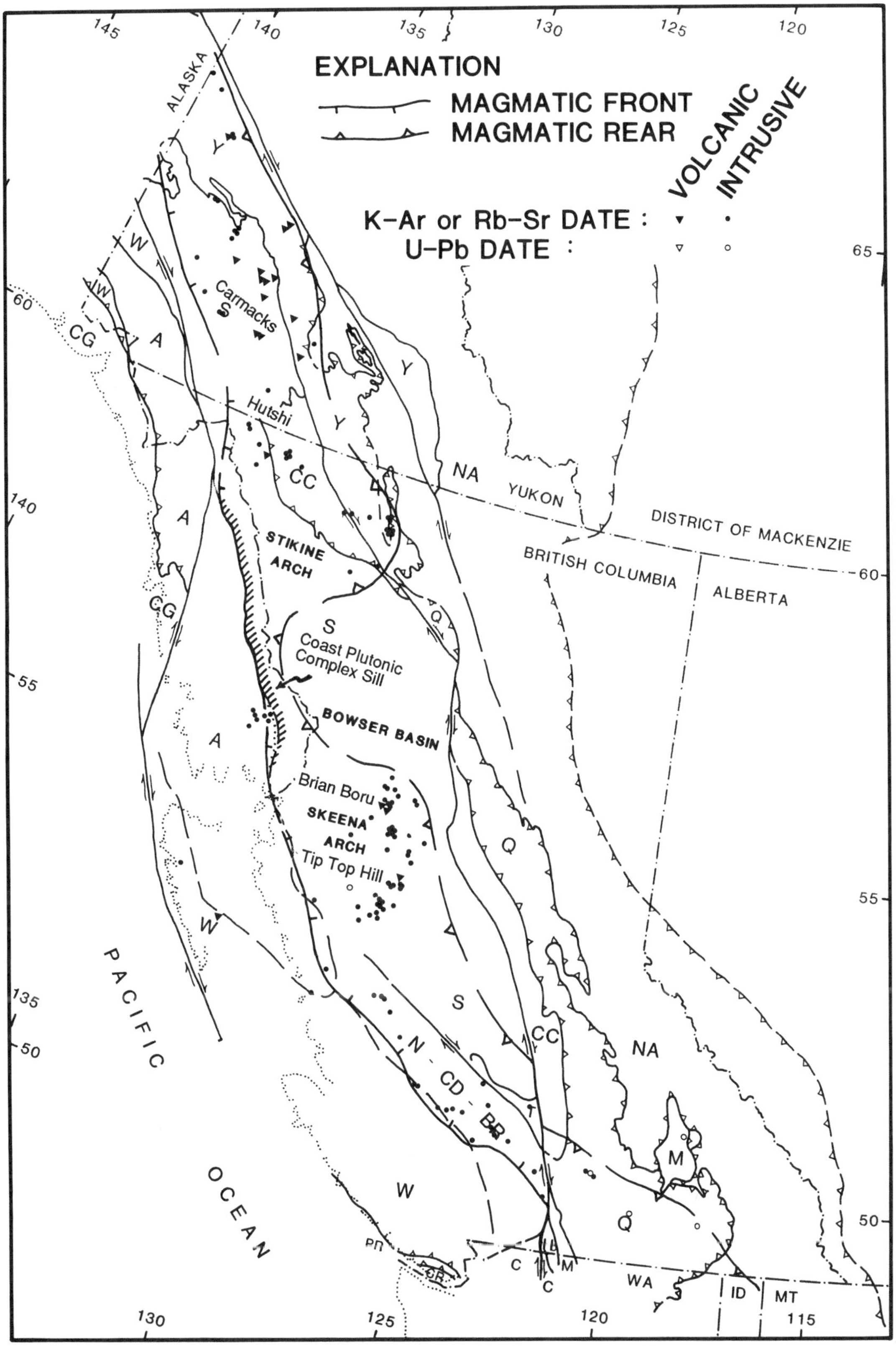

Figure 14. Map of the Canadian Cordillera showing distribution of dated magmatic rocks and outlines of magmatic belts for the time interval 84 to 64 Ma (middle Late Cretaceous to early Paleocene) showing the relatively narrow Late Cretaceous magmatic belt.

magmatic episode is a spectacular short-lived event with rapid onset and equally rapid decline. Most of the large body of relevant geochronometry lies within two analytical standard deviations of 50 Ma, so that the actual duration could not have been much more than 5 m.y. and was certainly less than 10 m.y. (predominantly 55 to 45 Ma). Any north to south transgression of the magmatism was small, the histogram peak being at about 52 Ma in the Yukon Territory and about 49 Ma at the Washington-Idaho border, a difference that is hardly resolvable and could be merely the result of greater depth of erosion in the south.

The lull that preceded this episode occurred from about 70 to 60 Ma. Very few igneous rocks of this age range have been dated. A few plutons exist in the Coast Plutonic Complex (Gehrels and others, 1984), on the Skeena Arch (Carter, 1982), and at the southern end of the Intermontane Belt (Preto and others, 1979). Cordilleran volcanism of latest Cretaceous–earliest Cenozoic age is confirmed by the bentonites used to date the Cretaceous-Tertiary boundary in Alberta and Montana (Baadsgaard and Lerbekmo, 1982).

Many names have been applied to the igneous rocks of the Early to Middle Eocene magmatic episode (Figs. 7, 8), partly because a paleontologic basis for precise correlation between separate areas has been lacking. Church (1970; and numerous later reports of the British Columbia Ministry of Energy Mines and Petroleum Resources) and Ewing (1980, 1981a,b) have discussed volcanic petrology, stratigraphy, and transtensional tectonics of southern British Columbia. No current synthesis exists for these rocks from central British Columbia to the Yukon—only studies of individual complexes (e.g., Lambert, 1974) or scattered mineral deposits (Carter, 1982; and numerous others). Plutonic rocks of this age are extensive in the Yukon Territory (Tempelman-Kluit and Wanless, 1975) and northern Coast Plutonic Complex (Hutchison, 1982; Brew and Morrell, 1983; Hudson, 1983; Gehrels and others, 1984).

The Eocene magmatic front (Fig. 15) is well defined and extends from the east edge of the southern Coast Plutonic Complex northward gradually across the Complex to near its western edge in Alaska. Granitic plutons occur in the area immediately east of the front. Volcanic rocks are found in the Intermontane Belt, together with numerous small intrusive bodies, but only the intrusive bodies (various siliceous stocks and lamprophyre dikes) occur in the Omineca Belt. This suggests a broad synformal warping of the Eocene volcanic cover, with consequent erosion from the uplifted Coast and Omineca belts during later Cenozoic time. Today's topography reflects this deformation pattern.

The chemical polarity of the Eocene volcanic arc is well known (Ewing, 1981a). Subalkaline quartz diorite bodies occur in the Coast Plutonic Belt, Intermontane rocks grade eastward from calc-alkaline to alkaline, and easternmost occurrences are invariably alkaline, which is the normal observation in wide subduction-related magmatic arcs (Gill, 1981).

Figure 15 also shows magmatic activity in four areas along the continental margin, well outboard of the main Eocene magmatic belt. At the southern end of Vancouver Island are the Metchosin Volcanics of Early to Middle Eocene age. These are interpreted to be a seamount that was rafted into place in approximately Late Eocene time (Muller, 1977, 1980). Also in southern Wrangellia the Catface magmatic episode of Vancouver Island is indicated by several dated rocks, but it continues to younger times. It may begin as long ago as 50 Ma, but most dates for Catface rocks lie in the range of 45 to 36 Ma, which makes this subsequent to the mainland Eocene event and not correlative. Available data do not rule out the possibility of two Eocene events here that bracket the inland volcanic episode. Metamorphic dates for the Leech River Schist of southern Vancouver Island, which lies between the Metchosin Volcanics and Jurassic and older basement, cluster tightly about 42 Ma. The cause for the shift from inland major arc to continent-margin igneous and metamorphic activity must lie with a dramatic rearrangement of plate boundaries at the end of Middle Eocene time. At this time, or soon thereafter, the modern plate pattern—with a triple junction in the vicinity of the north end of Vancouver Island—became established. This triple junction fixed the north end of the Cascade-Pemberton-Garibaldi arc in southern British Columbia for the remainder of the Cenozoic. A downgoing plate edge (Armstrong and others, 1985), passing triple junction (Marshak and Karig, 1977; Johnson and O'Neil, 1984), ridge subduction (Hill and others, 1981; Moore and others, 1983), or peculiarity of the end of subduction (Hudson and others, 1979) may explain the Catface episode. Cowan (1982) has argued that the Sitka Greywacke of Baranof Island is a displaced equivalent of the Leech River Complex, so it may be subject to the same interpretations.

Massett magmatism of the Queen Charlotte Islands is largely Oligocene and Miocene in age (Young, 1981). The geochronometry there is of poor quality and may include multiple rock groups under one heading, so a few 40- to 64-Ma dates for Massett rocks appear on Figure 15. These should be treated skeptically, but can only be rejected if more work is done in the area.

Early to Middle Eocene plutons, which occur in the Saint Elias–Gulf of Alaska region (Hudson, 1983), are also shown on Figure 15.

METAMORPHIC DATES

The CCGF dates excluded from Figures 9 through 15 were used to construct the mica chrontour map of Figure 16. In this way, the geographic position of all metamorphic and reset samples, and an indication of the dates obtained, can be displayed. The dates vary in a systematic manner and can be contoured to show the shape of the "metamorphic veil" (Armstrong, 1966). There is an inverse relationship between the veil and the pattern of igneous dates on maps showing magmatic episodes. Where the veil is as young as or younger than the magmatic event being illustrated, only those dates for that event that are very resistant to resetting will appear. On the metamorphic date map, the reset areas of the Canadian Cordillera are delineated. These corre-

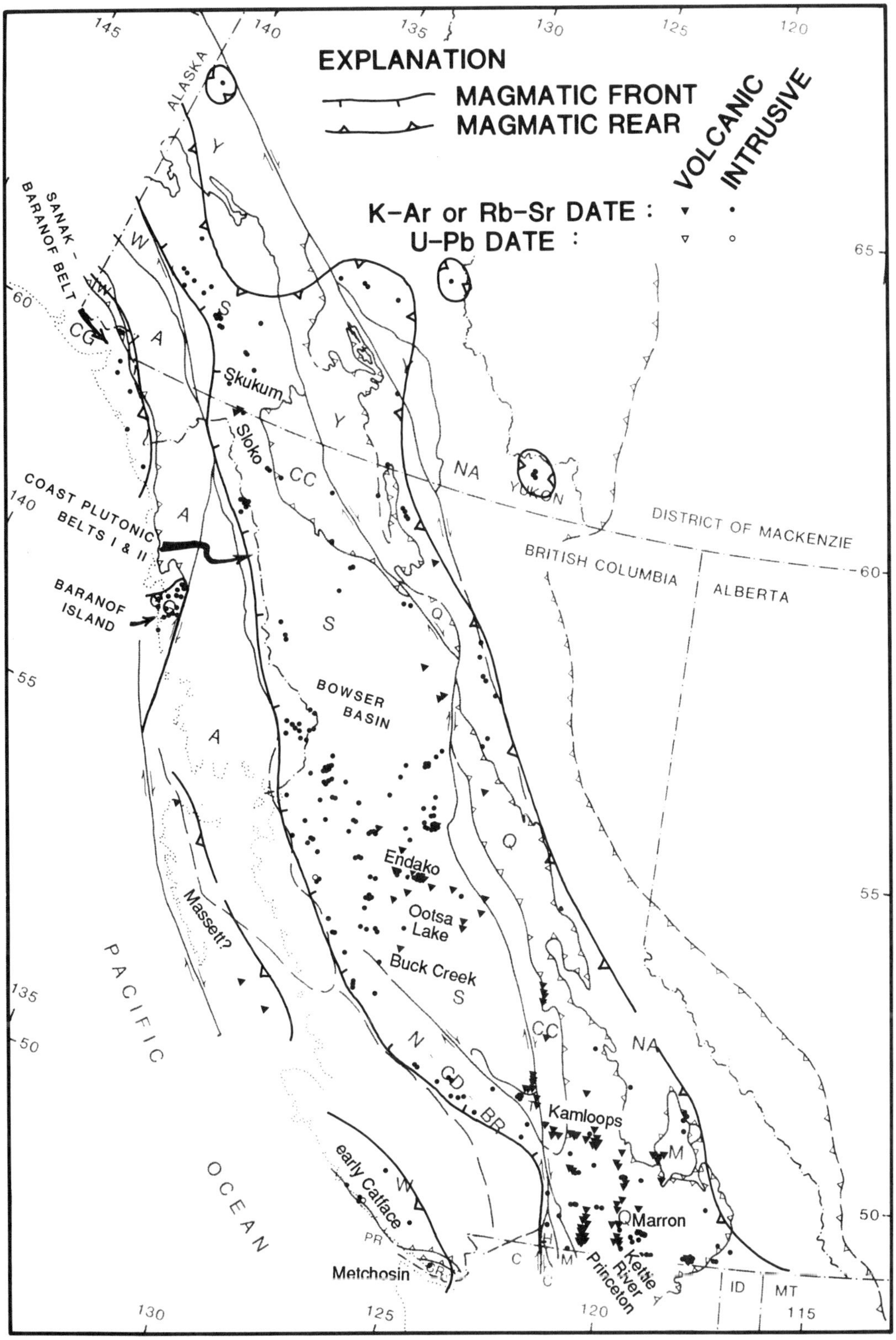

Figure 15. Map of the Canadian Cordillera showing distribution of dated magmatic rocks and outlines of magmatic belts for the time interval 64 to 40 Ma (Paleocene to end of Middle Eocene) showing the Early and Middle Eocene magmatic belt of the mainland, part of the Middle to Late Eocene magmatic belt of Vancouver and Baranof islands, and a hint of the mostly Oligocene and Miocene magmatism of the Queen Charlotte Islands.

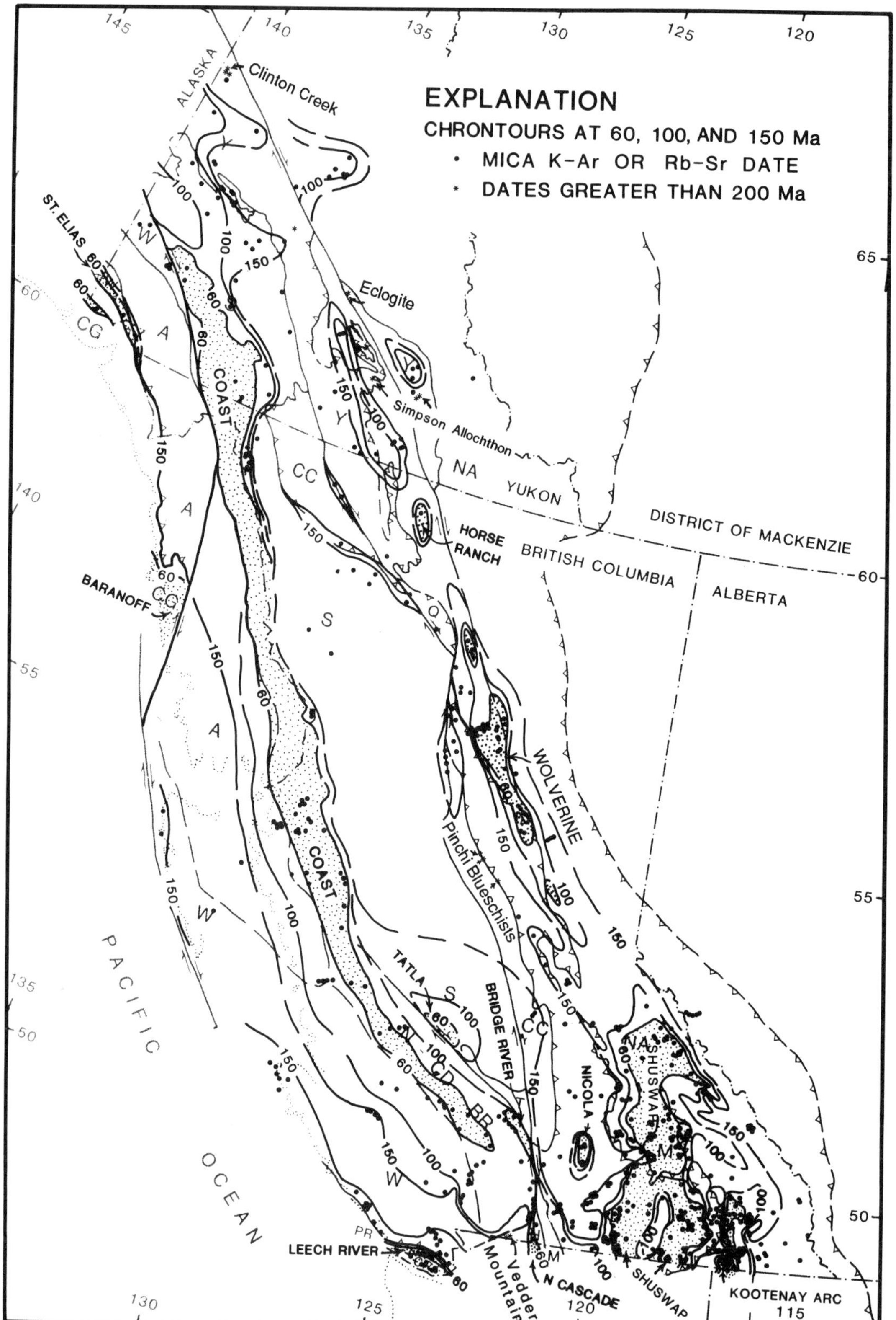

Figure 16. Metamorphic Veil of the Canadian Cordillera. Mica chontours represent a summary of metamorphic rock and reset magmatic rock dates that were excluded from Figures 9 through 15. To a first approximation this is also a map showing the two major metamorphic culminations—Coast and Omineca belts—of the Canadian Cordillera. Early Cenozoic reset areas are shaded.

spond in a general way to a map of the metamorphic belts of the Cordillera (Monger and Hutchison, 1971). In the Insular Belt, early Cenozoic resetting is observed in the Metchosin–Leech River area of southern Vancouver Island, on Baranoff Island, and in the Saint Elias region. Mesozoic resetting affects Pacific coastal areas of Vancouver Island and the Queen Charlotte Islands and Insular Belt Triassic and Jurassic magmatic rocks as the Coast Plutonic Complex is approached.

In the Coast Plutonic Complex there is progressive younging from a Cretaceous reset area in the west to an early Cenozoic reset area on the east. The belt of Cenozoic resetting and metamorphism extends into the southwestern Yukon Territory. The western limit of Cenozoic resetting is coincident with the Eocene magmatic front. Metamorphic resetting accompanied eastward migration of the magmatic front across the Coast Plutonic Complex. At the south end of the Coast Plutonic Belt, and offset en echelon from the main early Cenozoic reset core, is another early Cenozoic reset area (North Cascade–Bridge River) that has been offset by the Fraser strike-slip fault system in Middle to Late Eocene time.

In the Intermontane Belt, Mesozoic resetting is evident in rocks adjacent to the King Salmon, Kutcho, and Vital faults in northern British Columbia. Early Cenozoic resetting affects an area of poorly known gneisses that lie east of the Yalakom Fault in southern Stikinia and a small area containing the Nicola Batholith in southern Quesnellia.

The Omineca Belt metamorphic culmination shows a complex resetting pattern as a result of multiple, partly superimposed episodes of regional metamorphism and thermal events. In the Yukon Terrane some Early Jurassic or older dates have been reported, but, as already mentioned, the older dates are incompletely explained. Metamorphic rock fragments do occur in sediments as old as Late Triassic (Tempelman-Kluit, 1979a), and one major metamorphic event occurred in Late Triassic to Early Jurassic time (Tempelman-Kluit, 1976; Cushing and others, 1984). Otherwise, the major metamorphic event in central-southern Yukon Territory appears to have been a Cretaceous thermal overprint at the time of the mid-Cretaceous magmatic culmination. In British Columbia there is abundant evidence for Middle Jurassic regional metamorphism both in the Shuswap-Kootenay Arc region (Armstrong, 1982; Okulitch, 1984) and in the Wolverine Complex (Parrish, 1979). Some of the dates enclosed by the 150-Ma chrontour are from cooling following that metamorphic episode. Embedded within the Jurassic metamorphic areas of British Columbia are younger reset areas, all with early Cenozoic cores. This resetting event has been discussed in several papers (e.g., Parrish, 1979, 1981, 1984; Mathews, 1981, 1983; Archibald and others, 1984; Armstrong, 1982). Some resetting in southeastern British Columbia may also be a result of heating associated with the mid-Cretaceous magmatic culmination, but its effects are not clearly resolved. The early Cenozoic resetting is coincident with the Early to Middle Eocene magmatic event (Fig. 3). It is not, however, simply a contact metamorphic effect. Crustal extension, high heat flow, rapid uplift, erosion, and tectonic unroofing of metamorphic rocks are invoked to explain the high temperatures responsible for daughter product loss—even Ar loss from amphiboles has occurred over large areas (Medford, 1975). Locally, a ductile metamorphic strain has been imposed on Cretaceous or early Cenozoic plutonic igneous rocks (Parrish, 1984; Parkinson, 1985).

STRONTIUM ISOTOPIC COMPOSITION OF IGNEOUS ROCKS

A byproduct of the dating of rocks by the Rb-Sr method, and a closely related research activity, is the determination of initial $^{87}Sr/^{86}Sr$ ratios for igneous rocks. These provide a clue to the site of magma genesis or to processes of contamination of magmas (Leeman, 1982). In the United States, strontium isotopes have been shown to discriminate areas underlain by Precambrian basement from those accreted during Phanerozoic time (Kistler and Peterman, 1978; Armstrong and others, 1977). A ratio greater than 0.706 is taken as indicative of assimilation of, or genesis from, older crust or its underlying lithospheric mantle. Ratios in the range 0.703 to 0.704 are characteristic of magmas erupted in volcanic arcs and seamounts, and ratios of 0.702 to 0.703 are observed for depleted ocean floor basalts and considered characteristic of depleted mantle (Basaltic Volcanism Study Group, 1981).

Much of the Sr isotopic data shown on Figures 17 through 19 is unpublished. Petö (1974) discussed the few Sr initial ratios available up to 1973. Work at UBC began in the early 1970s; published UBC studies include LeCouteur and Tempelman-Kluit (1976), Petö and Armstrong (1976), Morrison and others (1979), Preto and others (1979), Wood and Armstrong (1982), and Pigage and Anderson (1985). The data presented here show the Sr isotopic signature to be characteristic of both tectonostratigraphic terrane *and* of time.

On the isochron-type diagrams of Figure 17, the initial ratio data have been sorted on the basis of tectonic belts, geography, and age. Initial ratios in the Insular, Coast, and Intermontane belts are commonly below 0.704, whereas those in the Omineca and Rocky-Mackenzie Mountains belts are invariably above 0.704, thus showing a fundamental geochemical difference between the eastern and western Cordillera, or miogeosyncline and eugeosyncline, respectively. Vancouver Island ratios are virtually all at or below 0.704, regardless of age. This is consistent with an arc or ocean island setting and relatively young basement. The southern Coast Plutonic Complex is identical to Vancouver Island in its low ratios, and is distinct from the northern Coast Complex, which has initial ratios of 0.704 to 0.706. The contrast coincides with the change from Wrangellia to Alexander Terrane basement and agrees with the known presence of older rocks in the Alexander Terrane (Brew and Morrell, 1983) and late Precambrian basement of the western Stikine Terrane (L. Werner and R. L. Armstrong, unpublished data). There does not appear to be an age effect in the Coast Plutonic Complex Sr isotope data.

Mesozoic rocks of the Intermontane Belt have low initial

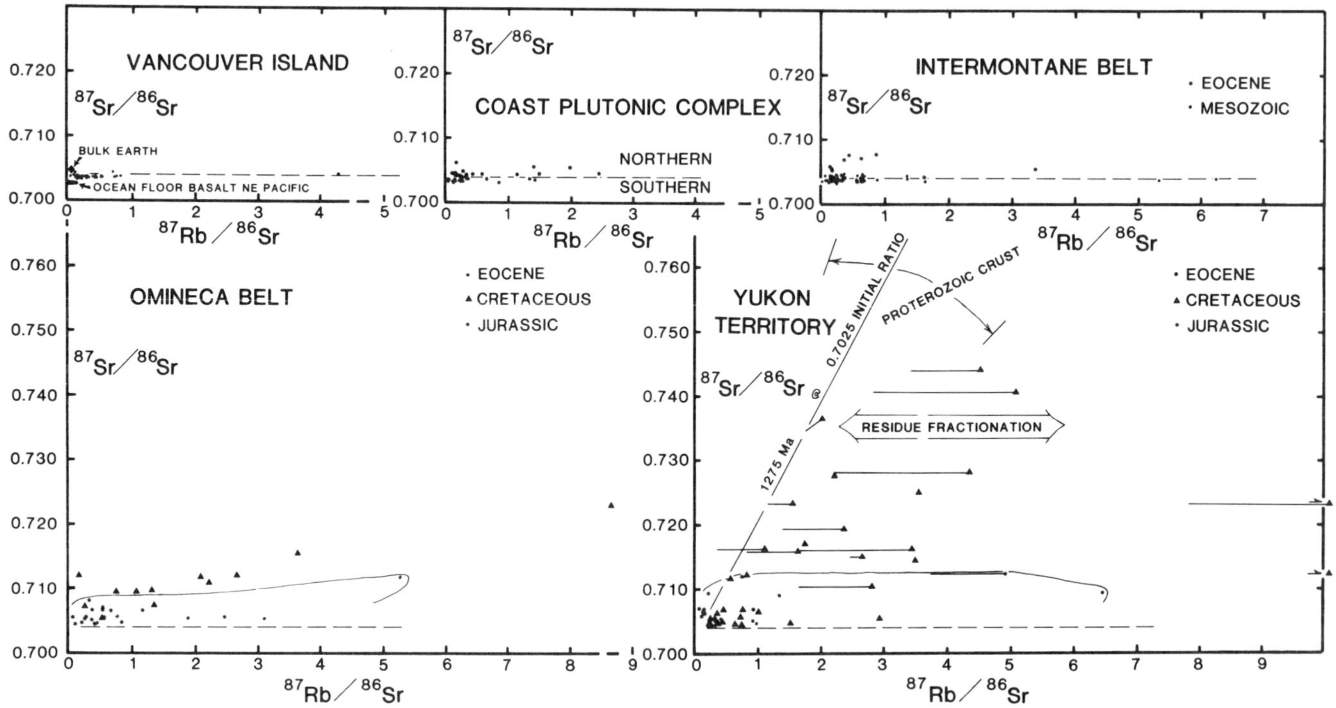

Figure 17. Compilation of $^{87}Sr/^{86}Sr$ *initial* ratios for different parts of the Canadian Cordillera on isochron-type diagrams. Symbols are mostly for average composition of multiple samples from single bodies, but in some cases they are for single samples. For Yukon Territory east of the Coast Plutonic Complex, the lines extend from the symbol for average values to the point of lowest value in the sample suite. The range in Rb/Sr ratio within a suite is created by igneous fractionation or mixing processes. In the case of mixing, pseudoisochron relationships are possible. To generate Yukon Territory Cretaceous magmas, a large component of old crust is required. This is true to a lesser degree for Omineca Belt Cretaceous magmas. Jurassic and Eocene magmas in both belts show a significant but small degree of crustal radiogenic Sr addition. Initial ratios are low, less than 0.704, in those areas underlain by Phanerozoic eugeosynclinal crust. This does not prove that there is no crustal component in the magmas, only that, if present, it if of minor importance and not sufficiently enriched in radiogenic Sr to have any measurable effect on the $^{87}Sr/^{86}Sr$ ratios of the mantle-derived magmas.

ratios like those of Vancouver Island, but the early Cenozoic rocks there are distinguished by more radiogenic Sr. The increase may be due to evolving crustal character—both thickness and age increasing through time. Only by early Cenozoic time was the lower crust radiogenic enough and sufficiently thick to be involved in magma evolution so that the effect on Sr isotopes became evident. Generally, high lower crustal temperatures associated with the Eocene magmatic event (recall the extensive early Cenozoic metamorphic reset areas) would encourage crustal incorporation into magmas and thus have been a third factor underlying the elevated ratios.

The Omineca Belt of British Columbia and of Yukon Territory east of the Coast Plutonic Complex show similar Sr isotopic character but have been separated to show the exaggerated pattern of crustal magma contributions in the Yukon (Fig. 17). In both areas all initial ratios lie above 0.704. Jurassic and early Cenozoic magmas show similar ranges of composition: ratios from 0.704 to 0.712, indicating a moderate and variable crustal contribution to the magmas. Only during the mid-Cretaceous magmatic episode do the effects of crustal involvement become spectacular. Cretaceous initial ratios are commonly above 0.710; some exceed 0.730. This may be the largest collection of exceptionally radiogenic magmas in the world. Most previous studies have reported the odd single example. Here the anomaly extends through entire magmatic suites over a strike distance of more than 2,000 km. Without going into detail here, it can only be said that these magmas involve large fractions of older crust. In some the amount may approximate 100 percent; a mantle component need not be involved. Most are probably mixtures of mantle-derived magma with crust that have been modified by fractionation. The belt of radiogenic Sr-enriched Cretaceous plutons can be followed south into Idaho (Armstrong and others, 1977), Nevada (Best and others, 1974; Kistler and others, 1981; Kistler, 1983; Lee and Christiansen, 1983), and California (Miller and others, 1982).

Anatexis, assimilation, or combined fractionation-assimilation processes have been invoked in petrologic explanations for high initial Sr ratios of magmas. The reason these processes are favored in Cretaceous time must lie in the thermal state of the crust. Otherwise it is much the same crust—before, during, and after the mid-Cretaceous event. A reasonable hypothesis is that lower crust temperatures were much higher during the mid-Cretaceous so that crustal contamination and anatexis were favored. A corollary of this would have been reduced lithosphere thickness and strength. A consequence would be localization of deformational crustal thickening in the area of radiogenic initial Sr magmas. This would all have been happening at the same time as much of the Rocky-Mackenzie Mountains deformation. Thus the isotopic composition of magmas through time in a given area can indicate changes in lithosphere thickness. In the Canadian Cordillera it is possible to distinguish multiple magmatic events affecting the same area during a time of changing tectonic conditions and thus to demonstrate the relationship between Sr isotopes and time, as well as with character of underlying crust.

The $^{87}Sr/^{86}Sr$ initial ratio maps of Figures 18 and 19 show the features already discussed in geographic context. The Mesozoic map shows the interconnected low-ratio area encompassing Vancouver Island, Stikinia, Cache Creek Terrane, and the Intermontane part of Quesnellia. A large prong with intermediate ratios essentially coincides with the western Stikine Terrane, where it lies next to the Alexander Terrane. Transitional ratios cover much of the southern Yukon Territory and eastern Quesnellia. In the Yukon Territory the transition zone coincides with regions containing Lower Paleozoic to Proterozoic clastic rocks but no known granitic Precambrian basement (Tempelman-Kluit, 1976, 1979b). The transition zone in southern Quesnellia, with ratios between 0.704 and 0.707, is a composite region containing scattered exposures of Middle Proterozoic or older basement, surrounded by allochthonous eugeosynclinal rocks, together engulfed in Mesozoic plutons. There is no systematic change of ratios across this transition zone, presumably a reflection of its complex geology. The high ratios, above 0.707, are confined to areas underlain by Purcell and older rocks of North America in part of the Omineca Belt and the western Mackenzie Mountains. Only in the south is the high-ratio belt bounded on the east by an occurrence of transitional-ratio rocks, the Howell Creek plutons of southeastern Alberta. These lay east of the Cretaceous thermal anomaly.

The Cenozoic map shows a similar but subdued pattern. The very low ratios, 0.704 or less, are more restricted in extent. Elsewhere the ratios are largely transitional, 0.704 to 0.707. The early Cenozoic magmas show some addition of crustal Sr in the Coast and Intermontane belts but very little crustal interaction farther east.

CHEMICAL TRENDS

The CCGF contains descriptive information about the rocks that have been dated, at least a rock name and a partial list of major minerals. This information has been used to create Figure 20, which shows some aspects of the petrology of igneous rocks plotted in the space-time framework of Figures 4 to 6. Aspects worth note are sodic versus potassic character, alkalinity, and distribution of hornblende- and muscovite-bearing plutons.

A traditional expression of the gradient from sodic magmas, near the continental margin, to potassic magmas, inland, is the quartz diorite line of Moore (1959) and Moore and others (1963). This line separated areas where quartz diorite predominated from areas where plutons were more often granodiorite or adamellite. On Figure 20 the occurrences of quartz diorite among the rocks dated is shown, as well as the position of the quartz diorite line from the map of Moore and others (1963). In southern British Columbia there is good agreement between current data for post-Jurassic plutons and the original quartz diorite line, but farther north the demarcation is less clear in CCGF data. In every magmatic episode the quartz diorites occur toward the west or are scattered throughout.

A different, and often contrasting, characteristic of magmas is alkalinity, usually expressed in the lithologic name used to describe the rocks (syenite, monzonite, lamprophyre, phonolite, etc.). The occurrence of alkaline magmas is symmetrically related in time and space to the mid-Cretaceous magmatic episode (Fig. 20). During the Jurassic, in southern and central traverses, alkaline rocks occur intermingled with calk-alkaline. This includes the well-known alkaline porphyry copper plutons (Ney and Hollister, 1976). The file contains only one alkaline Jurassic-dated body in the northern transect. During the Cretaceous, alkaline magmas are restricted to the most inland fringe of the magmatic belt in British Columbia and Alberta, but they occur throughout its eastern half in the northern, Yukon, traverse. During the Cenozoic, alkaline magmas occur much farther west again. They are particularly common in the east half of the Eocene field on the southern traverse and occur on the east in all Eocene magmatic fields. They occur even farther west in later Cenozoic time. The Jurassic magmatic fields are rather mixed in character and only weakly show a sodic to alkaline polarity from west to east, but from Cretaceous time onward this polarity is strong and not repeated. The evidence shown is consistent with subduction zone magmatism with an east-dipping Benioff Zone underlying the entire Cordillera after the end of the Jurassic. There is a possibility of two or more subduction zones during the Triassic and Jurassic (Griffiths, 1977; Griffiths and Godwin, 1983), providing magmas for the different Insular and Intermontane terranes. The pre-Cretaceous Intermontane subduction zones appear, from their chemistry, to have had the same east-dipping Benioff Zone polarity as the later and modern ones. Miller (1978) and Sylvester and others (1978) discussed equivalent Late Triassic–Jurassic alkaline plutons of the western U.S.; they concluded that the magmas were peculiar to earlier stages of the Cordilleran plutonic cycle and were of deeper origin than the calk-alkaline magmas that followed.

The eastward shift of alkaline magma occurrences in mid-Cretaceous time may be due to more rapid subduction, and consequent lower Benioff Zone dip, or it may be the result of steeper

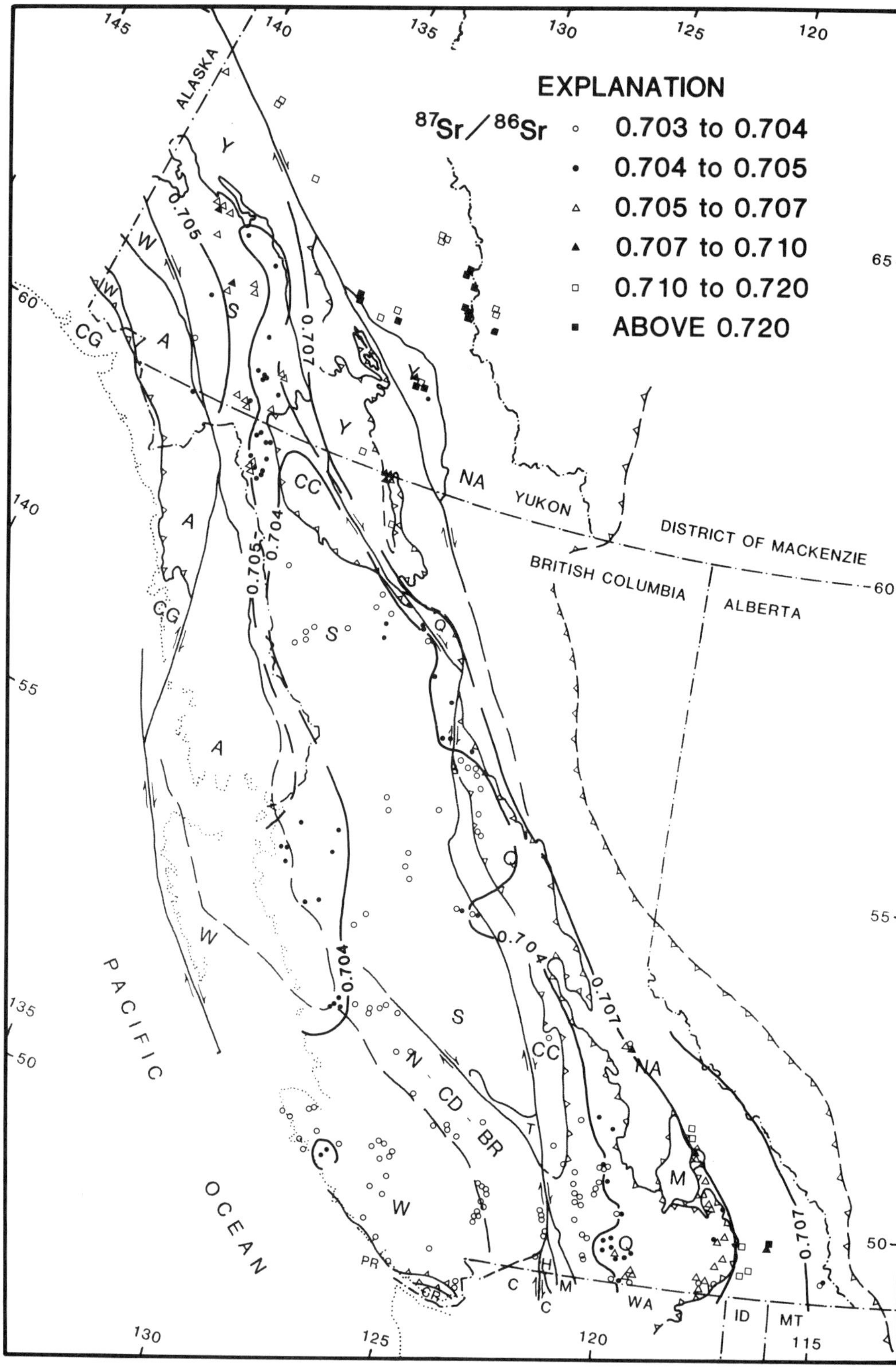

Figure 18. Map of the Canadian Cordillera showing Mesozoic $^{87}Sr/^{86}Sr$ initial ratios. Large area with ratios at or below 0.704 is underlain by ensimatic Phanerozoic crust. Areas with ratios greater than 0.707 are underlain by Precambrian crust that has been incorporated into younger magmas. Transitional areas contain scattered evidence of tectonically reworked Precambrian basement (southern Omineca Belt) or Proterozoic-age, continent-derived clastic sedimentary rocks (Yukon, northern Coast Plutonic Belt, and Omineca Belt).

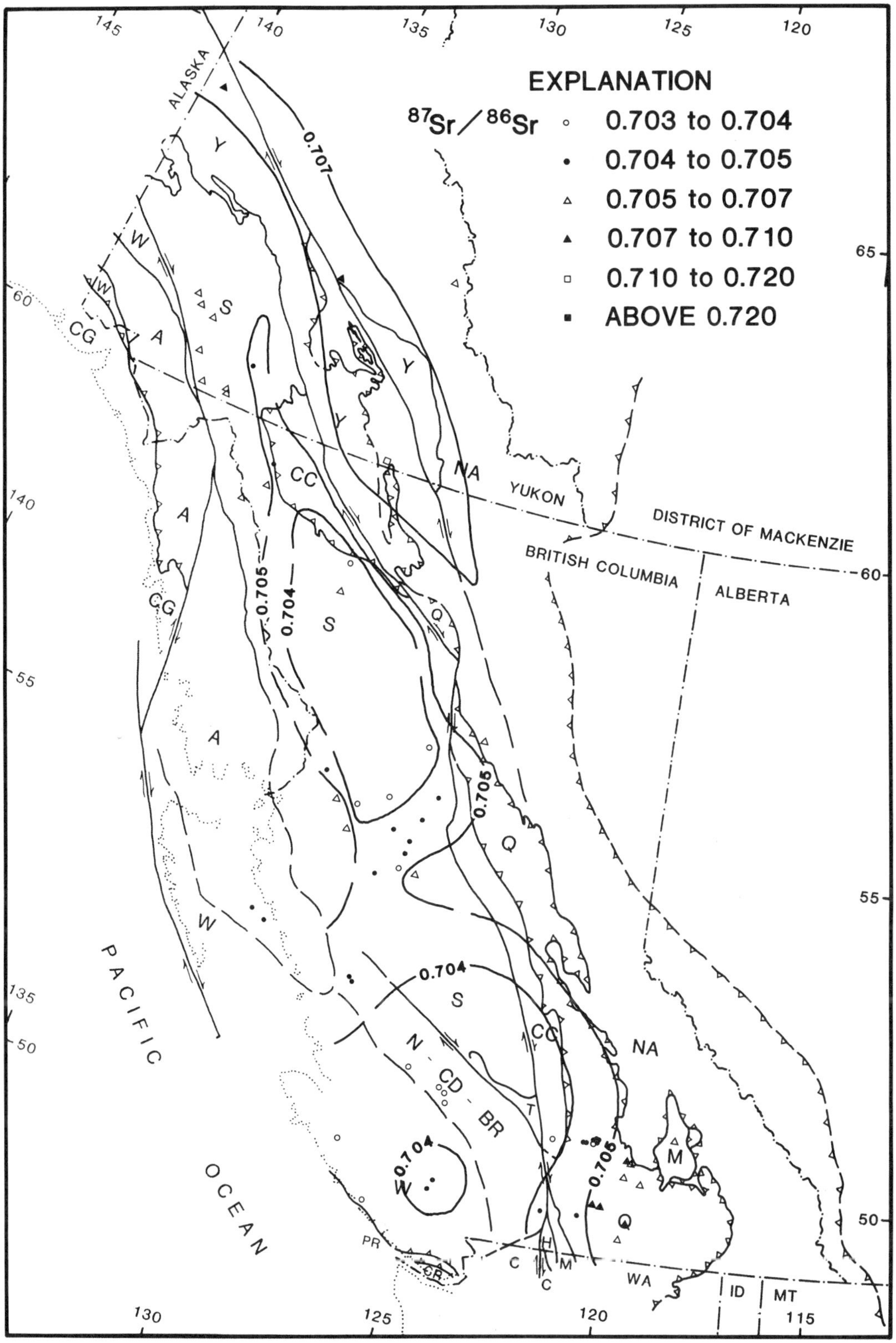

Figure 19. Map of the Canadian Cordillera showing early Cenozoic $^{87}Sr/^{86}Sr$ initial ratios. Pattern is a subdued version of the Mesozoic map (Fig. 18). Where Mesozoic ratios are low, the early Cenozoic ratios are slightly higher because the underlying crust has become more evolved. However, the area of very high ratios in Mesozoic rocks shows only transitional ratios (0.705 to 0.707) in early Cenozoic time, because continental crust is no longer making as large a contribution to the magmas.

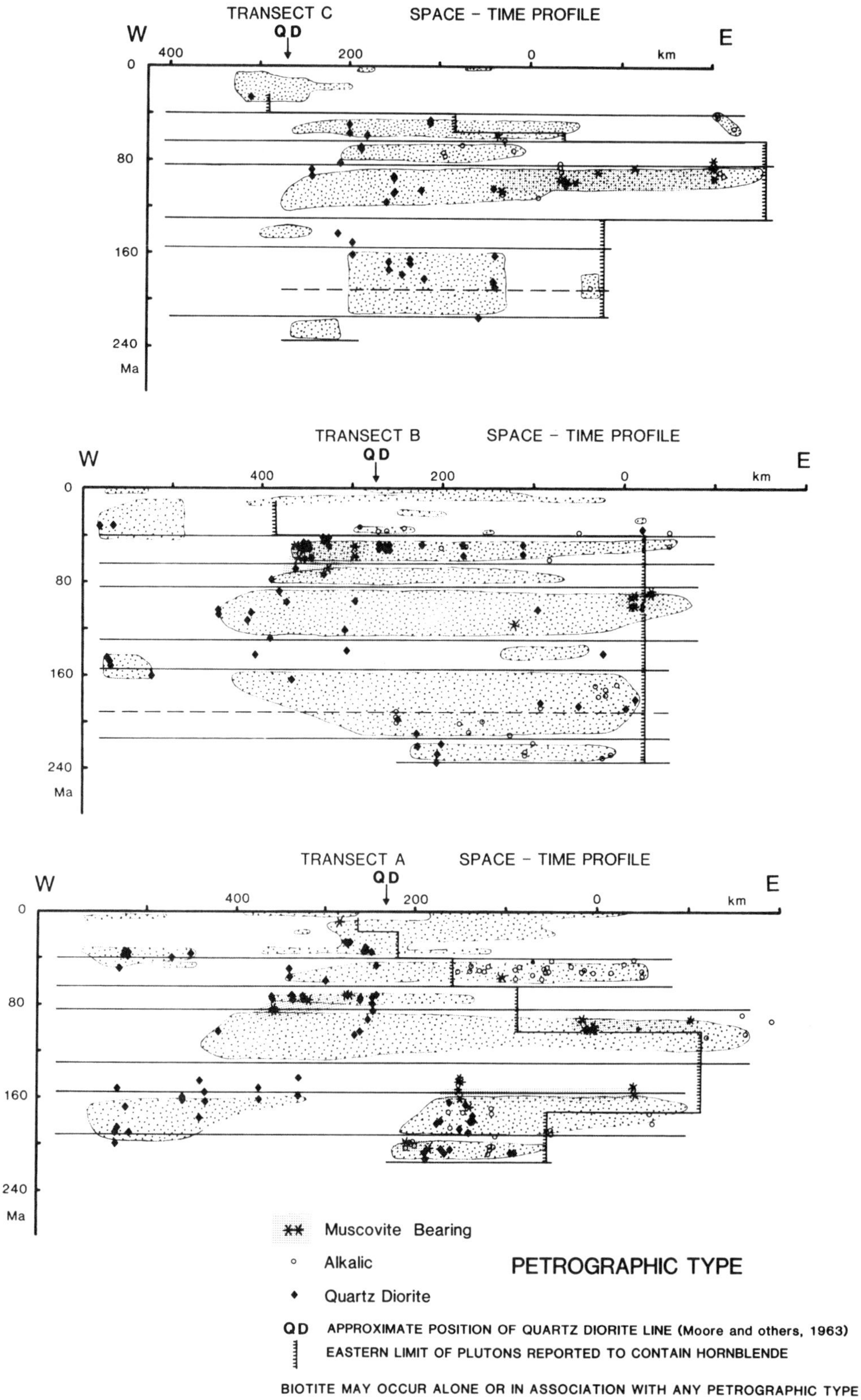

Figure 20. Space-time framework showing mineralogic-petrographic–type information in the Geochron File. In general, magmatic belts of Cretaceous and later time show a clear and singular chemical polarity: more sodic near the continental margin, alkalic far inland. Muscovite-bearing plutons are restricted to later parts of magmatic episodes and are especially abundant during the mid-Cretaceous episode in the Omineca Belt, implying an origin in thickened continental crust.

Cretaceous thermal gradients, which enhanced crustal interaction with mantle-derived magmas and thus suppressed alkalinity.

In accord with the eastward trend toward more potassic and alkaline magmas is the distribution of hornblende as shown on Figure 20. An eastern limit of hornblende-bearing igneous rocks is shown. Magmas farther east contain one or two micas or perhaps alkali amphibole or pyroxene as their mafic mineralogy.

The final petrographic aspect of interest in a space-time framework is the occurrence of muscovite in plutonic rocks. This can be seen on Figure 20 to be an exceptional feature of generally late appearance in some episodes and most common in the Cretaceous plutons of the Omineca and Rocky-Mackenzie Mountains belts. There are a couple of occurrences of muscovite in earliest Jurassic rocks in the southern transect. The Late Jurassic occurrences in that transect are instructive. The Oliver pluton on the west and Galena Bay pluton on the east are both muscovite- and garnet-bearing, and both the latest phases of Jurassic magmatism in their respective areas. Muscovite-bearing plutons of Cretaceous and early Cenozoic age in the Coast Plutonic Complex are likewise late bodies in the local intrusive sequence. Even in the eastern Cretaceous belt of muscovite-bearing plutons a relatively late age is indicated. In both central and southern transects the earlier plutons of the mid-Cretaceous episode in the east are hornblende and biotite bearing—but lack muscovite. Plutons of the eastern mid-Cretaceous belt with muscovite are also the ones most enriched in radiogenic Sr, and they show some of the characteristics of S-type plutons in Australia (Chappell and White, 1974; Archibald and others, 1983; Anderson, 1983b). All of this is consistent with a large crustal component in these magmas. The continuation of this eastern belt in the U.S. is discussed by Miller and Bradfish (1980), and details of the chemistry and petrology of muscovite-bearing plutons of southeastern California by Anderson and Rowley (1981).

The Jurassic and Coast Belt plutons with muscovite do not have markedly elevated initial $^{87}Sr/^{86}Sr$ ratios. Only their chemical composition suggests incorporation of sediment or lower crust into these magmas. Their source region may not have included old radiogenic crust—just crust of recent formation by tectonic and magmatic accretion processes—so there would have been no accumulation of radiogenic Sr prior to genesis of the aluminous magmas. Lack of radiogenic Sr enrichment does not rule out crustal involvement in these cases.

TECTONIC IMPLICATIONS

The position and polarity of magmatic arcs and the linkage of terranes by through-going magmatic belts are important elements of any Cordilleran tectonic synthesis. Figure 21 is a summary of the relationships between sutures separating terranes and the space-time framework of the Canadian Cordillera. Much of the following is a review of points documented in previous sections of this paper, so many citations are not repeated.

The oldest terrane linkage is between Cache Creek and Quesnel terranes. Rocks as old as Late Triassic, and of both oceanic and Quesnel arc affinity, are incorporated in the Cache Creek accretion wedge; blueschists of Late Triassic age are observed along the Pinchi Fault, and plutons of earliest Jurassic age intrude both terranes. Linkage was thus accomplished at the end of the Triassic.

Connections between Stikinia, Cache Creek Terrane, and northern Quesnellia are less certain, but a similar time of linkage has been inferred. Contrasting depositional environments existed until Late Triassic time, yet the Upper Triassic rocks of both Stikinia and Quesnellia are very similar, the name Takla Group being used in common. Early Jurassic plutons are similar across all three terranes.

Linkage of Anvil and Sylvester oceanic terranes with North America is not tightly bracketed (Tempelman-Kluit, 1979b). The contrast of depositional environments continued until Late Triassic time. Metamorphic rock clasts in Upper Triassic sediments of the Omineca Belt in the Yukon indicate that closure was in progress by that time. Telescoping along major thrust faults continued through Jurassic and into Early Cretaceous time so that suture closure and final emplacement of allochthons may be quite separate events. The upper bound on displacements is provided by the mid-Cretaceous plutons that cut autochthon and allochthon alike.

Emplacement of the Slide Mountain and Kaslo-Slocan-Rossland rocks along the eastern margin of Quesnellia is closely bracketed. Contrasting environments of accumulation existed until Toarcian time (late Early Jurassic). The oceanic rocks were then thrust large distances eastward over the western margin of North America before both autochthon and allochthon were pinned together by Middle Jurassic plutons. The structurally thickened zone experienced an important episode of regional metamorphism after imbrication and before emplacement of late Middle to early Late Jurassic plutons. A large fraction of the Canadian Cordillera was essentially in place against North America by the middle of Jurassic time, but perhaps not at its present latitude (Monger, 1984).

Figure 21 does not show the traditional suture between Wrangell and Alexander terranes because it is outside the compilation, in Alaska. That linkage is believed by some authors to postdate contrasting Upper Triassic rocks and predate the Late Jurassic to Early Cretaceous Gravina-Nutzotin overlap assemblage (Coney and others, 1980). Granitic plutons probably related to that assemblage occur in the Saint Elias region but they do not make a prominent contribution to the figures in this chapter. Work in progress on Paleozoic rocks suggests that there may be no need for a major suture between Wrangell and Alexander terranes.

Between the composite Insular terrane and the Intermontane terranes lay oceanic areas included in the Bridge River–Hozameen and Tyaughton terranes of southern British Columbia, and perhaps the Taku Terrane (Berg and others, 1978) of the Alaska panhandle. In the south, deep ocean basin environments lasted until Middle Jurassic time. Tipper (1984b) reported that shale of late Middle Jurassic age was common to Wrangel, Sti-

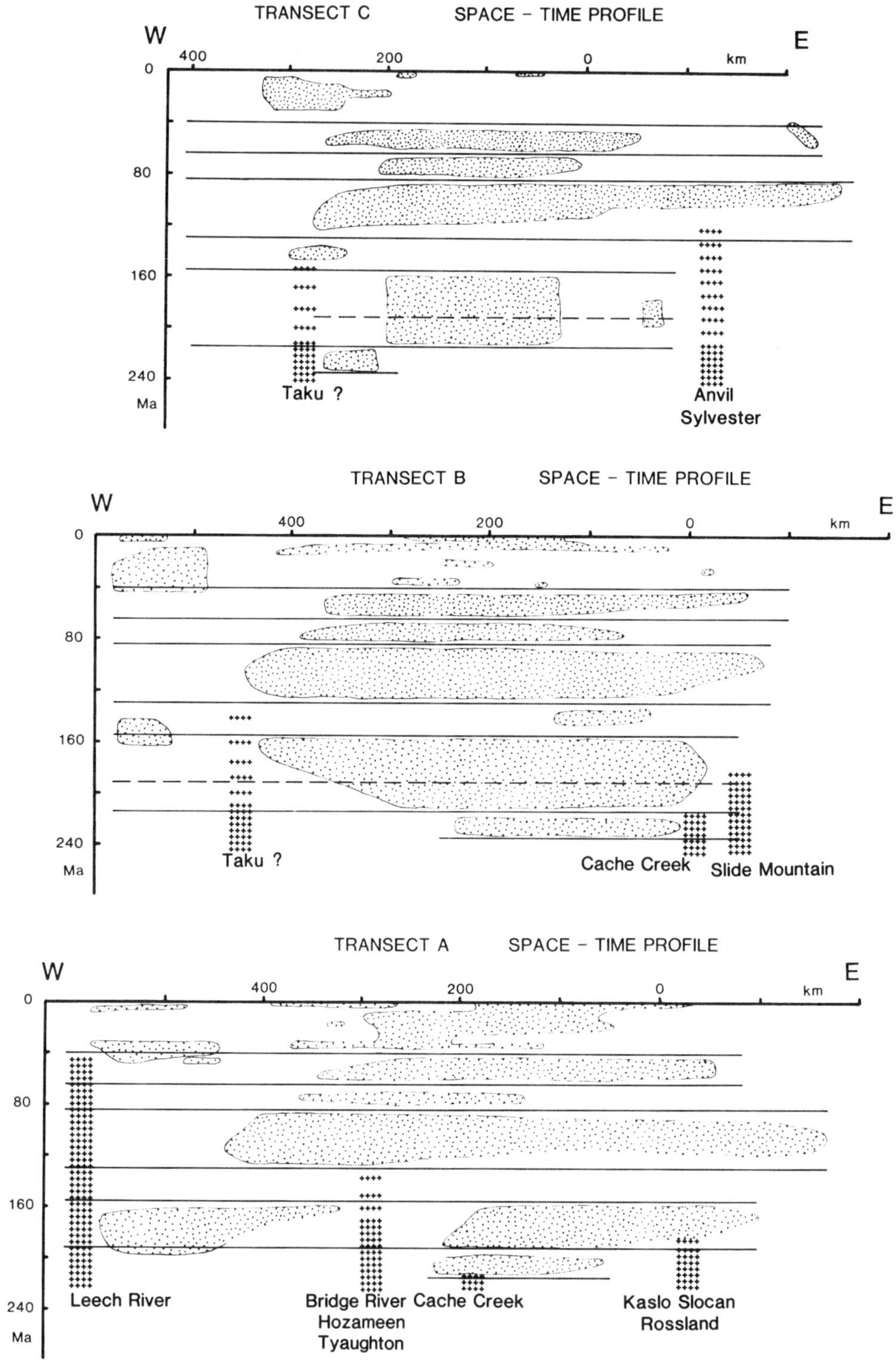

Figure 21. Relationship of major sutures between terranes of the Canadian Cordillera and magmatic belts. Where solid, the suture bands indicate certain separation; where broken, a linkage between terranes is possible; where missing, the linkage is firmly established. Assembly of terranes was largely accomplished during Late Triassic to the beginning of Early Cretaceous time. Mid-Cretaceous and Cenozoic magmatism occurred in an Andean continental margin setting. There were no post-140-Ma back-arc basins or multiple subduction zones. Strike-slip displacements younger than 140 Ma may be important (Monger, 1984) but cannot be specified from the evidence of magmatic history—they are neither required nor disproven.

kine, and Quesnel terranes; he suggested linkage by that time and wrote that in Late Jurassic time clastic debris was being shed into the northern Tyaughton trough from both sides. Potter (1983) and Haugerud (1984) agreed that closure of the Bridge River ocean was a late Middle to early Late Jurassic "Nevadan" event. Others reported that the Coast Plutonic source west of the Tyaughton Trough appeared only in mid-Cretaceous time (Cole and Tennyson, 1977; Tennyson and Cole, 1978; Monger and others, 1982; Kleinspehn, 1985). This difference of emphasis may partly depend on the latitude at which the observations were made. A southward transgressive zipping-up of the Jurassic ocean, and some delay between accretion and uplift of the Coast Mountains (delayed until the next magmatic episode commenced), may be agreeable to all.

In the north, the Insular-Intermontane suture is engulfed in the coast Plutonic Complex and thus largely unknown (Woodsworth and others, 1985). Brackets on its age are, on the older side, deformed Triassic sediments in the Taku Terrane and the overlying Upper Jurassic to Lower Cretaceous sediments of the Gravina-Nutzotin Belt and, on the younger side, the Coast Plutonic Complex itself, which started in mid-Early Cretaceous time. Lower Cretaceous rocks of the Gambier Group of the southern Coast Plutonic Complex may link both Wrangell and Stikine terranes. Well before the end of Early Cretaceous time, assembly of the Canadian Cordillera was largely complete.

The evidence of later Cretaceous and Cenozoic accretion is preserved only locally along the continental margin of Canada; this late accretion is much more evident in the western United States and Alaska. The Metchosin and Leech River–Pacific Rim units of Vancouver Island contain Triassic to early Cenozoic rocks emplaced in their present setting in early Cenozoic time (Muller, 1977; Brandon and others, 1983; Brandon, 1985). There is little accretion in Canada to relate to the Cretaceous and early Cenozoic magmatic arcs; either subduction erosion (decretion) was dominant over accretion, or the accretion wedges have been transported into the Gulf of Alaska.

COMPARISON WITH EVENTS OUTSIDE WESTERN CANADA

The magmatic belts of the Canadian Cordillera continue southward into the United States, without break. A few important comparisons are worth noting.

There are a few Triassic granitic plutons that mark the beginning of widespread Cordilleran magmatic arc activity in the western United States (Armstrong and Suppe, 1973; Walker, 1983; Dilles, 1983); their occurrence is sporadic and poorly known.

The episodic nature of Jurassic and Cretaceous plutonism in the Sierra Nevada and vicinity has long been under investigation. Evernden and Kistler (1970) reported a number of intrusive epochs at about 30 m.y. intervals starting at 215 Ma and ending at 82 Ma, but no later workers have agreed with the details of their interpretation. The current consensus is probably best expressed by the histogram of zircon U-Pb dates from Stern and others (1981) and Chen and Moore (1982) that is shown on Figure 22—a latest Triassic to earliest Jurassic episode at 215 to 200 Ma, a broad period of Jurassic activity, culminating at 160 to 155 Ma, a lull from 135 to 125 Ma, and a broad 125- to 80-Ma maximum of Cretaceous activity. The correspondence with southwestern Canada is excellent.

After 80 Ma, Sierra magmatism ceased, and a very different (Laramide) pattern appeared (Armstrong, 1974b). This major rearrangement did not affect western Canada. The Late Cretaceous magmatic belt of the Canadian Cordillera can be followed southward into Idaho and Montana (Armstrong and others, 1977; Hyndman, 1983; Marvin and others, 1983), as can the Eocene belt (Armstrong, 1974a).

In contrast to the similarities just described is magmatism during the last 80 m.y. in southwestern Alaska that is out of phase with that in Canada. The histogram of dates compiled by Wallace and Engebretson (1984) and presented on Figure 22 shows very clearly that Late Cretaceous to Oligocene magmatism on the west side of the Gulf of Alaska occurred when the Canadian Cordillera, as far north as the Yukon Territory, was relatively quiescent, and that Canadian magmatism occurred at times of quiescence in southwestern Alaska. This is a provocative observation. Since the areas trend at approximately a right angle to one another, it is tempting to speculate about switching back and forth of ocean-floor plate motion, relative to North America, from northwestward to northeastward. This would cause convergent and strike-slip movements to alternate between the two areas of contrasting magmatic history. This proposition is not pursued here. How the different and overlapping magmatic episodes interact within Alaska is a topic for future analysis; it probably requires more hard data, invites new syntheses, and will probably inspire much discussion and difference of opinion.

Another clear example of Mesozoic magmatism out of phase with the Canadian Cordillera is found in Japan, where the peaks of K-Ar date histograms (Fig. 22) occur at 120 to 115, 90 to 80, and about 60 Ma (Nozawa, 1983).

Implicit throughout this chapter is the concept that magmas arise from subduction of Pacific Ocean floor under North America, and that explanation for magmatic episodes must include changing geometry and rate of subduction. Convergence vectors for western North America can be derived, independent of local geology, from magnetic stripe maps and hot-spot traces, and two recent exercises of that sort are shown on Figure 22. The westward movement of North America relative to hot spots (Engebretson and others, 1984) shows variation broadly sympathetic with the tempo of magmatism. Likewise suggestive of a link of the westward drift of North America with magmatism are coincident magmatic events in the Atlantic realm. For example, volcanism offshore of eastern Canada at 125 to 105 Ma (Jansa and Pe-Piper, 1985) and in arctic Canada at about 50 Ma (Mitchell and Platt, 1983).

A specific plate-convergence event along the west coast of North America is the Farallon pulse (Jurdy, 1984) shown on

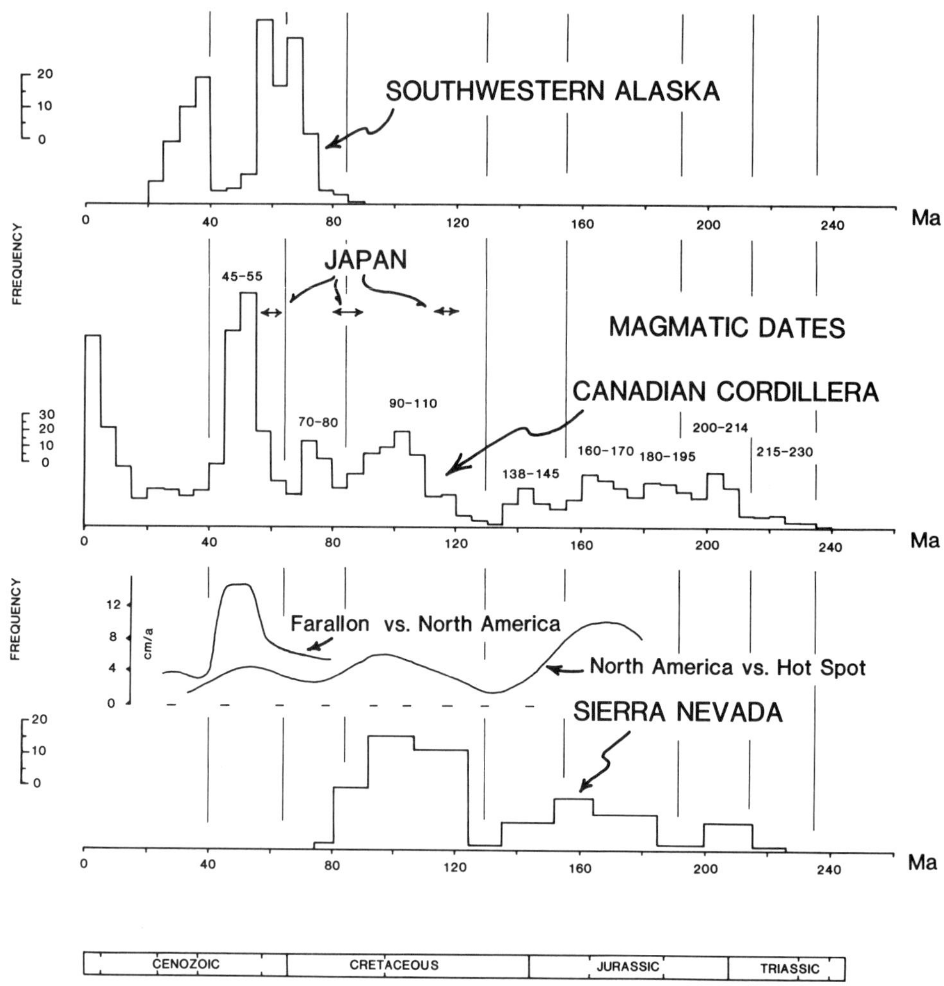

Figure 22. Comparison of Canadian Cordilleran magmatism with compilations from other areas and with Mesozoic-Cenozoic plate motions. Graphic framework and Canadian data from Figure 3. Southwestern Alaska geochronometry from Wallace and Engebretson (1984), Japanese magmatic culminations from Nozawa (1983), motion of Farallon Plate with respect to North America (with Antarctica treated as two plates) from Jurdy (1984), motion of North America with respect to hot spots from Engebretson and others (1984), and Sierra Nevada compilation of zircon U-Pb dates from Chen and Moore (1982).

Figure 22. It corresponds so exactly to the Eocene magmatic event as to make a cause-and-effect relationship seem inescapable. Significantly, this requires that the Farallon Plate extend as far north as the Gulf of Alaska as late as 45 Ma—however far north that area was then (but northward drift is a separate problem)—which is not the way the northeast Pacific is usually reconstructed (e.g., Riddihough, 1982).

Other coincidences between Pacific Ocean plate events and magmatism in the Canadian Cordillera include abundant mid-Cretaceous volcanic ash in western Pacific DSDP cores (Rea and Vallier, 1983), pulses of volcanism in the west Pacific at 110 to 85, 80 to 65, and around 50 Ma (Rea and Dixon, 1983), and accelerated ridge crest generation at 120 to 90 Ma (Watts and others, 1980). The correlation of events on a global scale gives encouragement to belief in plate tectonic models but indicates that any overall synthesis, and explanation, is beyond the scope of this paper.

CONCLUSIONS

Magmatism in the Canadian Cordillera has been more episodic than continuous or time-transgressive. The episodes vary in intensity, duration, extent, and character, and they appear to be linked to major changes in plate motion. The most notable time-transgressions are the rapid continentward advance of episode inception and the gradual retreat of the magmatic front over a much longer time span than individual magmatic episodes.

During Late Triassic and continuing through Jurassic time, the episodes were relatively terrane-specific, but from the Cretaceous onward, western Canada appears to have been the site of an Andean-type continent-margin magmatic arc. Sr isotopic composition of magmas is terrane-specific but is also affected by changing thermal conditions and crustal evolution. Chemical trends across magmatic belts reveal subduction polarity as well as changing thermal conditions. Eastward-descending subducted ocean crust appears to have been the general rule through Mesozoic time.

It will be interesting in the future to compare the Canadian Cordillera with other parts of the Circum-Pacific Mobile Belt, as their geochronometry becomes documented in equal detail. Eventually the assembled information on magmatic episodes and regional time-transgressive shifts of magmatism can be used to specify the history of continent-margin interactions with Pacific plates. This will reduce the freedom of extrapolation of plate boundaries and motions that now goes far beyond the preserved record of magnetic stripes and hot-spot traces.

In more local perspective, the space-time framework presented in this chapter can be used for future detailed discussion of igneous petrology, chemistry, and isotopes. The application of space-time information to Cordilleran metallogeny has already been recognized (Griffiths, 1977; Griffiths and Godwin, 1983) and will undoubtedly receive further attention. As time information from both paleontology and geochronometry accumulates, our presentation of geologic data will evolve from two-dimensional projections, such as familiar structure and tectonic maps, to increasingly detailed three-dimensional representations, such as time-interval maps and serial space-time projections. Likewise, we are forced into the world of data-base management to handle the results of our twentieth-century scientific activity explosion.

ACKNOWLEDGMENTS

Geochronometry at the University of British Columbia is funded by a Canadian Natural Sciences and Engineering Research Council operating grant and by Provincial and Federal government research contracts. J. Nelsen and E. Halpenny made large individual contributions toward organizing the CCGF, but the largest effort has been my own. The contributors of data to the CCGF are legion, as are the number of people who have engaged me in discussions of Cordilleran geology, but I must collectively recognize UBC graduate students and faculty and geologists of the Cordilleran Division of the Geological Survey of Canada. I thank George Gehrels for an especially thorough review of this manuscript; Peter Coney and B. C. Burchfiel also contributed reviews. K. K. Turekian initiated the writing process. This synthesis additionally discharges my principal obligation to the Canadian DNAG project. It is contributed to this volume to honor John Rodgers, whose influence on my career began 30 years ago at Yale University.

REFERENCES CITED

Anderson, J. L., and Rowley, M. C., 1981, Synkinematic intrusion of two-mica and associated metaluminous granitoids, Whipple Mountains, California: Canadian Mineralogist, v. 19, p. 83–101.

Anderson, P., 1976, Oceanic crust and arc-trench gap tectonics in southwestern British Columbia: Geology, v. 4, p. 443–446.

Anderson, R. G., 1983a, Geology of the Hotailuh Batholith and surrounding volcanic and sedimentary rocks, north-central British Columbia [Ph.D. thesis] Ottawa, Carleton University, 669 p.

——, 1983b, Selwyn plutonic suite and its relationship to tungsten skarn mineralization, southeastern Yukon and District of MacKenzie: Geological Survey of Canada Paper 83-1B, p. 151–163.

——, 1984, Late Triassic and Jurassic magmatism along the Stikine Arch and the geology of the Stikine batholith, north-central British Columbia: Geological Survey of Canada Paper 84-1A, p. 67–73.

Archibald, D. A., Glover, J. K., Price, R. A., Farrar, E., and Carmichael, D. M., 1983, Geochronology and tectonic implications of magmatism and metamorphism; Southern Kootenay Arc and neighbouring regions, southeastern British Columbia; Part I; Jurassic to mid-Cretaceous: Canadian Journal of Earth Sciences, v. 20, p. 1891–1913.

Archibald, D. A., Krogh, T. E., Armstrong, R. L., Farrar, E., 1984, Geochronology and tectonic implication of magmatism and metamorphism; Southern Kootenay Arc and neighbouring regions, southeastern British Columbia; Part II; Mid-Cretaceous to Eocene: Canadian Journal of Earth Sciences, v. 21, p. 567–583.

Armstrong, R. L., 1966, K-Ar dating of plutonic and volcanic rocks in orogenic belts, *in* Schaeffer, O. A., and Zahringer, J., eds., Potassium argon dating: Berlin, Springer-Verlag, p. 117–133.

——, 1974a, Geochronometry of the Eocene volcanic-plutonic episode in Idaho: Northwest Geology, v. 3, p. 1–15.

——, 1974b, Magmatism, orogenic timing, and orogenic diachronism in the Cordillera from Mexico to Canada: Nature, v. 247, p. 348–351.

——, 1982, Cordilleran metamorphic core complexes; From Arizona to southern Canada: Annual Review of Earth and Planetary Sciences, v. 10, p. 129–154.

Armstrong, R. L., and Suppe, J., 1973, Potassium-argon geochronometry of Mesozoic igneous rocks in Nevada, Utah, and southern California: Geological Society of America Bulletin, v. 84, p. 1375–1392.

Armstrong, R. L., Taubeneck, W. H., and Hales, P. O., 1977, Rb-Sr and K-Ar geochronometry of Mesozoic granitic rocks and their Sr isotopic composition, Oregon, Washington, and Idaho: Geological Society of America Bulletin, v. 88, p. 397–411.

Armstrong, R. L., Harakal, J. E., Brown, E. H., Bernardi, M. L., and Rady, P. M., 1983, Late Paleozoic high pressure metamorphic rocks in northwestern Washington and southwestern British Columbia; The Vedder Complex: Geological Society of America Bulletin, v. 94, p. 451–458.

Armstrong, R. L., Muller, J. E., Harakal, J. E., and Muhlenbacks, K., 1985, The Neogene Alert Bay Volcanic Belt of northern Vancouver Island: Journal of Volcanology and Geothermal Research, v. 26, p. 75–97.

Baadsgaard, H., and Lerbekmo, J. F., 1982, Abstracts for items NDS 126 and NDS 127, *in* Odin, G. S., ed., Numerical dating in stratigraphy: Chichester, Wiley-Interscience, v. 2, p. 795–797.

Baadsgaard, H., Folinsbee, R. E., and Lipson, J., 1961, Potassium-argon dates of biotites from Cordilleran granites: Geological Society of America Bulletin, v. 72, p. 689–702.

Basaltic Volcanism Study Project, 1981, Basaltic volcanism on the terrestrial planets: New York, Pergamon Press, 1,286 p.

Berg, H. C., Jones, D. L., and Richter, D. H., 1972, Gravina–Nutzotin belt; Tectonic significance of an upper Mesozoic sedimentary and volcanic sequence in southern and southeastern Alaska: U.S. Geological Survey Professional Paper 800D, p. 1–24.

Berg, H. C., Jones, D. L., and Coney, P. J., 1978, Map showing pre-Cenozoic tectonostratigraphic terranes of southeastern Alaska and adjacent areas: U.S. Geological Survey Open-File Report 78–1085, scale 1:1,000,000.

Best, M. G., Armstrong, R. L., Graustein, W. C., Embree, G. F., and Ahlborn, R. C., 1974, Mica granites of the Kern Mountains pluton, eastern White Pine County, Nevada; Remobilized basement of the Cordilleran miogeosyncline: Geological Society of America Bulletin, v. 85, p. 1277–1286.

Brandon, M. T., 1985, Pacific Rim Complex of western Vancouver Island; Tectonic evolution of a late Mesozoic active margin: Geological Society of America Abstracts with Programs, v. 17, p. 343.

Brandon, M. T., Cowan, D. S., Muller, J. E., and Vance, J. A., 1983, Pre-Tertiary geology of San Juan Islands, Washington and southeast Vancouver Island, British Columbia: Field Trip Guidebook, GAC-MAC-CGU Meeting, Victoria, 1983, no. 5, 65 p.

Brew, D. A., and Morrell, R. P., 1983, Intrusive rocks and plutonic belts of southeastern Alaska, U.S.A.: Geological Society of America Memoir 159, p. 171–193.

Brew, D. A., Ovenshine, A. T., Karl, S. M., and Hunt, S. J., 1984, Preliminary reconnaissance geologic map of the Petersburg and parts of the Port Alexander and Sumdum 1:250,000 Quadrangles, southeastern Alaska: U.S. Geological Survey Open-File Report 84-405, 43 p.

Brookfield, M. E., 1973, Lower Oxfordian (Upper Jurassic) sediments at Harrison Lake, British Columbia, and the age of the "Agassiz orogeny": Canadian Journal of Earth Sciences, v. 10, p. 1688–1692.

Carter, N. C., 1982, Porphyry copper and molybdenum deposits, west-central British Columbia: British Columbia Ministry of Energy Mines and Petroleum Resources Bulletin 64, 150 p.

Chappell, B. W., and White, A.J.R., 1974, Two contrasting granite types: Pacific Geology, v. 8, p. 173–174.

Chen, J. H., and Moore, J. G., 1982, Uranium-lead isotopic ages from the Sierra Nevada batholith, California: Journal of Geophysical Research, v. 87, p. 4761–4785.

Christopher, P. A., and Carter, N. C., 1976, Metallogeny and metallogenic epochs for porphyry mineral deposits in the Canadian Cordillera: Canadian Institute of Mining and Metallurgy Special Volume 15, p. 64–71.

Church, B. N., 1970, White Lake, *in* Geology Exploration and Mining in B.C., 1970: British Columbia Department of Mines and Petroleum Resources, p. 396–402.

Cole, M. R., and Tennyson, M. E., 1977, Comment *on* "Ocean crust and arc-trench gap tectonics in southwestern British Columbia": Geology, v. 5, p. 325–327.

Coney, P. J., Jones, D. L., and Monger, J.W.H., 1980, Cordilleran suspect terranes: Nature, v. 288, p. 329–333.

Cowan, D. S., 1982, Geological evidence for post-40 m.y. B.P. large-scale northwestward displacement of part of southeastern Alaska: Geology, v. 10, p. 309–313.

Cushing, G. W., Foster, H. L., Harrison, T. M., and Laird, J., 1984, Possible Mesozoic accretion in the eastern Yukon-Tanana Upland, Alaska: Geological Society of American Abstracts with Programs, v. 16, p. 481.

Dadisman, S. V., 1980, Radiometric ages of rocks in south-central Alaska and western Yukon Territory: U.S. Geological Survey Open-File Report 80–183, 80 p.

Dagenais, G. R., 1984, The oxygen isotope geochemistry of granitoid rocks from the southern and central Yukon [M.Sc thesis]: Emonton, University of Alberta, 168 p.

Daly, R. A., 1912, Geology of the North American Cordillera at the Forty-Ninth Parallel: Geological Survey of Canada Memoir 38, 3 v., 857 p.

Dickinson, W. R., 1973, Widths of modern arc-trench gaps proportional to past duration of igneous activity in associated magmatic arcs: Journal of Geophysical Research, v. 78, p. 3376–3389.

——, 1981, Plate tectonics and the continental margin of California, *in* Ernst, W. G., ed., The geotectonic development of California: New Jersey, Englewood Cliffs, Prentice-Hall, p. 1–28.

Dilles, J. H., 1983, Chronology of early Mesozoic plutonic and volcanism in the Yerington District, western Nevada: Geological Society of America Abstracts with Programs, v. 15, p. 383.

Eadie, E. T., 1976, K-Ar and Rb-Sr geochronology of the northern Hogem Batholith, B.C. [B.Sc. thesis]: Vancouver, University of British Columbia, 46 p.

Eisbacher, G. H., 1974, Evolution of successor basins in the Canadian Cordillera: Society of Economic Paleontologists and Mineralogists Special Publication 19, p. 274–291.

——, 1977, Mesozoic-Tertiary basin models for the Canadian Cordillera and their geological constraints: Canadian Journal of Earth Sciences, v. 14, p. 2414–2421.

Engebretson, D. C., Cox, A., and Thompson, G. A., 1984, Correlation of plate motions with continental tectonics; Laramide to Basin-Range: Tectonics, v. 3, p. 115–119.

Evernden, J. F., and Kistler, R. W., 1970, Chronology of emplacement of Mesozoic batholithic complexes in California and western Nevada: U.S. Geological Survey Professional Paper 623, 42 p.

Ewing, T. E., 1980, Paleogene tectonic evolution of the Pacific Northwest: Journal of Geology, v. 88, p. 619–638.

——, 1981a, Petrology and geochemistry of the Kamloops Group volcanics, British Columbia: Canadian Journal of Earth Sciences, v. 18, p. 1478–1491.

——, 1981b, Regional stratigraphy and structural setting of the Kamloops Group, south-central British Columbia: Canadian Journal of Earth Sciences, v. 18, p. 1464–1477.

Folinsbee, R. E., Baadsgaard, H., and Lipson, J., 1961, Potassium-argon dates of Upper Cretaceous ash falls, Alberta, Canada: New York Academy of Sciences Annals, v. 91, p. 352–363.

Gabrielse, H., and Reesor, J. E., 1964, Geochronology of putonic rocks in two areas of the Canadian Cordillera: Royal Society of Canada Special Publication 8, p. 96–138.

——, 1974, The nature and setting of granitic plutons in the central and eastern parts of the Canadian Cordillera: Pacific Geology, v. 8, p. 109–138.

Garnett, J. A., 1978, Geology and mineral occurrences of the southern Hogem batholith: British Columbia Ministry of Mines and Petroleum Resources Bulletin 70, 75 p.

Gehrels, G. E., and Berg, H. C., 1984, Geologic map of southeastern Alaska: U.S. Geological Survey Open-File Report 84–886, 28 p., scale 1:600,000.

Gehrels, G. E., Brew, D. A., and Saleeby, J. B., 1984, Progress report on U/Pb (zircon) geochronologic studies in the Coast plutonic-metamorphic complex east of Juneau, southeastern Alaska: U.S. Geological Survey Circular 939, p. 100–102.

Gill, J., 1981, Orogenic andesites and plate tectonics: New York, Springer-Verlag, 390 p.

Griffiths, J. R., 1977, Mesozoic–early Cenozoic volcanism, plutonism, and mineralization in southern British Columbia; A plate tectonic synthesis: Canadian Journal of Earth Sciences, v. 14, p. 1611–1624.

Griffiths, J. R., and Godwin, C. I., 1983, Metallogeny and tectonics of porphyry copper-molybdenum deposits in British Columbia: Canadian Journal of Earth Sciences, v. 20, p. 1000–1018.

Grond, H. C., Churchill, S. J., Armstrong, R. L., Harakal, J. E., and Nixon, G. T., 1984, Late Cretaceous age of the Hutshi, Mount Nansen, and Carmacks groups, southwestern Yukon Territory and northwestern British Columbia: Canadian Journal of Earth Sciences, v. 21, p. 554–558.

Hamilton, W., 1969, The volcanic central Andes; A modern model for the Cretaceous batholiths and tectonics of western North America: Proceedings of the Andesite Conference, Oregon Department of Geology and Mineral Industries Bulletin 65, p. 175–184.

Haugerud, R. A., 1984, Age, clinopyroxene chemistry, and tectonics of the

Hozameen Group, North Cascades, NW Washington and SW British Columbia: Geological Society of America Abstracts with Programs, v. 16, p. 532.

Hill, M., Morris, J., and Whelan, J., 1981, Hybrid granodiorites intruding the accretionary prism, Kodiak, Shumagin, and Sanak islands, southwest Alaska: Journal of Geophysical Research, v. 86, p. 10569–10590.

Htoon, M., 1979, Geology of the Clinton Creek asbestos deposit, Yukon Territory [M.S. thesis]: Vancouver, University of British Columbia, 194 p.

Hudson, T., 1983, Calc-alkaline plutonism along the Pacific rim of southern Alaska: Geological Society of America Memoir 159, p. 159–169.

Hudson, T., Plafker, G., and Peterman, Z. E., 1979, Paleogene anatexis along the Gulf of Alaska margin: Geology, v. 7, p. 573–577.

Hutchison, W. W., 1982, Geology of the Prince Rupert–Skeena map area, British Columbia: Geological Survey of Canada Memoir 394, 116 p.

Hyndman, D. W., 1983, The Idaho Batholith and associated plutons, Idaho and western Montana: Geological Society of America Memoir 159, p. 213–240.

Isachsen, C., Armstrong, R. L., and Parrish, R. R., 1985, U-Pb, Rb-Sr, and K-Ar geochronometry of Vancouver Island igneous rocks [abs.]: Geological Association of Canada Victoria Section Symposium, April 19, 1985, Sidney, B.C., Abstracts, p. 21–22.

Jansa, L. F., and Pe-Piper, G., 1985, Early Cretaceous volcanism on the northeastern American margin and implications for plate tectonics: Geological Society of America Bulletin, v. 96, p. 83–91.

Johnson, C. M., and O'Neil, J. R., 1984, Triple junction magmatism: A geochemical study of Neogene volcanic rocks in western California: Earth and Planetary Science Letters, v. 71, p. 241–262.

Jones, D. L., Silberling, N. J., and Hillhouse, J., 1977, Wrangellia; A displaced terrane in northwestern North America: Canadian Journal of Earth Sciences, v. 14, p. 2565–2577.

Jones, D. L., Howell, D. G., Coney, P. J., and Monger, J.W.H., 1983, Recognition, character, analysis of tectonostratigraphic terranes in western North America, *in* Hashimoto, M., and Uyeda, S., eds., Accretion tectonics in the Circum-Pacific Regions: Tokyo, Terra Scientific Publishing Company, p. 21–35.

Jordan, T. E., Isacks, B. L., Allmendinger, R. W., Brewer, J. A., Ramos, V. A., and Ando, C. J., 1983, Andean tectonics related to geometry of subducted Nazca plate: Geological Society of American Bulletin, v. 94, p. 341–361.

Jurdy, D. M., 1984, The subduction of the Farallon plate beneath North America as derived from relative plate motions: Tectonics, v. 3, p. 107–113.

Kistler, R. W., 1983, Isotope geochemistry of plutons in the northern Great Basin: Geothermal Resources Council Special Report 13, p. 3–8.

Kistler, R. W., and Peterman, Z. E., 1978, Reconstruction of crustal blocks of California on the basis of initial strontium isotopic compositions of Mesozoic granitic rocks: U.S. Geological Survey Professional Paper 1071, 17 p.

Kistler, R. W., Ghent, E. D., and O'Neil, J. R., 1981, Petrogenesis of garnet two-mica granites in the Ruby Mountains, Nevada: Journal of Geophysical Research, v. 86, p. 10591–10606.

Kleinspehn, K. L., 1985, Cretaceous sedimentation and tectonics, Tyaughton-Methow Basin, southwestern British Columbia: Canadian Journal of Earth Sciences, v. 22, p. 154–174.

Lambert, M. B., 1974, The Bennett Lake cauldron subsidence complex, British Columbia and Yukon Territory: Geological Survey of Canada Bulletin 227, 213 p.

Le Couteur, P. C., and Tempelman-Kluit, D. J., 1976, Rb/Sr ages and a profile of initial Sr^{87}/Sr^{86} ratios for plutonic rocks across the Yukon Crystalline Terrane: Canadian Journal of Earth Sciences, v. 13, p. 319–330.

Lee, D. E., and Christiansen, E. H., 1983, The granite problem as exposed in the southern Snake Range, Nevada: Contributions to Mineralogy and Petrology, v. 83, p. 99–116.

Leeman, W. P., 1982, Tectonic and magmatic significance of strontium isotopic variations in Cenozoic volcanic rocks from the western United States: Geological Society of America Bulletin, v. 93, p. 487–503.

Marshak, R. S., and Karig, D. E., 1977, Triple junctions as a cause for anomalously near-trench igneous activity between the trench and volcanic arc: Geology, v. 5, p. 233–236.

Marvin, R. F., Zen, E-an, Hammarstrom, J. M., and Mehnert, H. H., 1983, Cretaceous and Paleocene potassium-argon mineral ages of the northern Pioneer batholith and nearby igneous rocks in southwest Montana: Isochron/West, no. 38, p. 11–16.

Mathews, W. H., 1981, Early Cenozoic resetting of potassium-argon dates and geothermal history of north Okanagan area, British Columbia: Canadian Journal of Earth Sciences, v. 18, p. 1310–1319.

——, 1983, Early Tertiary resetting of potassium-argon dates in the Kootenay Arc, southeastern British Columbia: Canadian Journal of Earth Sciences, v. 20, p. 867–872.

Medford, G. A., 1975, K-Ar and fission track geochronometry of an Eocene thermal event in the Kettle River (west half) map area, southern British Columbia: Canadian Journal of Earth Sciences, v. 12, p. 836–843.

McCrossan, R. G., and Glaister, R. P., eds., 1964, Geological history of western Canada: Calgary, Alberta Society of Petroleum Geologists, 232 p.

Miller, C. F., 1978, An early Mesozoic alkalic magmatic belt in western North America, *in* Mesozoic paleogeography of the western United States: Pacific Section, Society of Economic Paleontologists and Mineralogists, p. 163–173.

Miller, C. F., and Bradfish, L. J., 1980, An inner Cordilleran belt of muscovite-bearing plutons: Geology, v. 8, p. 412–416.

Miller, C. F., Howard, K. A., and Hoisch, T. D., 1982, Mesozoic thrusting, metamorphism, and plutonism, Old Woman–Piute Range, southeastern California, *in* Frost, E. G., and Martin, D. L., eds., Mesozoic–Cenozoic tectonic evolution of the Colorado River region, California, Arizona, and Nevada: San Diego, Cordilleran Publishers, p. 561–581.

Mitchell, R. H., and Platt, R. G., 1983, Primitive nephelinitic volcanism associated with rifting and uplift in the Canadian Arctic: Nature, v. 303, p. 609–612.

Monger, J.W.H., 1984, Cordilleran tectonics; A Canadian perspective: Bulletin Societe géologique France, v. 26, p. 255–278.

Monger, J.W.H., and Church, B. N., 1977, Revised stratigraphy of the Takla Group, north-central British Columbia: Canadian Journal of Earth Sciences, v. 14, p. 318–326.

Monger, J.W.H., and Hutchinson, W. W., 1971, Metamorphic map of the Canadian Cordillera: Geological Survey of Canada Paper 70-33, 61 p. and supplement, 19 p.

Monger, J.W.H., and Price, R. A., 1979, Geodynamic evolution of the Canadian Cordillera; Progress and problems: Canadian Journal of Earth Sciences, v. 19, p. 770–791.

Monger, J.W.H., Souther, J. G., and Gabrielse, H., 1972, Evolution of the Canadian Cordillera; A plate tectonic model: American Journal of Science, v. 272, p. 577–602.

Monger, J.W.H., Price, R. A., and Tempelman-Kluit, D. J., 1982, Tectonic accretion and the origin of the two major metamorphic and plutonic welts in the Canadian Cordillera: Geology, v. 10, p. 70–75.

Moore, J. C., Byrne, T., Plumley, P. W., Reid, M., Gibbons, H., and Coe, R. S., 1983, Paleogene evolution of the Kodiak Islands, Alaska; Consequences of ridge-trench interaction in a more southerly latitude: Tectonics, v. 2, p. 265–293.

Moore, J. G., 1959, The quartz-diorite boundary line in the western United States: Journal of Geology, v. 67, p. 198–210.

Moore, J. G., Grantz, A., and Blake, M. C., Jr., 1963, The quartz diorite line in northwestern North America: U.S. Geological Survey Professional Paper 450E, p. 89–93.

Morrison, G. W., Godwin, C. I., and Armstrong, R. L., 1979, Interpretation of isotopic ages and $^{87}Sr/^{86}Sr$ initial ratios for plutonic rocks in the Whitehorse map area, Yukon: Canadian Journal of Earth Sciences, v. 16, p. 1988–1997.

Muller, J. E., 1977, Evolution of the Pacific margin, Vancouver Island, and adjacent regions: Canadian Journal of Earth Sciences, v. 14, p. 2062–2085.

——, 1980, Chemistry and origin of the Eocene Metchosin Volcanics, Vancouver Island, British Columbia: Canadian Journal of Earth Sciences, v. 17, p. 199–209.

Ney, C. S., and Hollister, V. F., 1976, Geologic setting of porphyry deposits of the

Canadian Cordillera: Canadian Institute of Mining and Metallurgay Special Volume 15, p. 21–29.

Norris, D. K., Stevens, R. D., and Wanless, R. K., 1965, K-Ar age of igneous pebbles in the McDougall-Segur conglomerate, southeastern Canadian Cordillera: Geological Survey of Canada Paper 65–26, 11 p.

Nozawa, T., 1983, Felsic plutonism in Japan: Geological Society of America Memoir 159, p. 105–122.

Okulitch, A. V., 1984, The role of the Shuswap Metamorphic Complex in Cordilleran tectonism; A review: Canadian Journal of Earth Sciences, v. 21, p. 1171–1193.

Orchard, M. J., 1981, Triassic conodonts from the Cache Creek Group, Marble Canyon, southern British Columbia: Geological Survey of Canada Paper 81-A, p. 357–359.

Palmer, A. R., 1983, The Decade of North American Geology [DNAG] 1983 Geologic Time Scale: Geology, v. 11, p. 503–504.

Parkinson, D., 1985, Geochronology of the western side of the Okanagan metamorphic core complex, southern B.C.: Geological Society of America Abstracts with Programs, v. 17, p. 399.

Parrish, R. R., 1979, Geochronology and tectonics of the northern Wolverine Complex, British Columbia: Canadian Journal of Earth Sciences, v. 16, p. 1428–1438.

——, 1981, Geology and regional tectonics of the Nemo Lakes Belt, northern Valhalla Range, British Columbia: Canadian Journal of Earth Sciences, v. 18, p. 944–958.

Parrish, R., 1984, Slocan Lake fault; A low angle fault zone bounding the Valhalla gneiss complex, Nelson map area, southern British Columbia: Geological Survey of Canada Paper 84-1A, p. 323–330.

Parrish, R., and Wheeler, J. O., 1983, A U-Pb zircon age of the Kuskanax batholith, southeastern British Columbia: Canadian Journal of Earth Sciences, v. 20, p. 1751–1756.

Paterson, I. A., and Harakal, J. E., 1974, Potassium-argon dating of blueschists from Pinchi Lake, central British Columbia: Canadian Journal of Earth Sciences, v. 11, p. 1007–1011.

Petö, P., and Armstrong, R. L., 1976, Strontium isotope study of the composite batholith between Princeton and Okanagan Lake: Canadian Journal of Earth Sciences, v. 13, p. 1577–1583.

Pigage, L. C., and Anderson, R. G., 1985, The Anvil Plutonic Suite, Faro, Yukon Territory: Canadian Journal of Earth Sciences, v. 22, p. 1204–1216.

Potter, C. J., 1983, Geology of the Bridge River complex, southern Shulaps Range, British Columbia; A record of Mesozoic convergent tectonics [Ph.D. thesis]: Seattle, University of Washington, 192 p.

Preto, V. A., Osatenko, M. J., McMillan, W. J., and Armstrong, R. L., 1979, Isotopic dates and strontium isotopic ratios for plutonic and volcanic rocks in the Quesnel Trough and Nicola Belt, south-central British Columbia: Canadian Journal of Earth Sciences, v. 16, p. 1658–1672.

Rahmani, R. A., and Lerbekmo, J. F., 1975, Heavy-mineral analysis of Upper Cretaceous and Paleocene sandstones in Alberta and adjacent areas of Saskatchewan: Geological Association of Canada Special Paper 13, p. 607–632.

Rea, D. K., and Dixon, J. M., 1983, Late Cretaceous and Paleogene tectonic evolution of the North Pacific Ocean: Earth and Planetary Science Letters, v. 65, p. 145–166.

Rea, D. K., and Vallier, T. L., 1983, Two Cretaceous volcanic episodes in the western Pacific Ocean: Geological Society of America Bulletin, v. 94, p. 1430–1437.

Read, P. B., and Okulitch, A. V., 1977, The Triassic unconformity of south-central British Columbia: Canadian Journal of Earth Sciences, v. 14, p. 606–638.

Riddihough, R. P., 1982, One hundred million years of plate tectonics in western Canada: Geoscience Canada, v. 9, p. 28–34.

Roddick, J. A., and Hutchinson, W. W., 1974, Setting of the Coast Plutonic Complex, British Columbia: Pacific Geology, v. 8, p. 91–108.

Ross, J. V., 1966, A note on histogram analysis of isotopic age data: Canadian Journal of Earth Sciences, v. 3, p. 259–262.

Saleeby, J. B., 1983, Accretionary tectonics of the North American Cordillera: Annual Review of Earth and Planetary Sciences, v. 15, p. 45–73.

Scholl, D. W., von Huene, R., Vallier, T. L., and Howell, D. G., 1980, Sedimentary masses and concepts about tectonic processes at underthrust ocean margins: Geology, v. 8, p. 564–568.

Souther, J. G., 1967, Acid volcanism and its relationship to the tectonic history of the Cordillera of British Columbia, Canada: Bulletin Volcanologique, v. 30, p. 161–176.

——, 1972, Telegraph Creek map-area British Columbia: Geological Survey of Canada Paper 71–44, 38 p.

——, 1977, Volcanism and tectonic environments in the Canadian Cordillera; A second look: Geological Association of Canada Special Paper 16, p. 3–24.

Steiger, R. H., and Jäger, E., 1977, Subcommission on geochronology; Convention on the use of decay constants in geo- and cosmochronology: Earth and Planetary Science Letters, v. 36, p. 359–362.

Stern, T. W., Bateman, P. C., Morgan, B. A., Newell, M. F., and Peck, D. L., 1981, Isotopic U-Pb ages of zircon from the granitoids of the central Sierra Nevada, California: U.S. Geological Survey Professional Paper 1185, 17 p.

Sutherland-Brown, A., 1968, Geology of the Queen Charlotte Islands, British Columbia: British Columbia Department of Mines and Petroleum Resources Bulletin 54, 226 p.

Sylvester, A. G., Miller, C. F., and Nelson, C. A., 1978, Monzonites of the White–Inyo Range, California, and their relation to the calc-alkalic Sierra Nevada batholith: Geological Society of America Bulletin, v. 89, p. 1677–1687.

Tater, J. M., 1965, Potassium-argon dates as provenance indicators for the Dunvegan Formation: Bulletin of Canadian Petroleum Geology, v. 13, p. 503–508.

Templeman-Kluit, D. J., 1976, The Yukon Crystalline Terrane; Enigma in the Canadian Cordillera: Geological Society of America Bulletin, v. 87, p. 1343–1357.

——, 1979a, Five occurrences of transported synorogenic clastic rocks in Yukon Territory: Geological Survey of Canada Paper 79–1A, p. 1–12.

——, 1979b, Transported cataclasite, ophiolite, and granodiorite in Yukon; Evidence of arc-continent collision: Geological Survey of Canada Paper 79–14, 27 p.

Tempelman-Kluit, D. J., and Wanless, R. K., 1975, Potassium-argon age determinations of metamorphic and plutonic rocks in the Yukon Crystalline Terrane: Canadian Journal of Earth Sciences, v. 12, p. 1895–1909.

——, 1980, Zircon ages for the Pelly Gneiss and Klotassin granodiorite in western Yukon: Canadian Journal of Earth Sciences, v. 17, p. 297–306.

Tennyson, M. E., and Cole, M. R., 1978, Tectonic significance of upper Mesozoic Methow-Pasayten sequence, northeastern Cascade Range, Washington and British Columbia, *in* Mesozoic paleogeography of the western United States: Society of Economic Paleontologists and Mineralogists, Pacific Section, p. 499–508.

Thorpe, R. S., Francis, P. W., Hammill, M., and Baker, M.C.W., 1982, The Andes, *in* Thorpe, R. W., ed., Andesites: New York, Wiley, p. 187–205.

Tipper, H. W., 1984a, The age of the Jurassic Rossland Group of southeastern British Columbia: Geological Survey of Canada Paper 84-1A, p. 631–632.

——, 1984b, The allochthonous Jurassic–Lower Cretaceous terranes of the Canadian Cordillera and their relation to correlative strata of the North American craton: Geological Association of Canada Special Paper 27, p. 113–120.

Tipper, H. W., and Richards, T. A., 1976, Jurassic stratigraphy and history of north-central British Columbia: Geological Survey of Canada Bulletin 270, 73 p.

Travers, W. B., 1978, Overturned Nicola and Ashcroft strata and their relation to the Cache Creek Group, southwestern Intermontane Belt, British Columbia: Canadian Journal of Earth Sciences, v. 15, p. 99–116.

von Huene, R., 1984, Tectonic processes along the front of modern convergent margins; Research of the past decade: Annual Review of Earth Planetary Sciences, v. 12, p. 359–381.

Walker, N. W., 1983, Pre-Tertiary tectonic evolution of northeastern Oregon and west-central Idaho; Constraints based on U/Pb ages of zircons: Geological

Society of America Abstracts with Programs, v. 15, p. 371.

Wallace, W. K., and Engebretson, D. C., 1984, Relationships between plate motions and Late Cretaceous to Paleogene magmatism in southwestern Alaska: Tectonics, v. 3, p. 295–315.

Wanless, R. K., compiler, 1970, Isotopic age map of Canada: Geological Survey of Canada Map 1256A, scale 1:5,000,000.

Watts, A. B., Bodine, J. H., and Ribe, N. M., 1980, Observations of flexure and the geological evolution of the Pacific Ocean basin: Nature, v. 283, p. 532–537.

Wheeler, J. O., and Gabrielse, H., coordinators, 1972, The Cordilleran structural province, *in* Price, R. A., and Douglas, R.J.W., eds., Variations in tectonic styles in Canada: Geological Association of Canada Special Paper 11, p. 1–81.

Wheeler, J. O., Charlesworth, H.A.K., Monger, J.W.H., Muller, J. E., Price, R. A., Reesor, J. E., Roddick, J. A., and Simony, P. S., 1974, Western Canada, *in* Spencer, A. M., ed., Mesozoic Cenozoic orogenic belts: Geological Society of London Special Publication 4, p. 591–623.

White, W. H., 1959, Cordilleran tectonics in British Columbia: American Association of Petroleum Geologists Bulletin, v. 43, p. 60–100.

White, W. H., Erickson, G. P., Northcote, K. E., Dirom, G. E., and Harakal, J. E., 1967, Isotopic dating of the Guichon batholith, B.C.: Canadian Journal of Earth Sciences, v. 4, p. 677–690.

White, W. H., Sinclair, A. J., Harakal, J. E., Dawson, K. M., 1970, Potassium-argon ages of Topley intrusions near Endako, British Columbia: Canadian Journal of Earth Sciences, v. 7, p. 1172–1178.

Wilson, F. H., Dadisman, S. V., and Herzon, P. L., 1979, Map showing radiometric ages of rocks in southeastern Alaska: U.S. Geological Survey Open-File Report 79-594, 33 p.

Wood, D. H., and Armstrong, R. L., 1982, Geology, chemistry, and geochronometry of the Cretaceous South Fork Volcanics, Yukon Territory: Geological Survey of Canada Paper 82-1A, p. 309–316.

Woodsworth, G. J., 1979, Metamorphism, deformation, and plutonism in the Mount Raleigh pendant, Coast Mountains, British Columbia: Geological Survey of Canada Bulletin 295, 58 p.

Woodsworth, G. J., Loveridge, W. D., Parrish, R. R., and Stephens, R. W., 1983, U-Pb dates from the Central Gneiss Complex and Ecstall pluton, Prince Rupert map area, British Columbia: Canadian Journal of Earth Sciences, v. 20, p. 1475–1483.

Woodsworth, G. J., van der Heyden, P., and Hill, M. L., 1985, Regional tectonic styles in the Coast Mountains between 52 and 55°N, B.C.: Geological Society of America Abstracts with Programs, v. 17, p. 418–419.

Young, I. F., 1981, Structure of the western margin of the Queen Carlotte Basin, British Columbia [M.Sc. thesis]: Vancouver, University of British Columbia, 380 p.

MANUSCRIPT ACCEPTED BY THE SOCIETY JUNE 1, 1987

Printed in U.S.A.

Geological Society of America
Special Paper 218
1988

A possible Jurassic-Cretaceous transform system in the Alps and the Carpathians

Rudolf Trümpy
Allmendboden 19, Küsnacht 8700, Switzerland

ABSTRACT

The west-east-directed part of the Penninic ocean basins, with their associated intraoceanic rises of the Alps and the western Carpathians, is considered as a complex Jurassic–Early Cretaceous transform zone that links the northern end of the Piemont–Ligurian ocean with the northern end of the Vardar ocean. The Maghreb transform belt may furnish an analogous example. Several traverses through the Alps are briefly described. Correlation between the Penninic zones of the Western and Eastern Alps is still controversial.

GENERAL FRAME

Since nappe structures and intense crustal shortening were recognized in the Alps at the beginning of this century, it has become clear that they implied relative movements on the order of hundreds, if not thousands, of kilometers between Eurasia (with its Iberian promontory or subcontinent) and Africa (with its Apulian promontory or subcontinent). But the amount, direction, and timing of these movements remained unclear, in spite of inspired guesswork by Argand (1924), Staub (1928), and Carey (1958).

With the advent of plate tectonics, it became possible to state certain boundary conditions, at least as far the relative positions of the three major plates—Africa, Europe, and North America—are concerned (Smith, 1976; Dewey and others, 1973; Dercourt and others, 1985; Le Pichon, 1982). This widely accepted scheme may undergo minor modifications as the ages of some of the more ancient magnetic anomalies become more precisely known, but the overall picture is fairly well established. The path of the two subordinate plates (European-related Iberia and African-related Apulia) is less certain, and the path of continental fragments, such as Corsica-Sardinia or Alborán, remains in dispute. More numerous and more critical paleomagnetic data should narrow our choice. Within the deformed belts, there is still considerable freedom in playing the game of palinspastic reconstructions, as long as the boundary marks—the relative location of the plates at each moment of geologic time—are respected.

Elsewhere in this volume, Le Pichon presents an outline of Mediterranean geometry and kinematics based on analysis of major plate movements and field geology. This chapter deals with one particular aspect of this synthesis, i.e., the existence of a Jurassic-Cretaceous transform zone north of the Apulian prong that links the northern end of the Piemont ocean with the northern end of the Vardar ocean in the wider sense. I believe the idea was first put forward by Biju-Duval and others (1977), then Bosellini (1981), and finally in a more explicit way by Lemoine (1984, 1985), Debelmas and others (1985), and Weissert and Bernoulli (1985). I agree fully with the general conclusion of these authors. While writing this paper, I received a preprint by J. Debelmas and M. Sandulescu (meanwhile published, 1987) and noted with pleasure the remarkable similarity of our independently derived conclusions. The trace of the northern transform zone, proposed here, is quite different from the one drawn by Dercourt and others (1985).

The western Mediterranean area belongs to the southern branch of the Variscan orogen. Late Variscan (late Carboniferous to mid-Permian) rifting and transcurrent faulting may be ascribed to a dextral shear between the closing Appalachian and Ural systems (Arthaud and Matte, 1975). This can also explain the local occurrence of compressional movements (e.g., in the Alès region of southern France: Gras, 1972).

Oceanic influences of the Triassic Tethys reach westward to Turkey and possibly into the Bükk hills in Hungary. Farther west, only narrow, deeper water channels (Hallstatt, Lagonegro) penetrate the extensive shallow-water platform of the western Mediterranean area. Shallow Triassic seas extended to Sicily and Tunisia, and to the Alborán platform, which lay at that time north of Tunisia.

After local, abortive rifting in mid-Triassic time, the carbonate platform was reestablished during the Carnian. Its definite

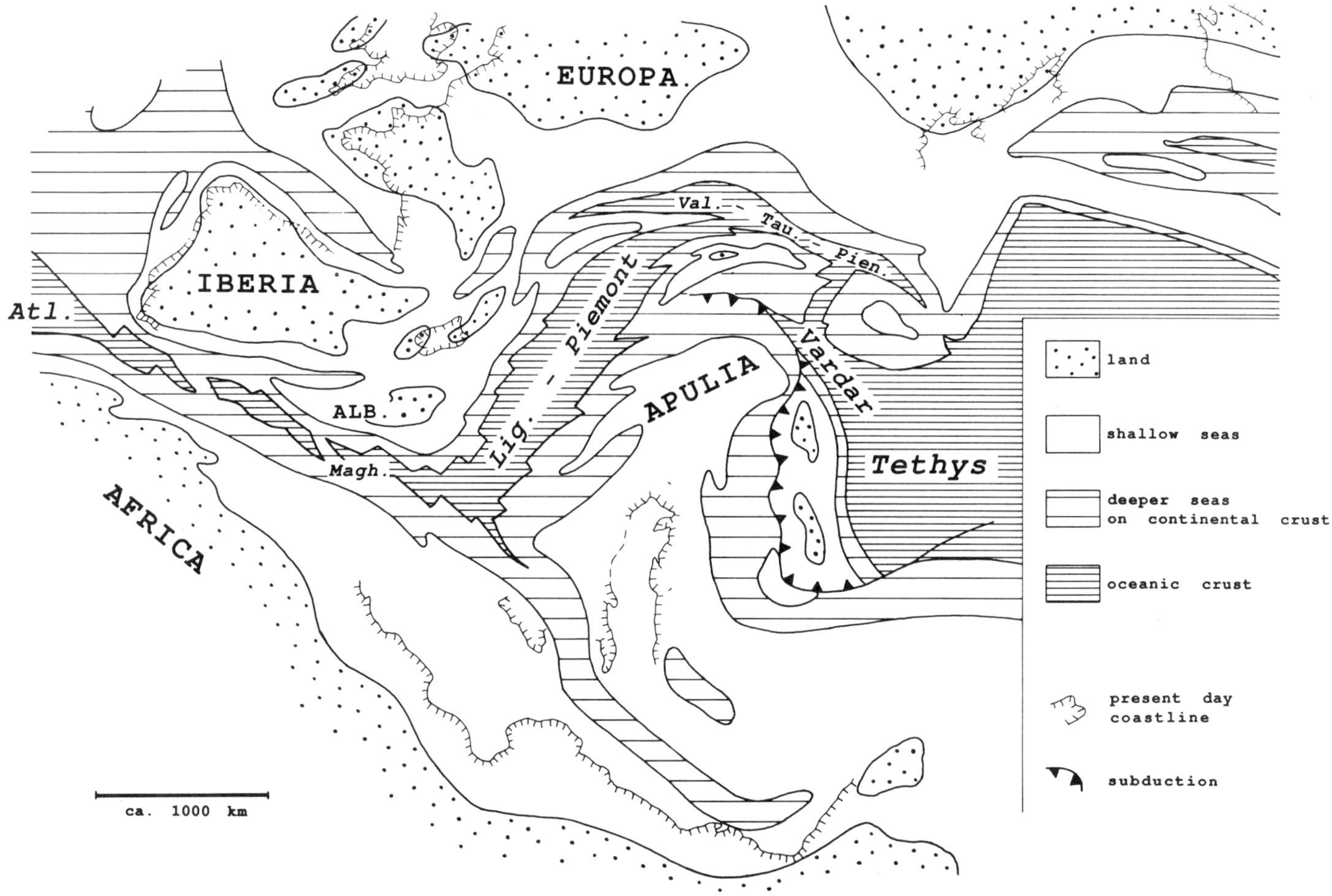

Figure 1. General situation of the future Mediterranean area at the Jurassic-Cretaceous boundary (after Dercourt and others, 1985, strongly modified for the Alpine-Carpathian area). Abbreviations: Atl. = Atlantic; Magh. = Maghreb transform zone; Lig. = Ligurian; Val.-Tau.-Pien. = Valais-Tauern-Pieniny transform; ALB. = Alboran block.

breakup started in the Norian but became accentuated during Hettangian and especially Sinemurian time. Early Jurassic and early Middle Jurassic normal faults are located along the margin of extra-Alpine Europe, on both sides of the Briançonnais rise, and along the Apulian margin, as well as along the Alborán platform (for a review, see Lemoine and Trümpy, 1986).

The oceanic opening of the central Atlantic, around 165 Ma or slightly earlier, produced a sinistral translation, of the order of 1,500 km, of Africa with respect to Europe and Iberia. This great Maghreb transform zone is briefly discussed later in this chapter. At its eastern end, it ran into the Ligurian-Piemont ocean or rhombochasm, which opened along preexisting rift lineaments at the same time as the first Atlantic Ocean did.

As the Piemont-Ligurian ocean can hardly correspond to the width of the contemporaneous Atlantic, there must be another transform zone joining the Piemont-Ligurian ocean with the eastern Vardar ocean, which included remnants of the Triassic–Early Jurassic eastern Tethys. A central Mediterranean Jurassic-Cretaceous transform can be sought either south or north of the Apulian block. The southern Calabrian path apparently did not exist during Jurassic time, when Apulia and Hyblean Sicily (presumably part of Africa) belonged to the same platform. Dercourt and others (1985) assumed an oceanic separation between Africa and Apulia only for mid-Cretaceous and later time. If this is true, a northern transform zone must have been active, at least during the Middle and Late Jurassic. The probability of such a northern, or Valais-Tauern-Pieniny, connection is discussed below.

For the general setting of the oceanic phase in the future western Mediterranean, readers are referred to Le Pichon (this volume). Figure 1 summarizes my concept of the location of the northern transform zone.

The Maghreb transform zone, where the dominance of lateral movements is beyond doubt, may serve as a model for paleotectonic belts of this kind.

REMARKS ON THE MAGHREB TRANSFORM ZONE

Review papers on the Alpine chains in the westernmost Mediterranean have been published by Egeler and Simon (1969), and, more recently, by Fontboté and Estevez (1980) for the Spanish part. Didon and others (1973) have written about the area around the straits of Gibraltar, and Wildi (1983), especially, has published a review paper for the African part.

From north to south, the following major zones can be distinguished:

1. Autochthonous and parautochthonous (Prebetic) cover of the Iberian Meseta.

2. Subbetic cover nappes, with continental/evaporitic/restricted marine Triassic overlain by Jurassic, Cretaceous, and Paleogene formations. The Jurassic-Cretaceous of the external (northern) Subbetic shows Prebetic affinities; the median part represents a pelagic furrow, with mid-Jurassic to Lower Cretaceous pillow basalts; the internal Subbetic (called Penibetic in the western part of the chain) corresponds again to a platform sequence.

3. Gneiss and schist nappes of the Sierra Nevada–Sierra de los Filabres, affected by Alpine metamorphism (including a phase of blueschist metamorphism), with scant Permian-Triassic(?) cover of Andalusian rather than Alpine type. The Sierra Nevada could be part of the exotic Internal Zones or a window pertaining to the Iberian continental margin (W. Wildi, personal communication, 1984).

4. Internal zones (in the narrow sense): subcontinental peridotites, pre-Ordovician basement, Paleozoic to Paleogene sediments; Alpine metamorphism increasing from higher to lower nappes. In the western parts, three units are distinguished: the Alpujarrides-Sebtides are overlain by the Malagides-Ghomarides, and the Mesozoic and Paleogene formations of the Rondaides-Tetuanides are derived from the cover of the Alpujarrides. In Algeria, the distinctions are less pronounced.

5. Maghrebine ("Mauretanian" and "Massylian") Flysch nappes; especially Lower Cretaceous and Eocene-Oligocene flysch, with few remnants of older rocks.

6. Allochthonous Rif and Tell nappes, with pelitic to arenaceous Middle Jurassic to Upper Cretaceous formations.

7. Atlas system: parautochthonous and autochthonous cover of the African continent.

Zones 1 and 2 (eventually also zone 3) belong to the Iberian zones, 6 and 7 to the African continental margin. The Jurassic-Cretaceous transforming plate boundary does not run through the Strait of Gibraltar (which is only affected by a quite subordinate, dextral post-Miocene fault), but instead is farther north; the southernmost part of the Iberian peninsula, the Campo de Gibraltar, was part of Mesozoic Africa. The Internal Zones (zone 4, eventually also zone 3) are exotic terranes, whose original position lay far to the east, north of present-day Tunisia (Andrieux and others, 1971; Tapponier, 1977). They differ from all surrounding areas by their weak Variscan overprint and their carbonatic ("alpine") Triassic formations. The so-called arc of Gibraltar corresponds to the western front of this Alborán terrane. As to the Flysch nappes (zone 5), their origin apparently lay between the Alborán terranes and the African margin; they are very probably connected with the Jurassic-Cretaceous transform boundary and have no typical analogs in Spain, outside the Campo de Gibraltar.

As the Maghreb transform zone provided the link between the Atlantic and the Alpine (Ligurian) oceans, it should be expected that at least parts of the zone had oceanic or paraoceanic character from mid-Jurassic to mid-Cretaceous time. There are indications (but no formal proof) of this premise, especially in the substratum of the more northerly ("Mauretanian") flyschs and along the southern margin of the Internal Zones. Radiolarian cherts of late Middle Jurassic and early Late Jurassic age are known from several localities (see, for example, Bouillin, 1977); they are underlain by pelagic *Bositra* shales and are overlain by micritic, Maiolica-type limestones. Flows and sills of basaltic volcanics are also present, but so far no typical ophiolite assemblage has been found, in spite of diligent search for such rocks. Furthermore, some outcrops of basaltic rocks have recently been dated as late Cretaceous (Gübeli and others, 1984; Ciszak and others, 1985) and cannot be considered as ophiolites. The post-Liassic sequence of the Internal Zones is also of pelagic, but not typically oceanic, character. It may be inferred that small pull-apart basins, with oceanic crust or anomalous continental crust, existed along the Maghreb zone, but there is little indication for major oceans. The whole situation is reminiscent of Carpathian rather than Alpine conditions.

The Lower Cretaceous (mainly Barremian and Aptian, more locally Valanginian, Hauterivian, and Albian) flyschs are another characteristic of the Maghreb scar. In spite of their relatively modest thickness (as much as 800 m), they contain a considerable volume of mature quartz sand. Their detritus is derived from northerly sources (Gübeli and others, 1984), but it is difficult to imagine transport from the Iberian landmass, across the intervening Subbetic trough. As with other flyschs in a transform rather than compressive setting, it is tempting to envisage crustal blocks caught between moving plates as source areas. This hypothesis, however, fails to account for the maturity of the components, unless the quartz is derived from older (Paleozoic?) sediments. The younger Paleogene to Lower Miocene flyschs, which occur mainly, but not exclusively, on the African side of the Internal Zones, can be interpreted as "normal" flyschs in a compressional context, except for the still-enigmatic Numidian Flysch.

The westward translation of the Alborán block, starting in Eocene and reaching its climax in Early to Middle Miocene time, was accompanied by a severe laceration of its continental crust. Reuber and others (1982) have shown that the roof contact of the Beni Bousera (or Beni Bouchera) peridotites of Morocco represented a ductile shear zone, involving extremely high geothermal gradients between peridotite and gneiss. The continental crust above this "tectonic Moho" is only a few kilometers thick. It is

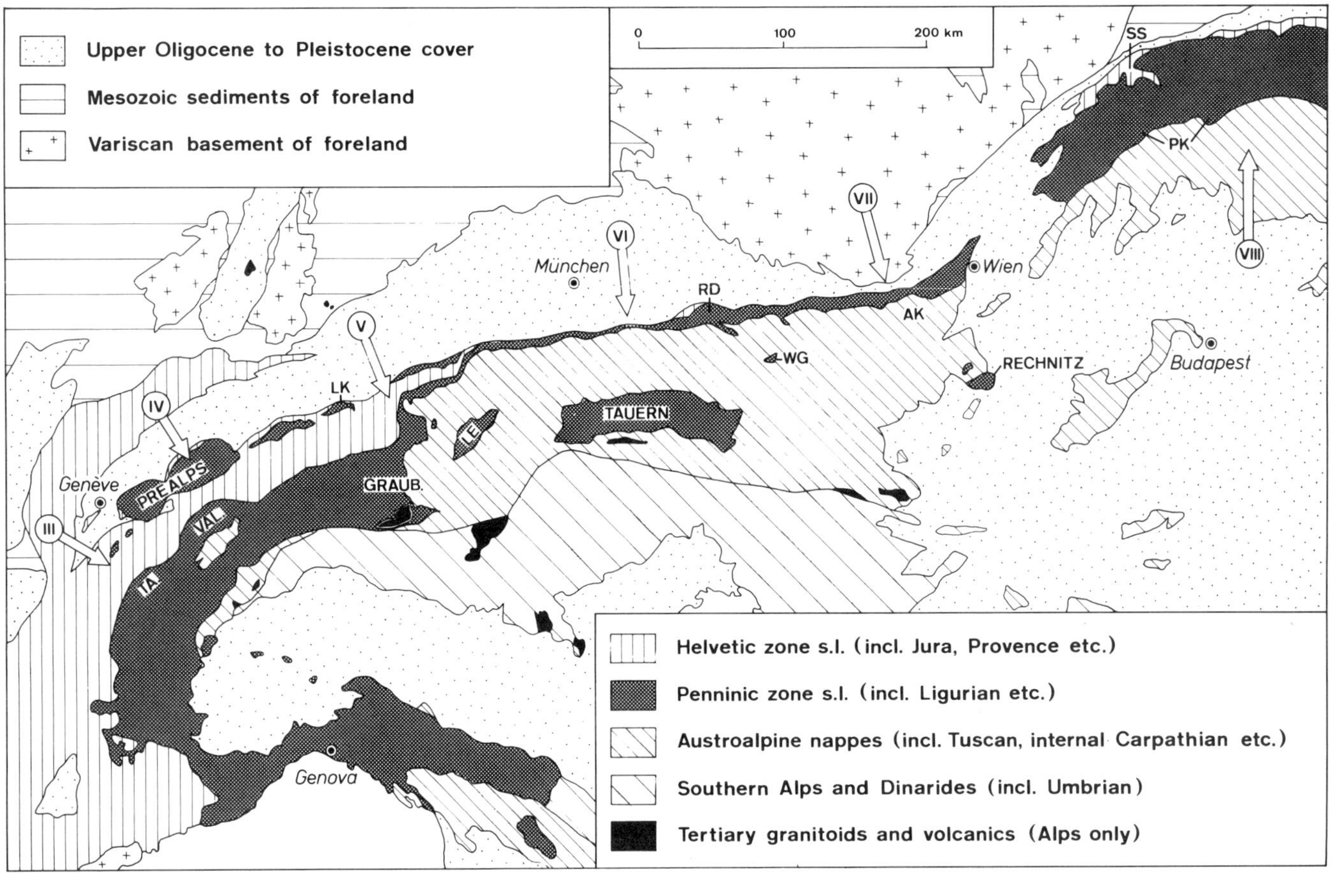

Figure 2. Sketch map of the Alps and surrounding areas. Abbreviations: TA. = Tarentaise; VAL. = Valais; LK = Klippen around Lake Lucerne; GRAUB. = Graubünden (Grisons); LE = Lower Engadine window; RD = Rheno-Danubic Flysch; WG = Windischgarsten window; AK = Austrian klippen; PK = Pieniny klippen; SS = Subsilesian units. Numbers III through VIII refer to the six traverses discussed in the text.

overlain by continental Permian redbeds, shallow-water Triassic carbonates, and thin pelagic (but not oceanic) Jurassic and Cretaceous formations. The character of this Permian and Mesozoic cover implies that the attenuation of the continental crust cannot be ascribed to Mesozoic listric faulting, but apparently occurred during the translation of the Alborán "plate"—or, rather, the Alborán flake. It is not yet certain whether the peridotite/gneiss contact represents a thrust or a lag (flat-lying normal fault). The same interpretation is valid for the contact of the Ronda peridotites, on the Spanish side, with the overlying gneisses (Felder, 1980). There, the cooling of the "hot contact" is dated at 22 to 19 Ma (Priem and others, 1979). The case of the Ronda and Beni Bousera peridotites is of special importance, as it underlines the prevalence of tectonic uncoupling between mantle and crust.

Thus, the Maghreb transform zone illustrates three characteristics of such belts: (1) development of deep-sea basins during the initial stage of translation, with extrusive and intrusive basalts ("ophiolitoids"): true ophiolites may have formed, but were not preserved at the surface; (2) deposition of flysch sediments during the final stage of translation, governed by a transform rather than compressional regime; and (3) anomalous thrust directions and tectonic dilaceration of the crust during the compressional stage.

THE VALAIS BELT IN SAVOY AND THE UPPER AOSTA VALLEY

Later in this chapter, the presumed northern transform zone (or "Swiss connection") is traced from west to east. Figure 2 provides a general overview of the locations. The units that can be assigned to the Valais belt in Savoy and in the Valais-Prealps traverse, where the classical west-Alpine paleogeographic pattern can still be applied, are discussed. In Graubünden, correlations are partly controversial, and still farther east, in the Tauern traverse (central part of the eastern Alps) as well as along the eastern edge of the chain, the facies belts have little in common with their west-Alpine analogs. Although the western (Slovakian-Polish) Carpathians are noted herein, a full discussion of the Carpathian part may be found in Debelmas and Sandulescu (1987).

In the major part of the French Alps, only one oceanic trench, the Piemont (-Ligurian) belt, is present, at least at the surface. The complex Briançonnais rise (including its westerly, Subbriançonnais subbelt) comes directly into tectonic contact with the basinal Jurassic and Cretaceous formations of the External (Dauphiné) Zones. In this part of the Alps, the Briançonnais belt is considered as a distal element of the European continental margin. Only to the north does it begin to be separated from the main part of the European margin by a complex basin, the Valais belt.

The westernmost part of the Alps in which the Valais belt is exposed lies in Savoy, in the Tarentaise Valley (headwaters of the Isère River). Pioneer work in this area was made by Barbier (1948), followed by Antoine (1971) and co-workers. Between the most internal part of the Dauphiné realm and the Subbriançonnais units, Barbier distinguished a series of digitations within a major nappe (nappe des Brèches de Tarentaise). The succession of the Tarentaise nappe, in the widest sense, comprises basement gneisses, Permian-Carboniferous continental detritals, thin shallow-water Triassic and Lower Jurassic formations, a Middle Jurassic(?) flysch-like formation including coarse conglomerates, and a transgressive (Lower and ?) Upper Cretaceous preflysch to flysch. Eocene flysch is present only in a very external and southerly digitation (Niélard), which might also be grouped with the more external ultradauphinois belt.

The abundance of scarp breccias and turbidites, as well as rapid lateral facies changes, attest to the considerable crustal mobility of the belt of the future Tarentaise nappe during mid-Jurassic to Late Cretaceous time. The more external (lower, westerly) digitations of the nappe were influenced by a complex Late Cretaceous to Eocene rise, the Cordillère Tarine of Barbier. The paleotectonic nature of this "marginal high" (Homewood, 1974) constitutes an important and still unresolved problem of Alpine geology. The higher digitations show more basinal characteristics, especially in late Cretaceous time.

The same units continue to the northeast, into the upper Aosta Valley of Italy. The most interesting difference concerns the presence of mafic (and, more rarely, ultramafic) rocks within the (Lower to "middle") Cretaceous pre-flysch sequence of the internal (Roignais-Versoyen) digitation. The tholeiitic metabasalts form sills and laccolite-type edifices within the Cretaceous formations (Loubat, 1968; Lasserre and Laverne, 1976). They can hardly be considered classical ophiolites; nevertheless, they attest to intensive impregnation of semi-consolidated sediments by tholeitic magmas and imply thinning of the continental crust even under this very westerly segment of the Valais realm.

South of Aime, the ancient capital of the Roman province of Alpes Graiae et Poeninae, the Valais belt is no longer visible at the surface. It is unresolved whether the southwestern termination of the exposed Valais belt is primary, or whether its prolongation was cut out by later (sinistral strike-slip?) movements. At any rate, a direct facies transition between the most internal part of the Dauphiné realm and the most external part of the Subbriançonnais belt, as drawn in many classical syntheses of the French Alps (e.g., Debelmas, 1983), is not exposed anywhere.

A connection between the Valais trough and the Cretaceous Vocontian trough of the external French Alps (Laubscher, 1971) is improbable, because of the quite different facies development and the presence of the Cordillère Tarine. The Vocontian facies belt does indeed continue into the Swiss Alps, but not into the Valais: the equivalent of its hemipelagic Lower Cretaceous formations is found in the "Néocomien à Céphalopodes" of the Ultrahelvetic nappes (see, for example, Guillaume, 1957). Likewise, the suggested link between the Briançonnais and the extra-Alpine Provence platforms (Laubscher, 1971; Ricou, 1980) is quite unconvincing. The only common point of these two highs is the presence of limestones (mainly pelagic in the Briançonnais, exclusively of shallow-water origin in Provence). Otherwise, their paleogeographic evolution is entirely different (e.g., in mid-Cretaceous time: pelagic sediments in one, bauxites in the other).

The situation of this westernmost part of the Valais belt may be summarized as follows: (1) the overall width of the belt was fairly modest (100 km?). (2) it comprised an external zone, with a Late Cretaceous–Early Paleogene ridge (the Cordillère Tarine) and an adjacent basin to the southeast; (3) subsidence of the basin began only in Cretaceous time: there is no indication of oceanic sediments, save for atypical black pelites; (4) mafic volcanics and subordinate ultramafic lenses were emplaced in the Aosta Valley region; and (5) the existence of a prolongation of the Valais belt south of Savoy can be neither proved nor disproved.

THE VALAIS BELT IN VALAIS AND THE SWISS PREALPS

In western Switzerland, the same paleogeographic situation as in the northern part of the French Alps prevails. The Briançonnais belt, in the wider sense, is well expressed and has a breadth of at least 50 km, more probably 100 km. Compared to the Piemont-Ligurian belt, which is still very wide and, in part at least, distinctly oceanic (Zermatt and Aosta Valley ophiolites), the Valais belt represents a narrower and "less oceanic" basin.

Nappes issued from the Valais domain are found in two positions: as isoclinal slabs east and south of the basement massifs and their cover (Montblanc to the west, Aar and Gotthard to the east), and as décollement sheets in the Prealps, especially between the Rhone Valley and Lake Thun (see Fig. 3).

In the southwestern part of Valais, the following units lie in proximal tectonic position, i.e., behind the Montblanc massif: (a) basal slices, traces of Triassic cornieule and slate with schists of uncertain age (see Fricker, 1960); (b) Ferret zone, thin Triassic sediments overlain by a distal flysch of uncertain age; (c) Versoyen-Roignais zone, recumbent folds composed of three Cre-

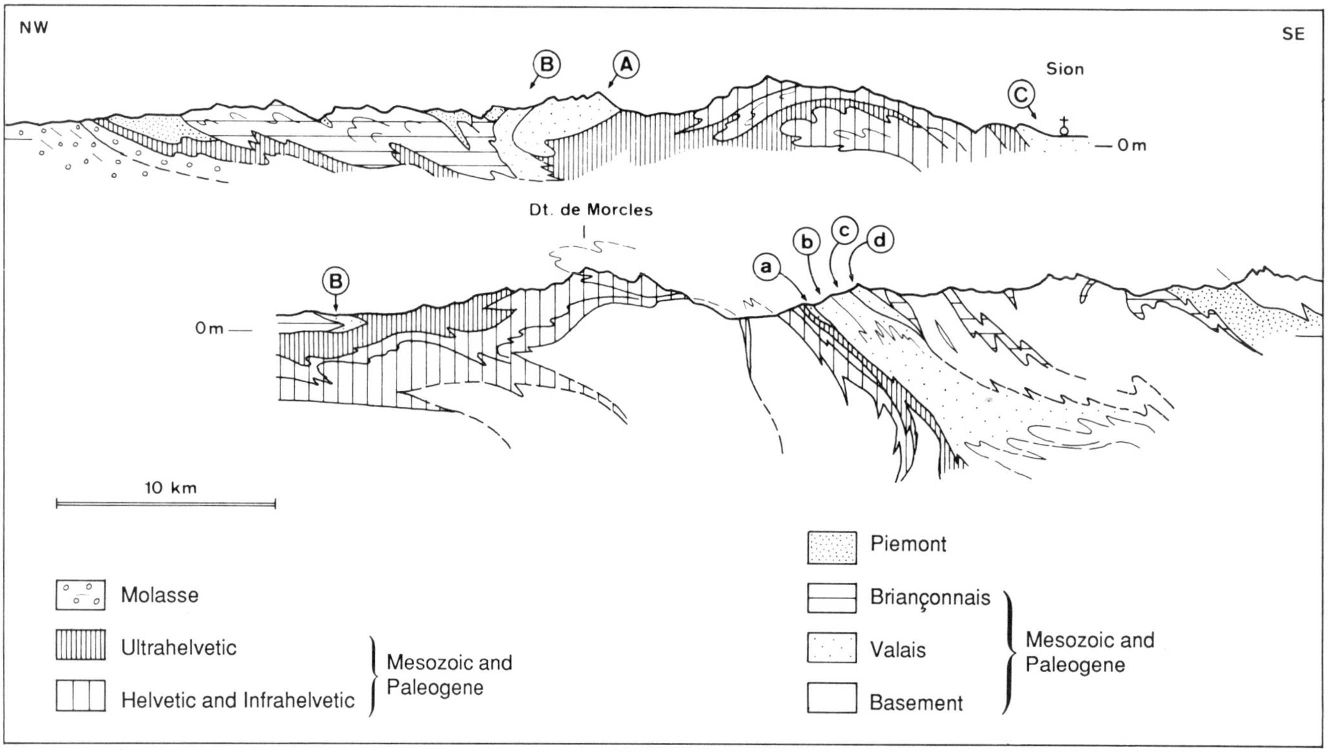

Figure 3. Schematic cross sections through the frontal Penninic and Prealpine nappes of western Switzerland. a-d and A-B: units derived from the Valais belt; see text.

taceous formations: the Aroley limestones and conglomerates, the Marmontains quartz sandstones and dark shales, and the St. Christophe Flysch; (d) Pierre Avoi zone: a Cretaceous(?) formation of conglomerates and black slates—the latter contain metabasaltic layers (ovardites). These rocks envelop slices and/or olistoliths of continental Carboniferous-Permian and shallow-water Triassic– Lower Jurassic carbonate rocks (see Burri and Jemelin, 1983).

In the distal Prealpine nappes, two units can be assigned to the Valais belt: (A) the Niesen nappe: basement slice, thin Triassic–Lower Jurassic carbonates, flysch-like Middle Jurassic, and a very thick Upper Cretaceous (essentially Maestrichtian) flysch, and (B) the Zone Submédiane: its significance has only recently been underlined by Weidmann and others (1976). It is interpreted as a tectonic mélange (with a matrix of shale, of gypsum or of cornieule) at the base of the Médianes nappe. Among the components, pelagic deep-water (oceanic?) sediments, such as cherty limestones with abundant radiolaria and maiolica-type limestones, as well as basaltic blocks are remarkable.

The correlation of units a through d and A through B is still being debated. Homewood and others (1984) have shown convincingly that the Niesen nappe is related to the Cordillère Tarine, and that its original position must have lain just to the southeast of the Ultrahelvetic domain. Plausible arrangements could be:

A - a - b - c - d - B   or   A - a - b - c - B - d.

The Submédianes mélange may have picked up different pieces from the Valais and adjoining frontal Subbriançonnais belt.

In the traverse under discussion, only traces of possible ophiolitic rocks from the Valais belt are preserved, whereas Piemontese ophiolites are abundant. Ophiolites do occur in a tectonically "low" position, i.e., below the basement rocks of the Grand St-Bernard and Monte Rosa nappes in the eastern Penninic Alps, especially in the Antrona Valley, below the east face of Monte Rosa. Most authors (e.g., Müller, 1983) now consider the Antrona rocks to be of Piemont origin, having been brought into their present position by a complex structural evolution. I still have some misgivings about this interpretation, especially since the Antrona ophiolites have a geochemical signature differing from that of the Piemontese ophiolites of the Zermatt region (Beccaluva and others, 1984). At any rate, the occurrence of Versoyen-type metabasalts in the Rhone Valley, below the front of the Grand St-Bernard nappe, is uncontested (Burri, 1979).

In the Prealps–western Valais traverse, the main characteristics of the Valais zone are: The belt is broader (about 150 km?) and more complex than in Savoy, although it is still subordinate with respect to the Piemont belt of the same traverse. It comprises several basins, with different development. Also, pelagic deep-sea sediments appear in Middle or Late Jurassic time, and the main infilling is Late Cretaceous. Finally, basaltic volcanics occur in the Pierre Avoi zone and in the Zone Submédiane; the tectonic position (Valais or Piemont?) of true ophiolites in the eastern Pennine Alps is uncertain.

The term of "géosynclinal valaisan" was proposed by Emile Haug in 1925, and then almost forgotten. I reintroduced the concept in 1955, as a synonym for "north-Pennic zone" (or belt, or geosyncline). In both Haug's view and my own, the term "Valais" covers the entire facies realm between the Ultrahelvetic and Subbriançonnais-Briançonnais belts. Restriction of the term to only those parts of the facies belt that contain the Aroley–Marmontains–St-Christophe trilogy (Bolli and others, 1980; Homewood and others, 1984) is unwarranted.

PENNINIC BELT IN GRAUBÜNDEN

The Gotthard traverse, in central Switzerland, furnishes few indications as to the paleogeographic pattern of the Penninic zone, as the Mesozoic sediments in the transverse uplift of the Toce and Ticino Valleys are strongly affected by Tertiary metamorphism and as only a few Penninic outliers (the original klippen) are preserved around Lake Lucerne. In eastern Switzerland and neighboring parts of Italy, Austria, and Lichtenstein, the Penninic nappes of the Western Alps plunge eastward under their Austroalpine cover. Penninic nappes are further exposed in the Lower Engadine window, and, as décollement sheets, in the Rheno-Danubic Flysch Zone near the northern margin of the Alps. For more detailed descriptions, refer to my review in Trümpy and Haccard (1969).

The following major paleogeographic units occur from north to south, according to one possible tectonic interpretation (see Figs. 4, 5).

1. Units pertaining to the northernmost part of the Penninic realm or to the Penninic-Ultrahelvetic resp. Infrapenninic margin: Jurassic Terri schists and associated formations of southeastern Graubünden; Upper Cretaceous to Lower Eocene Sardona Flysch of Glarus; Eocene boulder flysch with olistoliths of various rocks, including diabase, of the Feuerstätt nappe in Vorarlberg and Bavaria.

2. Great masses of Jurassic-Cretaceous Bündnerschiefer, with intercalations of basaltic sills and flows as well as fewer ultramafic lenses, in western Graubünden; several kilometers of Bünden (Prättigau schists); Paleocene and Lower Eocene flysch of Prättigau to Oberhalbstein. Bündnerschiefer (with metabasalts) and flysch of the Lower Engadine window occupy an analogous position. There, these sediments are overthrust by the Ramosch slice containing typical ophiolites (see Trümpy, 1972).

All these units noted in point 2 derive from the suture between the Adula and Tambo nappe, which also encloses the ophiolites of Chiavenna (Schmutz, 1976). The original width of the sedimentary domain was at least 150 km. Detrital influx is essentially from the north. A Jurassic high, the Adula swell, separates the units of 1 from those of 2 (see Probst, 1980).

3. Cretaceous Rheno-Danubic Flysch of Vorarlberg; substratum unknown, possibly oceanic or paraoceanic. Its original position must certainly be sought south of the Prättigau schists (Allemann, 1957). Detrital influx is essentially from the south and lengthwise. Hesse (1973) has shown that the Rheno-Danubic Flysch was laid down just to the north of the Falknis-Tasna high; this can be taken as one of the few well-established points in the reconstruction detailed in this chapter.

4. Falknis and Sulzfluh nappes of northern Graubünden, Tasna nappe in the western part of the Engadine window. Granitic substratum; thin Triassic dolomites. In the Late Jurassic and Early Cretaceous, the Falknis (and part of Tasna) nappes correspond to a north-facing scarp, the Sulzfluh nappe to a platform. As the succession in the easternmost outliers of the Préalpes Médianes nappe, northeast of Lake Lucerne, is practically identical to that of the Sulzfluh nappe, the Sulzfluh platform and a presumed platform south of the Tasna nappe can be considered as the easternmost, much reduced equivalents of the Briançonnais platform. Paleogene formations are present in units 1, 2, and 4; in eastern Austria they are also found in unit 3.

5. Schams nappes of southwestern Graubünden (Streiff et al., 1976): thick Middle Triassic carbonates, Jurassic partly of platform, partly of scarp (breccia and turbidite) facies, Cretaceous pelagic with detrital influence, presence of Tertiary formations possible but not proven.

6. Avers Bündnerschiefer (schistes lustrés) with metabasalts; to the south, Malenco ophiolites, with a large peridotite body.

7. Margna nappe: basement gneisses, Triassic carbonate rocks, thin Mesozoic formation including radiolarian cherts, but no ophiolites.

8. Arosa-Platta nappe: ophiolites, oceanic Middle Jurassic to Lower Cretaceous sediments; youngest formations of Cenomanian, eventually Lower Turonian age. Shards of sialic basement rocks and shallow-water Triassic carbonates are scattered through these units but may be of foreign origin (mélange components?); see Weissert and Bernoulli, 1985.

At the moment, this arrangement (hypothesis A) seems to be at least plausible. Units 1 through 3 would correspond to the Valais belt, units 4 and 5 to the Briançonnais platform, unit 6 to an external part and unit 8 to an internal part of the Piemont ocean, with the Margna nappe (unit 7) as an intervening intraoceanic mini-continent. I assume a position of the Schams south of Sulzfluh for two reasons: the sedimentologic link between the Rheno-Danubic Flysch and the Falknis-Tasna nappe, and the occurrence of Schams-type formations (Middle Jurassic sandstones) in a position above the Tasna nappe of the Lower Engadine window (see Oberhauser, 1980).

Other solutions are also possible. The still-undecided dilemma on the origin of the Schams nappes has a great influence on the paleogeographic reconstructions. It is by no means certain that Schams and Falknis-Sulzfluh come from the same high (Gruner, 1981). In that case, one might play with successions as:

1 - 2 - 3 - 4 - 6 - 5 = 7 - 8 (hypothesis B), or even
1 - 2 - 5 - 3 = 6 - 4 - 7 - 8 (hypothesis C)

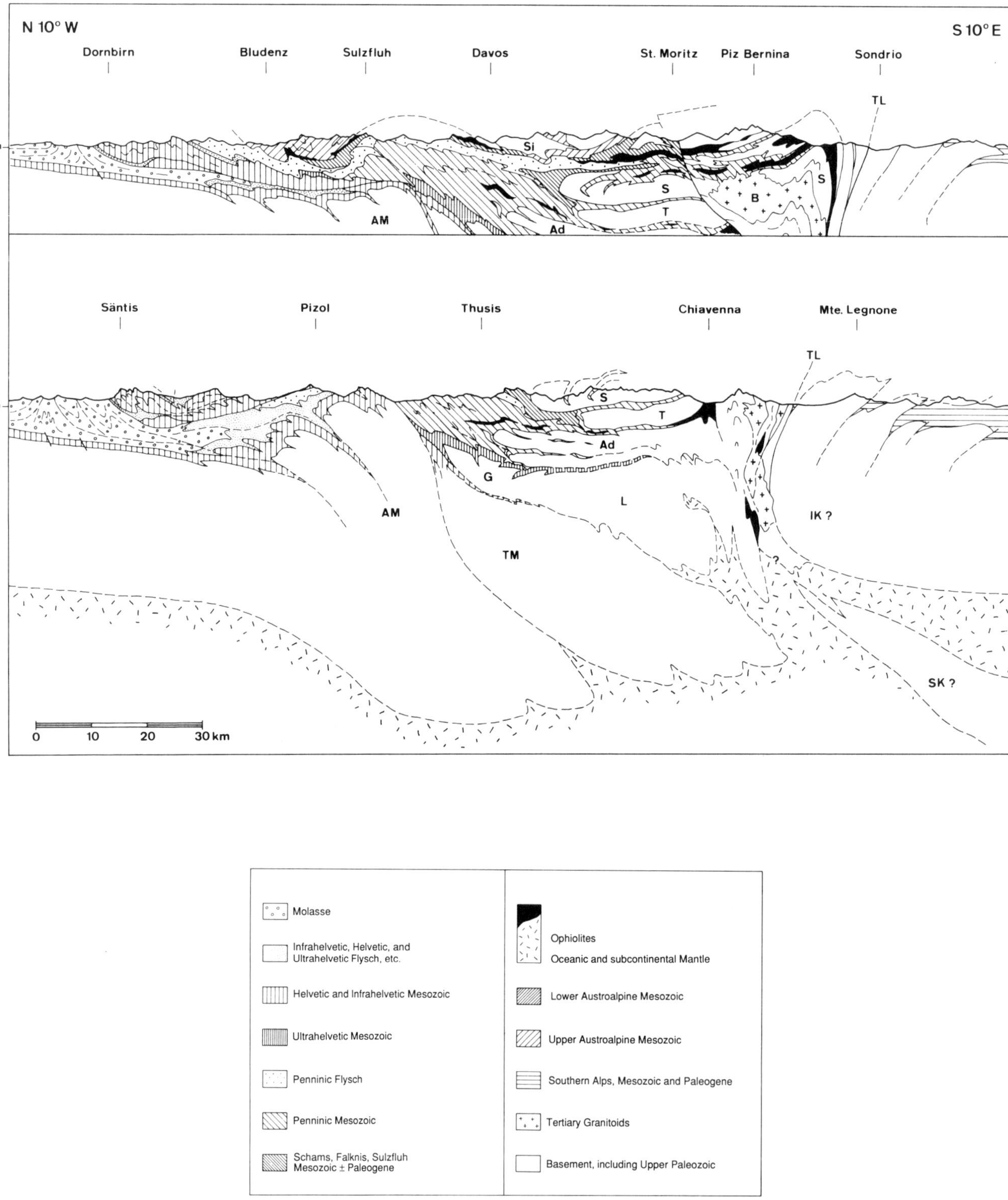

Figure 4. Sections through the Alps in the Graubünden traverse. Reproduced from R. Trümpy, 1985. AM = Aar massif; TM = Tavetsch massif; G = Gotthard "massif"; L = Leventina and associated gneiss nappes; Ad = Adula nappe; T = Tambo nappe; S = Suretta nappe; B = Val Bregaglia granitoids; TL = Tonale (Insubric s.l.) fault; IK = ? possible location of a prolongation of the Ivrea body; SK = ? possible subducted (eo- and meso-Alpine) continental and oceanic crustal fragments.

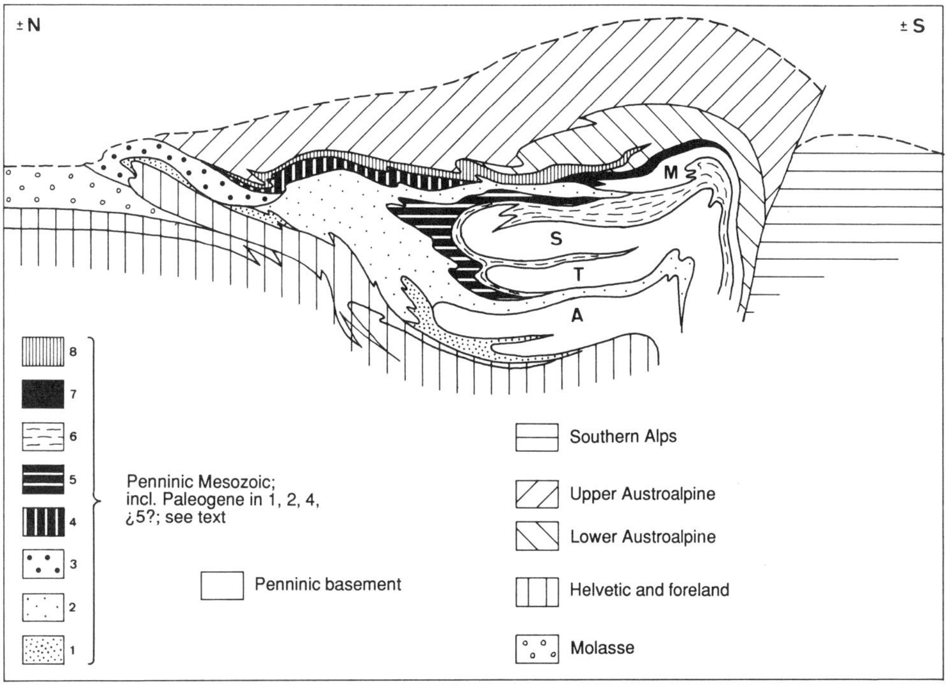

Figure 5. Highly schematic section through the Graubünden Alps, without scale, showing the relative position of tectonic units discussed in the text. For the "exact" position of the units, see Figure 4. A = Adula nappe; T = Tambo nappe; S = Suretta nappe; M = Margna nappe.

Some geologists (e.g., Staub, 1937) have derived both Schams and Falknis-Sulzfluh nappes from the Margna high, which would give:

1 - 2 - 3 - 6 = 6 - 4 + 5 = 7 - 8 (hypothesis D).

In the four hypotheses evoked so far, there are three ophiolitic sutures ("oceans") in the Graubünden traverse: Chiavenna ophiolites and Mesocco metabasalts between the Adula and Tambo basement nappes, Malenco ophiolites and Avers metabasalts between Suretta and Margna, and Platta ophiolites between Margna and Austroalpine nappes. In still another hypothesis, the Margna nappe is considered as a foreign (Lower Austroalpine?) body, emplaced between ophiolite nappes during the eo-alpine Cretaceous subduction. This would imply:

(1 - 5) - 6 - 8 - 7 (hypothesis E)

and reduce the number of ophiolite sutures to two.

Whatever solution is adopted, some points are evident:

1. The Valais belt has become very broad (200 to 250 km?), and increasingly complex. It comprises paraoceanic basins (for discussion and comparisons, see Kelts, 1981), some (narrow?) oceanic stretches (Chiavenna, Ramosch), and intrabasinal highs (e.g., Adula swell). Subsidence started at least in Middle Jurassic time (Pantić and Gansser, 1977).

2. By contrast, the Piemont belt is represented only by a few elements, especially in the Lower Engadine traverse. The older formations of the Tasna nappe, basement rocks, Permian volcano-sedimentary formations, and Triassic and Lower Jurassic carbonates, are very similar to those of the Lower Austroalpine nappes (Gruner, 1981; Wenk, 1962). Striking differences in the successions of the two units concern only post–mid-Jurassic formations. Up to that time, i.e., up to the onset of spreading, the Briançonnais belt of eastern Graubünden had more affinity with the Apulian than with the European continental margin, whereas the same belt has "European" affinities in the western Alps.

3. An apparent duplication of facies belts is notable: two "Briançonnais-type" platforms (Sulzfluh and Schams) and two "Piemont-type" oceans (Avers-Malenco and Platta-Arosa). The intercalation of kilometer-size fragments with granitic basement and shallow-water Triassic formations in ophiolite sheets, such as the Platta-Arosa nappe, is also a curious phenomenon.

Such anomalies have been attributed to tectonic imbrication during the Late Cretaceous lithospheric subduction and the Tertiary crustal subduction processes. Such a mechanism could indeed account for the small sialic slivers within oceanic units, but hardly for the triplication of ophiolitic sutures. Jurassic and Early Cretaceous transcurrent movements provide a more plausible explanation for the alternation of oceanic and continental fragments along one traverse.

PENNINIC ZONE OF THE TAUERN TRAVERSE

In the central segment of the Eastern Alps, Penninic rocks are exposed in two areas: in the Tauern window, where they appear underneath Austroalpine sheets, and in the Rheno-Danubic Flysch belt, north of the Austroalpine Northern Calcareous Alpes. Many of the correlation problems stem from the fact that the connection between the two areas is buried below Austraoalpine nappes.

The Tauern window has an east-west length of some 160 km. Its core is composed of first-order nappes (recumbent folds) of granodioritic and tonalitic gneisses, the Zentralgneiss, with an envelope of pre-Triassic metasediments and metavolcanics, which are partly older and partly younger than the emplacement of the Zentralgneiss metagranitoids. These basement rocks are overlain, especially in the western part of the window, by the thin "autochthonous" Hochstegen sequence. It comprises Permian(?) clastics, Triassic quartzites and dolomites, thin black metapelites with quartzitic bands (Lower or Middle Jurassic), and as much as 200 m of limestones with chert bands, the Hochstegenkalk. These limestones have furnished the (almost) unique determinable post-Triassic fossil in the Tauern window, a *Perisphinctes* sp. of early Late Jurassic (or ±Oxfordian) age. A hanging formation of schists with sandstones and breccias (Kaserer Formation) may, at least in part, represent younger (Cretaceous) rocks of the Hochstegen sequence. The Zentralgneiss nappes with their Hochstegen cover are overridden by thick, allochthonous sheets of Jurassic-Cretaceous Bündnerschiefer, with gneiss lamellae, and, locally, several hundred meters of Triassic carbonates (Seidlwinkel Triassic). Stratigraphic and structural analyses of the Tauern Bündnerschiefer are only beginning (e.g., Frasl and Frank, 1966; Frisch, 1980; Tollmann, 1975). A lower sequence (Brennkogel type) is characterized by quartzites and arkosic schists, a middle one (Glockner type) mainly by calc-schists, and a higher one (Fusch type) by black pelitic schists. Metabasalts are common, especially in the Glockner schists; serpentinites are subordinate. No complete ophiolite sequence is preserved. Age and original sedimentologic characteristics of the Tauern Bündnerschiefer remain uncertain. Locally, for instance, above the tunnel piercing the watershed on the Grossglockner road, they recall certain types of Cretaceous "preflysch." Flysch, or at any rate flysch-like turbidites, occupies a high position within the Tauern edifice along its southern margin (Baggio, 1969), and especially in the Klammkalk zone along its northern edge (Prey, 1977b).

The Flysch zone along the northern edge of the Alps comprises essentially the thick Cretaceous Rheno-Danubic Flysch. Pre-Flysch and flysch sedimentation continued through a long time span, from Aptian to Maestrichtian. Except for some Lower Cretaceous pelagic limestones, the original substratum of the flysch sediments is unknown in this traverse; they are overthrust on the so-called "Helvetic" Gresten-Buntmergel units (see below). Thus, we find, along the Tauern traverse: an area with reduced Jurassic sedimentation—the Hochstegenkalk high; to the south, the (partly) oceanic, or at least paraoceanic, basin of the Tauern Bündnerschiefer; and either north or south of the Hochstegen high, the trough of the Rheno-Danubic Flysch.

There are two fundamental problems. Do the Tauern Bündnerschiefer belong to the Valais or to the Piemont belt of Switzerland, or to both? And does the Rheno-Danubic Flysch come from a basin to the north or to the south of the Hochstegen rise?

Tollmann (1965) has correlated the Hochstegen high with the Briançonnais rise, or, even more specifically, with the Acceglio zone of the French-Italian Alps. Many geologists, including Dercourt and others (1985), have followed this view, which I find difficult to accept.

When last seen, coming from the west, in the western part of the Engadine window, the Briançonnais remnant (Tasna nappe) occupied a very "high" position within the Penninic edifice, overlying all the Bündnerschiefer units and overlain only by thin slices of the Arosa zone, which does not contain Bündnerschiefer. The stratigraphic affinities of the pre–mid-Jurassic sequence of the Tasna nappe are "Apulian" (Lower Austroalpine). The Hochstegen zone, on the contrary, has a very "low" position, below the Tauern Bündnerschiefer and in stratigraphic contact with gneiss nappes, which, if they did reach the Engadine area, would lie far below its surface. Furthermore, the facies development of the Hochstegen sequence has a "European" character. The distance between the eastern end of the Engadine window and the western end of the Tauern window is only 60 km. It is true that the Engadine window is slightly more external, especially if the sinistral displacement by the Engadine fault is taken into account (Trümpy, 1977). Still, it is difficult to see how such a fundamental change of position could be realized.

From Savoy to Graubünden, the Valais belt becomes successively broader, more complex, and more oceanic from west to east, while the Piemont basin narrows in the same direction. If this tendency is extrapolated to the Tauern traverse, it would be quite surprising to see the Piemont basin expanding again to encompass the entire mass of the Tauern Bündnerschiefer.

There is no certain equivalent of the Briançonnais platform in the Tauern window, unless the Seidlwinkel Triassic carbonates are a distant analog of those of the Schams, nor even in the eastern, Tyrolese part of the Engadine window. Oberhauser (1980) has shown that the Prutz nappe in the latter area cannot be correlated with the Tasna nappe of the western part of the window. It is probable that the west-Alpine Briançonnais does not exist in the Austrian Alps, and that the Valais and Piemont belts are no longer differentiated there.

Apart from the presence of Upper Jurassic limestones, the Hochstegen succession shows little resemblance to typical Briançonnais development. Frisch (1975), and especially Lammerer (1986, 1987), have recently restudied the Hochstegen sequence. They have insisted on its "northern" affinities, with scanty Triassic and quartz detritus in the Lower or Middle Jurassic. The latter feature recalls the Gresten beds of the Klippen zones. Whether the Hochstegen rise is really "Helvetic" (i.e., part of the proximal European margin) or whether it constitutes an intra-Valais platform analogous to the Adula rise, is difficult to decide. Our

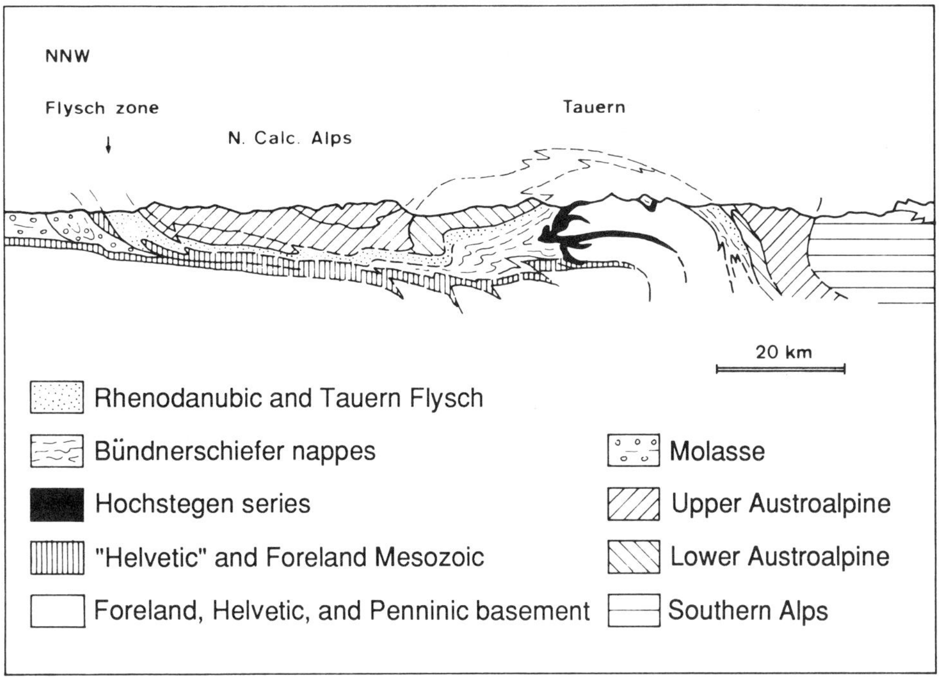

Figure 6. Highly schematic section through the central part of the Eastern Alps.

reconstruction (Fig. 7) uses the latter solution. At any rate, analogies between the Hochstegen sequence and those of the "true" Helvetic facies belt in Switzerland are very tenuous. The comparison between the Tauern gneiss nappes and the external massifs of Switzerland, especially with the Aar massif (Frisch, 1976), is quite improbable, because of the position, the rock content, the style of deformation, and the grade of Alpine metamorphism; such a radical change of all features over an east-west distance of only 150 km seems out of the question. I much prefer to compare the Zentralgneiss nappes with the Lower Penninic and Infrapenninic (Adula and "Lepontinic") nappes of Switzerland, where there is agreement on all the four above-mentioned criteria: situation along the same axis, gneiss cores with marble cover, recumbent folds, and amphibolite-grade Tertiary metamorphism.

The origin of the Rheno-Danubic Flysch is a controversial question. Geologists looking at the problem from the east (e.g., Prey, *in* Oberhauser, 1980) tend to assign a very northerly position to the Flysch Zone, north of the Hochstegen rise and close to the "Helvetic" Gresten-Buntmergel belt (see also Del Negro, 1970). "Western" (Swiss and other) geologists, on the contrary, have insisted on the southerly position of the Rheno-Danubic Flysch within the Valais belt, south of the great mass of the Prättigau Bündnerschiefer and a little to the north of the northern scarp of the Briançonnais (Falknis-Tasna) platform.

Recently, geologists from Fribourg (Homewood and Caron, 1982; Winkler and others, 1985) have gone even further. Noting the great similarities of post-Turonian sediments in the ultra-Briançonnais flyschs (Helminthoid Flysch, Schlieren-Gurnigel Flysch, etc.) of Switzerland and in the Rheno-Danubic Flysch, they have (more or less cautiously) suggested that all these flyschs were laid down in the same basin. The similarity also concerns the heavy mineral spectrum, which differs markedly from that of undoubted north-Penninic flyschs, such as Niesen or Prättigau. This raises some difficulties insofar as the Rheno-Danubic Flysch is probably in citra-Briançonnais position (i.e., north of the Briançonnais rise; Hesse, 1973). On the Briançonnais rise itself, the Upper Cretaceous to Lower Eocene formations are represented by the pelagic limestones of the Couches Rouges. At present, I see no easy way out of the dilemma, even if the Briançonnais of eastern Switzerland was considered as a displaced terrane of west-Alpine origin.

In the Tauern traverse, the question is, does the Rheno-Danubic Flysch originate from an area to the north of the Tauern (including the Hochstegen belt), now hidden underneath the Northern Calcareous Alps, or was it stripped from the Tauern window (Hochstegen and/or Bündnerschiefer belts) itself?

The second solution (more plausible to the west-Alpine geologist) is possible only if the Tauern area or part of it was open to sedimentation up to Late Cretaceous time. Opinions differ widely as to the date at which the Tauern window was covered by the Austroalpine nappes: Tollmann (1963) would place it in the mid-Cretaceous; Oberhauser (1968), only in the Tertiary. At any rate, the Penninic rocks of the Tauern window have so far furnished only Tertiary ages (Oxburgh and others, 1966), whereas the overlying Austroalpine nappes give Cretaceous dates.

We need to know the age of the youngest, flysch-like rocks

in the Tauern window; we should also be sure whether an earlier, Cretaceous metamorphism in the Tauern window was overprinted by the Tertiary greenschist to amphibolite-grade event. And, finally, we should continue persuading oil companies that gas may be found below the Northern Calcareous Alps.

PENNINIC AND RELATED BELTS AT THE EASTERN EDGE OF THE ALPS

Here again, Penninic and related rocks are exposed in the flysch belt to the north, especially in the Wienerwald hills west of Vienna, and in the Rechnitz window, 70 km to the south, on the Austro-Hungarian border. The schists of the Rechnitz window were formerly considered as being of Paleozoic age. W. J. Schmidt (1956) first pointed out their Penninic character, and Schönlaub (1973) brought the proof by determining sponge spiculae of Early Cretaceous or early Late Cretaceous age. The Rechnitz schists are mainly calc-schists of greenschist grade; they contain breccias with dolomite components and are associated with metadiabase and serpentinite. There can be little doubt that the Rechnitz rocks are eastern equivalents of the Tauern Bündnerschiefer. Their presence on the edge of the Pannonian plain is of prime importance for demonstrating the far-travelled allochthony of the Austroalpine nappes, even this far east, and of the internal nappes of the western Carpathians.

Compared to outcrop areas farther west, the Flysch Zone of the Wienerwald is broader, tectonically more complex—it can be divided into three nappes or digitations—and stratigraphically more complete, reaching up into the Middle Eocene or early Upper Eocene.

One of the difficult and controversial problems of Alpine geology concerns the "klippen" in the Flysch belt of lower and upper Austria. These are "Slovak-type klippen" in the sense of Andrusov and Scheibner (1968), i.e., inliers of older and/or tectonically lower formations, in contrast to the "Swiss-type klippen," which are outliers of higher nappes. Outcrop conditions are poor, but painstaking work (e.g., Prey, 1975, 1977a, 1979, 1983; Schnabel, 1979, 1980) has clarified many problems.

A northern string of klippen (Gresten, or main klippen zone) contains sandy Lower Jurassic formations (with arkose and conglomerate), Middle Jurassic shales (including *Bositra* marls), radiolarian chert, Upper Jurassic–Lower Cretaceous limestones and shales, and Upper Cretaceous to Upper Eocene variegated, hemipelagic marl shales (Buntmergel) that are rich in pelagic foraminifera. Its contact with the flysch nappes is clearly tectonic, all the more as the lowest flysch is of late Early to early Late Cretaceous age; it belongs to a unit originally north of the Flysch basin.

The southern St. Veit or Ybbsitz klippen zone has an even more pelagic development of the Jurassic and Lower Cretaceous formations, including radiolarian chert and aptychus limestones (Maiolica type); it comprises only pre-Albian formations. Both klippen zones are strongly dilacerated, and stratigraphic contacts between the formations are rarely seen. The southern (St. Veit) klippen zone may represent part of the substratum of some flysch nappes. It has generally been compared with the Pieniny klippen zone of the Carpathians (but see the cautionary words of Prey, 1975, p. 23).

If we unravel the autochthonous substratum, the "Helvetic" units including the Gresten klippen zone and the Flysch nappes including the St. Veit klippen zone, this takes us well back to the Tauern window or the Rechnitz window (del Negro, 1970). The question is whether a Tertiary crustal subduction zone with the "root" of the Flysch and St. Veit nappes lies hidden below the Northern Calcareous Alps, north of the Tauern and Rechnitz uplifts.

Terms such as Helvetic, south-Helvetic, and northern and southern Ultrahelvetic are used for various units, including the Gresten-Buntmergel slices, along the front of the Eastern Alps. But the true Helvetic zone in Switzerland is characterized as a proximal continental margin or unstable shelf, with thick, mainly calcareous Jurassic and Lower Cretaceous formations. This shelf narrows eastward and is practically nonexistent south of the Bohemian massif. At the longitude of Salzburg, the Helvetic facies belt "goes autochthonous," and its typical Upper Cretaceous formations are encountered in boreholes below the Molasse of the foreland basin (Janoschek, 1957). Only far to the northeast, in the Subsilesian units of the Polish Carpathians, does an analogous—but by no means homologous—shelf reconstitute itself. In my opinion, the term "Helvetic," including its derivatives, should *not be used east of Salzburg.*

The Jurassic and Lower Cretaceous units of the Ybbsitz–St. Veit klippen zone, which may represent part of the substratum of the Flysch Zone, contain deep-sea sediments and may have been of oceanic or paraoceanic nature. This oceanic character is reinforced by the presence of serpentinite, serpentinite breccias, and picritic basalts, which are associated with rocks of late Early Cretaceous to early Late Cretaceous age (Prey, 1977).

Schnabel (1979, p. 33) has pointed to certain similarities between the succession of the Ybbsitz klippen and those of the Arosa zone of Graubünden (Piemont belt). In my opinion, this analogy merely implies that both sequences were formed in more or less oceanic basins.

The stratigraphic successions of each of the two klippen zones also have some points in common. Their paleogeographic relation to the Hochstegen high of the Tauern window can be interpreted in different ways: both to the north of the Hochstegen high, Gresten to the north, and Ybbsitz–St. Veit to the south, or both to the south. The second solution seems the most probable one—if the Hochstegen rise, with its thick Upper Jurassic limestones, existed at all east of the Tauern. Comparable limestones are not quite absent from the Gresten klippen zone (Arzberg limestones; see Trauth, 1921).

PENNINIC EQUIVALENTS IN THE WESTERN CARPATHIANS

In this section, I briefly comment on the prolongation of Alpine paleogeographic units into the External Zones of the western Carpathians. Readers are referred to reviews by Andrusov (1968), Mahel (1974), Sandulescu (1985), and especially to the article by Debelmas and Sandulescu (1987).

The Internal Zones, consolidated in Cretaceous time and roughly corresponding to the Austroalpine sheets or terranes of the Alps, will not be treated here, except for their most external part, the Tatrides s.l. The External Zones, with Tertiary deformation, are much wider than in the Eastern Alps; they comprise the broad Flysch Zone to the north and the narrow but very complex Pieniny Klippen Zone to the south.

The Flysch Zone consists of several nappes of Cretaceous to Oligocene flysch formations, deposited in at least two large basins; the aggregate minimum width of these basins (not counting the hypothetical "cordilleras" furnishing detritus) is of the order of 200 km. Drilling for oil and gas, as well as seismic exploration, has shown beyond doubt that the entire flysch mass is allochthonous, and that it overlies the gently south-dipping autochthonous sediments of the foreland. Little is known about the pre–mid-Cretaceous formations of the Flysch Zone. A thin belt of more or less oceanic Jurassic and Lower Cretaceous formations, including Callovian to early Upper Jurassic radiolarites, along the southern edge of the zone in Poland is considered to represent the substratum of the most internal (Magura) nappe of the flysch belt (Birkenmajer, 1977). This would imply that the flysch basins, or at least their southern part, developed on a belt with thinned continental, if not oceanic, crust. The Magura nappe can be followed into the Rheno-Danubic Flysch belt. Although the Klippen Zone is only some hundreds of meters to a few kilometers wide, it contains several quite distinct facies belts made up of rocks of the same (Jurassic to Eocene) age and must have had an original width of at least 100 km, more probably 200 km. Summaries by Andrusov (in Mahel, 1974) for the Slovakian and by Birkenmajer (1977) for the Polish part give a good idea of the stratigraphic composition. An external zone, the Czorstyn unit, comprises Middle to Upper Jurassic crinoidal, nodular, and micritic limestones; the Lower Cretaceous is often missing, and red pelagic Upper Cretaceous marly limestones (Couches Rouges) rest directly on the Czorstyn limestones. South of the Czorstyn belt, the Kysuca (or Pieniny, in the narrow sense) belt shows basin facies, with *Bositra* shales, radiolarian cherts, and pelagic limestones. Its oldest formation in most outcrops are Aalenian black shales with sandstone beds (Aalenian "flysch" of many Carpathian authors); older formations (Upper Triassic in Carpathian Keuper facies, Liassic Fleckenmergel) are rarely seen.

The most southerly elements of the klippen belt, in the broad sense, are only locally exposed, especially in Slovakia. Pebbles in Cretaceous conglomerates indicate the presence of a pronounced, emergent high south of the Kysuca zone, Trauth's (1921) Ultrapienidic ridge. This may well have been a sort of island arc exposing basement rocks and Cretaceous volcanics (Mišik and Sýkora, 1981). Some pebbles show traces of high-pressure metamorphism. The deeper basin of the Manin and of the controversial Kostelec units may represent a sort of back-arc basin, between the Ultrapienidic ridge and the Apulian front.

In a much simplified manner, the Jurassic to mid-Cretaceous sequence of the western external Carpathians can be interpreted to show two deep basins, presumably underlain by anomalous (thinned) continental crust, separated by the Czorstyn high and limited to the south by the presumed Cretaceous island arc of the Ultrapienidic ridge. The Carpathian Flysch belt and the major part of the Pieniny zone correspond more or less to the Flysch belt and to the klippen (at least the St. Veit klippen) of the Eastern Alps. There is no obvious prolongation of the Tauern (and Rechnitz) Bündnerschiefer and ophiolites visible at the surface. Two possibilities can be considered: either the Tauern-Rechnitz basin becomes starved toward the east and merges with Pieniny-type basins, or it continues eastward below the Central Carpathians and becomes subducted, presumably in Cretaceous time.

Seismic profiles show that the European basement, deformed by normal faulting, but apparently not by thrusting or folding, can be followed southward to the vicinity of the Pieniny klippen belt. The substratum of the Flysch Zone (about 200 km or more) and of the Klippen Zone (about 150 km or more) must have disappeared in depths somewhere to the south, either in a crustal subduction zone below the Klippen belt itself, or, more probably, below the Internal Zones. This implies that the Internal Zones, with their Cretaceous deformation, have been displaced (in Tertiary time) by at least 400 km, possibly much more, and that they are entirely allochthonous. Their place of origin, with respect to stable Europe, would lie somewhere under the southern part of the Pannonian plain.

In my opinion, the Apulian edge is marked by the Mesozoic high of the Tatra (Andrusov, 1968). Yet Tollmann (1963b), and following him, Dercourt and others (1985), have linked the Tatra with a "Briançonnais" belt, and have sought the continuation of a "Piemont" belt in the somewhat deeper basin of the Krizna nappe, to the south of the Tatra. I do not see any evidence of an oceanic basin south of the Tatra; the deepest (paraoceanic or locally oceanic) basins definitely lay to the north of the Tatra high.

The Klippen Zone (at least its Kysuca unit) can be followed eastward into the Ukrainian Carpathians. Does the paraoceanic basin continue southward, underneath the cover of Neogene sediments and volcanics, to join the frankly oceanic belt of the southern Apuseni Mountains, presumably a northern extension of the Vardar ocean? What is the origin of the Transylvanian nappes? Does a branch of the west Carpathian deep basins continue to the southeast, into the external Dacides? Carpathian geologists are seeking answers for all these questions (e.g., Sandulescu, 1984; Debelmas and Sandulescu, 1987).

A tempting hypothesis might involve the existence of a "Ukrainian triple junction" between the Tauern-Pieniny (para-) oceanic belt to the west, a Dacide basin to the southeast, and an Apuseni-Vardar Ocean to the south. It might constitute a sort of mirror image to a "Tyrolese triple junction" of the Alps, between the Tauern-Pieniny belt to the east, the Valais belt to the west, and the Piemont-Ligurian Ocean to the south.

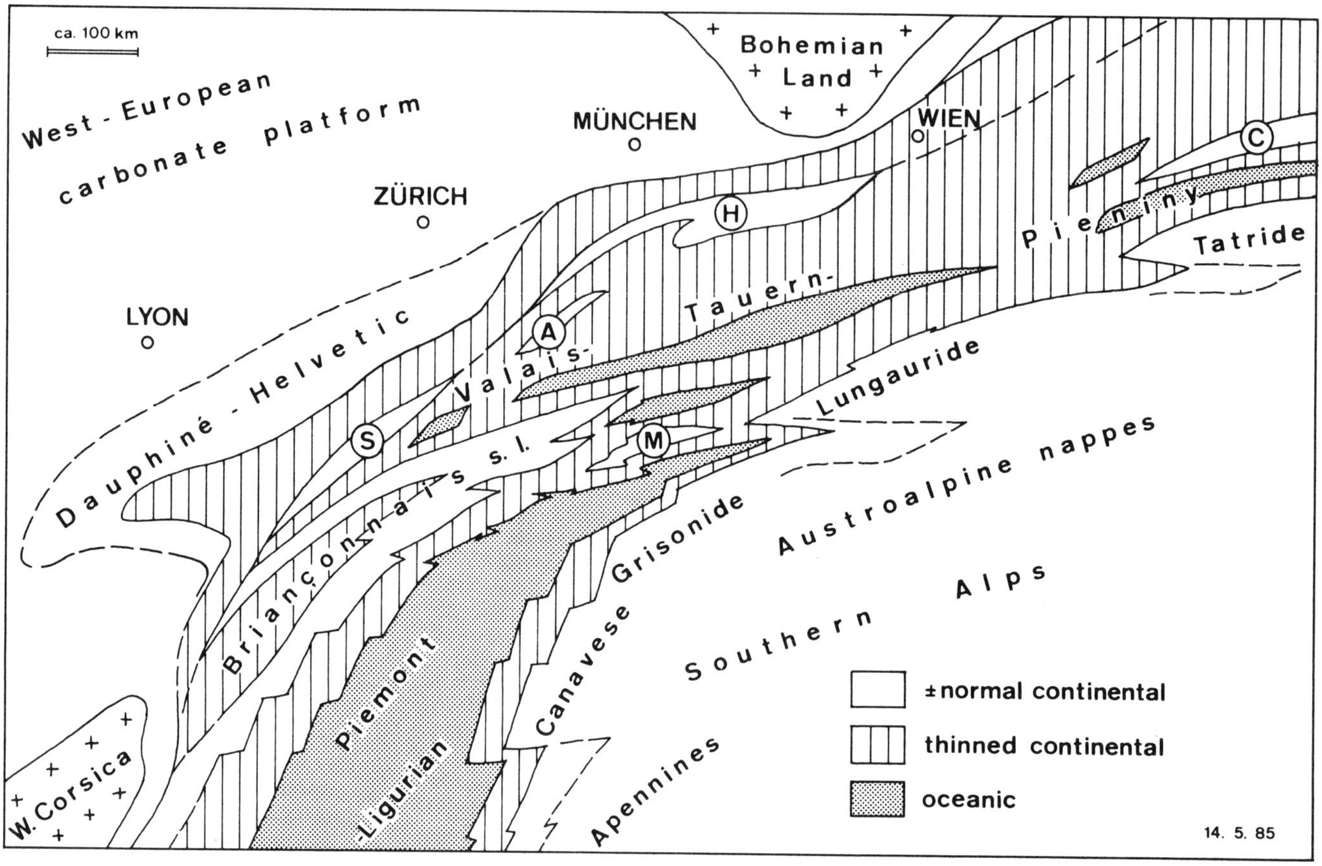

Figure 7. Tentative palinspastic map of the Alps in Late Jurassic time (reproduced from Lemoine and Trümpy, 1986). S = Marginal high and Cordillère Tarine; A = Adula rise; M = Margna rise; H = Hochstegen rise; C = Czorstyn rise.

CONCLUSIONS

I have tried to follow the east-west–trending segment of the Alpine oceans from Savoy to Slovakia. From Savoy over Valais to Graubünden, the north Pennininic or Valais belt increases in width, complexity, and partly oceanic character. The Briançonnais platform, separating the Valais from the Piemont ocean, dwindles eastward and cannot be followed into the Austrian Alps. In its western and main part, its affinities are definitely European; at its eastern end they are rather Apulian. In the Tauern and Rechnitz-Vienna traverses, Valais and Piemont belts are no longer distinct.

During Late Jurassic and Cretaceous time, parts of these basins became starved and received only thin sediments of hemipelagic or pelagic character. Upper Jurassic radiolarian cherts and maiolica-type limestones are widespread (Zone Submédiane, klippen belts of the eastern Alps and of the western Carpathians). Other basins were reached by fine-grained detrital influx; they were filled by thick Jurassic and especially Cretaceous Bündnerschiefer, comprising distal, commonly calcareous turbidites (Kelts, 1981). In the course of Late Cretaceous time, Bündnerschiefer graded into flysch.

Basalts resp. metabasalts are common, especially in the Bündnerschiefer basins, where they constitute flows or sills. Ultramafic rocks also occur, but their context is often unclear. Ophiolite sequences are rather subordinate but not quite absent (Chiavenna, Ramosch, Glockner zone of the Tauern). At any rate, there are few complete and typical ophiolitic successions in the west-east segment of the Alpine oceans; these are mainly restricted to the north-south– or southwest-northeast–trending segments (Piemont-Ligurian and Vardar-Apuseni oceans).

Platforms with continental basement lie within the more or less oceanic domain. These platforms are not continuous; thus, the Briançonnais belt disappears eastward in western Tyrol and southward between Liguria and Corsica. Quite similar facies associations are found in other traverses (Penibetic of southern Spain; Czorstyn rise of the Carpathians). These platforms were certainly not connected. In other cases, a duplication of facies belts within one traverse seems probable (especially in Graubünden).

Island arcs are rare and atypical. Only the Cordillère Tarine of Savoy and the Ultrapienidic rise of Slovakia are possible candidates.

Paleogeographic reconstructions in the Penninic belt are strongly influenced by the solution adopted as to the correlation of units along different traverses. The question of the origin of the Prealpine nappes was resolved around 1950 (especially by Ellenberger, 1958). I feel fairly confident about the correlation of the gneiss nappes of the Hohe Tauern with the Adula and lower Penninic nappes of Graubünden, which are proposed in this chapter. On the other hand, the provenance of the Schams nappes and of the Rheno-Danubic Flysch nappes remains controversial.

The Apulian margin toward the Penninic oceans, i.e., the "Lower Austroalpine," is probably discordant to the facies belts within the Apulian platform and within the Penninic oceanic domain. Thus, the facies succession is quite different in the Carpathians (Tatrides), in the frame of the Tauern window (Lungaurides), in the Upper Engadine (Grisonides), and in the westernmost southern Alps (Canavese). Grisonides and Canavese show the strongest affinities to each other. Occasionally, oceanic rocks occur in an anomalous position south of Lower Austroalpine units (Reckner serpentinites southeast of Innsbruck; see Enzenberg, 1969).

All these features can be explained more easily by a transform régime, predominantly sinistral, than by a mechanism of orthogonal spreading (Kelts, 1981; see also Weissert and Bernoulli, 1985). This applies not only to the Valais-Tauern-Pieniny belt, but also to the northeastern part of the Piemont belt.

Thus, I would explain the origin of the Penninic oceanic and paraoceanic belts between Savoy and Slovakia as the result of a transform system, active from Middle Jurassic to mid-Cretaceous time and already prefigured by Early Jurassic rifting, linking the northern end of the Piemont ocean to the northern end of the Vardar ocean. The intraoceanic platforms could be transtensional features and might have undergone considerable lateral shift with respect to the European foreland and/or the Apulian prong. The duplication of facies belts might point in the same direction. Needless to say, such possibilities do not really make palinspastic reconstructions easier. Many alpine nappe complexes can be considered as tectono-sedimentary "terranes", although their exotic character can reach only modest proportions, with relative displacement of a few hundred kilometers at most.

If such a hypothesis is considered, the Valais belt of Savoy and western Switzerland and the external Dacide belt of Romania might be interpreted as failed arms of a Piemont-Vardar transform system, west of a "Tyrolean triple junction" and southeast of a "Ukrainian triple junction." This could explain their curious termination on the Isère to the west and on the Ialomita to the southeast.

We must abandon the cylindristic concept of parallel facies belts. Basins and swells may change their character along strike. Major paleogeographic elements, such as the Briançonnais rise, may wedge out and be replaced, in another traverse, by a similar but not homologous swell. This behavior is also exemplified in the most plausible present-day analogue of the Mesozoic Alpine basins, the seas and islands between Indonesia and Australia (Hamilton, 1979).

I have insisted on the uncertainty of certain correlations. A hypothesis proposed by an elderly professor is not necessarily the true one or the only possible one. The upcoming IGCP project 198, on the northern margin of the Tethys, should provide the opportunity for young geologists from the Rhone Valley to the Caspian Sea to work together to discover the means whereby the puzzle of the old Mediterranean oceans can be put on a more solid base of facts and a more inspiring set of fancies.

REFERENCES CITED

Allemann, F., 1957, Geologie des Fürstentums Liechtenstein (Südwestlichster Teil), unter besonderer Berücksichtigung des Flyschproblems: Jahrbuch des Historischen Vereins für das Fürstentum Liechtenstein, Vaduz, v. 56, 244 p.

Andrieux, J., Fontboté, J. M., and Mattauer, M., 1971, Sur un modèle explicatif de l'arc de Gibraltar: Earth and Planetary Science Letters, v. 12, no. 2, p. 191–198.

Andrusov, D., 1968, Grundriss der Tektonik der Nördlichen Karpaten: Vyd. Slovak Academy of Science, Bratislava, p. 1–188.

Andrusov, D., and Scheibner, E., 1968, Classification of "Klippes" or "Klippen": 23d International Geological Congress, Prague, v. 3, p. 92–102.

Antoine, P., 1971, La zone des Brèches de Tarentaise entre Bourg-Saint-Maurice (Vallée de l'Isère) et la frontière italo-suisse: Thèse Université Grenoble, 367 p.

Argand, E., 1924, La tectonique de l'Asie: 13th Congrès géologique international, Bruxelles, 1922, p. 171–372.

Arthaud, F., and Matte, P. H., 1975, Les décrochements tardihercyniens du sudouest de l'Europe, et essai de reconstruction des conditions de la déformation: Tectonophysics, v. 25, p. 139–171.

Baggio, P., 1969, La formazione mesozoica pennidica dei calcescisti con ofioliti delle valli di Vizze, Fundres e Valles, nell'Alto Adige Orientale: Memorie del Museo Tridentino di Scienze naturali, Trento, v. 17, no. 3, p. 205–244.

Barbier, R., 1948, Les zones ultradauphinoise et subbriançonnaise entre l'Arc et l'Isère: Mémoires pour servir à l'explication de la Carte géologique de la France, 291 p.

Beccaluva, L., dal Piaz, G. V., and Macciotta, G., 1984, Transitional to normal MORB affinities in ophiolitic metabasites from the Zermatt-Saas, Combin, and Antrona units, Western Alps; Implications for the paleogeographic evolution of the western Tethyan Basin: Geologie en Mijnbouw, v. 63, no. 2, p. 165–177.

Biju-Duval, B., Dercourt, J., and Le Pichon, X., 1977, From the Tethys Ocean to the Mediterranean Seas; A plate tectonic model of the evolution of the Western Alpine System: International Symposium on Structure and History of Mediterranean Basins, Split 1976, Paris, Technip, p. 143–164.

Birkenmajer, K., 1977, Jurassic and Cretaceous lithostratigraphic units of the Pieniny Klippen Belt, Carpathians, Poland: Studia geologica Polonica, v. 45, 159 p.

Bolli, H., Burri, M., Isler, A., Nabholz, W., Pantic, N., and Probst, Ph., 1980, Der nordpenninische Saum zwischen Westgraubünden und Brig: Ecologae geologicae Helvetiae, v. 73, no. 3, p. 779–795.

Bosellini, A., 1981, The Emilia Fault; A Jurassic fracture zone that evolved into a Cretaceous-Paleogene sinistral wrench fault: Bollettino della Societè geologica Italiana, v. 100, p. 161–169.

Bouillin, J. P., 1977, Géologie de la Petite-Kabylie dans les régions de Collo et d'El Milia (Algérie): Travaux du Laboratoire de Géologie méditerranéenne, associé aux CNRS, Paris, 509 p.

Burri, M., 1979, Les formations valaisannes dans la région de Visp: Eclogae geologicae Helvetiae, v. 72, no. 3, p. 789–802.

Burri, M., and Jemelin, L., 1983, Notice explicative, Feuille 77, Sembrancher, de l'Atlas géologique Suisse: Comm. géol. Suisse, 51 p.

Carey, S. W., 1958, A tectonic approach to continental drift, *in* Symposium on Continental Drift 1956: Hobart, University of Tasmania, Geology Department, p. 177–349.

Ciszak, R., Bandet, Y., Gonord, H., and Rolet, J., 1985, Sur la datation radiométrique de "roches vertes" intrusives dans les massifs à schistosité, para-autochtones, de l'Oranie (Algérie occidentale); Implications: Comptes rendus hebdomadaires des séances de l'Académie des Sciences, v. 300/II/7, p. 291–294.

Debelmas, J., 1983, La formation des Alpes: Paris, La Recherche, M. 150, p. 1542–1552.

Debelmas, J., and Sandulescu, M., 1987, Transformante nord-penninique et le problême des corrélations paléogéographiques entre les Alpes et les Carpathes: Bulletin de la Sociéte géologique de France, v. 3, no. 8, p. 403–408.

Debelmas, J., Escher, A., and Trümpy, R., 1985, Profiles through the Western Alps: International Geodynamics Project, Final report W.G. 9, subgroup 1.

Del Negro, W., 1970, Salzburg, Verhandlungen der Geologischen Bundesanstalt.

Dercourt, J., and 16 others, 1985, Présentation de 10 cartes paléogéographiques au 1/20,000,000[a] s'étendant de l'Atlantique au Pamir pour la période du Lias à l'actuel. Bulletin de la Société géologique de France, v. 8/I/5, p. 637–652.

Dewey, J. F., Pitman, W. C., Ryan, W.B.F., and Bonnin, J., 1973, Plate tectonics and the evolution of the Alpine system: Geological Society of America Bulletin, v. 84, p. 3137–3180.

Didon, J., Durand-Delga, M., and Kornprobst, J., 1973, Homologies géologiques entre les deux rives du détroit de Gibraltar: Bulletin de la Société géologique de France, v. 15, no. 7, p. 77–105.

Egeler, C. G., and Simon, O. J., 1969, Sur la tectonique de la Zone Bétique (Cordillère Bétique, Espagne); Étude basée sur les recherches dans le secteur compris entre Alméria et Vélez Rubio: Verhandelingen der Koninklijke Nederlandse Akademie van Wetenschappen, Amsterdam, v. I, no. 25, p. 3.

Ellenberger, F., 1958, Etude géologique du pays de Vanoise: Mémoires pour servir à l'explication de la Carte géologique de la France, 561 p.

Enzenberg, M., 1969, Die Geologie der Tarntaler Berge (Wattener Lizum), Tirol: Mitteilungen der Gesellschaft der Geologie und Bergbaustudenten in Österreich, Wien, v. 17, p. 5–50.

Felder, T. E., 1980, Geologic evolution of the westernmost part of the Internal Betic Zone (Betic Cordilleras, southern Spain): Geologische Rundschau, v. 69, no. 1, p. 131–148.

Fontboté, J. M., and Estevez, A., 1980, Geologia de la Cordillera Bética: Boletín del Instituto de Geología y Mineralogía, Granada, v. 41, no. 2, p. 249–292.

Frasl, G., and Frank, W., 1966, Einführung in die Geologie und Petrographie des Penninikum im Tauernfenster, *in* Metz, R., ed., Zur Mineralogie und Geologie des Landes Salzburg und der Tauern: Aufschluss, v. 15, p. 30–58.

Fricker, P. E., 1960, Geologie der Gebirge zwischen Val Ferret und Combe de l'A (Wallis): Eclogae geologicae Helvetiae, v. 53, no. 1, p. 33–132.

Frisch, W., 1975, Hochstegen-Fazies und Grestener Fazies—ein Vergleich des Jura: Neues Jahrbuch für Geologie und Paläontologie, v. 2, p. 82–90.

——, 1976, Ein Modell zur alpidischen Evolution und Orogenese des Tauernfensters: Geologische Rundschau, v. 65, no. 2, p. 375–393.

——, 1980, Post-Hercynian formations of the western Tauern window; Sedimentological features, depositional environment, and age: Mitteilungen der österreichischen geologischen Gesellschaft, v. 71/73, p. 49–63.

Gras, H., 1972, Etude géologique détaillée du Bassin houiller des Cévennes (Massif Central Français): These docteur S.C., Université Clermont-Ferrand, 300 p.

Gruner, U., 1981, Die jurassischen Breccien der Falknis-Decke und altersäquivalente Einheiten in Graubünden: Beiträge zur geologischen Karte der Schweiz, N.F., v. 154, 136 p.

Gübeli, A. A., Hochuli, P. A., and Wildi, W., 1984, Lower Cretaceous turbiditic sediments from the Rif chain (northern Morocco); Palynology, stratigraphy, and palaeogeographic setting: Geologische Rundschau, v. 73, no. 3, p. 1081–1114.

Guillaume, H., 1957, Geologie du Montsalvens: Beiträge zur geologischen Karte der Schweiz, N.F. 104, 170 p.

Hamilton, W., 1979, Tectonics of the Indonesian region: U.S. Geological Survey Professional Paper 1078, 345 p

Haug, E., 1925, Contribution à une synthèse stratigraphique des Alpes Occidentales: Bulletin de la Société géologique de France, v. 4, no. 3, p. 97–244.

Hesse, R., 1973, Flysch-Gault und Falknis-Tasna-Gault (Unterkreide); Kontinuierlicher Uebergang von der distalen zur proximalen Flyschfazies auf einer penninischen Trogebene der Alpen: Geologica et Palaeontologica, v. SB2, 53 p.

Homewood, P., 1974, Le flysch du Meilleret (Préaples romandes) et ses relations avec les unités l'encadrant: Thèse D.S., Université Lausanne.

Homewood, P., and Caron, C., 1982, Flysch of the western Alps, *in* Hsu, K. J., ed., Mountain building processes: London, Academic Press, p. 157–168.

Homewood, P., Ackermann, T., Antoine, P., and Barbier, R., 1984, Sur l'origine de la nappe du Niesen et la limite entre les zones ultrahelvétique et valaisanne: Comptes rendus hebdomadaires des séances de l'Académie des Sciences, v. 299, no. 2, p. 1055–1059.

Janoschek, R., 1957, Die Molassezone, *in* F. Bachmayer, ed., Erdoel in Oesterreich: Wien Natur and Technik, p. 75–85.

Kelts, K., 1981, A comparison of some aspects of sedimentation and tectonics from the Gulf of California and the Mesozoic Tethys, northern Penninic Margin: Eclogae geologicae Helvetiae, v. 74, no. 2, p. 317–388.

Lammerer, B., 1986, Das Autochthon im westlichen Tauernfenster: Jahrbuch der Geologischen Bundesanstalt, Wien, v. 112, p. 153–196.

——, 1987, Thrust-regime and transpression-regime tectonics in the Tauren Window (eastern Alps): Geologische Rundschau (in press).

Lasserre, J.-L., and Laverne, Ch., 1976, Le volcanisme tholéiitique de la zone du Versoyen (Alpes franco-italiennes); Minéralogie, pétrographie, et géochimie: Thèse 3'cycle, Grenoble, 252 p.

Laubscher, H. P., 1971, The large-scale kinematics of the western Alps and the northern Apennines and its palinspastic implications: American Journal of Science, v. 271, p. 193–226.

Lemoine, M., 1984, La marge occidentale de la Téthys ligure et les Alpes occidentales, *in* Boillot, G., ed., Les marges continentales actuelles et fossiles autour de la France: Paris, Masson, p. 155–248.

——, 1985, Structuration jurassique des Alpes occidentales et palinspastique de la Téthys ligure: Bulletin de la Société géologique de France, v. I, no. 1, p. 126–137.

Lemoine, M., and Trümpy, R., 1987, Pre-oceanic rifting in the Alps: Tectonophysics, v. 133, p. 305–320.

Le Pichon, X., 1982, Land-locked oceanic basins and continental collision; The eastern Mediterranean as a case example, *in* Hsü, K., ed., Mountain building processes: London, Academic Press, p. 201–211.

Loubat, H., 1968, Etude pétrographique des ophiolites de la "Zone du Versoyen," Savoie (France), Prov. d'Aoste (Italie): Archives des Sciences, Genève, v. 21, no. 3, p. 265–457.

Mahel, M., ed., 1984, Tectonics of the Carpathian Balkan regions: Bratislava, Geological Institute Dionyz Stur, 453 p.

Mišik, M., and Sýkora, M., 1981, Der pieninische exotische Rücken, rekonstruiert aus Geröllen karbonatischer Gesteine kretazischer Konglomerate der Klippenzone und der Manin Einheit (in Slovakian, with extensive summary in German), *in* Západné Karpaty, geológia 7: Bratislava, Geologicky Ustav Dionyza Stura, 149 p. and XXVIII p.

Müller, R., 1983, Die Struktur der Mischabelfalte (Penninische Alpen): Eclogae geologicae Helvetiae, v. 76, no. 2, p. 391–416.

Oberhauser, R., 1968, Beiträge zur Kenntnis der Tektonik und der Paläogeographie während der Oberkreide und dem Paläogen im Ostalpenraum: Jahrbuch der Geologischen Bundesanstalt, Wien, v. 111, p. 115–145.

——, ed., 1980, Der geologische Aufbau Oesterreichs: Wien/New York, Geology Bundesanstalt/Springer, 699 p.

Oxburgh, E. R., Lambert, R., Baadsgaard, H., and Simons, J., 1966, Potassium-Argon age studies across the southeast margin of the Tauern window, the eastern Alps: Verhandlungen der Geologischen Bundesanstalt, Wien, p. 17–33.

Pantić, N., and Gansser, A., 1977, Palynologische Untersuchungen in Bündnerschiefern: Ecologae geologicae Helvetiae, v. 70, no. 1, p. 59–81.

Prey, S., 1975, Neue Forschungesergebnisse über Bau und Stellung der Klippenzone des Lainzer Tiergartens in Wien (Oesterreich): Verhandlungen des Geologischen Bundesanstalt, Wien, v. 2-3, p. 1–25.

——, 1977a, Der Serpentin von Kilb in der Flysch-Mittelkreide am Nordende des Flysch zone (Niederösterreich): Verhandlungen der Geologischen Bundesanstalt, Wien, v. 3, p. 271–277.

——, 1977b, Flyscherscheinungen in den "Flysch-artigen Serien" der östlichen Tauernnordrandes: Verhandlungen der Geologischen Bundesanstalt, Wien, v. 3, p. 313–320.

——, 1979, Der Bau der Hauptklippenzone und der Kahlenberger Decke im Raume Purkersdorf-Wienerwaldsee (Wienerwald): Verhandlungen der Geologischen Bundesanstalt, Wien, v. 2, p. 205–228.

——, 1983, Das Ultrahelvetikum-Fenster des Gschliefgrabens südöstlich von Gmunden (Oberösterreich): Jahrbuch der Geologischen Bundesanstalt, Wien, v. 126, no. 1, p. 95–127.

Priem, N.H.A., Boelrijk, N., Hebeda, E., Oen, I. S., Verdurmen, E., and Verschure, A., 1979, Isotopic dating of the emplacement of the ultramafic masses in the Serranía de Ronda, southern Spain: Heidelberger Beiträge zur Mineralogie und Petrographie, v. 70, no. 1, p. 103–109.

Probst, Ph., 1980, Die Bündnerschiefer des nördlichen Penninikums zwischen Valser Tal und Passo di San Giacomo: Beiträge zur geologischen Karte des Schweiz, N.F. 153, 63 p.

Reuber, J., Michard, A., Chalouan, A., Juteau, T., and Jermoumi, B., 1982, Structure and emplacement of the Alpine-type peridotites from Beni Bousera, Rif, Morocco; A polyphase tectonic interpretation: Tectonophysics, v. 82, p. 231–251.

Ricou, L. E., 1980, La zone subbriançonnaise des Alpes Occidentales interprétée comme la trace d'un ample décrochement senestre sub-méridien: Comptes rendus hebdomadaires des séances de l'Académie des Sciences, Paris, v. 290, sér. D, p. 835–838.

Sandulescu, M., 1984, Geotectonica României: Bucuresti, ed. Technica, 334 p.

Schmidt, W. J., 1956, Die Schieferinseln am Ostrand der Zentralalpen, *in* Führungen und Fachausflüge 1953: Mitteilungen der Geologischen Gesellschaft in Wien, v. 47, p. 359–366.

Schmutz, H.-U., 1976, Der Mafitit-Ultramafitit-Komplex zwischen Chiavenna und Val Bondasca (Prov. Sondrio/Italien; Kt. Graubünden/Schweiz): Beiträge zur geologischen Karte des Schweiz, N.F. 149, 73 p.

Schnabel, W., ed., 1979, Blatt 71, Ybbsitz der Geologischen Karte der Republik Oesterreich 1:50,000: Arbeitstagung der Geologischen Bundesanstalt, Wien.

——, 1980, Die Geologie des Voralpengebiets im Abschnitt Waidhofen an der Ybbs-Ybbsitz-Gresten: Waidhofener Heimatblätter, v. 6, p. 13–27.

Schönlaub, H. P., 1973, Schwamm-Spiculae aus dem Rechnitzer Schiefergebirge und ihr stratigraphischer Wert: Jahresberichte der Geologischen Bundesanstalt, Wien, v. 116, p. 35–49.

Smith, A. G., 1976, Plate tectonics and orogeny; A review: Tectonophysics, v. 33, p. 215–285.

Staub, R., 1928, Der Bewegungsmechanismus der Erde, dargelegt am Bau der irdischen Gebirgssysteme: Berlin, Bornträger, 270 p.

——, 1937, Gedanken zum Bau der Westalpen zwischen Bernina und Mittelmeer, I. Teil: Vierteljahrsschrift der Naturforschenden Gesellschaft in Zürich, v. 82, 140 p.

Streiff, V., Jäckli, H., and Neher, J., 1976, Geologischer Atlas Schweiz, Blatt 56, Andeer: Bern, Kümmerly and Frey, 106 p.

Tapponnier, P., 1977, Evolution tectonique du système alpin en Méditerranée; Poinçonnement et écrasement rigide-plastique: Bulletin de la Socété géologique de France, v. 7, no. 19, p. 437–460.

Tollmann, A., 1963a, Ostalpen-Synthese: Wien, Deuticke, 256 p.

——, 1963b, Résultats nouveaux sur la position, la subdivision et le style structural des zones Helvétiques, Penniques et Austro-Alpine des Alpes Orientales: Société géologique de France, Livre à la mémoire de Paul Fallott, v. 2, p. 477–490.

——, 1965, Die Fortsetzung des Briançonnais in den Ostalpen: Mitteilungen der Geologischen Gesellschaft in Wien, v. 57, no. 2, p. 469–478.

——, 1975, Ozeanische Kruste im Pennin des Tauernfensters un die Neugliederung des Deckenbaues der Hohen Tauern: Neues Jahrbuch für Geologie und Paläontologie, v. 148, no. 3, p. 286–319.

Trauth, F., 1921, Ueber die Stellung der pieninischen Klippenzone und die Entwicklung des Jura in den niederösterreichischen Voralpen: Mitteilungen der Geologischen Gesellschaft in Wien, v. 14, p. 105–265.

Trümpy, R., 1972, Zur Geologie des Unterengadins, *in* Oekologische Untersuchungen im Unterengadin, Lieferung I/4: Ergebnisse der wissenschaftlichen Untersuchungen im schweizerischen Nationalpark, v. 12, p. 71–87.

——, 1977, The Engadine line; A sinistral wrench fault in the central Alps: Memoir of the Geological Society of China, v. 2, p. 1–12.

——, 1985, Die Plattentektonik und die Entstehung der Alpen: Neujahrsblatt der Naturforschenden Gesellschaft in Zürich, 47 p.

Trümpy, R., and Haccard, D., 1969, Réunion extraordinaire de la Société Géologique de France: Les Grisons: Bulletin de la Société géologique de France, v. 9, p. 330–396.

Weidmann, M., Homewood, P., Caron, C., and Baud, A., 1976, Réhabilitation de la "Zone Submédiane" des Préalpes: Eclogae geologicae Helvetiae, v. 69, no. 2, p. 265–277.

Weissert, H. J., and Bernoulli, D., 1985, A transform margin in the Mesozoic Tethys; Evidence from the Swiss Alps: Geologische Rundschau, v. 74, no. 3, p. 665–679.

Wenk, E., 1962, Einige Besonderheiten des unterostalpinen Kristallins im Unterengadin: Eclogae geologicae Helvetiae, v. 55, no. 2, p. 457–460.

Wildi, W., 1983, La chaîne tello-rifaine (Algérie, Maroc, Tunisie); Structure, stratigraphie et évolution du Trias au Miocène: Rev. Geogr. phys. Geol. dyn., v. 24, no. 3, p. 201–297.

Winkler, W., Wildi, W., van Stuijvenberg, J., and Caron, Ch., 1985, Wägital-Flysch et autres flyschs penniques en Suisse Centrale; Stratigraphie, sédimentologie, et comparaisons: Eclogae geologicae Helvetiae, v. 82, no. 1, p. 1–22.

MANUSCRIPT ACCEPTED BY THE SOCIETY JUNE 1, 1987

Printed in U.S.A.

Geological Society of America
Special Paper 218
1988

Plate kinematics and tectonics leading to the Alpine belt formation; A new analysis

Xavier Le Pichon
Département de Géologie, Ecole Normale Supérieure, 24 rue Lhomond, 75231 Paris Cedex 05, and C.N.R.S., UA 215, France
Françoise Bergerat and Marie-José Roulet
Département de Géotectonique, Université Pierre et Marie Curie, 4 Pl. Jussieu, T.26, ler ét., 75252 Paris Cedex 05, and C.N.R.S., UA 215, France

ABSTRACT

We use the paleogeographic reconstructions of the Tethyan belt made during a recent French-Soviet cooperative program (Dercourt and others, 1985, 1986) to consider the relationship between plate kinematics and plate tectonics. These reconstructions are based on kinematic and paleomagnetic syntheses. They enable us to obtain a quantitative estimate of the total amount of shortening along the Tethyan belt, both in subduction and in collision. We propose that, during Mesozoic time, the basic system was a three-plate system that involved the northward migration of a mid-ocean ridge across the ocean, its subduction beneath Eurasia, and the subsequent formation of a new Rift along the African margin. This might explain consecutive northward migrations of microcontinents across the Tethys. Our quantitative evaluation of continental collision indicates that about half the total amount of continental crust involved must have disappeared into the mantle. We suggest that this is because only the upper brittle part of the crust is involved in the stacking of nappes, whereas the lower ductile crust disappears into the mantle. As a result, the Tethyan continental collision resulted in a 75-m lowering of sea level. A discussion of the paleo-stress trajectories over the West European platform since the Eocene shows that Oligocene time was dominated by east-west extension related to relative motion between western Europe and central Eurasia. This relative motion is confirmed by the difference between the plate kinematics deduced from Arctic magnetic lineations and those deduced from North Atlantic magnetic lineations. The paleo-stress trajectories are remarkably parallel over most of the West European platform at all stages. However, since latest Miocene time, there is an outward fanning of the compressive stress trajectories in a zone 300 to 150 km wide, along the forefront of the Alps. The significance of this zone of fanning stress trajectories is not clear.

INTRODUCTION AND DISCUSSION OF KINEMATIC CONSTRAINTS

A recent French-Soviet cooperative program led to the establishment of nine paleogeographic maps of the Tethyan belt, from the Atlantic to the Pamirs, between Pliensbachian (190 Ma) and present times, at a scale of 1:20,000,000 (Dercourt and others, 1985, 1986). The maps are based on kinematic (Savostin and others, 1986), paleomagnetic (Westphal and others, 1986), and geologic syntheses (Ricou and others, 1986; Kazmin and others, 1986a,b; Zonenshain and Le Pichon, 1986; Knipper and others, 1986), all published in the same special issue of *Tectonophysics* (v. 123, March 1986). In this chapter, we use these syntheses and paleogeographic reconstructions to discuss in more detail the relationship between plate kinematics and tectonics. We present quantitative evaluations of the amounts of oceanic plate subduction and continental collision through time, and we discuss the

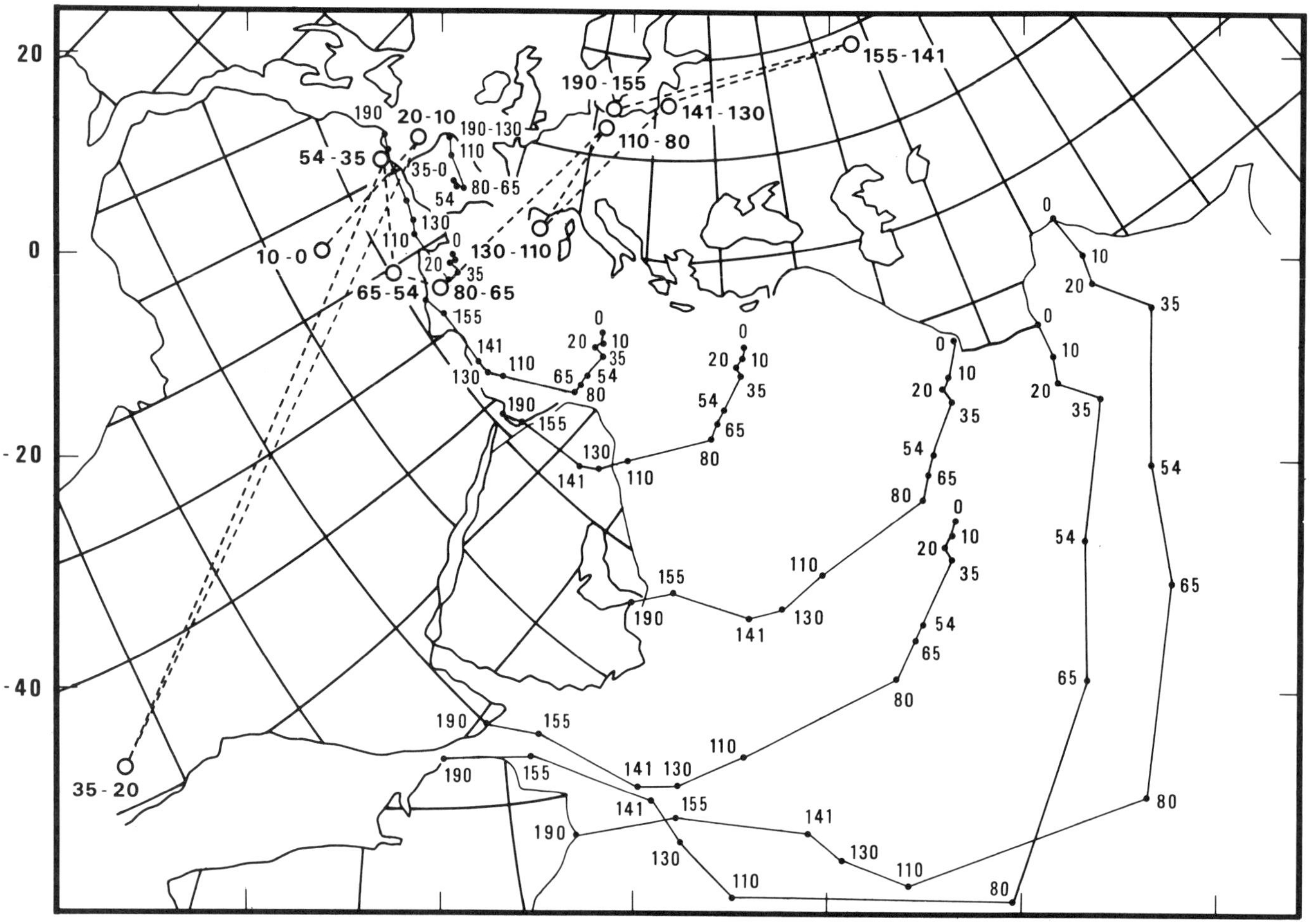

Figure 1. Relative motion history of Africa, Iberia, Arabia, and India with respect to Eurasia from Early Jurassic to Present time. Oblique Mercator projection with North Pole at 50°N, 155°W. Position of continents at 190 Ma. Stages in Ma. Present latitudes and longitudes every 10° on continents. Interpolated flow-lines: continuous lines. Pole positions describing motion of Africa with respect to Eurasia from stage to stage are also shown (circles connected by dashed lines with parameters of Savostin and others, 1985).

significance of these evaluations. Finally, we discuss within this framework the evolution of the West European paleo-stress fields between Eocene time and the present, obtained through quantitative analyses of fault populations by Bergerat (1985).

We do not try here to justify the proposed reconstruction of the evolution of the Tethyan belt; the reader is referred to the 1986 *Tectonophysics* special issue and to Dercourt and others (1985). We do, however, briefly describe some of its main characteristics. The primary basis of the reconstruction is a kinematic analysis of the relative motions of the America, Eurasia, Africa, Arabia, and India plates based on identification of the Vine and Matthews lineations in the oceans. For the central Atlantic, Savostin and others (1986) principally used the recent synthesis of Olivet and others (1984). For the North Atlantic, however, Savostin and others (1986) have retained the solution of Savostin and Karasik (1981) and Savostin (1981) for anomaly 24 to the present. This is because the Savostin analysis is based on Arctic data that are well constrained by their proximity to the Eurasia–North America pole of rotation and are presumably more representative of the motion of the bulk of Eurasia than are the more southerly anomalies. This choice introduces some discrepancy in the fit of the magnetic anomalies in the North Atlantic, which suggests some limited relative motion between westernmost Europe and the main part of Eurasia (e.g., Unternehr, 1982); this point is discussed later. Another problem comes from the fact that Olivet and others (1984) did not give parameters for anomalies 5 (Tortonian), 6 (Aquitanian), and 30 (K/T boundary). Consequently, the corresponding positions had to be interpolated, thus introducing larger uncertainties for these stages.

In the Indian ocean, Savostin and others (1986) mostly used the kinematic analysis of Patriat and others (1982).

Figure 1 shows the resulting relative motions of Africa, Ibe-

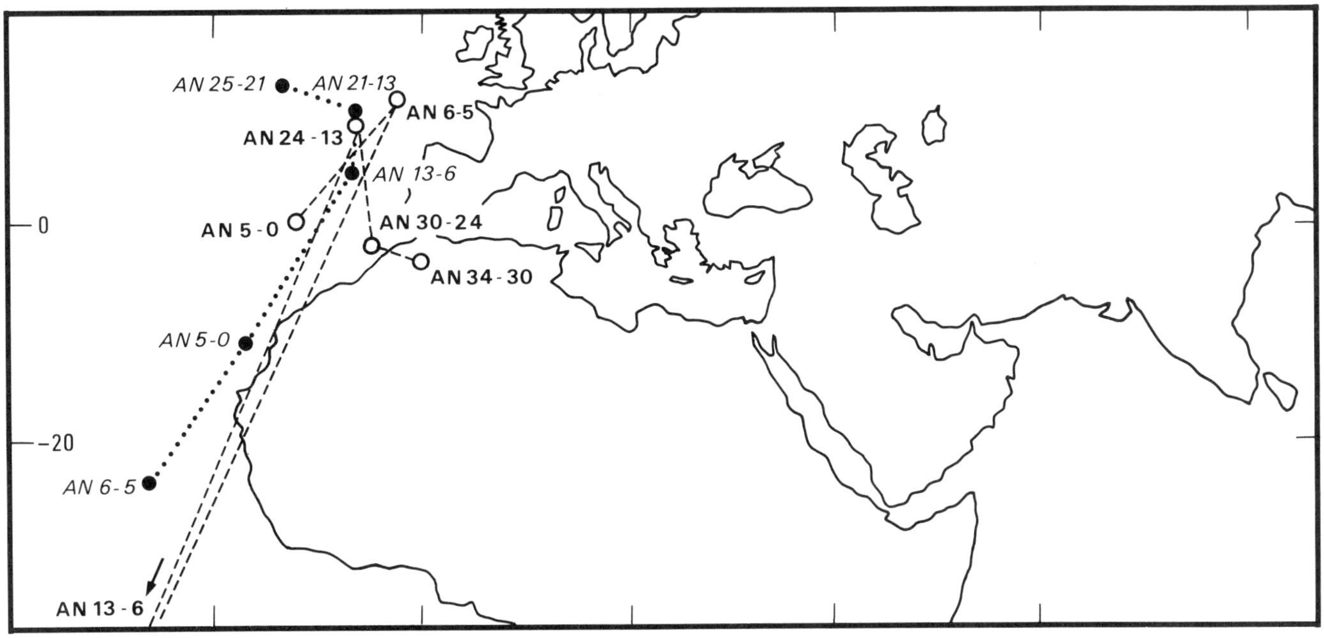

Figure 1b. Comparison of poles of Savostin and others (1986) (open circles) with poles of Klitgord and Schouten (filled circles). Present-day position of continents.

ria, Arabia, and India with respect to Eurasia in the oblique Mercator projection. (Used throughout this project, Mercator projection provides minimal distortion along the Tethyan belt. The pole of the projection is at 50°N, 155°W.) Although there are still considerable uncertainties in the relative positions of the continents, it is felt that significant progress has been made since the early kinematic solutions, such as the one used by Biju-Duval and others (1977). This progress is reflected in the relatively smooth trajectories followed by the different continental pieces in Figure 1. In particular, the spectacular Late Cretaceous hook, proposed by Pitman and Talwani (1972)—which resulted in a change from sinistral to dextral motion of Africa with respect to Eurasia—has disappeared.

Recently, Klitgord and Schouten (1985) published a kinematic solution which is very similar to the one published by Olivet and others (1984) but more complete. This solution was not available to Savostin and others (1986) at the time they established the reconstruction used here. Figures 2 through 10 show the vectors of motion of the main plates with respect to Eurasia, which is arbitrarily maintained in its present position. These vectors correspond to the solution used by Dercourt and others (1985, 1986), based on the kinematic solution of Savostin and others (1986) and on an adopted independent motion for Apulia, between 130 and 80 Ma, based on paleomagnetics (as discussed later). Figures 7 through 10 also show the Africa-Eurasia vectors of motions with the parameters of Klitgord and Schouten (1985). For Figures 7 and 10, the differences are probably within the present uncertainties of this type of determination. The differences in the orientations of the vectors of course become very large near the pole of rotation (see Fig. 1). However, for pre-Aquitanian (Fig. 8) and pre-Tortonian time (Fig. 9), the differences are very large and beyond the uncertainties of the fits. During this period, the use of the Arctic isochrons gives a significantly different result from that obtained using the Atlantic isochrons.

It is interesting to investigate the possible significance of this difference: consider first Figure 8. The vectors of Savostin and others (1986) resulting from the Arctic anomaly fits, can be considered representative of the main part of Eurasia. They indicate an east-southeast motion of Main Eurasia with respect to Africa. Conversely, the vectors of Klitgord and Schouten (1985), resulting from the Atlantic anomaly fits, are representative of westernmost Eurasia. They indicate a southward motion of westernmost Eurasia with respect to Africa. Thus, if these vectors are correct, some limited extension must have been present during pre-Aquitanian time between westernmost Eurasia and the main part of Eurasia. The situation there was nearly exactly reversed (Fig. 9) during Early and Middle Miocene time, when some limited compression should have been present. This is geologically quite plausible, as is discussed in the last section of this paper. Thus, it might have been better to use Atlantic-type fits for reconstruction of westernmost Eurasia during the Cenozoic.

Although the positions of the major plates identified in Figure 1 are entirely fixed by the kinematic solution of Savostin and others (1986), the positions of the smaller plates within the Tethyan belt are not. The paleomagnetic synthesis of Westphal and others (1986) leads to a very significant constraint on the relative motion of Apulia, both with respect to Eurasia and to Africa.

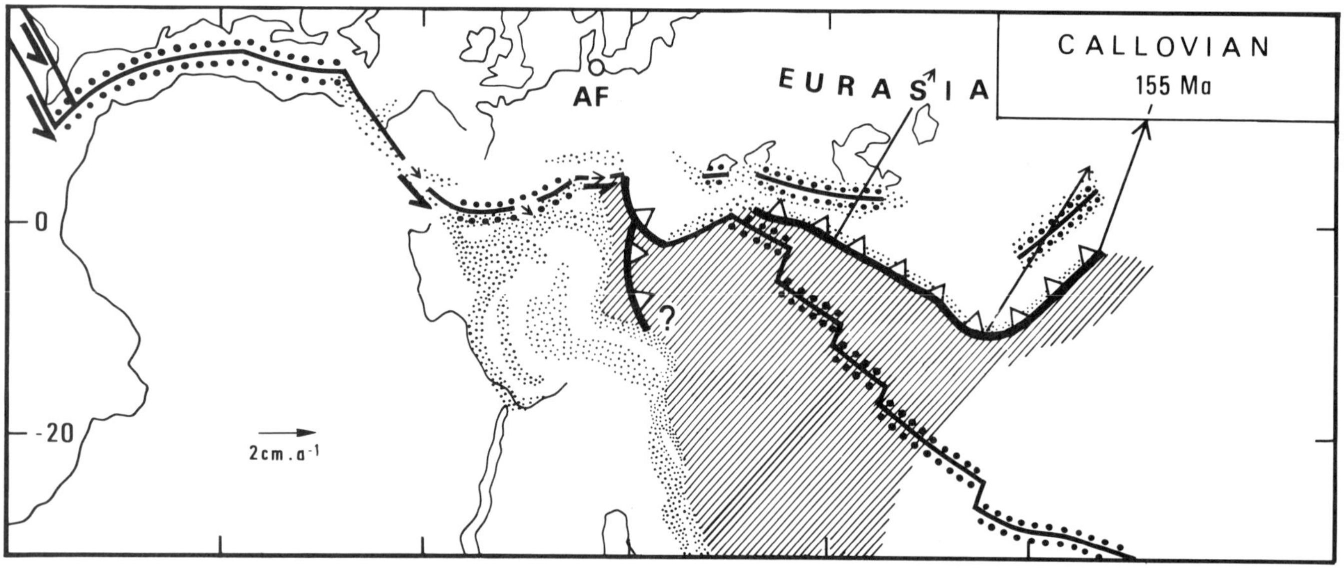

Figure 2. Simplified plate tectonic scheme; Blake Spur Anomaly, 155 Ma Callovian. Eurasia is in its present position. Relative motion vectors in centimeters per annum along northern boundary of Tethys are computed from previous stage (190 Ma) to this stage. Position of poles of rotation of plates shown with respect to Eurasia from previous stage to this stage. Schematic plate boundaries indicated by continuous line; transform; open triangles, oceanic subduction; closed triangles, continental collision; double dotted line, oceanic accretion; hachured pattern, oceanic crust; dotted pattern, thinned continental crust.

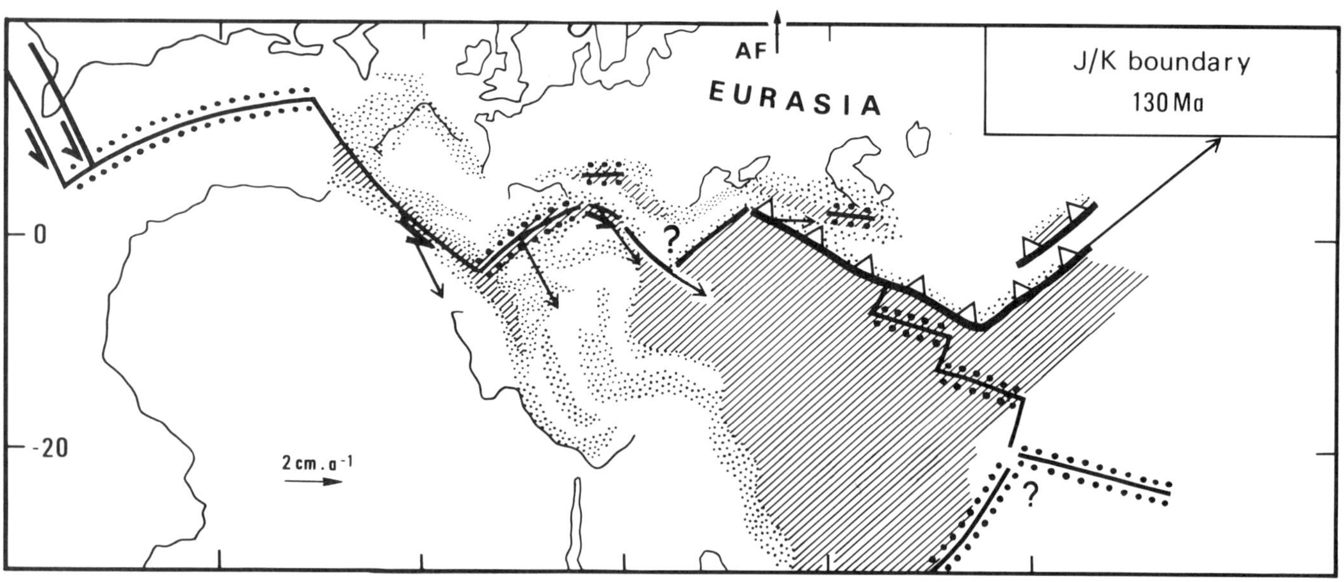

Figure 3. Jurassic/Cretaceous boundary, 130 Ma. Connection with Caribbean after Pindell, 1985. Same legend as in Figure 2.

Westphal and others (1986) considered that the Appennines are allochthonous units that were rotated counterclockwise with respect to Apulia during the Miocene. They consequently considered that the most critical paleomagnetic data for Apulia are those coming from the Gargano peninsula (Van den Berg, 1983), which they interpreted as indicating a counterclockwise rotation of Apulia with respect to Africa during Late Cretaceous time. This Late Cretaceous rotation of about 30° between 130 and 80 Ma led to the formation of the Eastern Mediterranean ocean. Biju-Duval and Dercourt (1980) and Le Pichon (1984) have both argued that geologic data point to a Middle to Late Cretaceous time of origin of the Eastern Mediterranean sea.

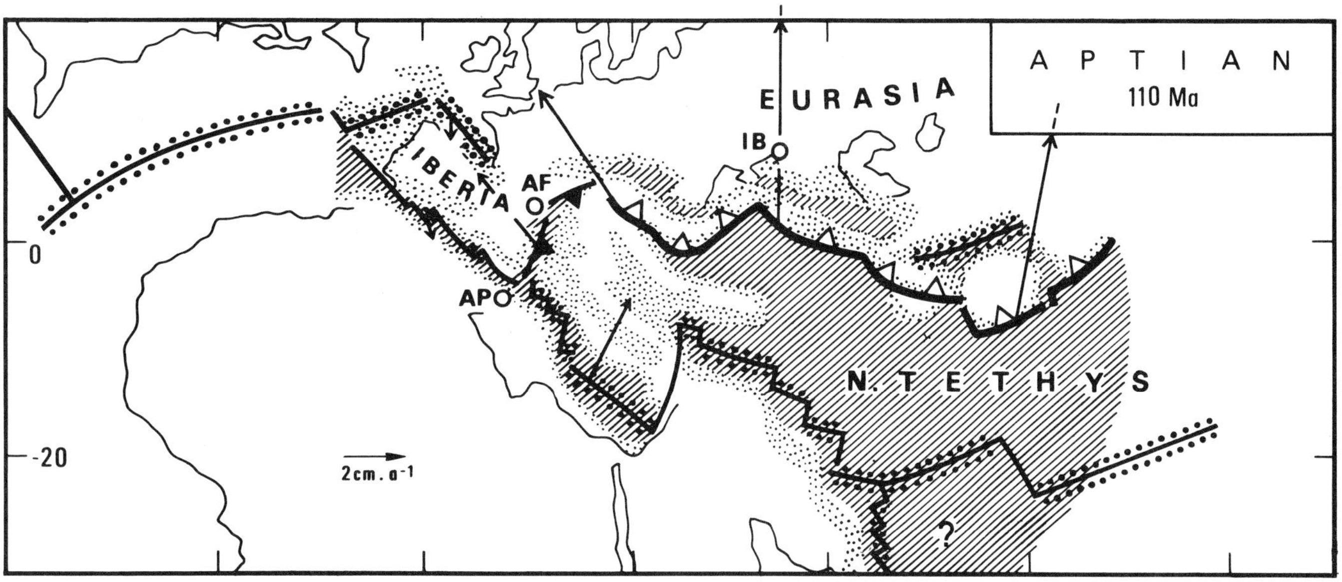

Figure 4. Magnetic anomaly M0, Aptian time, 110 Ma. Same legend as in Figure 2.

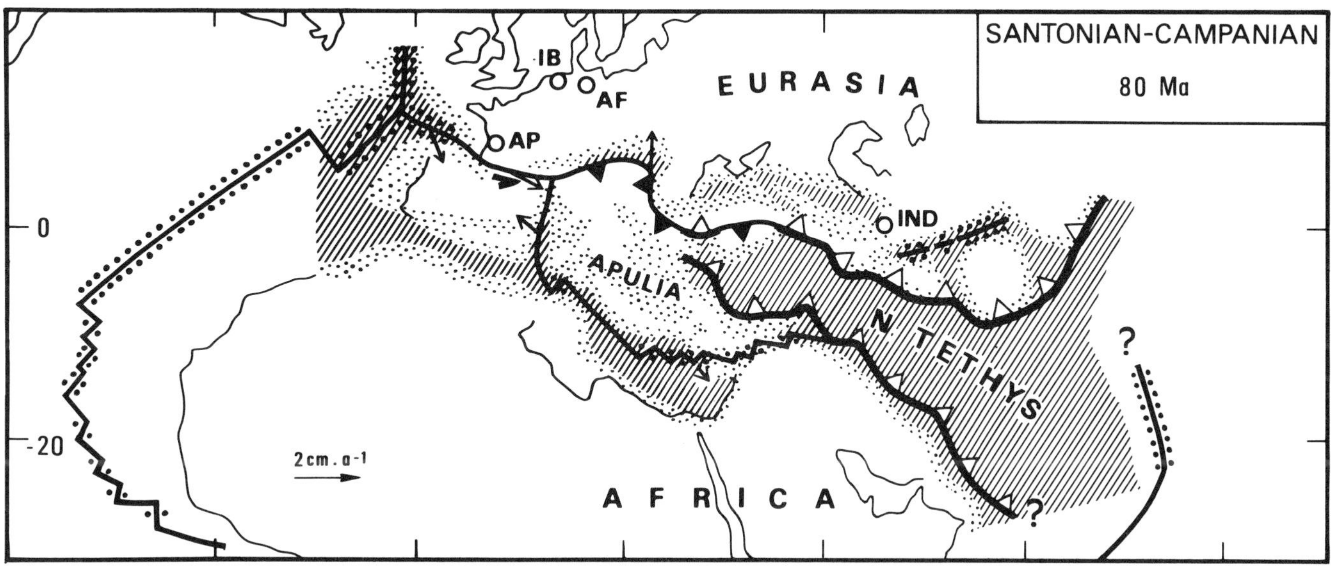

Figure 5. Magnetic anomaly 34, Santonian-Campanian, 80 Ma. Same legend as in Figure 2.

Two other important paleomagnetic constraints in the analysis of Westphal and others (1986) are a post-Oligocene 50° clockwise rotation of western Greece, about 20° in Early Miocene and 30° in post-Medial Miocene, (Epirus: Kissel and others, 1985) and a large—30° to perhaps 90°—not precisely dated, counterclockwise rotation of the Lut-Tabbas block in central Iran. The 20° Early Miocene rotation of western Greece was assumed to result from a "roller-bearing" type of mechanism (Mann and Burke, 1984). A 90° Tertiary rotation was adopted for the Lut block.

A difficult problem with any attempt of this type is the chronology. The Odin time scale (Odin and Curry, 1981; Odin and Kennedy, 1982) has been adopted for the correspondence between isotopic ages and geologic stages. A simplified comparison with another time scale is given in Table 1. The correspondence between magnetic anomalies and geologic stages is from Vogt and Einwich (1979) for Jurassic time, Keating and Helsley (1978) for anomaly 34, Hardenbol and Berggren (1978) for anomalies 24 and 13, and Vincent (1977) for anomalies 5 and 6.

Given these constraints on the positions of the major plates,

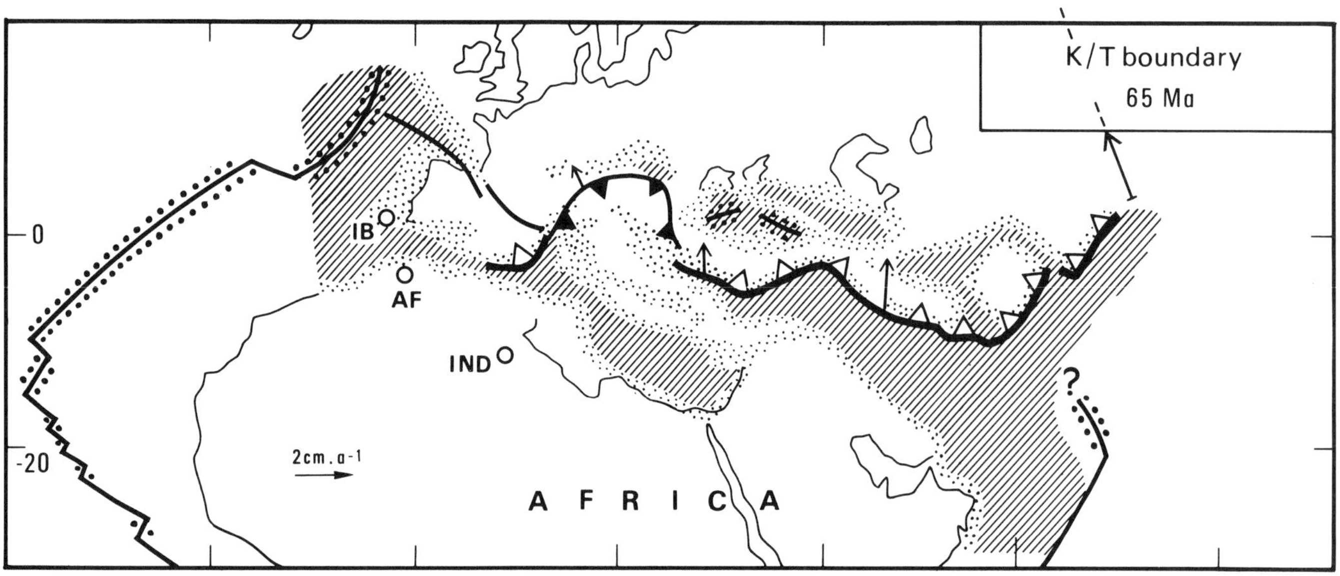

Figure 6. Cretaceous/Tertiary boundary, 65 Ma. Same legend as in Figure 2.

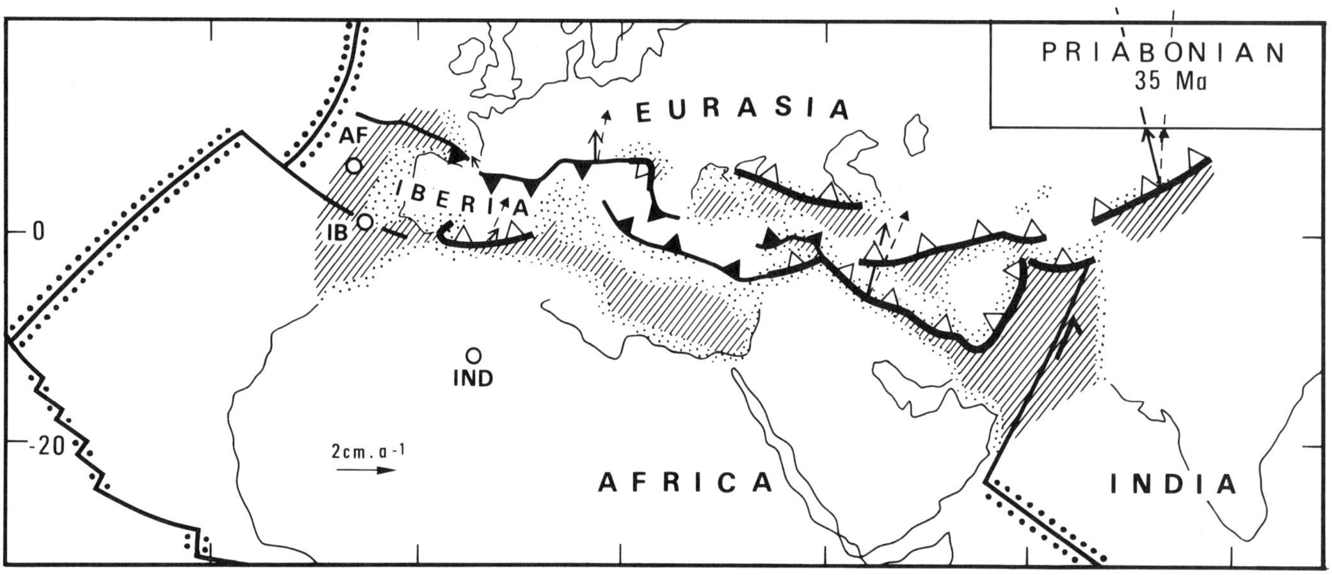

Figure 7. Magnetic anomaly 13, Priabonian time, 35 Ma. Same legend as in Figure 2. Relative motion vectors in centimeters per annum computed with parameters of Klitgord and Schouten are also represented in this figure and in following figures as dashed vectors.

a trial and error method was followed by Dercourt and others (1985, 1986), going through the whole set of maps in order to account for the geologic data. Any change for one single part of a map implies changes over a wider area and through the preceding and following maps. This is especially true for the peri-Mediterranean chains, which form a single entity, without significant oceanic separation. Dercourt and others (1986) discovered that there is very little freedom to adopt significantly different solutions in the western part of the Tethyan belt, once the general constraints are accepted. Thus, unless these constraints are grossly in error, one might hope that the different solutions that might be adopted would all be parts of the same family, with the same general plate tectonic characters.

Figures 2 through 9 depict schematically the main features of the plate tectonic solutions adopted by Dercourt and others (1985, 1986) in their set of paleogeographic maps. These sketches

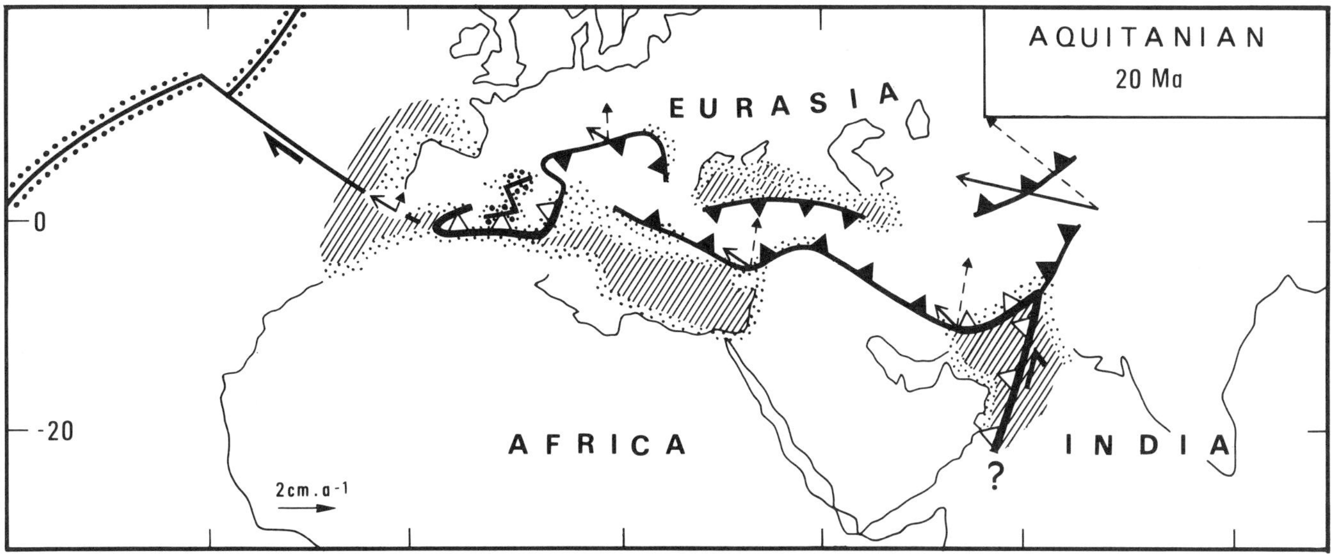

Figure 8. Magnetic anomaly 6, Aquitanian time, 20 Ma. Same legend as in Figures 2 and 7.

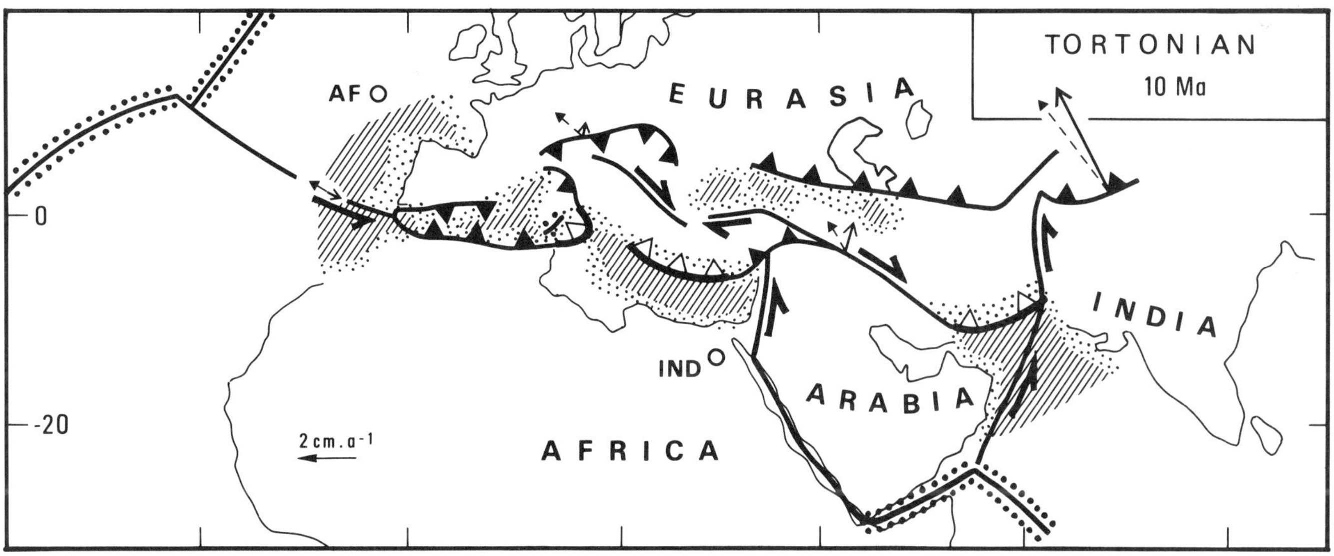

Figure 9. Magnetic anomaly 5, Tortonian time, 10 Ma. Same legend as in Figures 2 and 7.

are sufficient for the purpose of the general plate tectonic discussion below, but note that it would be better to look at these maps while reading this chapter.

EASTERN NEO-TETHYS AND THE MIGRATING ACCRETING PLATE BOUNDARY

Figures 2 through 6 confirm the existence of a wide ocean east of Apulia during the Mesozoic. Its width was about 3,000 km at the longitude of central Iran in medial Jurassic time, about 155 Ma (Fig. 2). It had decreased to less than 1,000 km at the Cretaceous-Tertiary boundary 65 Ma ago (Fig. 6). Thus, the average shortening rate was about 2 cm. This shortening had to be accomplished through subduction. It is now generally accepted that the geologic data observed in the Sanandaj Sirjan chain in Iran are best explained by the existence of subduction zones along the northern margin of Neo-Tethys (Kazmin and others, 1986b). Conversely, the southern margin of Neo-Tethys seems to have behaved as a passive margin (Kazmin and others, 1983a), except locally, during uppermost Cretaceous (Knipper

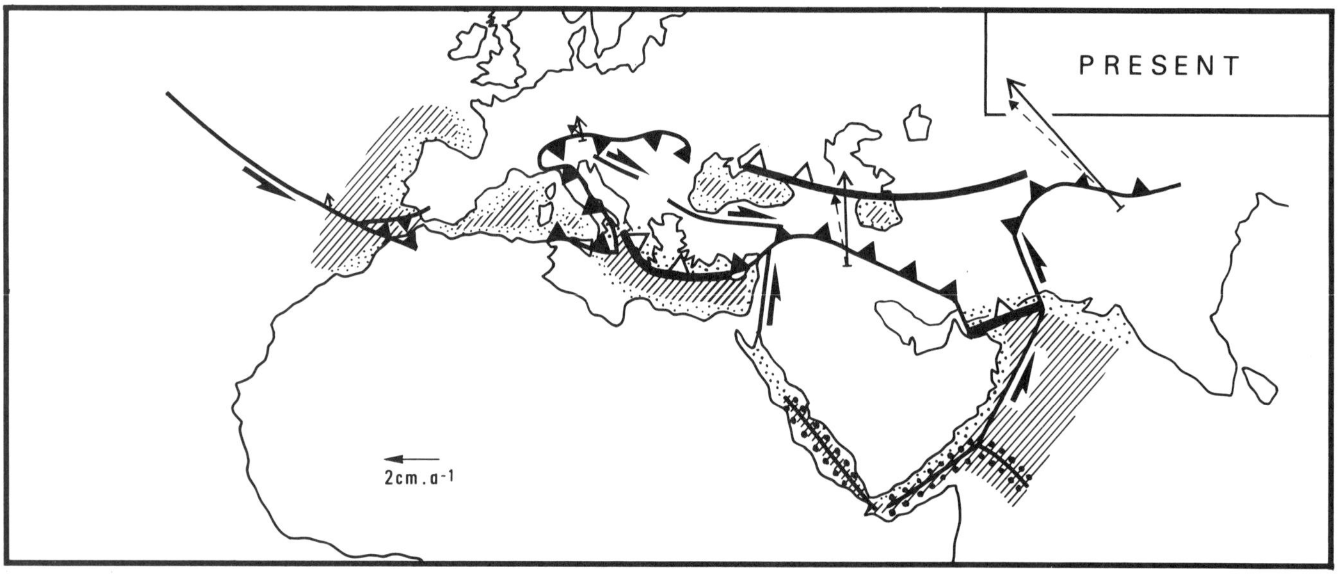

Figure 10. Present-day position of continents. Same legend as in Figures 2 and 7.

and others, 1986; see Fig. 5). 2 cm/yr is small for a subduction rate. Trench pull generally tends to establish higher subduction rates, closer to 6 to 8 cm/yr.

A simple way of establishing a higher subduction rate is to have an accreting plate boundary (a mid-ocean ridge) that either separated the subduction zone from the northern Eurasian continent within a marginal basin or separated the subduction zone from the southern (African) continent, creating an independent entirely oceanic plate. The relatively large width of the Neo-Tethys ocean and its proven existence since at least earliest Jurassic time, during about 150 m.y., suggest that an active accreting plate boundary existed in its midst, as depicted in Figure 2; this is compatible with the relatively young ages of the ophiolites on its northern margin (Knipper and others, 1986).

Consider such a three-plate system (Fig. 2). Because of the existing subduction along the northern margin of the ocean, but not along the southern margin, the accreting boundary migrates northward, with respect to Africa, at a velocity equal to the spreading rate. It also migrates with respect to Eurasia at a velocity equal to the spreading rate plus the Africa-Eurasia convergence rate (assuming, for simplicity of discussion, that all vectors are oriented north-south). The ridge will migrate until it becomes subducted (as in Fig. 3), at which time the three-plate system will change to a two-plate system.

We have seen that the average Africa-Eurasia convergence rate along this part of the boundary is 2 cm/yr. Assuming an average spreading rate of 3 cm/yr, the subduction rate is 8 cm/yr, and the northward migration velocity of the accreting plate boundary is 5 cm/yr. In other words, 60 m.y. are necessary for the accreting plate boundary to cross the whole 3,000-km width of the ocean, but only 30 m.y. are necessary to reach the subduction zone, if the accreting plate boundary started in a median position.

TABLE 1. CORRESPONDENCE BETWEEN TIME SCALES OF ODIN AND PALMER

Magnetic Anomalies	Age	Odin* (Ma)	Palmer† (Ma)
BSMA	Callovian	155	166
M 17	Jurassic/Cretaceous boundary	130	144
M 0	Aptian	110	116
An 34	Santonian-Campanian	80	81
An 30	Cretaceous/Tertiary boundary	65	66.4
An 13	Priabonian	35	38.3
An 6	Aquitanian	20	22.8
An 5	Tortonian	10	8.9

*From Odin and Curry (1981) and Odin and Kennedy (1982).
†From Palmer (1983).

Once the accreting plate boundary has reached the subduction zone, accretion stops and the subduction rate abruptly decreases from 8 to 2 cm/yr. As the subducted plate is very young, it is consequently very hot, light, and short (Molnar and others, 1979). The slab-pull force becomes insignificant. It takes at least 20 m.y. of subduction (or 400 km) to increase the age of the subducted plate sufficiently for the slab-pull force to become significant again. From then on, the slab-pull force will continu-

TABLE 2. STAGE POLE IN EURASIA FRAME

Period	Apulia/Africa	Apulia/Eurasia	Apulia/Iberia	Iberia/Africa
130/110	29°33N 1°95E 27°1	31°47N 3°14E 33°5	30°14N 0°35W 30°8	30°57N 18°22W-3°8
110/80	36°78N 124°37E-3°0	44°23N 0°33E 14°3	60°37N 90.74E-3°8	
65/35				45°73N 20°82W-3°8

ously increase and may well start a new accreting plate boundary—but where? As continental lithosphere is mechanically less resistant than oceanic lithosphere because of the presence of the continental crust (Molnar and Tapponnier, 1975), the splitting is more likely to occur within the continental crust (Vink and others, 1984). If the splitting occurs within the southern African continent, perhaps within the weakened continental margin, the detached narrow band of continent will travel across the ocean at a velocity of 8 cm/yr.

We suggest that this process explains the successive northward migration of microcontinents between a northern Paleo-Tethys and a southern Neo-Tethys in pre-Liassic time (Boulin, 1981; Tapponnier and others, 1981; Dercourt and others, 1985, 1986). Assume, for example, that a new accreting plate boundary was formed within the African margin at 220 Ma; this would lead to collision of a microcontinent with the subduction zone at 180 Ma, and to subduction of the spreading ridge at 150 Ma.

In Figure 4, the geologic scenario proposed by Dercourt and others (1985, 1986) for Aptian time is the formation of a new accreting plate boundary next to the African continental margin, about 20 m.y. after subduction of the preceding spreading ridge, which occurred in latest Jurassic time (Figs. 2, 3). The geologic evidence for the presence of this new ridge next to the African margin consists of the peri-Arabian ophiolites, which were emplaced as oceanic crust in Medial Cretaceous time, 100 to 110 Ma, began to be deformed 90 m.y. ago and then were overthrust on the African margin 80 m.y. ago (Knipper and others, 1986; Dercourt and others, 1986). We believe, unlike Dercourt and others (1985, 1986), that the new ridge was not formed within oceanic crust, but instead, was probably formed within the African continental margin complex.

However, the development of a subduction zone close to this new accreting plate boundary, as indicated by the presence of calc-alkaline volcanic rocks (Coleman, 1981) and the following obduction, indicate a complication in the simple scenario developed above. The new ridge system aborted; from then on, the situation was a simple two-plate system, perhaps because of the small width of the remaining ocean.

QUANTITATIVE EVALUATION OF SUBDUCTION AND COLLISION

In discussing the relative role of subduction and collision in the formation of the Alpine mountain belt, we note that quantitative evaluation of the amount of surface destroyed by subduction and collision is needed. Although there are still many uncertainties in the reconstructions of Dercourt and others (1985, 1986), they can be used relatively simply to obtain a quantitative evaluation of the amount of subduction and collision, provided their estimates of the times at which continental collision began are correct.

It is very simple to evaluate the net amount of shortening, between Iberia-Africa-Arabia-India and Eurasia, from the Atlantic to Pamir (the Himalayas excluded), using the parameters published by Savostin and others (1986). The amount of shortening corresponds to the surface outlined by the trajectories in Figure 1. However, we want to be able to separate the shortening due to collision from the shortening due to subduction. To make the problem tractable, we have neglected the surface created within marginal basins along the southern border of Eurasia, which results in an underestimate of the amount of subduction. In addition, we have assumed that all the shortening occurs along one single boundary on the southern border of Eurasia. The parameters for Iberia and Apulia, which were not given by Savostin and others (1986), have been computed and are given in Table 2.

The rotation of Apulia was schematically represented by Dercourt and others (1985, 1986) as resulting from two successive finite rotations, between 130–110 and 110–80 Ma. These were chosen in such a way that they can account for the repeated overthrust of Austro-Alpine material above the European margin, and such that the total rotation with respect to Africa is slightly less than 30°, based on paleomagnetic data (see Westphal and others, 1986). The total equivalent rotation is actually 27.5°, which is perhaps a little bit larger than indicated by paleomagnetics. The passage from subduction to collision is as indicated on the maps of Dercourt and others (1985, 1986), and is represented in a schematic fashion in Figures 4 through 9. The only remaining major uncertainties are the direction and amount of accretion along the oceanic accreting plate boundary shown in Figures 2 through 4. In the absence of any information, we have assumed very arbitrarily a constant total accretion rate of 6 cm/yr parallel to the subduction vector; that is, we have added 6 cm/yr to the subduction vectors situated north of the oceanic accreting boundary to take into account the effect of increased subduction rate. The corresponding vectors of motion—shown in Figures 2 through 9, Tables 3 and 4, and Figures 11, 12, and 13—recapitulate the main results of these computations. The fact that we do not know the actual spreading rate along the oceanic ridge

TABLE 3. SUBDUCTION ALONG THE EU/AF-AR BOUNDARY*

Period (Ma)	Surface Disappearance (10^6 km^2)	10^3 km^2/m.y.	Length of Boundary (km)	Rate (km^2/m.y.)
190-155	7.9	225	5270	43
155-130	1.9	75	3660	21
130-110	6.3	316	6140	52
110-80	3.0	100	4650	22
80-65	3.6	242	6010	40
65-35	6.3	209	6260	33
35-20	0.22	15	3350	4.4
20-10	0.21	21	2230	9.4
10-0	0.47	47	1860	25

*Total surface subducted is 30 x 10^6 km^2. Total length subducted is 5,700 km.

TABLE 4. COLLISION ALONG THE EU/AF-AR BOUNDARY AND ALONG THE AP/EU BOUNDARY*

Period (Ma)	Surface Disappearance (10^6 km^2)	10^3 km^2/m.y.	Length of Boundary (km)	Rate (km^2/m.y.)
110-80	0.68 (0.32)	23 (11)	2980 (1550)	7.6 (6.8)
80-65	0.11 (0.11)	7.4 (7.4)	1800 (1800)	4.1 (4.1)
65-35	1.2 (0.36)	40 (12)	3720 (1000)	11 (12)
35-20	1.3 (0.21)	85 (14)	7000 (2050)	12 (7.0)
20-10	0.75 (0.086)	75 (8.7)	7500 (2170)	10 (4.0)
10-0	1.2 (0.19)	120 (19)	8930 (2910)	14 (6.7)

*First figure: EU/AF-Ar boundary. Figure in parantheses: AP/EU boundary.

Notes:
Total surface disappeared by collision: 5.3 x 10^6 km^2 (1.3 x 10^6).
Total length disappeared by collision: 1020 km since 110 Ma; 740 km since 65 Ma.

crest results in a large uncertainty in the evaluation of the total amount of surface subducted, but it does not affect in any way the evaluation of the amount of surface destroyed by collision. This is because the collision rate estimate is independent of the spreading rate estimate within the ocean.

To understand the significance of these computations, it is necessary to first consider the continental collision in the Alps, north of Apulia. There was no indication of collision at 130 Ma at the Jurassic-Cretaceous boundary, and oceanic crust was still present to the north of Apulia, although collision appears to have begun to the northeast and east (Dercourt and others, 1985, 1986). By 110 Ma, collision in the Alps was clearly occurring over the whole northern margin, as the Apulian margin was overthrust on the Eurasian margin (Tollman, 1966; Dercourt and others, 1985, 1986; Geyssant, 1980). Thus, collision started 110 Ma at the latest, and subduction ended, at the earliest, 130 Ma. Let us assume conservatively that collision started 110 Ma. Then, the average amount of continental shortening since Aptian time in the Alps has been somewhat more than 800 km, corresponding to an average shortening rate of 0.7 cm/yr, the rate being maximum during Eocene time (1.2 cm/yr). The present average width of the Alps is about 150 km; thus the shortening factor is about six. However, the 800 km of shortening include the Apulian and south European margins, as well as the Valais trough, which may have reached the oceanic stage, at least in its eastern portion. Thus about 400 km, or possibly somewhat more, of these 800 km of shortening correspond to thinned continental margin crust.

By early Oligocene, 35 Ma, the Valais trough had been

totally crushed. So, at least, the 200 km of collision since that time must have been "true" continental collision of normal thickness crust. We conclude that sometime during the Eocene, the transition from thinned continental crust collision to "true" collision occurred. Note that this estimate of 200 km of shortening since early Oligocene is about three times larger than the estimate of 55 to 75 km made by Trümpy (1973, 1975, 1980) on the basis of structural considerations. His conservative method apparently seriously underestimated the total amount of shortening.

As pointed out by Gillet and others (1986), the metamorphic facies indicate that deep crust was involved in the early African thrusts of the Austro-Alpine units, as shown by the widespread occurrence of pre-Alpine granulitic rocks. Later thrusting of the European basement involves only the upper crust. Finally, the latest thrusting in the external zones involves mostly the post-Hercynian sedimentary cover. We propose that this observation can be explained by assuming that only the brittle portion of the crust is involved in the stacking of nappes, and that the décollement occurs at the brittle ductile boundary, or occasionally at a higher lithologic interface. The brittle-ductile transition would be expected to occur at a temperature of about 350°C in continental-type material or at a temperature of about 650°C in mantle-type material. In a "normal" temperature and heat-flow situation, the brittle crust is unlikely to exceed a thickness of about 15, or perhaps 20, km. On the other hand, even in a "hot" continental crust, it is unlikely to be less than 7 km thick. Thus, in the early thrusting, when thinned, lower continental margin–type crust is involved, the deep granulitic facies is part of the brittle crust. But later, during the thrusting of normal thickness upper continental margin- to shelf-type crust, the lower crust belongs to the ductile domain. Thus only the upper continental crust is involved in the stacking. *The key to our hypothesis is that thrusting always involves approximately the same thickness above the brittle-ductile transition and that consequently the lower ductile portion disappears into the mantle.* This proposal—a completely different interpretation from the one proposed by Gillet and others (1986)—is supported by volumetric estimates of the amount of crust involved in collision. In any case, note that the metamorphic evidence they discussed implies that a great amount of lower continental crust had been lost during the stacking of the European units.

The total excess volume of continental crust to dispose of can be estimated in a crude fashion. Again, to be on the conservative side, we assume that before 35 Ma, the collision involved only thinned continental crust having an average thickness of only half the normal 35-km thickness: this is equivalent to a volume of about 14×10^6 km^3 of crust. Since 35 Ma, the continental crust involved had its normal 35-km thickness; the corresponding volume of crust is 17×10^6 km^3, so the total amount is about 30×10^6 km^3. Compare this with the volume of 4 to 5×10^6 km^3 presently stored in the roots of the Alps. Where are the missing 25×10^6 km^3? Could they have been eroded and disposed of as sediments in the adjacent basins? We have seen above that metamorphic evidence indicates that the lower portion of the continental crust was never involved in the stacking of the European units and that consequently it must have disappeared in the mantle. This is confirmed by the fact that the largest available reservoirs of sediment are the Mediterranean basins that contain a maximum of about 8 to 10×10^6 km^3 (using the figures of Le Pichon, 1984). Yet, a good part of the Mediterranean sediments came from sources other than the Alps. Even if we take into account the large North Sea, Ligurian, and Pannonian basins, the total amount of eroded sediments coming from the Alps is unlikely to exceed about 10×10^6 km^3.

We have not yet taken into account the lateral expulsion of the Alps, toward the west in the western Alps and toward the east in the Carpathian belt. Such lateral expulsion has certainly been significant since Early Miocene time. If we assume that lateral expulsion accounts for a maximum of half the total shortening since the Early Miocene along the whole length of the Apulian collision boundary, then the volume of crust involved in the shortening is reduced by 5×10^6 km^3. Thus about 5×10^6 km^3 are stored in the excess crust of the Alps, not more than 10×10^6 km^3 have been eroded away, and not more than 5×10^6 km^3 can be accounted for by lateral expulsion. As a result, at least 10×10^6 km^3 are missing and consequently must have disappeared into the mantle in agreement with our earlier conclusion.

To avoid the problem of lateral expulsion, the same quantitative estimates have been made by C. Grandjacquet (in preparation, 1986), both along sections and over the maps, for the central portion of the Alps, which cannot have been involved in any lateral expulsion. The estimates of Grandjacquet reach conclusions similar to ours. About half the crust involved in the shortening cannot be accounted for. We believe that this missing half is the lower part of the crust, which, outside the Austro-Alpine units, is not involved in the stacking of the nappes. Note that the nappes do not have thicknesses exceeding 10 to 15 km, which is less than half the thickness of the continental crust.

Having evaluated the significance of these results, we return to a more general discussion of Tables 3 and 4 and Figures 11 through 13. Consider first subduction, with Table 3 and Figure 11. Prior to Oligocene time, the subduction rate oscillated between 0.3 and 0.1 km^2/yr (compared to a total subduction rate of 3.45 km^2/yr at present; Sclater and others, 1980; Le Pichon and others, 1985). The maxima in Early Jurassic and Early Cretaceous time correspond to the presence of an accreting plate boundary within the Neo-Tethys ocean (three-plate system), whereas the minima in Late Jurassic and Late Cretaceous time are related to a two-plate system without an active Tethys ridge. Finally, the Early Tertiary maximum is related to the highly increased subduction along what is now Pakistan, opposite the Indian plate.

Figure 11 indicates that subduction becomes insignificant after Eocene time and cannot possibly be considered to have contributed significantly to the dynamics of the Tethyan belt west of Pamir in the last 35 m.y. This is because India was already colliding with Eurasia and because Arabia was also entering into collision with Eurasia. The remaining flysch basin in Zagros was

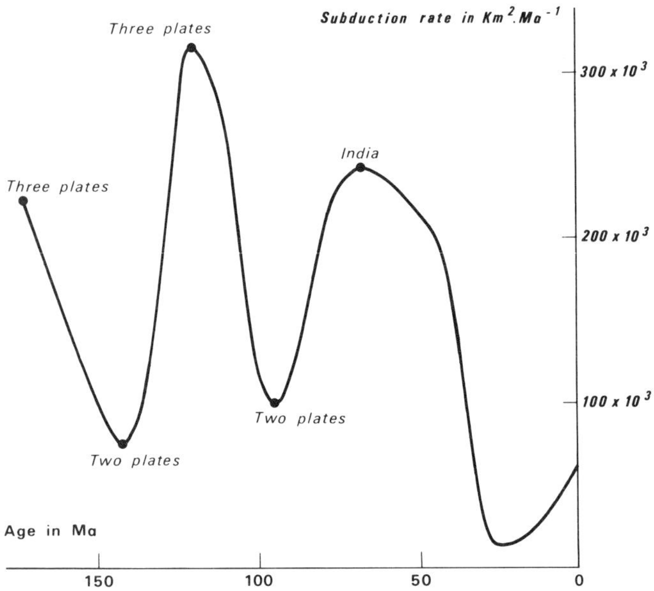

Figure 11. Variations in rate of subduction along northern boundary of Tethys.

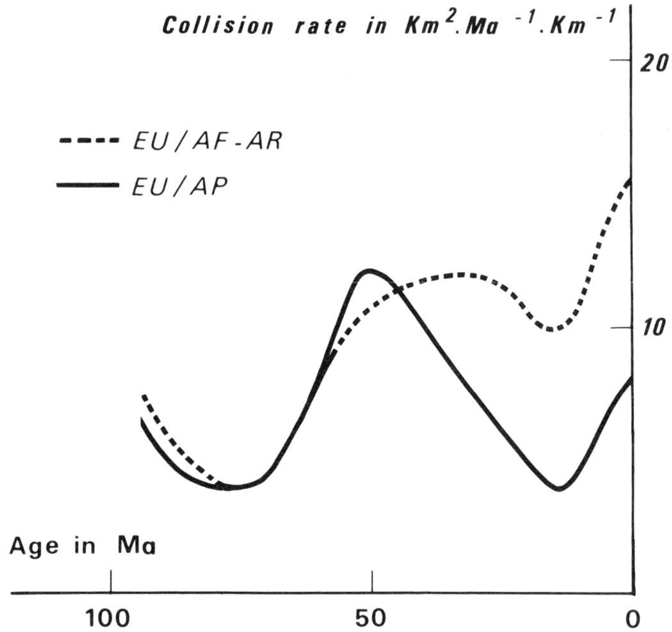

Figure 12. Variation in rate of collision along the northern boundary of Tethys and along the Alpine frontier.

finally filled and crushed before Aquitanian time (Kazmin and others, 1986a). Although it is clear, for reasons discussed earlier, that the subduction rate estimates are highly inaccurate, these figures establish that subduction was both significant and quite variable prior to Oligocene time and that it has not been significant since. The amount of surface subducted was several times larger than the total surface of the Neo-Tethys ocean in Lower Jurassic time.

Consider now the collision results in Table 4 and Figures 12 and 13. The present average collision rate is 0.12 km^2/yr, a figure that compares closely with the one given by Parsons (1981) of 0.1 km^2 for the Eurasia collision with Arabia and Africa. Our estimate includes the part of the collision between India and Eurasia west of Pamir, which explains why it is larger than Parsons's figure. The main indication of Figures 12 and 13 and Table 4 is that the collision rate increases very rapidly and nearly linearly since Late Cretaceous time because collision progresses from west to east. The major change happens between Eocene and Oligocene time, mostly because of the active continental collision between India and Eurasia since 35 Ma. The total surface area of continent that has disappeared is about 5×10^6 km^2. The portion corresponding to the Himalayas, east of Pakistan, is an approximately equivalent amount since 35 Ma. So the total amount of surface disposed of by continental collision along the Tethys belt is of the order of 10×10^6 km^2. Some 2.4 to possibly as much as 4×10^6 km^2 corresponds to the present area of double-thickness continental crust (Allègre, 1985; Harrison and

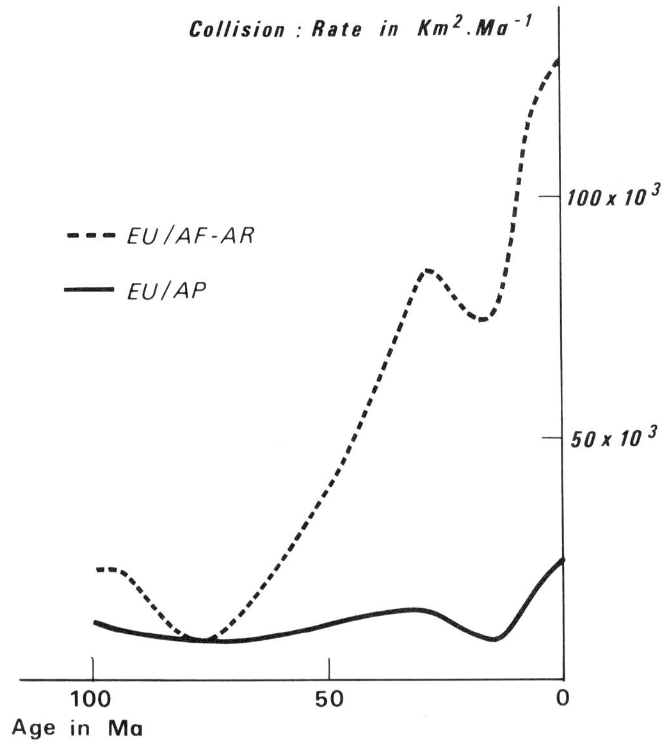

Figure 13. Same legend as in Figure 12.

others, 1983). The remaining 7×10^6 km^2 probably corresponds to about 185×10^6 km^3 of crust (half the 35-km thickness and half the 17.5-km thickness). This is compared to the amount of sediments present in the Bay of Bengal (8×10^6 km^3; Harrison and others, 1985) or the total amount of sediments trapped in the world continental margins (44×10^6 km^3).

Part of this surface actually corresponds not to surface destruction but to a transfer through lateral escape of continental crust (Molnar and Tapponnier, 1975). It is difficult to estimate the amount of lateral escape of Asiatic continental crust toward the Pacific. Lyon-Caen and Molnar (1985) recently argued, assuming steady-state flexure of the basement below the Ganga plain south of the Himalayan range, that the rate of convergence during the last 15 to 20 m.y. was only 10 to 15 mm/yr instead of 50 mm/yr, as given by plate kinematics. As Molnar and Deng (1984) had previously argued that the rate of crustal shortening in Tien Shan appears to be only 10 ±5 mm/yr, Lyon-Caen and Molnar (1985) concluded that half or more of the present convergence rate occurs by squeezing parts of Tibet and areas north and east of it eastward. Lyon-Caen and Molnar, however, have implicitly assumed that the relative plate velocity is perpendicular to the local orientation of the Ganga plain. This is not correct, as it makes a large angle of 50 to 70° with this perpendicular along the western and central sets of drilling holes they used. Thus the rate of convergence obtained, using their assumption, should be 30 mm/yr compared to about 45 mm/yr for the plate kinematic velocity. Consequently, if their hypothesis is correct, these data actually suggest that little velocity is absorbed by lateral expulsion. We believe that the best estimate is that approximately half the total volume of shortened continental crust probably disappeared into the mantle, about one-fourth was eroded away or was absorbed by lateral crust expulsion, and the remaining one-fourth is stored in the roots of mountain belts. The part of the crust that has disappeared corresponds to the lower ductile continental crust.

WESTERN EUROPEAN PALEO-STRESS FIELD AND CONTINENTAL COLLISION

The preceding section has emphasized the significant transition that occurred between 80 and 35 Ma, from a subduction-dominated Tethyan belt to a collision-dominated one. This evolution in western Europe, between Late Eocene and Present, is now considered in more detail, based on quantitative analyses of fault populations recently made by one of us (Bergerat, 1985).

The study of fault populations was made in zones that have been little affected by the folding that occurred in the forefront of the Alpine belt. Special studies were made along the major Pfahl and Franconia lineaments and along the west European Rift. In most cases, faults of differing ages have been found on the same sites. The separation of successive tectonic episodes and the establishment of their chronology were first achieved by field observations. Then, the main geometric characteristics of the mean stress field for each tectonic episode were established (Angelier, 1979, 1984). Several thousand field measurements were supplemented by previously published data on the least deformed portions of the Iberian and North African platforms, as well as on the intracratonic belts. The data are quite coherent and have been used to reconstruct the mean paleo-stress fields in Late Eocene time (Fig. 14), Oligocene (Fig. 15), Early Miocene (Fig. 16), Late Miocene (Fig. 17) and present (Fig. 18).

Late Eocene, about 40 Ma

During late Eocene time, the West European paleo-stress field was remarkably homogeneous, with an approximately north-south horizontal compression that resulted in a system of large strike-slip faults (Fig. 14). The major fault system at this time was the "German-Czech triangle" (Bergerat and Geyssant, 1980). This "triangle" is limited, to the east, on the border of the Bohemian Massif, by right-lateral strike-slip faults (Pfahl and Franconia lineaments), oriented 130 to 150°. To the west, it is limited by left-lateral strike-slip faults, oriented 0° to 20° to 45°, which will give rise to the future West European Rifts. Note that this is the time at which the transition occurred in the Alps from thinned continental crust collision to "true" collision (see above).

In the Iberic and Catalan chains within Iberia, the north-south Late Eocene to Oligocene compression results in left-lateral reverse NE–SW faults and in folds (Guimera, 1983). Lepvrier and Mougenot (1983, 1984) have shown that this compressive episode results in strike-slip faults in "several regions of Portugal."

The whole West European platform was thus affected by a constant north-south horizontal σ_1. At this time, the pole of rotation of Africa with respect to Eurasia was situated northwest of Portugal, according to both Savostin and others (1986) and to Klitgord and Schouten (1985, see Figure 1b). Thus, the relative Africa-Eurasia motion is also oriented approximately north-south, if the boundary on which this motion is computed approximately coincides with the Pyreneo-Alpine boundary.

Oligocene, about 35 to 30 Ma

There is a remarkable change during Oligocene time in the paleo-stress field of the West European platform (Fig. 15). The tectonic regime changes from compression to extension, which results, particularly on the western border of the German Czech triangle, in the formation of the West European Rift system. The first phase of extension of the Ligurian sea may have been initiated at this time (e.g., Rehault, 1981; Burrus, 1984; Le Pichon, 1984). Extension can also be observed at this time along the Pfahl lineament and in the vicinity of major faults, such as the Lauchert graben and the Bohemian troughs. The direction of extension is generally approximately east-west, except in the southern part of the West European Rift, where it becomes NW–SE.

Within the Alps, large right lateral strike-slip faults to the north (Insubrian and Gail faults) and left-lateral to the south (Ligurian fault) could have been related to the westward motion of an Insubrian block (Laubscher, 1971).

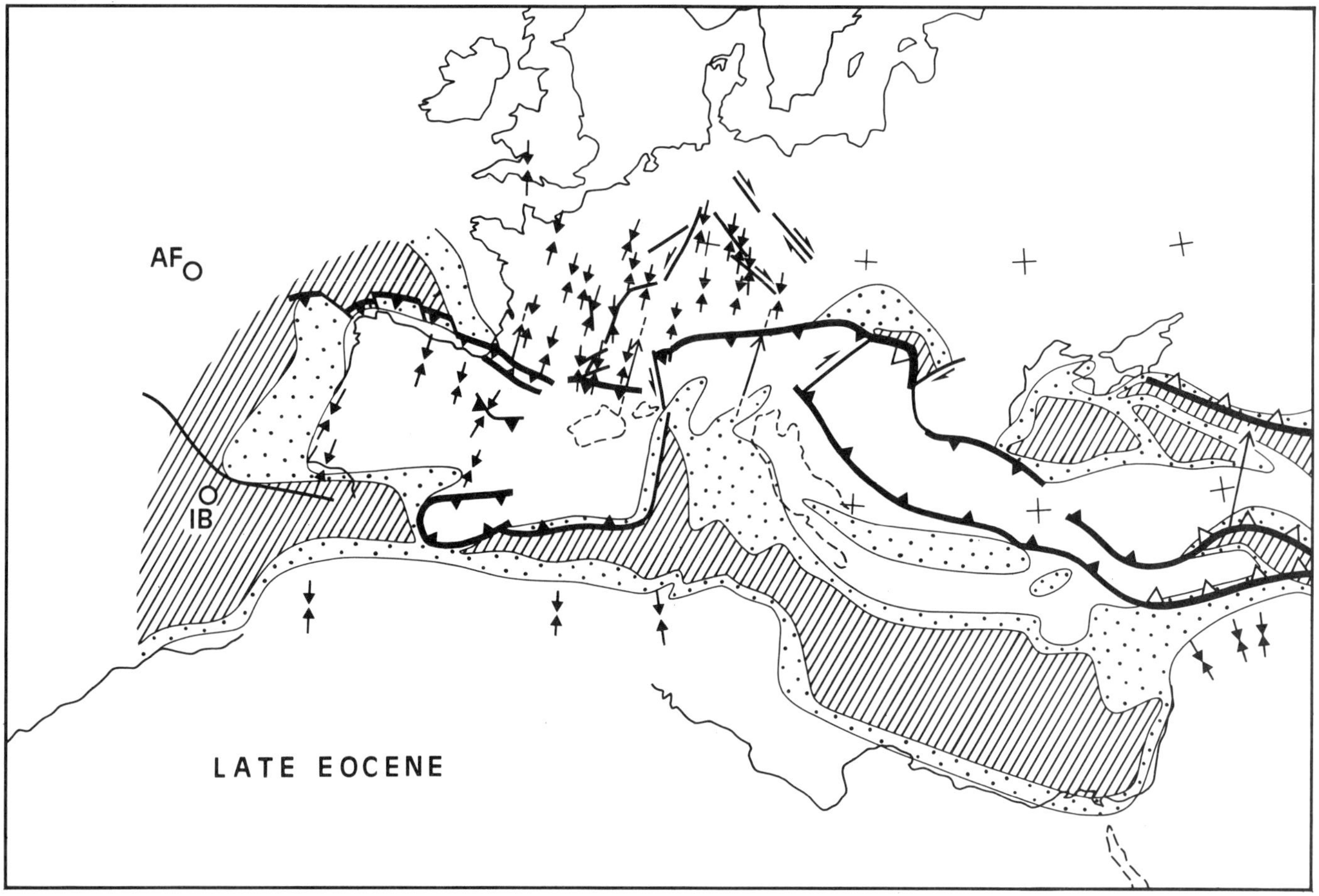

Figure 14. Reconstruction of late Eocene stress field and geodynamic evolution of European platform in Cenozoic Africa/Eurasia collision framework. Eurasia is fixed. AF indicates rotation pole Africa/Eurasia; IB, rotation pole Iberia/Eurasia; AR, rotation pole Arabia/Eurasia. Structural and paleogeographic framework after Dercourt and others (1985). Paleo-stress data from Bergerat (1985) (European platform), Letouzey and Trémolières (1980) (North Africa, Arabia), Guimerà (1983, 1984), Viallard (1979), Lepvrier and Mougenot (1984) (Iberian peninsula). See map legend Figure 15. Rotation poles and motion vectors between 54 and 35 Ma.

We have already discussed relative motion of Africa with respect to Eurasia in Oligocene time (Figs. 1, and 8). There is a large discrepancy between the relative motion based on Arctic magnetic lineation data (as given by Savostin and others, 1986), which results in a relative southeastward motion of Eurasia, and the relative motion based on Atlantic magnetic lineation data (as given by Klitgord and Schouten, 1985), which results in a relative southward motion of Eurasia. Assuming that this large difference represents the difference in motion between the main part of Eurasia and westernmost Eurasia, the resulting relative motion is indeed extensional between west Eurasia and central Eurasia, accounting for the change in tectonic regime within Eurasia, and with compression still being prevalent within the Alpine belt. This is clearly shown in Figure 19, where the positions of western Eurasia are derived after Klitgord and Schouten (1985), whereas the positions of the main part of Eurasia are after Savostin and others (1986).

Everything happens, at this time, as if west Europe had begun an evolution similar to the one followed by Iberia, which had been transferred from the Eurasian to the African plate. A new plate boundary began to be formed along preexisting weakened lineaments (Rheno-Rhodanian fault system), but the evolution aborted in Early Miocene.

Early Miocene, about 22 to 20 Ma

In Early Miocene time (Fig. 16), all the large west European fault systems are reactivated by a NE–SW compression. The

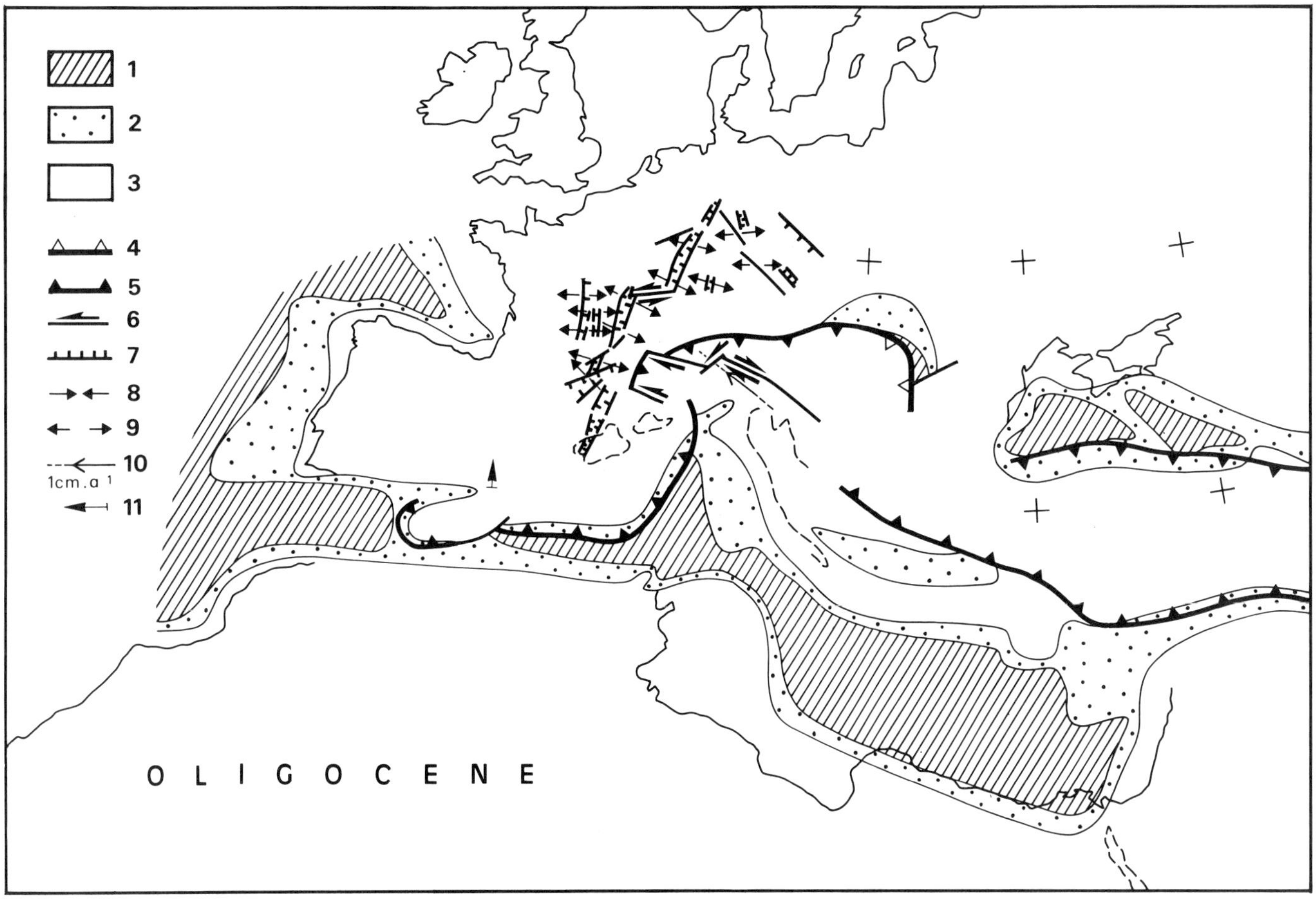

Figure 15. Reconstruction of Oligocene stress field and geodynamic evolution of European platform in Cenozoic Africa/Eurasia collision framework. 1 indicates oceanic crust; 2, thinned continental crust; 3, continental crust; 4, subduction; 5, overthrust; 6, strike-slip fault; 7, normal fault; 8, azimuth of main maximal stress σ_1; 9, azimuth of main minimal stress σ_3; 10, relative vector motion Africa/Eurasia (centimeters per annum), from Savostin and others (1986). 11, relative vector motion Africa/Eurasia (centimeters per annum), from Klitgord and Schouten (1985). Rotation pole and motion vectors between 35 and 20 Ma.

Pfahl and Franconia lineaments are affected by reverse faulting; the border faults of the Rhine graben are affected by right-lateral strike-slip faulting. In the Iberic chain, the major shortening phase is Late Oligocene to Early Miocene. It resulted in 120 to 130 folds with stacking and thrusting toward the southwest, as well as in left-lateral strike-slip faults (Viallard, 1979). This NE–SW compression was also described by Lepvrier and Mougenot (1984) in the Mesozoic and Cenozoic cover of Portugal, where it produced left-lateral faults trending ENE–WSW to ESE–WNW. It was dated by these authors as being close to the Oligo-Miocene boundary or Early Miocene time.

Consequently, the West European platform was affected by NE–SW horizontal compression during Early Miocene time. The direction of σ_1 is quite constant over the whole platform, around 45°.

As discussed above, the kinematic situation is indeed exactly reversed with respect to Oligocene time (Fig. 19). The solution of Savostin and others (1986), presumably representative of western Eurasia with respect to Africa, indicates approximately north-south motion. The solution of Klitgord and Schouten (1985), presumably representative of western Europe with respect to Africa, indicates NW–SE compression. The difference between the two motions is such that compression should exist between western Europe and central Europe, independent of the general Africa-Eurasia shortening.

Late to Post-Miocene, about 7 to 4 Ma

North of the Alps, the main fault systems are situated immediately before the Alpine chain, in the Jura and Septentrional

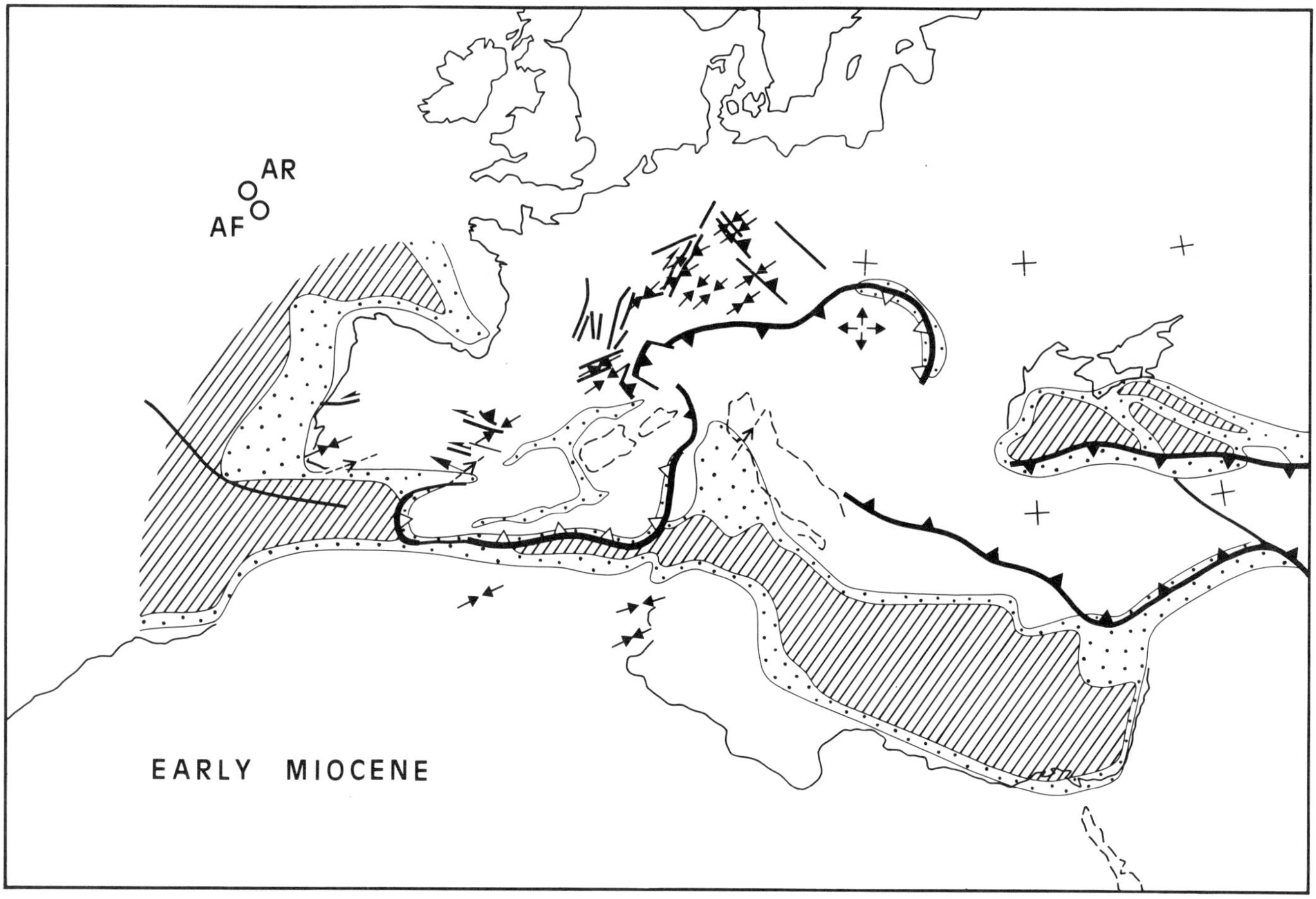

Figure 16. Reconstruction of early Miocene stress field and geodynamic evolution of European platform in Cenozoic Africa/Eurasia collision framework. Same legend as in Figure 15. Rotation pole and motion vectors between 20 and 10 Ma.

Subalpine chains. A remarkable outward fanning of the σ_1 trajectories characterizes this new pattern (Fig. 17), in complete contrast to the earlier constant patterns discussed above.

In the French Massif Central, numerous 8- to 6-Ma dikes indicate the existence of a NW–SE compression (Féraud and Campredon, 1983; Gastaud and others, 1983).

In Portugal, Lepvrier and Mougenot (1984) attributed to Late Miocene time a NNW–SSE compression, which reactivates earlier left-lateral faults. In Spain, a NW–SE shortening is dated as post-Burdigalian by Letouzey and Trémolières (1980).

To summarize, the part of the West European platform closest to the Alpine front was characterized by an outward fanning of σ_1 (from NW–SE to the east, to E–W to the west, and N–W to the south). Farther away from the Alps, σ_1 was close to NW–SE.

Present

The Late Miocene situation is very similar to that in the present (Fig. 18). In situ stress measurements and earthquake fault-plane mechanisms have given results quite consistent with those coming from structural data (Philip, 1980, 1983). A NNW–SSE horizontal compression over most of the West European platform is replaced, close to the Alpine front, by an outward fanning of the compressive stress trajectories.

The general NNW–SSE compression over the platform is in fair agreement with the relative direction of motion between Africa and Eurasia, both in the solutions of Savostin and others (1986) and Klitgord and Schouten (1985) (Fig. 10). However, the outward fanning of the stress trajectories, immediately before the Alpine front, is a new post–Lower Miocene feature that must result from a change in the collision regime within the Alpine belt.

DISCUSSION AND CONCLUSIONS

Despite the many uncertainties remaining in the reconstructions of the evolution of the Tethyan belt, the main characteristics

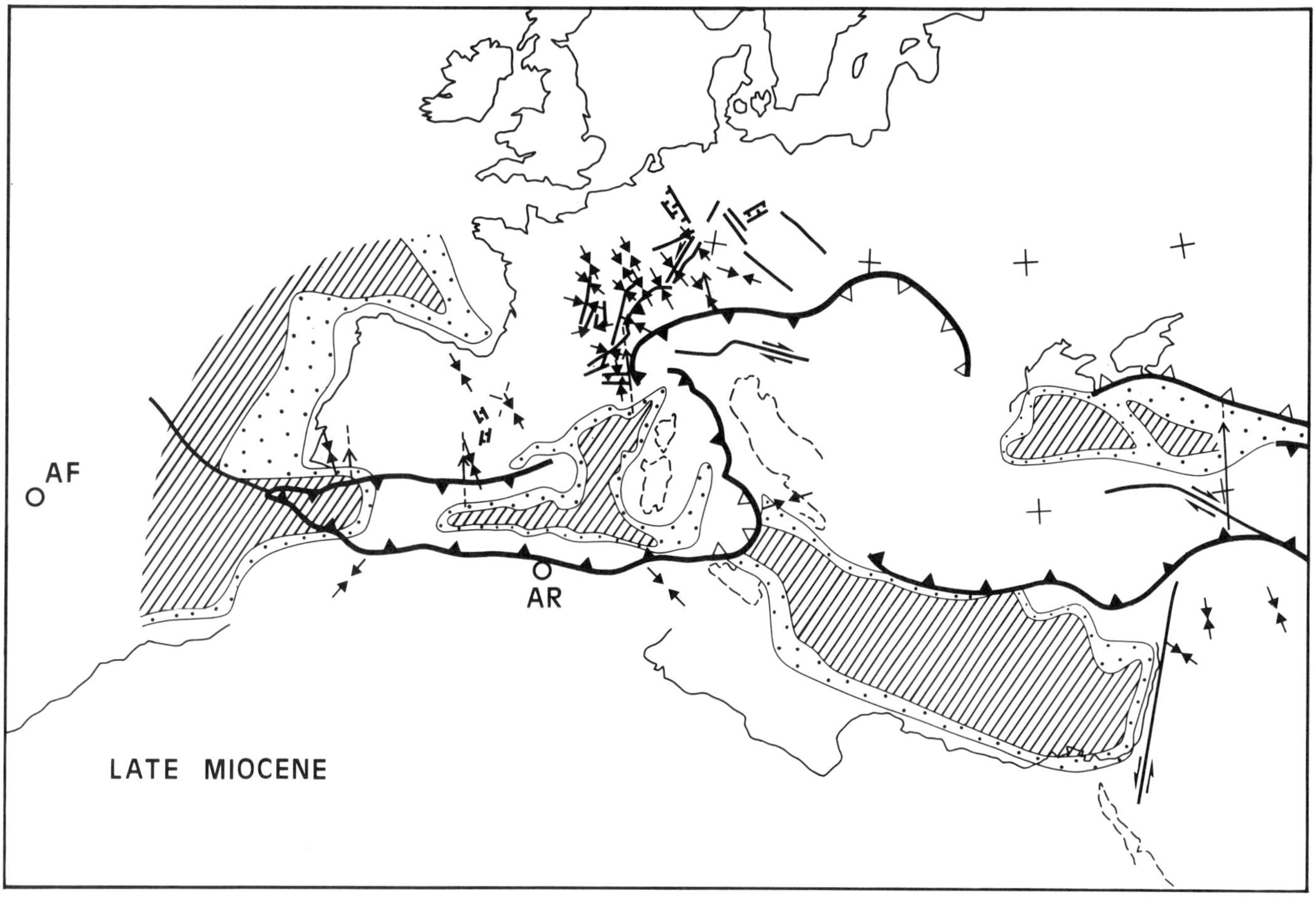

Figure 17. Reconstruction of late to post-Miocene stress field and geodynamic evolution of European platform in Cenozoic Africa/Eurasia collision framework. Same legend as in Figure 15. Rotation pole and motion vectors between 10 and 0 Ma.

of the plate tectonic system can now be outlined and some quantitative considerations made.

The first characteristic is the change from a subduction-dominated system to a collision-dominated one. This change occurred earlier in the west (between 100 and 80 Ma) than in the east (35 Ma) and migrated eastward with time. It was essentially complete at 35 Ma, with the collision between India and Eurasia, which finally closed the Neo-Tethys open oceanic space to the east.

During the Mesozoic, the subduction-dominated system oscillated between two modes, a three-plate mode and a two-plate mode, depending on the existence of an accreting plate boundary within the closing ocean. Because of the inherent asymmetry of such an ocean—bounded on the north by a subduction zone and on the south by a passive margin—the accreting plate boundary migrated northward toward the south Eurasian subduction zone. Once this mid-ocean ridge became subducted, the subduction rate abruptly decreased until rifting could detach a new piece of African margin to the south, starting a new ocean south of the first one. We suggest that this process explains the northward migration of microcontinents across the Tethys realm.

A discussion of the amount of surface that has disappeared through continental collision along the Tethyan belt leads to the conclusion that about half the amount of continental crust involved in the collision must have disappeared in the mantle. *We suggest that this is because only the brittle part of the crust is involved in the stacking, whereas the lower ductile crust disappears into the mantle.*

Thus, the isotherm at the base of the part involved in the stacking would typically be about 350°C for continental-type material and about 650°C for mantle-type material. In the early stages of collision, only lower continental margin, highly thinned crust participates in the tectonics. The whole thinned crust is in the brittle mode and is preserved in the stacking. Later on, less thinned upper continental margin to normal thickness continental crust participates in the collision. The lower part of the crust being in the ductile mode disappears in the mantle, and only the upper brittle crust is preserved in the stacking.

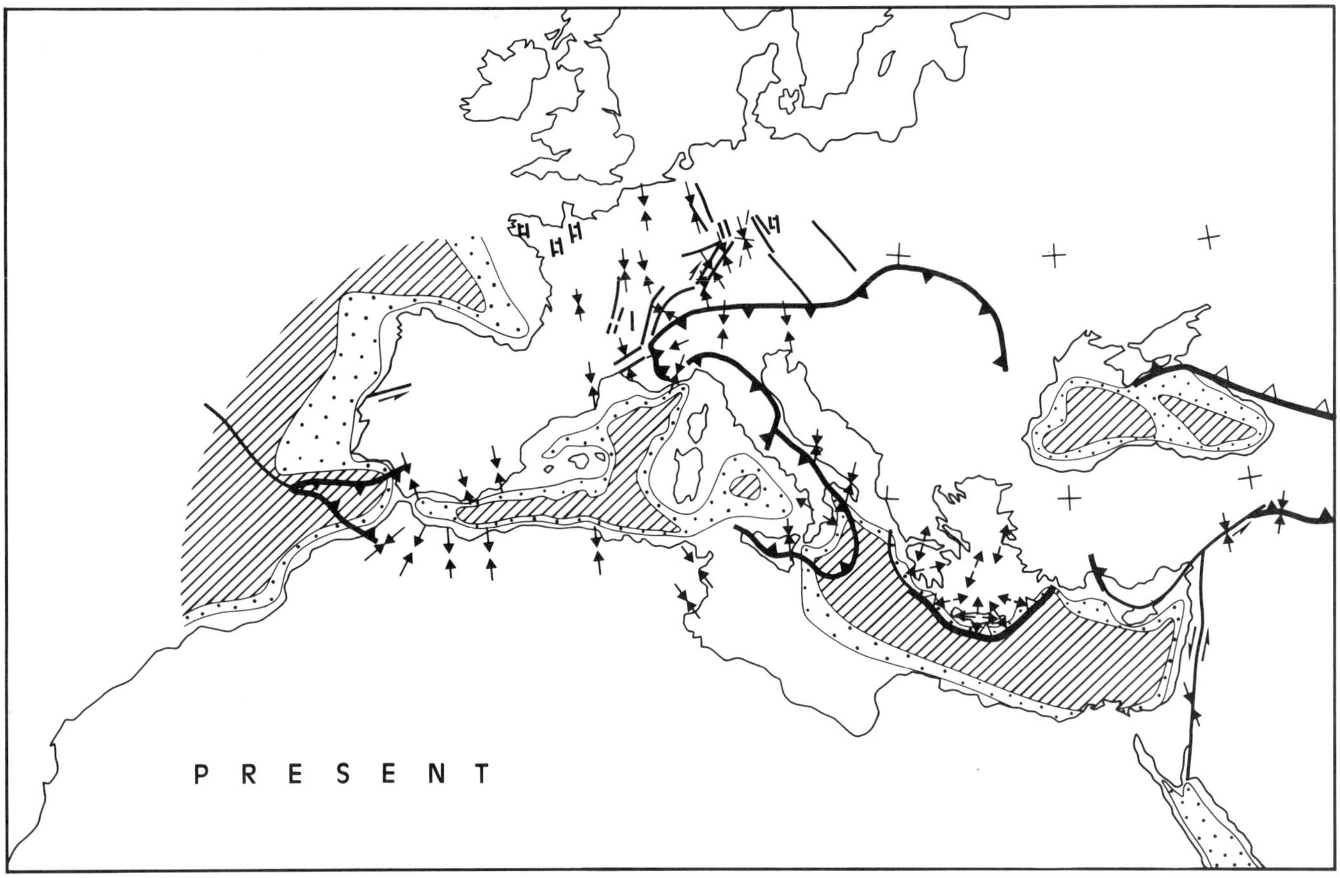

Figure 18. Reconstruction of present stress field and geodynamic evolution of European platform in Cenozoic Africa/Eurasia collision framework. Present stress data from in situ measurements, focal mechanisms of earthquakes and Philip (1980, 1983). Same legend as in Figure 15.

The total amount of continental crust that disappeared along the Tethyan belt is probably about 2 percent of the total present volume of continental crust, most of it since 35 Ma. The present rate of disappearance of continental crust in the mantle is about 3 km^3/yr, three times larger than the rate of subduction of pelagic sediments along trenches. However, this collision process is episodic, being controlled by the occurrence of large-scale continental collision orogenies.

The amount of sea-level regression, related to the continental surface shortening produced by this collision, has been about 75 m. This is somewhat less than the amount computed by Le Pichon and others (1985) in part because of more accurate revised rates of collision based on the reconstructions of Dercourt and others (1985, 1986), mostly because about one-fourth of the shortened crust was eroded and sedimented in the oceans. But the amount of regression produced is three times larger than the amount proposed by Harrison and others (1985) and should be taken into account in global sea-level budgets.

The reconstruction of the paleo-stress trajectories over the West European platform from the Eocene to the present gives additional information on the collisional regime. Two remarkable observations can be made.

1. Paleo-stress trajectories appear to be essentially parallel over most of the west European platform during each successive tectonic stage. The direction appears to be controlled by the Africa-Eurasia relative motion vector along the main African-Eurasian boundary. However, during Oligocene time, the principal essentially east–west stress trajectories are extensional and appear to be controlled by east–west extension motion between west Europe and central Eurasia, as given by the difference between the Arctic magnetic lineation (Savostin and others, 1987) and the North Atlantic ones (Klitgord and Schouten, 1985). In the same way, during Early Miocene time, the principal NE–SW compressional stress trajectories may be controlled by a reversal in the relative motion between west Europe and central Eurasia, as given by the same two sets of magnetic anomalies.

2. A peculiar stress pattern existed along the forefront of the Alpine belt during the last 7 to 8 Ma. Both in Latest Miocene and Present time, there is an outward fanning of the compressive stress trajectories in this area, with a total angle of 135° between

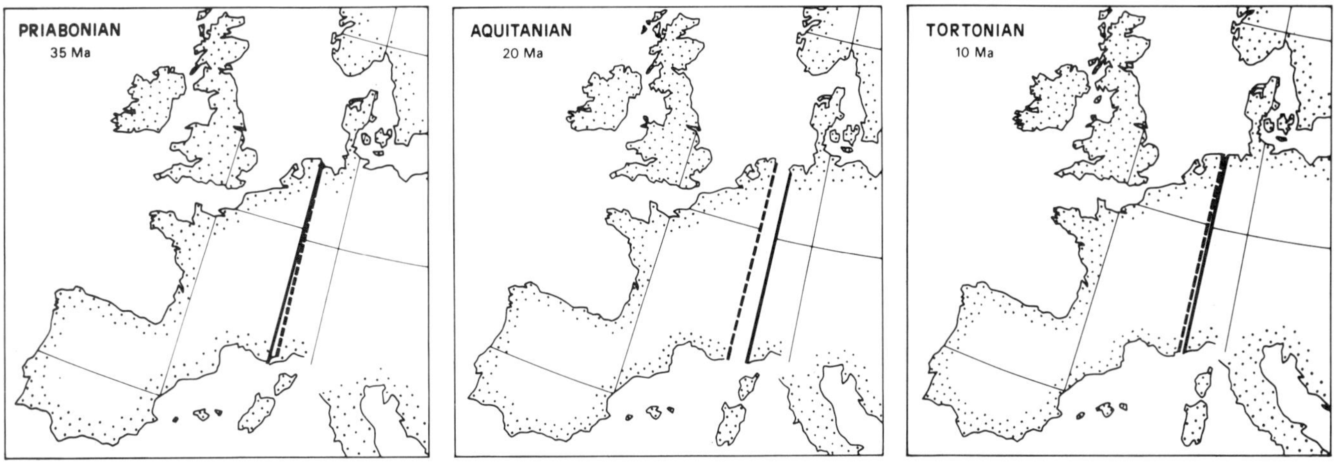

Figure 19. Positions of western Europe with respect to main Europe in Priabonian, Aquitanian, and Tortonian time, using the parameters of Klitgord and Schouten (1985) for western Europe and those of Savostin and others (1986) for main Europe. Note extension between Priabonian and Aquitanian time, followed by equivalent compression between Aquitanian and Tortonian time. Amount of extension is too large, but precision of kinematics is too low for result to be discussed quantitatively.

the two extremities of the fan. The width of this zone of fanning stress trajectories is about 300 km in Uppermost Miocene and appears to be only 150 km at present. Such a pattern would be expected in the indenter model (Tapponnier, 1977) or in the body forces model (Fleitout, 1984). But why would the indenting begin only in Late Miocene time, whereas continental collision began much earlier? Or why would the Alps begin to rise to high altitudes only in Late Miocene, whereas a phase of very rapid uplift appears to have been detected in Late Oligocene–Early Miocene time (Bradbury and Nolen-Hoeksema, 1985)?

One possible reason might be the progressive change in deformation regime of the whole Alpine domain at the end of Oligocene if, as indicated by Gillet and others (1986), the deformation became more homogeneous under greenschist to amphibolitic conditions, perhaps because of a more homogeneous thermal structure. The rapidly stacked cold thrusts would begin to heat, and the whole domain would be less rigid and closer to a plastic-type material allowing indenting and volume body forces to produce their outward fanning effect.

Another possible reason might be the existence of significant relative motion, extensional in Oligocene time and compressional in Early Miocene time, between western Europe and central Eurasia. The resulting stress pattern would then override the one resulting from the Alps indenting or the Alpine body forces.

ACKNOWLEDGMENTS

This paper was prepared at the invitation of Sydney P. Clark for the John Rodgers symposium. The first author is honored and grateful to have been chosen to participate. The French-Soviet Tethys program was supported by the French Ministry of Foreign Affairs, by IFREMER and by CNRS, and by the USSR Academy of Sciences. We appreciate the constructive reviews and comments of Bert Bally, Kevin Burke, Jean Dercourt, Claude Grandjacquet, and Luc-Emmanuel Ricou.

REFERENCES CITED

Allegre, C. J., and Jaupart, C., 1985, Continental tectonics and continental kinematics: Earth and Planetary Science Letters, v. 74, p. 171–186.

Angelier, J., 1979, Determination of the mean principal directions of stress for a given fault population: Tectonophysics, v. 56, p. T17–T26.

—— ,1984, Tectonic analysis of fault data sets: Journal of Geophysical Research, v. 89, p. 5835–5848.

Bergerat, F., 1985, Deformations cassantes et champs de contrainte tertiaires dans la plate-forme européenne [thèse d'état]: Paris, Mem. Sc. Terre Univ. Pierre et Marie Curie, 315 p.

Bergerat, F., and Geyssant, J., 1980, La fracturation tertiaire de l'Europe du Nord; Resultat de la collision Afrique Europe: Paris, Comptes Rendus de l'Académie des Sciences, v. 290, p. 1521–1524.

Biju-Duval, B., and Dercourt, J., 1980, Les bassins de la Méditerranée orientale reposent-ils sur les restes d'un domaine océanique, la Mesogée, ouvert au Mésozoique et distinct de la Tethys?: Bulletin de la Société Géologique de France, v. 7, no. 21, p. 43–60.

Biju-Duval, B., Dercourt, J., and Le Pichon, X., 1977, From the Tethys ocean to the Mediterranean seas; A plate tectonic model of the evolution of the western Alpine system, *in* Biju-Duval, B., and Montadert, L., eds., Histoire structurale des bassins mediterraneens: Split, Yougoslavie, Technip ed., Paris,

p. 143–164.

Boulin, J., 1981, Afghanistan structure, Greater India concept and eastern Tethys evolution: Tectonophysics, v. 72, p. 261–287.

Bradbury, H. J., and Nolen-Hoeksema, R. C., 1985, The Lepontine Alps as an evolving metamorphic core complex during A-type subduction; Evidence from heat flow, mineral cooling ages, and tectonic modeling: Tectonics, v. 4, no. 2, p. 187–211.

Burrus, J., 1984, Contribution to the geodynamic synthesis of the Provençal Basin (north-western Mediterranean): Marine Geology, v. 55, p. 247–270.

Coleman, R. G., 1981, Tectonic setting for ophiolite obduction in Oman: Journal of Geophysical Research, v. 86, p. 2497–2508.

Dercourt, J., and 16 others, 1985, Presentation de 9 cartes paléogéographiques au 1:20,000,000e s'étendant de l'Atlantique au Pamir pour la période du Lias à l'Actuel: Bulletin de la Société Géologique de France, v. 1, no. 5, p. 637–653.

Dercourt, J., and 18 others, 1986, Geological evolution of the Tethys belt from Atlantic to the Pamirs since Liassic: Tectonophysics, v. 123, p. 241–315.

Féraud, G., and Campredon, R., 1983, Geochronological and structural study of Tertiary and Quaternary dikes in southern France and Sardinia; An example of the utilization of dike swarms as paleostress indicators: Tectonophysics, v. 98, p. 297–325.

Fleitout, L., 1984, Modélisation des contraintes tectoniques et des instabilites thermo-mécaniques dans la lithosphère [thèse sc. nat.]: Université Paris XI, Orsay, 433 p.

Gastaud, J., Campredon, R., and Féraud, G., 1983, Les systèmes filoniens des Causses et du Bas-Languedoc (Sud de la France); Géochronologie, relations avec les paléocontraintes: Bulletin de la Société Géologique de France, v. 7, no. 5, p. 737–746.

Geyssant, J., 1980, Correlations péri-adriatiques le long des Alpes orientales; Rapports entre domaines austro-alpin et sud-alpin et tectogenèse crétacée: Bulletin de la Société Géologique de France, v. 7, no. 22, p. 31–42.

Gillet, P., Choukroune, P., Ballevre, M., and Davy, P., 1986, Thickening history of the western Alp: Earth and Planetary Science Letters, v. 78, no. 1, p. 44–52.

Guimera, J., 1983, Evolution de la deformation alpine dans le Nord-Est de la Chaine Ibérique et dans la Chaîne cotière Catalane: Paris, Comptes Rendus de l'Académie des Sciences, v. 297, no. 2, p. 425–430.

——, 1984, Palaeogene evolution of deformation in the northeastern Iberian Peninsula: Geological Magazine, v. 121, no. 1, p. 413–420.

Hardenbol, J., and Berggren, W. A., 1978, A new Paleogene numerical time scale, *in* Cohee, G. U., Glaessner, M. F., and Hedberg, H. D., eds., The geologic time scale: American Association of Petroleum Geologists Studies in Geology, v. 6, p. 213–234.

Harrison, C.G.A., Miskell, K. J., Brass, G. W., Saltzman, E., and Sloan, J., II, 1983, Continental hypsography: Tectonics, v. 2, p. 357–377.

Harrison, C.G.A., Brass, G. W., Miskell-Gerhardt, K. J., and Saltzman, E., 1985, Reply *to* Comment *on* "Continental hypsography": Tectonics, v. 4, p. 257–262.

Kazmin, V. G., Sbortshikov, I. M., Ricou, L. E., Zonenshain, L. P., Boulin, J., and Knipper, A. L., 1986a, Volcanic belts as markers of Mesozoic-Cenozoic active margin of Eurasia: Tectonophysics, v. 123, p. 123–152.

Kazmin, V. G., Ricou, L. E., and Sbortshikov, I. M., 1986b, Structure and evolution of the passive margin of the eastern Tethys: Tectonophysics, v. 123, p. 153–179.

Keating, B., and Helsley, C. E., 1978, Magnetostratigraphy of Cretaceous age sediments from Sites 361, 363, and 364, *in* Initial reports of the Deep Sea Drilling Project: Washington, D.C., U.S. Government Printing Office, v. 40, p. 459–468.

Kissel, C., Laj, C., and Muller, C., 1985, Tertiary geodynamic evolution of north-western Greece; Paleomagnetic results: Earth and Planetary Science Letters, v. 72, no. 2–3, p. 190–204.

Klitgord, K. D., and Schouten, H., 1985, Plate kinematics of the central Atlantic, *in* Vogt, P. R., and Tucholke, B. E., The western North Atlantic: Boulder, Colorado, Geological Society of America, The Geology of North America, v. M, p. 351–378.

Knipper, A. L., Ricou, L. E., and Dercourt, J., 1986, Ophiolites as indicators of the geodynamic evolution of the Tethyan Ocean: Tectonophysics, v. 123, p. 213–240.

Laubscher, H. P., 1971, The large scale kinematics of the western Alps and the northern Apennines and its palinspastic implications: American Journal of Science, v. 271, p. 193–226.

Le Pichon, X., 1984, The Mediterranean Seas, *in* Origin and history of marginal and inland seas, Proceedings of the 227th International Geological Congress: VNU Science Press, v. 23, p. 189–222.

Le Pichon, X., Huchon, P., and Barrier, E., 1985, Pangea, geoid, and the evolution of the western margin of the Pacific Ocean, *in* Nasu, N., ed., Formation of active ocean margins: Tokyo, Terra Scientific Publishing Co., p. 3–42.

Lepvrier, C., and Mougenot, D., 1983, Evidence microtectonique de la compression pyrénéenne dans l'ensemble de la couverture mésozoique du Portugal: Paris, Comptes Rendus de l'Académie des Sciences, v. 296, no. 2, p. 737–740.

——, 1984, Deformations cassantes et champs de contrainte post-hercyniens dans l'Ouest de l'Iberie (Portugal): Revue de Géographie Physique et de Géologie Dynamique, v. 25, no. 4, p. 291–305.

Letouzey, J., and Trémolières, P., 1980, Paleo-stress fields around the Mediterranean since the Mesozoic derived from microtectonics; Comparisons with plate tectonic data, *in* Géologie des chaînes alpines issues de la Téthys: 26th International Geological Congress, Paris, Mem. B.R.G.M., v. 115, p. 261–273.

Lyon-Caén, H., and Molnar, P., 1985, Gravity anomalies, flexure of the Indian plate, and the structure, support, and evolution of the Himalaya and Ganga Basin: Tectonics, v. 4, no. 6, p. 513–538.

Mann, P., and Burke, K., 1984, Neotectonics of the Caribbean: Reviews of Geophysics and Space Physics, v. 22, no. 4, p. 309–362.

Molnar, P., and Denq Qidong, 1984, Faulting associated with large earthquakes and the average rate of deformation in central and eastern Asia: Journal of Geophysical Research, v. 87, no. B7, p. 6203–6227.

Molnar, P., and Tapponnier, P., 1975, Cenozoic tectonics of Asia; Effects of a continental collision: Science, v. 189, p. 419–426.

Molnar, P., Freedman, D., and Shih, J. S., 1979, Lengths of intermediate and deep seismic zone and temperatures in downgoing slabs of lithosphere: Geophysical Journal of the Royal Astronomical Society, v. 56, p. 41–54.

Odin, G. S., and Curry, D., 1981, L'échelle numérique des temps paléogènes en 1981: Paris, Comptes Rendus de l'Académie des Sciences, v. 293, p. 1003–1006.

Odin, G. S., and Kennedy, W. J., 1982, Mise à jour de l'échelle des temps mésozoiques: Paris, Comptes Rendus de l'Académie des Sciences, v. 294, p. 383–386.

Olivet, J. L., Bonnin, J., Beuzart, P., and Auzende, J. M., 1984, Cinématique de l'Atlantique nord et central: Paris, CNEXO, Rapports Scientifiques, 108 p.

Palmer, A. R., 1983, The Decade of North American Geology 1983 Geologic Time Scale: Geology, v. 11, p. 503–504.

Parsons, B., 1981, The rates of plate consumption and creation: Journal of Geophysical Research, v. 67, p. 437–448.

Patriat, P., and 7 others, 1982, Les mouvements relatifs de l'Inde de l'Afrique et de l'Eurasie: Bulletin de la Société Géologique de France, v. 24, p. 363–373.

Philip, H., 1980, Tectonique récente et sismicité de la France; Caractéristique géodynamiques, *in* Image géophysique de la France: 26th International Geological Congress, Paris, Mem. B.R.G.M., v. 107, p. 42–50.

——, 1983, La tectonique actuelle et récente dans le domaine mediterranéen et ses bordures, ses relations avec la sismicité [these sc. nat.]: Universite de Montpellier, 240 p.

Pindell, J. L., 1985, Alleghenian reconstruction and subsequent evolution of the Gulf of Mexico, Bahamas, and Proto-Caribbean: Tectonics, v. 4, no. 1, p. 1–39.

Pitman, W. C., and Talwani, M., 1972, Seafloor spreading in the North Atlantic: Geological Society of America Bulletin, v. 83, p. 619–646.

Rehault, J. P., 1981, Evolution tectonique et sédimentaire du bassin ligure (Méditerranée occidentale) [these d'etat]: Mem. Sc. Terre, Univ. P. and M. Curie,

Paris, 132 p.

Ricou, L. E., Dercourt, J., Geyssant, J., Grandjacquet, J. C., Lepvrier, C., and Biju-Duval, B., 1986, Geological constraints on the alpine evolution of the Mediterranean Tethys: Tectonophysics, v. 123, p. 83–122.

Savostin, L. A., 1981, Cenozoic plate tectonics of the Arctic region, north-eastern and inner Asia and paleogeodynamic reconstructions in the global scale [thesis]: Moscow, Science Institute of Oceanology, 498 p. (in Russian).

Savostin, L. A., and Karasik, A. M., 1981, Recent plate tectonics of the Arctic basin and north-eastern Asia: Tectonophysics, v. 74, p. 111–145.

Savostin, L. A., Sibuet, J. C., Zonenshain, L. P., Le Pichon, X., and Roulet, M. J., 1986, Kinematic evolution of Tethys Belt from the Atlantic Ocean to the Pamirs since Triassic: Tectonophysics, v. 123, p. 1–35.

Sclater, J. G., Royden, L., Horvath, F., Burchfiel, B. C., Semlen, S. C., and Stegena, L., 1980, The formation of the intra-Carpathian basins as determined from subsidence data: Earth and Planetary Science Letters, v. 51, p. 139–162.

Tapponnier, P., 1977, Evolution tectonique du systême alpin en Mediterranée; Poinconnement et écrasement rigide plastique: Bulletin de la Société Géologique de France, v. 7, p. 437–460.

Tapponnier, P., Mattauer, M., Proust, F., and Cassaigneau, C., 1981, Mesozoic ophiolites, sutures, and large scale tectonic movements in Afghanistan: Earth and Planetary Science Letters, v. 52, p. 355–371.

Tollman, A., 1966, Die alpidischen Gebirgsbildung Phasen in der Ostalpen und Westkarpathen: Stuttgart, Geotecktonische Forschungen, v. 21, 156 p.

Trümpy, R., 1973, L'évolution de l'orogénèse dans les Alpes Centrales; Interprétation des données stratigraphiques et tectoniques: Eclogae Geologicae Helvetiae, v. 66, no. 1, p. 1–10.

—— , 1975, Penninic-Austroalpine boundary in the Swiss Alps; A presumed former continental margin and its problems: American Journal of Science, v. 275A, p. 209–238.

—— , 1980, An outline of the geology of Switzerland, 26th International Geological Congress, Schweizer Geol. Komm. Ed.: Basel and New York, Wepf and Co., Publishers, 334 p.

Unternehr, P., 1982, Etude structurale et cinématique de la mer de Norvège et du Groenland; Evolution du microcontinent de Jan Mayen [thesis]: Brest, Université Bretagne Occidentale, 228 p.

Van den Berg, J., 1983, Reappraisal of paleomagnetic data from Gargano (south Italy): Tectonophysics, v. 98, p. 29–41.

Viallard, P., 1979, La chaîne ibérique; Zone de cisaillement intracontinentale pendant la tectogénèse alpine: Paris, Comptes Rendus de l'Académie des Sciences, v. 289, p. 65–68.

Vincent, E., 1977, Indian Ocean Neogene planktonic foraminiferal biostratigraphy and its paleo-oceanographic implications, *in* Heirtzler, J. R., Bolli, H. M., Davies, T. A., Saunders, J. B., and Sclater, J. G., eds., Indian Ocean geology and biostratigraphy: American Geophysical Union, p. 469–584.

Vink, G. E., Morgan, W. J., and Zhao, W. L., 1984, Preferential rifting of continents; A source of displaced terranes: Journal of Geophysical Research, v. 89, no. B12, p. 10072–10076.

Vogt, P. R., and Einwich, A. M., 1979, Magnetic anomalies and sea-floor spreading in the western North Atlantic and a revised calibration of the Keathley (M) Geomagnetic reversal chronology, *in* Initial reports of the Deep Sea Drilling Project: Washington, D.C., U.S. Government Printing Office, v. 43, p. 857–876.

Westphal, M., Bazhenov, M. L., Lauer, J. P., Pechersky, D. M., and Sibuet, J. C., 1986, Paleomagnetic implications on the evolution of the Tethys Belt from the Atlantic Ocean to the Pamirs since the Triassic: Tectonophysics, v. 123, p. 37–82.

Zonenshain, L. P., and Le Pichon, X., 1986, Deep basins of the Black and Caspian Seas as remnants of Mesozoic back-arc basins: Tectonophysics, v. 123, p. 181–211.

Manuscript Accepted by the Society January 15, 1987

Printed in U.S.A.

Geological Society of America
Special Paper 218
1988

Detachment faulting in continental extension; Perspectives from the Southwestern U.S. Cordillera

Gregory A. Davis
Department of Geological Sciences, University of Southern California, Los Angeles, California 90089-0741
G. S. Lister*
Bureau of Mineral Resources, Geology and Geophysics, Canberra City, 2601 Australia

ABSTRACT

Cordilleran detachment faults, as defined here, are extensional faults of low initial dip, probably less than 30°, and subregional to regional scale. Some detachment faults have large translational displacements, i.e., in excess of several tens of kilometers. First interpreted as Tertiary extensional structures in the eastern Great Basin by Armstrong (1972), they are now known to be widespread throughout those Cordilleran regions that have undergone greatest Cenozoic extension. Detachment faults are commonly, but not necessarily, associated with lower-plate mylonitic gneisses that compose the so-called "metamorphic core complexes." Probably nowhere in the U.S. Cordillera are detachment faults more widely and spectacularly developed than in the region that borders the lower Colorado River in southernmost Nevada, southeastern California, and southwestern Arizona. We believe that our studies and those of numerous other workers in this region, the Colorado River extensional corridor of Howard and John (1987), provide a number of new perspectives on the origin, geometry, and evolution of Cordilleran detachment faults.

Detachment faults are best explained as evolving shallow-dipping shear zones that have accommodated Tertiary crustal extension (Wernicke, 1981). The fault zones are believed to root at midcrustal or lower upper crustal depths into broad zones of intracrustal flow, the tectonic regime in which mylonitic gneisses form. At their upper ends, major detachment faults either reach the surface directly or terminate at shallow depth into pull-apart complexes of closely spaced normal faults. Along these evolving shear zones, lower-plate mylonitic gneisses are drawn upward and out from beneath upper-plate rocks. As footwall gneisses rise structurally upward, they are retrograded, sheared, and shattered at progressively colder and shallower crustal levels to form the chloritic breccias and microbreccias characteristic of many major detachment faults. At advanced stages of detachment fault evolution, lower-plate mylonitic gneisses formed at depths >12 km are tectonically juxtaposed beneath unmetamorphosed supracrustal rocks and exposed at the surface through combinations of crustal upwarping, tectonic denudation, and erosion.

Contrary to popular belief, the master detachment faults exposed today are probably not in their entirety those faults that formed at the start of extensional deformation, but rather are only the youngest in a succession of major detachment faults. Detachment faults undergo warping at high angles to the direction of crustal extension, probably in large part related to isostatically induced distortions of originally more planar faults. Such warping leads to the development of younger, more planar fault splays that either

*Present address: Department of Earth Sciences, Monash University, Clayton, Victoria, 3168 Australia

cut upward into former upper-plate rocks (excisement) or downward into former lower-plate rocks (incisement).

Recognition of such geometric complexities offers fresh insights into deciphering the evolving strain patterns within major detachment terranes. Studies in the Whipple Mountains region of southeastern California indicate that: (1) detachment faults have formed by both excisement and incisement tectonics; (2) northeast-southwest-trending "folds" of major detachment faults, oriented parallel to the direction of extension, are in reality primary corrugations or flutes in the fault surface (a conclusion previously reached by other workers in nearby areas); (3) most normal faults in the upper plates of major detachments originally had listric geometries before losing their flattened lower segments as the consequence of excisement tectonics; and (4) detachment faults can transect upper crustal rocks as primary, low-dipping shear zones without pre-existing, shallow-dipping structural controls (e.g., thrust faults) on their localization; the northeast-southwest-trending curviplanar geometry of the Whipple fault does, however, seem to mimic preexisting fold structure in lower-plate mylonitic gneisses crossed by the fault. Finally, the rate of translation along master faults of some evolving detachment systems apparently can be very rapid (>1 cm/yr), much faster than rates that we (and perhaps other workers) once deemed reasonable. A very good case can be made on the basis of geochronologic and field studies that footwall mylonitic gneisses were transported upward along the Whipple detachment system from lower upper crustal depths to near surface levels in less than 2 m.y. (between 18 and 20 Ma).

INTRODUCTION

With few exceptions, information published prior to 10 years ago on extension of the continental crust focused attention on moderately to steeply dipping normal faults as the fundamental structures of extended terranes. Steeply dipping swarms of igneous dikes were recognized as the consequences of limited crustal extension in some areas (e.g., feeder dikes for Miocene Columbia Plateau basalts in northeastern Oregon and west-central Idaho). Some investigators of the Basin and Range province of the U.S. Cordillera (e.g., Thompson, 1959; Stewart, 1971) realized that the widely spaced (ca. 20–30 km) and relatively steeply dipping (>50°) range-bounding faults of that province could produce only limited crustal extension, and that below some critical depth (ca. 8–15 km) the rheology of the hotter than normal Basin and Range crust should not permit extensional strain by normal faulting. Estimates of total extensional strain in the province by these geologists were geometrically constrained to low values, e.g., 10 to 15 percent, by the wide spacing and steep dips of observed range-front faults.

Other workers, however, were troubled by such low estimates. Hamilton and Myers (1966), impressed by geophysical data indicating that the Basin and Range crust was considerably thinner than that beneath adjacent provinces to the west and east, concluded that Cenozoic extension had probably been on the order of 50 to 100 percent. Subsequent field-based studies, such as those of Anderson (1971), Armstrong (1972), and Davis and Burchfiel (1973), indicated that certain areas within the Basin and Range province had indeed undergone extension compatible with Hamilton and Myers's estimates for province-wide strain. Davis and Burchfiel, for example, analyzed geometric relations within a geologic terrane offset by the major northeast to east-striking Garlock fault of southern California. They concluded that extension within the 200-km-wide Great Basin area north of the fault had exceeded extension in Mojave Desert areas to the south and east by approximately 100 percent, and they interpreted the Garlock fault as an intraplate transform bounding the southern limit of the Basin and Range province in California.

Cordilleran geologists now agree that the Basin and Range province (Fig. 1) has indeed experienced larger (ca. 50–100%) extensional strains than those envisioned by most early workers, although such strains are nonuniformly distributed across the province (e.g., Stewart, 1978; Miller and Gans, 1984). Such agreement has come about through a growing awareness of two major aspects of basin-and-range geology: (1) that crustal extension occurred widely throughout the province prior to the late early to middle Miocene (locally, Pliocene) formation of present basin-and-range topography (e.g., Loring, 1972, 1974; McDonald, 1976); and (2) that low-angle normal faults are widespread and extremely important structural elements of those Cordilleran regions that have experienced the greatest extensional strains.

The second of these geologic aspects is the principal topic of this paper. Shallow-dipping normal faults were until recently the neglected structures of the U.S. Cordillera. In eastern Nevada and western Utah, areally extensive faults of this type were first described by Peter Misch and his students in the late 1950s and 1960s (e.g., Snelson, 1957; Misch, 1960; Nelson, 1966, 1969; Thorman, 1970). These faults, generally developed within pre-Jurassic strata, were interpreted as updomed, hinterland exposures of Mesozoic thrust faults belonging to the foreland fold and thrust belt to the east (Misch, 1960, 1971). Geometrically similar faults, clearly of Cenozoic age because they truncate inclined

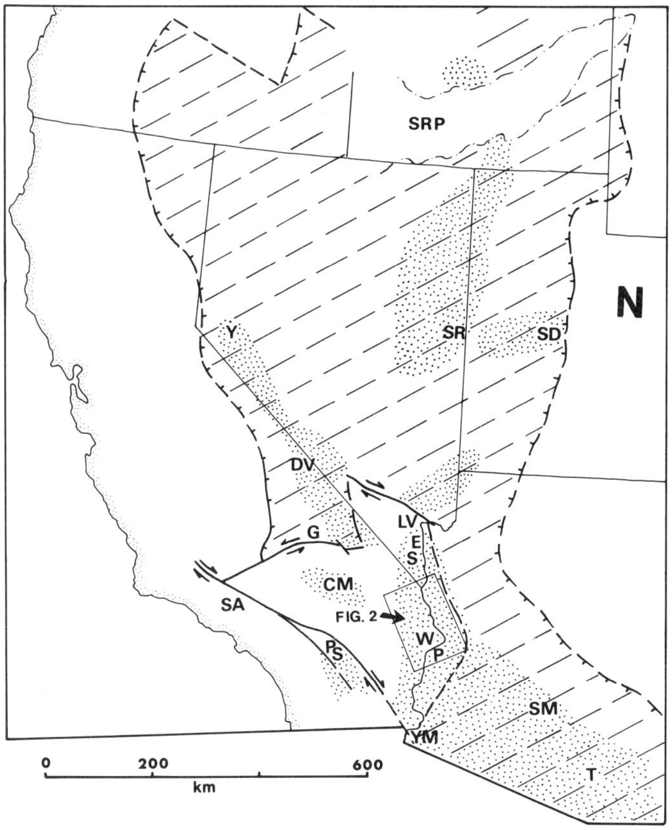

Figure 1. Location map of southern U.S. Cordillera showing area of Basin and Range province (diagonally ruled pattern) and locations of detachment fault complexes (stippled pattern). Geographic localities and geologic features cited in the text (in alphabetical order): CM = central Mojave Desert; DV = Death Valley; E = Eldorado Mountains; G = Garlock fault; LV = Las Vegas; P = Parker; PS = Palm Springs; S = Searchlight; SA = San Andreas; SD = Sevier Desert; SM = South Mountain; SR = Snake Range; SRP = Snake River Plain; T = Tucson; W = Whipple Mountains; Y = Yerington; YM = Yuma.

Tertiary strata in their upper plates, are also widely exposed in the Death Valley–eastern Mohave region and in an arcuate belt extending southward from Las Vegas along the lower Colorado River and curving southeastward into southern Arizona. However, they too, were generally interpreted as thrust faults (e.g., in the Death Valley–eastern Mohave area: Noble, 1941; Hewett, 1956; Hunt and Mabey, 1966; in the lower Colorado River area: Ransome, 1931; Coonrad, 1960; Wilson, 1962; Terry, 1972) or as unconformities (Kemnitzer, 1937; Terry, 1972; and Dibblee, 1971, for the central Mohave region). In retrospect, it is difficult to understand why the low-angle faults in these geologic terranes were so widely mistaken for thrust faults, given that they characteristically juxtapose younger over older rock units, and unmetamorphosed or low-grade metamorphic rocks over higher grade metamorphic and/or mylonitic rocks.

Armstrong (1972) was the first to reinterpret the low-angle faults in areas studied by Misch and his students as the consequence of major regional Tertiary extension, although Lee and others (1970) had earlier recognized Tertiary movement on the Snake Range décollement from Tertiary K-Ar dates in fault zone rocks. Terming such structures "denudation faults," Armstrong believed that they were the product of regional extension, beginning about 40 Ma, of a crust thickened during Mesozoic compression. He recognized the likelihood that some low-angle normal faults might reactivate pre-existing thrust faults, and he suggested that faults formed at depth during early stages of extension might be cut by steeper, younger faults of higher crustal position as tectonic denudation continued.

Some Tertiary low-angle faults described in the Basin and Range area during the 1970s were initially interpreted as extension faults, rather than being misidentified as thrust faults, because of observable offsets of Tertiary strata in their hanging and foot walls. Such faults, with low dips attributable to primary listric geometry or to rotation of once-steep faults, are not as cryptic as the major low-angle "denudation" faults of the eastern Great Basin, or those of the lower Colorado River–southern Arizona area where Tertiary strata are present only in upper-plate positions. Ernest Anderson's recognition (1971) of downward-flattening (listric) late Miocene normal faults in the Eldorado Mountains, southern Nevada, was a major contribution to Cordilleran extensional tectonics for at least two reasons: (1) it documented the existence of closely spaced listric normal faults, which were interpreted as flattening downward into either a subhorizontal basal fault surface or inflating plutonic complex below which extension by normal faulting had not occurred; and (2) it demonstrated that the listric geometry of the imbricate Eldorado Mountains faults and the consequent down-dip rotation of late Tertiary strata along them had produced large extensions of supracrustal rocks. As the Eldorado normal faults become subhorizontal, some steeply dipping strata truncated by them are subhorizontally translated by amounts equal to the cumulative dip-slip on the entire family of listric structures. Hence, Anderson's paper was among the first to demonstrate the efficacy of low-angle normal faults in producing large extensional strain in the crust transected by them.

Proffett's study of multiple phases of normal faulting in the Yerington area of west-central Nevada (1977; Proffett and Dilles' detailed map, 1984) was another landmark contribution. In that area, very shallow dipping normal faults displace steeply dipping Tertiary strata and basement rocks for distances up to several kilometers. Proffett interpreted the shallow-dipping faults as originally much steeper, moderately listric structures that cut subhorizontal Tertiary strata and its underlying basement. His explanation for present geometric relations relied on downward rotation of hanging wall units along early formed listric faults, followed by the repeated rotation of these faults and their wall rocks along younger generations of listric faults. Proffett's thorough documentation of multiple, overprinted episodes of closely spaced normal faulting helped direct geological attention away from the simple, wide-spaced range-bounding fault system of the Great Basin. It also illustrated that rotated low-angle normal faults, however formed, occur far from Anderson's Eldorado Mountains and are extensional phenomena that must be reckoned with in analyses of province-wide strain.

Fault geometries quite comparable to those described by Proffett have been mapped above the Snake Range décollement (detachment) in eastern Nevada (Gans and Miller, 1983; Miller and others, 1983), but there, domino-style rotation of originally moderately dipping (50° to 60°) planar faults of two generations is interpreted as the mechanism for producing very low dipping normal faults and extensions locally as great as 450 to 500 percent. Rotation is restricted, however, to the upper plate of the Snake Range décollement.

Low-angle faults in the Eldorado and Yerington areas owe their low dips to various combinations of original listric geometry and/or rotation of listric or planar normal faults. But such faults are not the detachment faults of principal concern to this chapter, although it is likely that they have developed in the extending upper plates of unexposed detachment systems. It is to such detachment fault systems that we now turn.

DETACHMENT FAULTS

The detachment faults of the Cordillera have known many names: décollements, abscherung zones, denudational faults, dislocation surfaces, and LANFs (low-angle normal faults; Wernicke, 1981; Brun and Choukroune, 1983), but the term detachment fault (Carr and Dickey, 1976; Davis and others, 1979) has gained most common usage. Two characteristics of detachment faults have already been cited—that they juxtapose younger over older or structurally higher over structurally lower rocks, and that they commonly separate upper-plate unmetamorphosed or low-grade metamorphic rocks from lower-plate crystalline rocks of higher metamorphic grade; the latter commonly have mylonitic gneissic fabrics. Four other characteristics can be added: (1) the master faults of detachment complexes are developed on a regional to subregional scale (e.g., the Whipple detachment fault system of the lower Colorado River area, California, Arizona, and southernmost Nevada, appears to have once underlain at least 15,000 km^2; the northern Snake Range décollement extends for at least 50 km in an east-west dimension); (2) the upper plates of detachment faults are typically extended by one or more generations of normal faults that either merge downward into the detachment or end abruptly at it without a shallowing of dip; similar faults are absent or less well-developed in lower plates; (3) several lines of evidence (e.g., telescoping of metamorphic facies, juxtaposition of deep structural levels beneath much shallower levels, and difficulty in matching displaced upper- and lower-plate units) suggest that some detachment faults have very large displacements; (4) detachment faults are commonly underlain by a distinctive sequence of first ductilely, then brittlely deformed, rocks, with each younger superposed member of the sequence developed in progressively narrower zones. The most common sequence, oldest to youngest, is (a) mylonitic gneisses (not always present); (b) sheared, retrograded mylonitic gneisses (or, in their absence, other crystalline basement rocks) widely termed "chloritic breccias"; (c) pseudotachylites and flinty cataclasites or microbreccias, sometimes with injection vein geometries from layers resembling melt generation surfaces (Sibson, 1977); and (d) fault gouge.

We present evidence elsewhere in this chapter that detachment faults develop initially through the upper crust with low angles of dip, probably less than 25° to 30°. In our opinion, low-angle faults that owe their present shallow dip to the rotation of originally steeper faults (e.g., the Singatse and parallel faults of the Yerington area: Proffett, 1977), or that constitute the lower portion of listric faults that steepen upward (e.g., some faults in the Eldorado Mountains: Anderson, 1971) should not be termed detachment faults. *The essential elements of extensional detachment faults,* as the term is used here, *are low angle of initial dip, subregional to regional scale of development, and large translational displacements, certainly up to tens of kilometers in some instances.*

DETACHMENT FAULTING IN THE COLORADO RIVER REGION SOUTH OF LAS VEGAS

Probably nowhere in the U.S. Cordillera are detachment faults more widely and spectacularly developed than in the region that borders the lower Colorado River between Las Vegas, Nevada and Yuma, Arizona, and a wide contiguous terrane that extends eastward from the river across the southern third of Arizona and into Sonora, Mexico (Fig. 1). This Oligo-Miocene detachment terrane is quite likely offset by the San Andreas fault system near Yuma, Arizona (Garner and others, 1982). Portions of the displaced terrane may be present in the eastern Peninsular Ranges west of the Salton Sea and south of Palm Springs (Wallace and English, 1982; Engel and Schultejann, 1984).

This discussion emphasizes detachment faulting in the northern lower Colorado River (NLCR) region, i.e., from the approximate latitude of Parker, Arizona, northward to the southern tip of Nevada (Figs. 1, 2). The terms "Colorado River extensional corridor" (Howard and John, 1987) and "lower Colorado River detachment terrane" refer to the same region. Most of the studies of the past decade in this region have been conducted by three groups of researchers: by the U.S. Geological Survey, largely under the leadership of Keith Howard, but including important independent studies by Will Carr, Ivo Luchitta, Neil Suneson, and Barbara John among others; by J. Lawford Anderson and G. A. Davis and their students at the University of Southern California; and by Eric Frost and his students at San Diego State University. As one measure of the intensity of geologic studies in the NLCR region, the results of 18 separate investigations in that area by more than 30 earth scientists were published in 1982 (Frost and Martin) in a volume commemorating the pioneering Colorado River area studies of Ernest Anderson and Warren Hamilton. The results of this collective

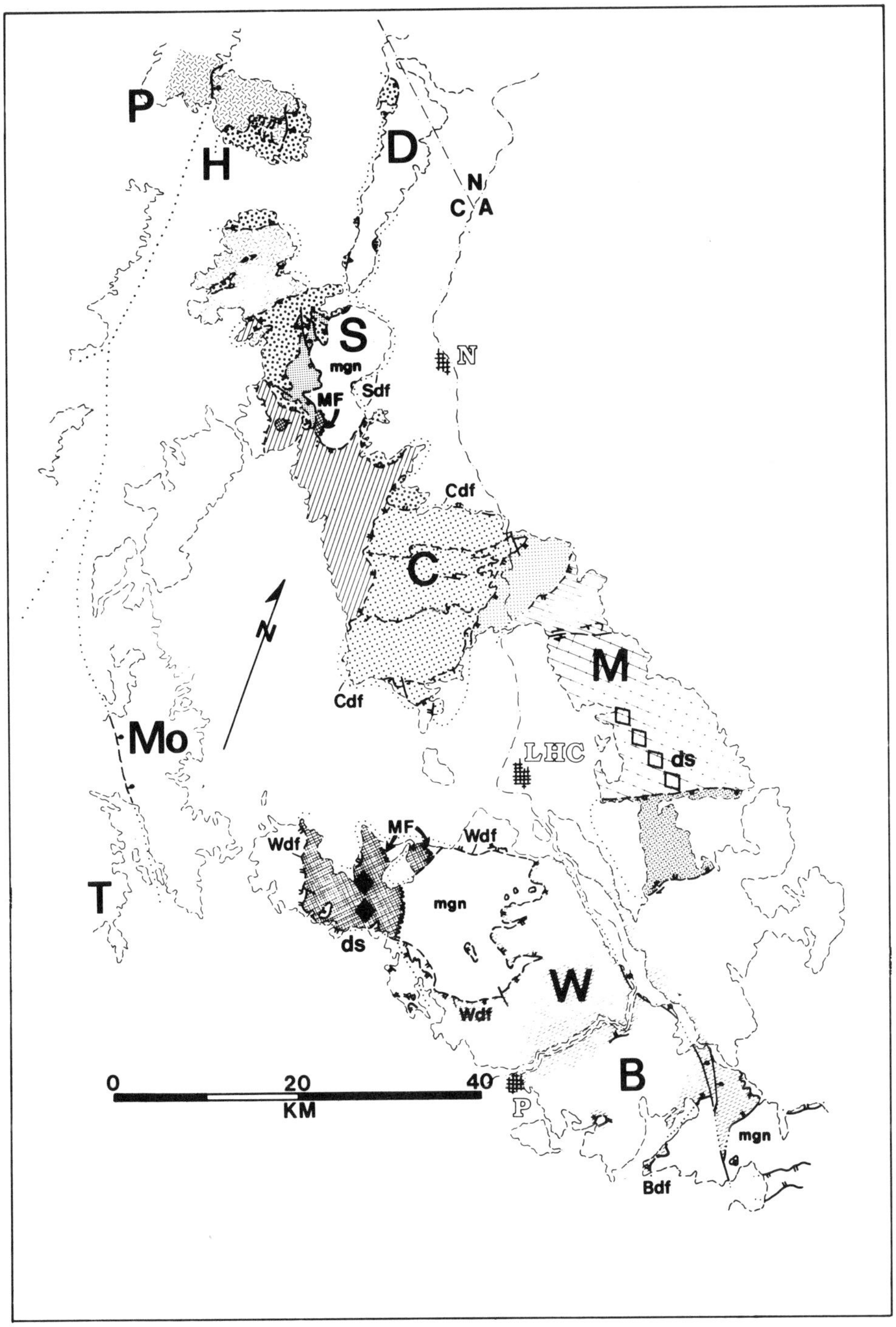

Figure 2. Simplified geologic map of lower Colorado River region of southernmost Nevada (N), southeastern California (C), and western Arizona (A). Map emphasizes location of major faults, primarily low-angle detachment faults. Various detachment-bounded plates are designated by different patterns. Cities from north to south (open letters): N = Needles; LHC = Lake Havasu City; P = Parker. Mountain ranges (in alphabetical order): B = Buckskin; C = Chemehuevi, D = Dead; H = Homer; M = Mohave; Mo = Mopah; P = Piute; S = Sacramento; T = Turtle; W = Whipple. Geologic abbreviations: Wdf = Whipple detachment fault; Sdf = Sacramento detachment fault; MF = mylonitic front; mgn = mylonitic gneiss; ds = dike swarm, offset and displaced by faults of the Whipple detachment system. Data for map (north to south): Spencer and Turner (1982); Spencer (1985); McClelland (1982, 1984); Howard and others (1982b); G. A. Davis and colleagues (unpublished data); and Wilkins and Heidrick (1982).

geological effort in the Colorado River extensional corridor are briefly summarized below in order to define the environment of detachment faulting. The summary includes a number of generalizations with which, we believe, most of the area's workers would agree—a statement not meant to imply that there are no significant differences of opinion about major aspects of the area's evolution.

GEOLOGIC SUMMARY OF THE NORTHERN LOWER COLORADO RIVER REGION

This region is underlain by Precambrian crystalline basement rocks with well-established ties to the North American continent. This basement consists of a heterogeneous assemblage of high-grade orthogneisses, paragneisses, and amphibolites metamorphosed approximately 1.7 Ga; and younger igneous intrusions of diverse age and composition. Among the latter components of the Proterozoic basement assemblage are (1) synkinematic to late kinematic, foliated granitic rocks (ca. 1.7 Ga), (2) postorogenic ("anorogenic") granites (ca. 1.4 to 1.45 Ga; see Anderson, 1983, for description and appropriate references), and (3) diabase dikes and sills (1.1 to 1.2 Ga?).

The Precambrian basement of the region was once covered by a thin (ca. 2 to 3 km) platform sequence of Cambrian through Triassic sedimentary rocks and Jurassic volcanic and clastic rocks related to development of an Andean-type continental margin in the southwestern United States (Burchfiel and Davis, 1972, 1975). Scattered plutons of Jurassic and Cretaceous age are an additional expression of arc development. The Paleozoic and Mesozoic cover was eroded from the NLCR region north of Parker, Arizona, prior to late Oligocene(?) and Miocene deposition of terrestrial sedimentary and volcanic rocks. Pre-Tertiary basement and supracrustal units were strongly folded, thrust-faulted, and metamorphosed during late Jurassic and/or Cretaceous time in areas southwest (e.g., Harquahala Mountains), south (e.g., Big Maria and Riverside Mountains), and west (e.g., Old Woman and Arica Mountains) of the NLCR region shown in Figure 2. The extensional corridor north of Parker (Fig. 2) was in general little affected by this deformation since there is widespread preservation throughout its crystalline basement of complex Precambrian structural and intrusive relations, e.g., in the eastern Whipple Mountains (Davis and others, 1980), the Turtle Mountains (Howard and others, 1982a), and the Buck Mountains east of the Mojave Range, Arizona (Howard and others, 1982b). Precambrian crystalline rocks in at least the southern half of the NLCR region (Fig. 2) did not escape, however, a regional Mesozoic thermal event that reset K-Ar clocks in Precambrian hornblende and biotite to ages typically between 135 and 160 Ma (Anderson and Frost, 1981). Locally, portions of the Precambrian NLCR terrane experienced penetrative strain, possibly during this thermal event. Plutons of the anorogenic 1.4-Ga suite and younger Precambrian diabase sills and dikes exhibit mylonitic shear zones and penetrative foliation development of tectonic origin at some localities in the NLCR region (Howard and others, 1982b; J. L. Anderson, personal communication, 1986). The age or ages, geometry, and extent of post-1.4-Ga deformation are poorly known.

By Oligocene time the lower Colorado River region north of Parker had been denuded of its Paleozoic and Mesozoic cover. During an episode of Oligocene(?) and Miocene arc volcanism, areas both north and south of Parker experienced profound crustal extension. This extension was largely manifested by the development of an east-rooting system of detachment faults in the NLCR region as far north as the town of Searchlight in southern Nevada (Fig. 3). To the north of Searchlight (Fig. 1) lies a separate Miocene extensional terrane, studied by Anderson (1971) and mentioned briefly above. The geometry of fault structures within this northern terrane, in the Eldorado Mountains, the Highland Spring Range to the west, and the McCullough Mountains still farther west, is compatible with the existence of a west-rooting detachment system beneath it (S. Davis, 1984), although neither the inferred detachment nor rocks that would be lower plate to it are known to be exposed.

Multiple low-angle faults have been recognized throughout most of the detachment terrane south of Searchlight, both in upper- and lower-plate positions (Fig. 2). Interpretations vary as to whether all such faults developed with shallow dips or represent, in some cases, initially steep normal faults that have experienced rotational shallowing of dip. Upper-plate rocks above the major detachment faults in the Colorado River extensional corridor have characteristically been distended by numerous normal faults, the great majority of which dip northeastward at angles less than 60° (Fig. 3). Complex relationships exist in the detachment terrane between multiple generations of extension faults that now dip both steeply and gently; an interpretation of these relationships is one of the objectives of this chapter.

Lower-plate rocks for the east-dipping detachment system include Precambrian metamorphic rocks generally similar to the 1.7-Ga high-grade gneisses and amphibolites of the upper plate, Cretaceous plutons, commonly sheetlike or tabular in form, and Tertiary dikes and plutons. A steep, nonmylonitic, northeast-striking foliation is discernible within some of the Cretaceous (89 ± 3 Ma) plutons in the central Whipple Mountains, and closely parallels what appears to be Precambrian compositional layering and foliation in enclosing gneisses. Its tectonic significance is not known.

Lower-plate crystalline rocks in some ranges within the NLCR region experienced penetrative early(?) to mid-Tertiary deformation within a crustal shear zone. This event was characterized by the development of a shallow-dipping mylonitic foliation and a regionally consistent northeast–southwest-trending stretching lineation in the mylonitic rocks. The uppermost structural level of penetrative mylonitic gneiss development is the "mylonitic front" of Davis and others (1980). It is exposed in both the Whipple and southern Sacramento Mountains (McClelland, 1982; Figs. 1, 2) as a west-southwest–dipping gradational zone several meters to several tens of meters wide (Fig. 2). Exposures of the mylonitic fronts in the two ranges are in a north-

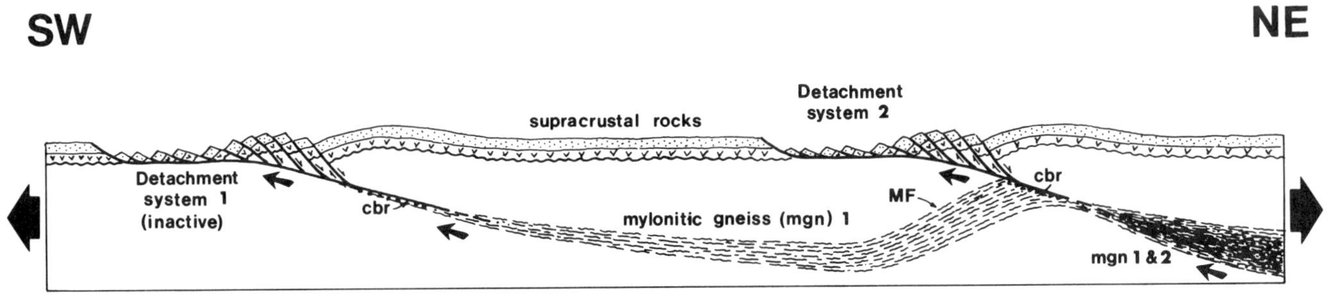

Figure 3. Interpretation of Cordilleran detachment faults as evolving shear zones in an extending continental crust. Width and depth of cross section are approximately 170 and 20 km, respectively. The detachment fault system 1 on the left (southwest) side of the diagram roots at depth into a thick zone of intracrustal flow. Lower levels of mylonitic gneisses formed within that zone are transported upward in the footwall of the detachment fault system where they become retrograded, sheared, and shattered (= chloritic breccias [cbr]) at progressively higher and colder crustal levels. A younger, more easterly splay of the evolving shear zone, detachment system 2, has the geometry of system 1, but captures in its footwall the previously formed mylonitic gneisses of system 1. In this case, the entire kinematically "dead" mylonitic sequence is transported upward; its structural top is mylonitic front (MF) as seen in the lower plates of the Whipple and Sacramento detachment faults, southeastern California (see text). Mylonitic gneisses that form at depth along detachment system 2 presumably overprint mylonitic gneisses formed during earlier phases of deep crustal shear within the evolving zone.

northwest–south-southeast alignment. We interpret the absence of the front and underlying Tertiary mylonitic rocks in the intervening Chemehuevi Mountains (John, 1987) as an indication that levels of exposure within that range are not as deep as they are in the adjacent Whipple and Sacramento mountains. In other words, the mylonitic front lies hidden beneath the Chemehuevis. A corrollary to this conclusion is that the several detachment fault-bounded allochthons of the Chemehuevi Range all lie structurally above the Whipple fault and its presumed Sacramento Mountains equivalent. Mylonitic gneisses now present beneath major detachment faults exhibit shearing and chloritic alteration through thicknesses of as much as 300 m. As discussed below, we interpret both the mylonitization and subsequent shearing and alteration to be phases in a deformational continuum related to Tertiary crustal extension.

Exposures of lower-plate rocks along the extensional corridor from southern Nevada into western Arizona are controlled by two structural elements: (1) a broad (100 ± 30 km) northwest-southeast–trending regional arch; and (2) numerous (ca. 30) northeast-southwest–cross-trending "antiforms" and "synforms" with an average wavelength of approximately 8 km (Frost, 1984) and "limb" dips varying from only a few degrees to as much as 20° to 25°. Lower-plate rocks are exposed in the cores of the doubly plunging "antiformal" structures along the broad hinge of the regional arch. Most recent workers (e.g., Davis and Coney, 1979; Spencer, 1982, 1984, 1985; Howard and others, 1982a) call upon differential isostatic uplift resulting from tectonic denudation of the upper crust as an explanation for the broad northwest-southeast arching at high angles to the direction of northeast-southwest extension.

The most prominent "folds," however, trend parallel to the regional extension direction. An interpretation gaining favor for these northeast-southwest–trending structures is that they are not folds at all, as initially assumed by most workers in the extensional corridor, but gigantic primary flutings or corrugations of the detachment faults oriented parallel to displacements along them (Woodward and Osborne, 1980; Woodward, 1980; Osborne, 1981; Wilkens and Heidrick, 1982; John, 1987). One of us (Davis, *in* Davis and others, 1980, 1982, Fig. 1) has until recently regarded such structures in the Whipple Mountains as late-formed folds because the largest "antiform" in the Whipple detachment fault (Figs. 4, 5) is defined by the geometry of folded, lower-plate sheetlike granitic plutons and mylonitic foliation as well.

However, the limbs of the northernmost Whipple antiform as defined by mylonitic foliation dip more steeply (by 5° to 25°) than do limbs of the antiform defined by dips on the detachment fault (Fig. 5). An early, pre-detachment phase of lower-plate folding is indicated by this relationship, and supported by foliation relationships in the west-central Whipple Mountains. Here, the antiformal geometry of mylonitic foliation is not expressed by the more planar orientation of northwest-striking, southwest-dipping foliation directly beneath the Whipple mylonitic front (Figs. 4, 5). Thus, antiformal arching of lower-plate plutonic sheets and mylonitic foliation seem to have predated a higher level of transpositional foliation development directly beneath the mylonitic front. The front is in turn cut discordantly by the War Eagle detachment fault and the still-younger Whipple fault. It seems appropriate to conclude that the curviplanar geometry of the Whipple fault is not due to folding per se. Instead, the shape

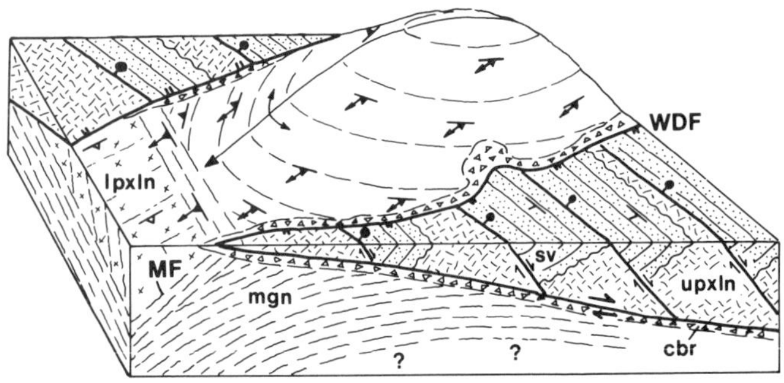

Figure 4. Diagrammatic representation of geologic relations, Whipple Mountains, southeastern California, viewed to north. Width of block diagram is approximately 30 km; vertical topographic relief, ca. 1.1 km, is highly exaggerated in the diagram. Structural features designated by symbols: WDF = Whipple detachment fault; MF = mylonitic front. Rock units: lpxln = lower plate crystalline rocks (predominantly Precambrian gneisses); mgn = undifferentiated mylonitic gneisses (see text); cbr = chloritic breccias; upxln = upper plate crystalline rocks (predominantly Precambrian, but not correlative with "lpxln" within area of figure); sv = Miocene sedimentary and volcanic rocks.

of the fault can be described as corrugated, with the axes of the corrugations oriented parallel to the well-documented direction of fault displacement. This corrugated geometry may have been strongly influenced by preexisting patterns of foliation-defined folds in mylonitic gneisses cross-cut by the fault (i.e., the form of the fault mimicked to some extent the curviplanar structure in the gneisses). John (1987) reached a similar conclusion in her excellent treatise on detachment faults in the Chemeheuvi Mountains, noting that, "There appears to be some influence of pre-existing structures on the overall geometry of the fault system," and the strength of her arguments led us to reevaluate our former position on folded detachment faults.

The tectonic implications of the juxtaposition of upper- and lower-plate units in the lower Colorado River detachment terrane deserve special attention. Mylonitic gneisses of Tertiary age lie in the footwalls of some of the region's major detachment faults, commonly directly below Tertiary strata of broadly equivalent age. Thus, Oligo(?)-Miocene sedimentary and volcanic rocks were being deposited on Precambrian and younger basement rocks while mylonitic gneisses were forming by ductile shear at greater depth in that basement. As explained in greater detail below, J. L. Anderson (1981, 1985, 1987) has undertaken thermobarometric studies of mylonitic gneisses from several areas in the Mojave-Sonora Desert region of California and Arizona. His most detailed investigation has been of samples collected in the Whipple Mountains, where such gneisses have a thickness in excess of 3.5 km. According to his analysis, mylonitic gneisses now in the upper two-thirds of the section, formed at an average depth of approximately 16 km ± 4 km (4.4 ±1.1 kbar).

There has been relatively little erosion of the upper-plate crust in the Whipple Mountains since Miocene time, ca. 20 Ma, because thin (ca. 1–3 km) Miocene surficial deposits are still widely preserved at the earth's surface. It seems clear that the telescoping of upper midcrustal mylonites against coeval supracrustal rocks has occurred primarily by the upward displacement of the former in the footwalls of master detachment faults (Fig. 3). For most faults, arguments about which wall or plate was active and which was passive are purely relative. This is not the case for the major detachment faults of the Colorado River extensional corridor. Using the earth's surface (Miocene and present) as a crude reference datum, the active plate during crustal extension and detachment faulting in the corridor south of Searchlight, Nevada, was the lower plate. Although some vertical movement of the upper plate cannot be discounted, the present juxtaposition of structural levels once up to 16 ± 4 km apart (in a vertical sense) has been accomplished primarily by the transport of lower plate rocks upward and southwestward out from beneath North America, the upper plate. This is an important point in understanding the nature of detachment faulting during crustal extension, and one agreed upon by most present workers in the area (cf. Luchitta and Suneson, 1981; Reynolds and Spencer, 1985; Howard and John, 1987). It makes no sense geologically to conclude that Miocene surficial deposits were actively downfaulted 16 km with respect to sea level or some absolute earth reference surface, e.g., the geoid, in order to now lie atop midcrustal rocks (unless one is willing to propose that the Whipples in Miocene time had an average elevation of 16,000 m compared with their present average of 700 to 800 m).

The duration of detachment faulting activity in the Colorado River extension corridor is imperfectly known; typically, it is estimated indirectly by studies of the depositional and structural history of upper-plate Tertiary strata. Throughout an area greater than 15,000 km^2, such strata have been block-faulted and rotated, and now generally dip southwestward at variable angles. Angular unconformities within detached Tertiary sections indicate multiple episodes of block faulting, stratal rotation, and presumably, detachment of upper-plate units. Although dips in rotated Tertiary strata generally increase northeastward across the Colorado River detachment terrane (Howard and others, 1982a; Howard and John, 1987), many exceptions point to significant complexities in temporal and spatial patterns of detachment faulting (Davis, 1986a). The analysis of stratal tilting is complicated by uncertainties in correlation and dating of many Tertiary units, although the regionally widespread Peach Springs Tuff (ca. 18.3 Ma, Glazner and others, 1986) provides a critical datum for such analysis. Otton (1982) concluded that Tertiary strata in west-central Arizona (Date Creek Basin) had undergone southwestward tilting of as much as 70° prior to 21 to 24 Ma. If Otton's age determinations are correct, then detachment faulting can be assumed to have also begun prior to this time. However, in most areas farther west (Fig. 2), the first major episode of tilting occurred either just before or just after deposition of the Peach Springs Tuff (Nielson and Glazner, 1986). Spencer (1985) reported that detachment-related tilting of Tertiary units in the

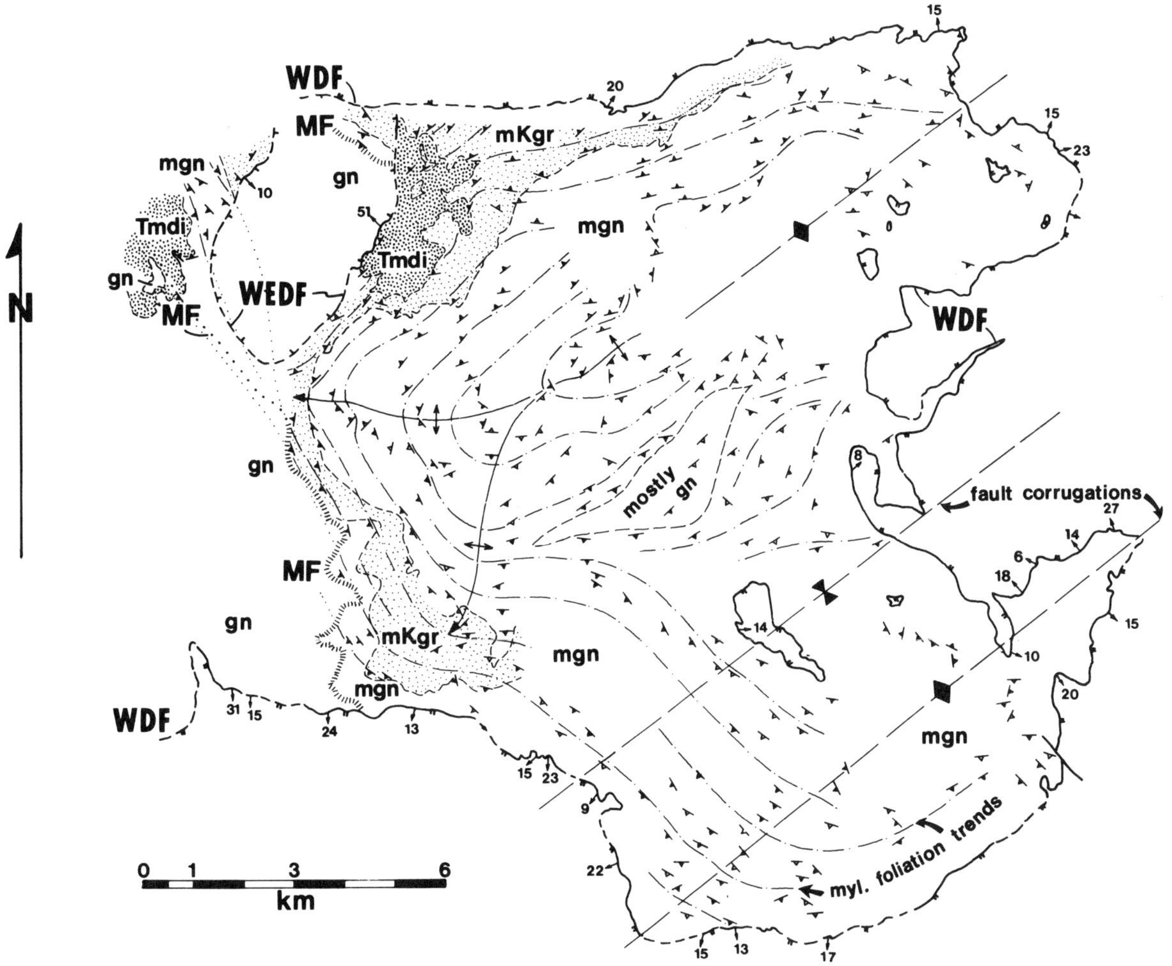

Figure 5. Geologic map of the lower plate of the Whipple detachment fault (WDF), central and eastern Whipple Mountains, California. The map emphasizes the contrasting geometry of the corrugated Whipple fault and folded mylonitic rocks below the Whipple mylonitic front (MF). Foliation trends are shown in mylonitic gneisses (mgn) and a mylonitized composite Cretaceous granitic pluton (mKgr, light stippled pattern), the latter with a sheetlike geometry (now folded). Foliation attitudes depict dip: open barbs, <25°; closed barbs, >25°. Nonmylonitized gneisses (gn) structurally overlie the mylonitic front, which is intruded by a composite Miocene diorite/gabbro pluton (Tmdi, heavy stippled pattern). War Eagle detachment fault (WEDF) offsets the mylonitic front ca. 4.5 km in a N30°E direction. The War Eagle allochthon lies within a "synformal" corrugation with the same trend. Displacement along the still-younger Whipple fault, upper plate with respect to lower plate, was N50°E and parallel to the major wavelike corrugations of the fault surface.

northern Sacramento Mountains ceased between 14 and 15 Ma. Collectively, available data from the Tertiary record indicate that detachment faulting had begun prior to 21 Ma and persisted until shortly after 15 Ma. Geologic and geochronologic relations in the Whipple Mountain region, amplified below, indicate that major detachment fault activity within presently exposed structural levels of the range occurred after 19 to 20 Ma. There is little direct evidence on the age of the detachment fault itself. One reset K-Ar whole-rock age for a cataclasized mylonitic gneiss lying 35 m below the Whipple fault is 15.3 ± 0.5 Ma (Davis and others, 1982).

GEOLOGY OF THE WHIPPLE MOUNTAINS

The geology of the Whipple Mountains is generally similar to that of several other ranges within the NLCR region, but the range, with a relief of one kilometer and many deep canyons that cut across the Whipple fault, provides some of the most spectacular exposures of detachment fault–related tectonics in the U.S. Cordillera (Fig. 6). Figure 4 is a diagrammatic block diagram, viewed to the north, of the central and eastern Whipple Mountains. It illustrates the major structural features of the range, including: (1) the Whipple Mountains antiform, a major foliation

Figure 6. Low-angle aerial view of the Whipple Mountains, California, from westernmost Arizona; the Colorado River (Lake Havasu) along the boundary between the two states is in the middle ground. View is to the west. The crest of the Whipple Mountains rises 1.1 km above the river and its coincident with foliation-defined antiform (Figs. 4, 5) in the lower plate of the Whipple fault. The isolated dark hills above light-colored, lower-plate mylonitic gneisses are klippen above Whipple fault. The prominent contact between light-colored rocks just west of the Colorado River and darker, cliff-forming rocks is a nonconformity separating upper-plate Precambrian crystalline rocks from SW-tilted Oligo(?)-Miocene sedimentary and volcanic rocks.

arch in thick (>3.5 km) lower-plate mylonitic gneisses (mgn); (2) the southwest-dipping mylonitic front (MF), which separates lower-plate crystalline rocks (lpxln) from their structurally deeper, mylonitic counterparts; (3) the Whipple detachment fault (WDF) and the chloritic breccias (cbr) developed beneath it; and (4) the characteristic pattern of upper-plate faulting, a series of closely spaced (1 to 2 km), northeast-dipping normal faults that repeatedly offset and rotate to the southwest upper-plate Tertiary sedimentary and volcanic units (sv) and their largely Precambrian crystalline basement (upxln). The figure illustrates the strong discordance seen in the range between the Whipple fault and both lower-plate and upper-plate structures (Fig. 7).

Protoliths of the mylonitic rocks are not differentiated in Figure 4, but consist primarily of Precambrian gneiss (1.7 Ga), Cretaceous peraluminous (89 ± 3 Ma) and metaluminous (73 ± 3 Ma) plutons—most of them shallow dipping and sheetlike in form, and thin (<3 m), ubiquitous Oligo-Miocene (26 ± 5 Ma) biotite tonalite sills and dikes (Wright and others, 1986). Mylonitization of Whipple Mountains lower-plate rocks occurred under conditions of upper greenschist to lower amphiolite grade. Two feldspar and amphibole-plagioclase thermometries from Cretaceous plutons in the upper two-thirds of the mylonitic section yield temperatures increasing with depth from 458 ± 35 to 535 ± 44°C (Anderson, 1981, 1985, 1987; Anderson and Rowley, 1981; Davis and others, 1982).

Data on the depth of Tertiary mylonitization, and, by infer-

a

b

Figure 7. a, Structural discordance between steeply southwest-tilted Tertiary strata and the underlying, gently southwest-dipping Whipple detachment fault. This aerial view of Savahia Peak in the southwestern Whipple Mountains (courtesy of John Shelton) is to the southeast. Lower-plate Precambrian crystalline rocks at this locality are above the Whipple mylonitic front. Savahia Peak rises approximately 330 m above the lower terrain in the foreground. b, Structural discordance in the south-central Whipple Mountains between southwest-dipping mylonitic gneisses and a lower-plate detachment fault developed at the base of resistant, cliff-forming chloritic breccias. View is to the north-northwest. The Whipple fault (above the ground surface here) and chloritic breccias below it have a discordant relationship to lower-plate mylonitic foliation throughout most of the range (Figs. 4, 5, 9).

ence, the depth of rooting of detachment fault systems beneath the California-Arizona detachment terranes are still sparse. Anderson (1985, 1987, *in* Wright and others, 1986), using existing calibrations for GAR-PL-BIO-MU, MU-BIO-KSP-QZ, and KSP-PL equilibria, and the P-T limits of alumina solubility in hornblende, has estimated P-T conditions for mylonitization of granitic rocks in the Sacramento and Whipple Mountains, California, and the Santa Catalina Mountains north of Tuscon, Arizona (Fig. 1). Pressure estimates at the time of Tertiary mylonitization in the three areas are, respectively, 2.0, 4.4 ± 1.1, and a poorly constrained 3.8 ± 0.9 to 5.2 ± 1.0 kbar. The pressure range for the Whipple mylonitic rocks corresponds to a depth range of approximately 12 to 20 km; much shallower levels of formation indicated for the Sacramento mylonites appear to us as anomalous, but stem from significant differences in Whipple and Sacramento mylonite mineralogy (Anderson, 1987).

The mylonitic front in west-central portions of the range dips 25° to 50° to the west-southwest. At the front, structurally higher, northeast-striking gneisses, amphibolites, and foliated Cretaceous plutonic rocks (Fig. 4) become abruptly rotated, transposed, and mylonitized (cf. Davis and others, 1980, Figs. 15, 16; Davis, 1987). Prior to mylonitization and at levels below the present mylonitic front, steeply dipping Precambrian gneisses and amphibolites had been intruded by discordant, subhorizontal plutonic sheets of Cretaceous age. The plutons became preferential loci of shear strain during mylonitization, quite likely because of their favorable low-dipping geometry and their quartz-rich mineralogy. Panels or lenses of gneisses with relict steeply dipping foliation are locally preserved between plutonic sheets (cf. Davis and others, 1982, Fig. 2). These relict structural domains are as much as 1 km thick (Fig. 5, see domain labeled "mostly gn" below the mylonitic front). They terminate upward and downward into zones of rotation and transposition commonly associated with the highly foliated contacts of the plutonic sheets. More uniform rotation and transposition of the older gneisses has occurred in lower-plate areas below the front where Cretaceous plutonic sheets were not present and could not, therefore, act as preferential strain guides. Mylonitization did occur locally hundreds of meters above the front (cf. Davis and others, 1980, Fig. 16), especially, but not exclusively, in shallow- to moderate-dipping Tertiary dikes that were intruded during late stages of more pervasive mylonitization at deeper structural levels.

We have collected over 100 oriented samples of mylonitized crystalline rocks from several traverses that in the aggregate cross most of the 3.5+-km-thick mylonitic gneiss section. Fabric and microstructural analyses of these samples (Lister, unpublished manuscript) reveal a variety of kinematic indicators. Sense of shear in most samples could typically be determined by two or three indicators, including S-C fabric relations, oblique foliation in dynamically recrystallized quartz aggregates, asymmetric mica "fish," and asymmetric pressure shadows. These analyses demonstrate that approximately 65 percent of the samples collected formed by northeast-directed shear (higher structural levels relative to lower) parallel to the penetrative mylonitic lineation (N45 ± 10E) and to the sense and direction of transport along the Whipple detachment fault; 18 percent exhibit evidence for southwest-directed shear, and 17 percent were not kinematically definitive. We (Lister and Davis, 1983) have interpreted our data as indicating formation of the Whipple mylonitic rocks in a zone of intracrustal laminar flow. Within the zone, large shear strains took place by dynamic recrystallization (e.g., quartz, micas), with minor diffusional mass transfer (e.g., secondary quartz, K-feldspar), and cataclasis (plagioclase, hornblende, garnet). This conclusion supersedes the earlier interpretation by Davis (Davis and others, 1982) that mylonitization occurred during coaxial strain characterized by the flattening in a subhorizontal plane of preexisting thermally weakened rocks with attendant northeast-southwest extensional flow. However, evidence favoring such pure shear deformation is locally present within at least one of the relict structural domains bounded by well-developed mylonitic shear zones (cf. Davis and others, 1982, Fig. 6D, p. 418–419).

Mylonitization throughout the lower plate of the Whipple Mountains is of Tertiary age, possibly in part Oligocene and clearly of Early Miocene age. Davis and others (1980, 1982) concluded erroneously that the gneisses had formed during Cretaceous time. This age, correctly considered by some of our Arizona colleagues to be unreasonably old, based on their field and geochronologic studies in southwestern and south-central Arizona (cf. G. H. Davis, 1980; Rehrig and Reynolds, 1980; Reynolds, 1982), was based on several lines of evidence, among them: (1) concordant biotite K-Ar (78.5 ± 5.5 Ma) and sphene fission track (82.9 ± 3.0 Ma) ages from a mylonitic plutonic gneiss found as clasts in an upper-plate Tertiary debris flow; (2) geologic relations and isotopic age data indicating Cretaceous mylonitization in areas surrounding the Whipple Mountains, e.g., the Iron and Chemehuevi Mountains, California (Miller and others, 1981; John, 1982, 1987, respectively); and (3) our early interpretation that the Cretaceous plutonic sheets (89 Ma) had been intruded synkinematically during mylonitization. The latter interpretation was subsequently invalidated by finding a steep, premylonitic (and nonmylonitic) foliation in at least two of the 89-Ma plutons.

A Tertiary age for Whipple mylonitization has now been thoroughly documented by recent geochronologic and field studies, although the time of its inception has not been closely constrained. Fine-grained, porphyritic (biotite, plagioclase) tonalite sills and dikes occur throughout the thick section of mylonitic gneiss. Field relations indicate that their intrusion was synkinematic to mylonitization. For example, some biotite tonalite dikes were intruded, mylonitized with their country rocks, folded, and then intruded by identical (in the field) planar dikes along the axial surfaces of the folds. The younger dikes were also mylonitized and have a northeast-southwest–stretching lineation that is parallel to that in the older dikes. Zircons from one of the biotite tonalite dikes have been dated as 26 ± 5 Ma, an Oligo-Miocene age (Wright and others, 1986).

The cessation of lower-plate mylonitization is now rather well constrained. Tertiary dikes of the Chambers Well dike swarm (Davis and others, 1982, Figs. 1, 5) were intruded across the mylonitic front. Below the front, andesite and dacite dikes of the swarm always exhibit mylonitic fabrics, but younger diabase dikes that intrude them are typically not mylonitized or display only schistose fabrics within their chilled marginal zones. One such lower-plate mafic dike, collected above the mylonitic front, yields an undisturbed ^{40}Ar-^{39}Ar hornblende age of 21.5 ± 0.7 Ma (E. DeWitt, written communication, 1984). All Chambers Well dikes and the mylonitic front are intruded by a postkinematic (i.e., nonmylonitic) hornblende quartz-diorite to olivine-clinopyroxene gabbro pluton (Tmdi, Fig. 5). Zircons from the dominant quartz diorite phase yield concordant U-Pb ages of 19 ± 2 Ma (Wright and others, 1986). Actinolite hornblende from the same body has a ^{40}Ar/^{39}Ar plateau age of 19.8 ± 0.1 Ma (E. DeWitt, written communication, 1986). Sparse, thin (<0.25 m) aplite dikes that intrude the diorite do have mylonitic fabrics, but the field relations described here indicate that the major episode of lower-plate mylonitization had ceased by 21.5 Ma. All lower-plate rocks described here—the mylonites, all dikes of the Chambers Well swarm, and the diorite/gabbro pluton which intrudes them—are truncated discordantly by the Whipple detachment fault. Its age in the central portion of the range is, therefore, younger than 19 to 20 Ma.

DETACHMENT ZONE ROCKS AND A MODEL FOR DETACHMENT FAULT ORIGIN

Physical descriptions of detachment faults in the southern U.S. Cordillera have been published by many authors, most notably in the Geological Society of America Memoir on *Cordillera Metamorphic Complexes* (Crittenden and others, 1980) and in the volume of papers dealing with the Colorado River region mentioned earlier (Frost and Martin, 1982). Figure 8 illustrates some of the characteristics of such faults. They are typically very planar, but characteristically exhibit domal and basinal or antiformal and synformal geometries on the scale of individual mountain ranges. Fault surfaces, although locally polished, generally exhibit a dull orangish-brown to reddish-brown limonitic(?) patina. This surface is often underlain by dark-colored (black, brown, dark green) layer(s) of flinty cataclasite or microbreccia, usually containing angular, nonoriented porphyroclasts of lower-plate rocks. The layer is generally less than 0.2 m thick and forms a resistant ledge across the topography (cf. Davis and others, 1980, Fig. 24). Phillips (1982) has studied the microbreccia by scanning electron microscopy (SEM) and observed that its matrix is composed of very small (average, 0.004–0.006 mm), equant, interlocking quartz and feldspar grains showing little signs of deformation. Field relationships, especially injection veins into underlying breccias, suggest that these cataclasites flowed during their formation. Phillips suggests that these random fabric fault rocks might have formed initially by cataclastic flow, but that structural superplasticity became the dominant deformational mechanism after grain size had been reduced. Further SEM studies of the cataclasite (H. Green and G. Lister, in progress) reveal evidence of multiple phases of silicification subsequent to repeated episodes of cataclastic flow.

The detachment fault microbreccias and below them sparse, thin (<3 mm) layers resembling pseudotachylite (cf. Fig. 2a of Davis and others, 1986) are overprinted across much thicker (5 to 100 m) "chloritic breccias" (Fig. 4, cbr) which have been derived from lower-plate crystalline rocks. Their development is clearly tied to the detachment fault, since the breccias always lie directly below it and are found both above and below the mylonitic front. These breccias show their thickest development when developed across mylonitic protoliths. Upper-plate rocks have never been observed to be included within the chloritic breccias. Rocks within the breccia zone have variably experienced the effects of locally profound shearing (in extreme development to aphanitic, chert-like cataclasites), in situ shattering, southwest-tilting along northeast-dipping normal faults, and pervasive alteration. The alteration that accompanied shearing and cataclasis is probably lower greenschist in grade, involving as it does the growth of chlorite, epidote, and sericite. The bottom of the chloritic breccia zone is a low-angle fault throughout much of the Whipple Mountains (Fig. 7b), although in some areas the effects of shearing, brecciation, and alteration simply die out downward. Normal faults that cut the breccias are of more than one generation. Some flatten with listric geometries into the fault at the base of the breccias. Other faults cut both the breccias and their underlying low-angle fault, but are themselves truncated upward by the Whipple fault with its associated microbreccias (Fig. 9).

Davis and others (1983, 1986) have interpreted the progressively overprinted sequence of premylonitic crystalline rocks, mylonitic gneisses, chloritic breccias, and flinty cataclasites or microbreccias as indicating that deep-seated rocks have been transported upward from crustal depths of 10 to 15 km or greater along evolving shallow-dipping shear zones (Fig. 3) of the general type envisioned by Wernicke (1981, 1985), Reynolds (1982), and G. H. Davis (1983). The deformational behavior of rocks now exposed in the footwalls of major detachment faults of the NLCR region has changed with time as follows: (1) penetrative deformation (mylonitization) of preexisting crystalline rocks in low-dipping zones of intracrustal laminar flow; (2) upward passage of mylonitic gneisses in the footwalls of detachment fault zones with retrograde formation of chloritic breccias in zones of intense shearing and cataclasis; (3) development of pseudotachylite along narrow (<3 mm) layers cross-cutting the breccias, probably as fault-generated melts formed during intervals of seismic slip; and (4) late-stage formation of discrete detachment faults with associated microbreccias, and, locally, younger fault gouge. These relations suggest progressive deformation of footwall rocks at decreasing crustal depths and temperatures during a Tertiary period of unknown duration. During crustal extension the originally deep lower-plate rock assemblages of the detachment terrane are drawn upward and outward from beneath the brittlely extending upper plate. Lower-plate rocks become exposed at the surface through variable combinations of regional and subregional warping, tectonic denudation, and erosion

a

b

Figure 8. a, Exhumed Whipple detachment fault surface, south-central Whipple Mountains. Cholla cactus in a pained Gordon Lister's left hand provides scale. The detachment fault surface is underlain by a resistant microbreccia at this locality. b, Exposure of Whipple detachment fault on south side, Whipple Wash. Very planar, gently dipping fault separates lower-plate chloritic breccias from subhorizontal upper-plate Tertiary sedimentary and volcanic section (Tmvs, Fig. 13). Neither the prominent northeast-dipping normal fault (on right) or antithetic southwest-dipping normal fault (to left) flatten into the detachment fault. Height of exposure is approximately 35 m.

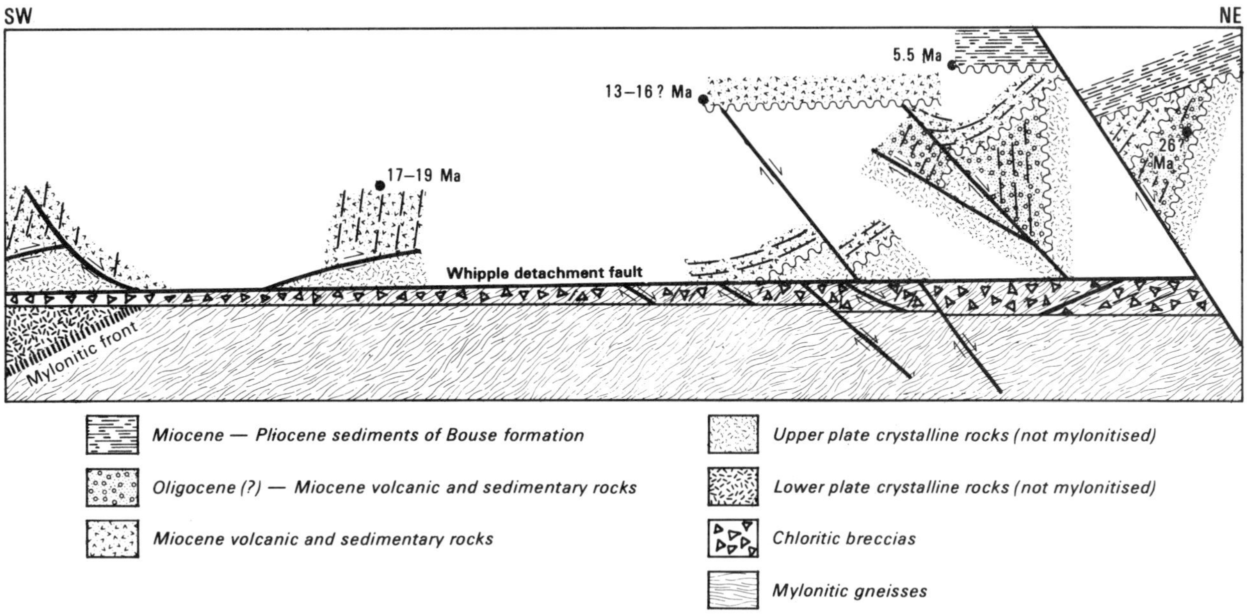

Figure 9. Diagrammatic northeast-southwest cross section through the Whipple Mountains, southeastern California. Section illustrates late age of Whipple detachment fault (WDF) with respect to multiple generations of preexisting upper- and lower-plate structures. Ages are given for most critical stratigraphic units. Listric upper-plate fault (far left) cuts shallow-dipping upper-plate detachment faults and is presumably younger than steeper upper-plate faults (on right). These faults are interpreted as having once been listric with respect to earlier, structurally deeper generations of Whipple detachment faults (see text). The Havasu Springs fault on the far right rotates hanging wall strata of the Mio-Pliocene Bouse Formation and is believed to be a post-detachment normal fault of Basin-Range type. Figure 7b illustrates the angular relationship, shown on this section, between southwest-dipping mylonitic gneisses and the cross-cutting detachment fault at the base of the chloritic breccias.

(Fig. 3). This model accounts for the abrupt contrast in metamorphic grade and deformational history of upper- and lower-plate rocks now juxtaposed across the major detachment faults.

The evolving shear zone model for crustal extension in the NLCR region receives support from geochronologic studies of lower-plate rocks in the Whipple Mountains, and from field studies that indicate the timing and magnitude of displacement along the Whipple fault. Geochronologic studies strongly support our tectonic model by confirming that Miocene mylonitic gneisses experienced very rapid cooling, between roughly 19 to 20 and 18 Ma, presumably during rapid uplift and tectonic denudation. This conclusion was first reached by Dokka and Lingrey (1979) from fission track dating of mylonitic rocks collected from just below the Whipple fault. Five age determinations (three from zircons, one each using apatite and sphene) from three rock samples yielded concordant ages that varied from 17.9 to 20.4 Ma with an overlap of error bars between 18.4 and 19.5 Ma. Because the three analyzed mineral species have widely disparate fission track retention temperatures, their concordant ages led Dokka and Lingrey to conclude that the mylonitic gneisses had experienced a significant temperature drop (>80°, <220°C) between 18 and 20 Ma. $^{40}Ar/^{39}Ar$ age-spectrum dating of Whipple mylonitic rocks by E. DeWitt and J. Sutter (written communication, 1986) offers impressive confirmation of the Dokka and Lingrey study. Tertiary mylonitization (26 ± 5 Ma) of a Cretaceous hornblende quartz diorite in the Whipple Mountains occurred at a temperature of 535° ± 44°C (Anderson, 1987). By 19.2 ± 0.2 Ma, neomineralized hornblende (formed during mylonitization of the diorite) had cooled below about 450°C. Neomineralized muscovite from a closely adjacent wall rock mylonite has a plateau age of 18.0 ± 0.1 Ma, indicating that the mylonitic assemblage "had cooled below about 275°C only 1 m.y. after it was at more than 450°C" (E. DeWitt, written communication, 1986). Orthoclase from a structurally higher mylonitized granitic rock collected elsewhere in the range has a near plateau age of 18.5 Ma. This age indicates cooling of the rock below 150°C, only about 1 m.y. after the hornblende in deeper rocks had cooled below 450°C. Lower plate mylonitic rocks reached the earth's surface during detachment faulting. Tertiary mylonitic clasts are present in some tilted upper-plate Miocene fanglomerates that are older than 16 Ma.

Rapid upward transport of mylonitic gneisses in the lower plate of the Whipple detachment fault system, suggested by the rapid cooling of these rocks, is supported by field studies indicating high collective rates of slip along Whipple system faults after 20 Ma. As mentioned above, the Whipple fault truncates a lower-plate assemblage of rocks near the mylonitic front that includes the Chambers Well dike swarm and a somewhat younger (19.8 Ma), cross-cutting hornblende quartz diorite pluton (Fig. 5). An upper-plate dike swarm, which is almost certainly the offset equivalent of the Chambers Well swarm, is exposed in the Mohave Mountains to the northeast (Fig. 2, Nakata, 1982; Howard and others, 1982a). If this correlation of dike swarms is valid, at least 40 km of offset (horizontal component) has occurred between rocks of the two ranges after approximately 20 Ma (the age of the diorite pluton). If a constant rate of slip is assigned to the Whipple fault system between 20 and 15 Ma, a slip rate (horizontal component) of 0.8 cm/yr (8 km/m.y.) is indicated. However, much higher slip rates are probable for the period 18 to 20 Ma in order to account for the rapid cooling of footwall mylonites during fault-controlled uplift within that time interval. Most, although not all, major tilting of Tertiary strata in the Whipple-Mohave mountains region occurred just before or soon after deposition of the Miocene Peach Springs Tuff (mean sanidine K-Ar age of 18.3 Ma, Glazner and others, 1986; Nielson and Glazner, 1986; Davis, 1986).

FOOTWALL MYLONITIC GNEISSES

The temporal and spatial relations between detachment faults and the mylonitic gneisses exposed below them need amplification. Structural relations throughout the detachment terranes of the southwestern Cordillera indicate kinematic continuity between the Tertiary period of deformation in which the now-exposed mylonitic gneisses formed and the period during which detachment faulting occurred (Davis and others, 1983, 1986). At any given time during their development, major Tertiary detachment faults, such as the Whipple fault, were probably continuous down-dip into progressively wider and deeper zones of brecciation, shearing, and mylonitization (e.g., detachment system 1, Fig. 3; cf. Wernicke, 1981; Reynolds, 1982; G. H. Davis, 1983; Davis and others, 1983, 1986). But lower-plate mylonitic gneisses in the Whipple Mountains are, as one example, somewhat older than the Whipple fault below which they are now exposed. This age relationship has been documented by geochronologic studies reviewed above and is demonstrated by the angular discordance of the Whipple mylonitic front (and most mylonitic foliation below it) with the overlying Whipple fault (Figs. 4, 7b, 9; Davis and others, 1980, 1982, Figs. 1, 2). The Whipple mylonitic front leaves the detachment fault and dips below the ground surface to the southwest (Fig. 4). This is a geometric necessity since only directly beneath and near the capturing detachment fault are footwall rocks, including those at the mylonitic front, elevated to surface or near-surface structural levels (Fig. 3). It appears likely that the front is "seen" in CALCRUST seismic reflection profiles at a depth from 3 to 4 sec, respectively, 16 to 60 km west of the surface trace of the mylonitic front (Davis, 1986b; Okaya and Frost, 1986).

From the relationships described above, it is probable that the Whipple fault did not exist at the time of the penetrative ductile deformation that formed the mylonitic gneisses now exposed in its lower plate. Two extensional tectonic settings are possible for these pre–Whipple fault mylonitic gneisses. One is that they formed at depth along the roots of a more westerly detachment fault system (e.g., detachment system 1, Fig. 3). If so, these early Miocene mylonitic gneisses were presumably crosscut and "captured" by the younger, most easterly Whipple detachment fault and then carried rapidly upward (ca. 18 to 20 Ma) in the footwall of that fault (e.g., detachment system 2, Fig. 3). Dokka and Woodburne (1986) and Dokka and Baksi (1986) have documented the existence of a major detachment fault system in the central Mojave region to the west (Fig. 1). This system was active between about 24 and 20 Ma and may have been kinematically linked with the lower Colorado River detachment system (R. K. Dokka, 1986, personal communication). An alternative tectonic setting is that the mylonitic gneisses formed in an intracrustal zone of noncoaxial laminar flow (a zone of midcrustal delamination) without direct or obvious connection to structural levels now exposed in the southwestern Cordillera (Davis, 1987).

Mylonitic gneisses are not present beneath all major Cordilleran detachment faults, nor should they be. According to the crustal shear zone model described above, major detachment faults lacking lower-plate mylonitic rocks probably either did not have displacements large enough to carry deep-seated mylonitic gneisses to upper-crustal levels, or alternatively (and less likely), the faults were generated at crustal levels above those where mylonitic gneisses could have formed during Tertiary crustal extension (Fig. 3). The Chemehuevi fault in the NLCR mountain range of that name is a case in point of a major detachment fault (displacement >8 km; John, 1987) that lacks Tertiary mylonites in its footwall; Mesozoic mylonitic rocks are present, but are intruded by a late kinematic and postkinematic plutonic suite, a younger phase of which yields a 64-Ma biotite K-Ar cooling age (John, 1987). John has correlated the Chemehuevi fault, which carries southwest-tilted Tertiary strata in its upper plate, with the Whipple and Sacramento faults in areas to the south and north, respectively. Alternatively, we consider the Chemehuevi fault, lacking Tertiary mylonites in its footwall, to lie structurally above the Whipple fault and to have a much smaller displacement. It is possible that the Chemehuevi fault is an upper-plate splay off the Whipple fault and that both northeast-dipping faults merge at depth. Because Tertiary mylonitic gneisses are not exposed in the Chemehuevi Mountains. Howard and John (1987) conclude that in this range "the middle Tertiary crustal extension was not accommodated in a major way by either igneous intrusion or ductile distension at any crustal levels now exposed." We agree with this conclusion for the Chemehuevi Mountains, but believe that it is not applicable to those ranges in the lower Colorado River detachment terrane where deeper levels of Tertiary structural development are exposed (e.g., Sacramento, Whipple, Buckskin mountains with their mylonitic lower plates, Fig. 2).

The geometry at depth of detachment zones and their relationship to spatially associated and broadly coeval mylonitic gneisses is uncertain, and a matter of much disparate opinion among Cordilleran workers. We will not attempt here a discussion of tectonic models that are alternatives to inclined crustal shear zone models, e.g., the coaxial ("pure shear") strain model proposed by Miller and others (1983) for the Snake Range of Nevada; such discussion is undertaken elsewhere (Lister and Davis, 1987). Wernicke (1981, 1985) has postulated that detachment zones may cross the entire lithosphere and root into underlying asthenospheric rocks. We believe, however, that the common occurrence of mylonitic gneisses in the footwalls of major detachment faults argues for rooting of the faults into those lower upper crustal or midcrustal levels (see below) where such gneisses can form. Mesoscopic and microscopic analyses of mylonitic gneisses from several "core complexes" in Arizona, California, and Nevada strongly indicate that they have formed in shallow-dipping ductile shear zones, with a sense of shear the same as the relative sense of displacement inferred for spatially associated detachment faults (e.g., Lister and Davis, 1983). S-C mylonite types I and II (Lister and Snoke, 1984) are common throughout the mylonitic sequences. Kinematic complexities do exist (e.g., subordinate antithetic shear zones, mylonite-bounded relict lenses exhibiting internal coaxial strain effects), but they do not appear to detract from the general conclusion that the movement picture inferred for most mylonitic gneiss terranes is compatible with simple shear tectonic models. This conclusion does not negate the possibility that elements of pure shear were involved in the evolution of some detachment fault/mylonitic gneiss terranes, especially at mid- to lower crustal depths (e.g., Hamilton, 1982; Miller and others, 1983; Lee and others, 1987; Lister and Davis, 1987).

COMPLEXITIES IN THE EVOLVING GEOMETRY OF DETACHMENT FAULTS

As described above in the evolving shear zone model (Fig. 3), the lower plate of detachment zones is drawn upward and out from beneath the upper plate as high levels of the lithosphere are distended. Figure 3 is oversimplified, however, because it suggests that the initial detachment fault that propagated upward from midcrustal levels remains operative throughout the entire history of lithosphere extension. Field relations in the northern lower Colorado River (NLCR) region indicate that this is not the case and that the geometric evolution of detachment fault zones is quite complex. For example, the Whipple detachment fault, the most spectacular of the structural elements in the Colorado River extensional corridor, is unequivocally the youngest major structure of the region in which it is developed (with the exception of postdetachment antiformal warping along the regional northwest-southeast trend). The fault truncates discordantly steeper upper- and lower-plate normal faults that formed during earlier phases of detachment tectonics (Fig. 9), a relationship that

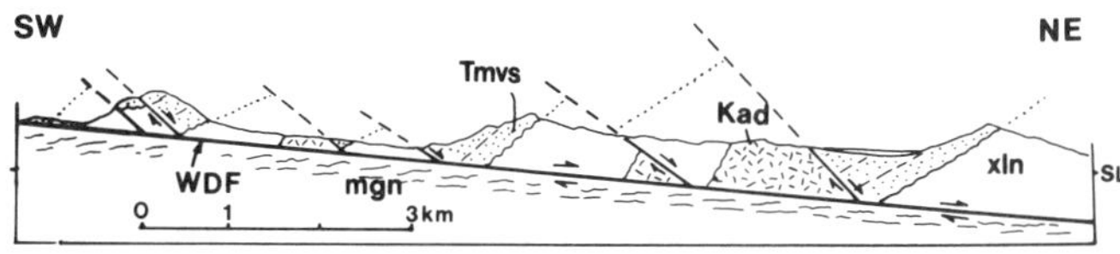

Figure 10. Geologic cross section through upper plate of Whipple detachment fault (WDF), eastern Whipple Mountains (from Frost, *in* Anderson and Frost, 1981). Gene Canyon reservoir lies above easternmost exposures of southwest-dipping Miocene sedimentary and volcanic strata (Tmvs). Other upper-plate units include nonmylonitized, largely Precambrian metamorphic and plutonic rocks (xln) and disrupted portions of a Cretaceous adamellite pluton (Kad). Lower-plate chloritic breccias are present but are not illustrated. Attempts to balance this section, e.g., by realigning the multiply offset Miocene nonconformity below the Tertiary section reveals that large volumes of former upper-plate rocks are now missing. Domino-style rotation above the present Whipple fault of the northeast-tilted fault blocks seen here would produce such major space problems that this mode of rotation appears to be geometrically precluded. No vertical exaggeration.

Davis, his co-workers, and other investigators did not fully appreciate in earlier phases of Whipple studies.

There was an earlier tendency on our parts to regard the internal fault structures of the two juxtaposed plates as somehow developing wholly with the plate in which they are now found. For example, at one stage of our studies it was assumed that upper-plate normal faults were originally planar, that they had originally terminated downward at the Whipple detachment fault and that rotation of tilted Tertiary strata in upper-plate blocks bounded by them had occurred domino-style during displacement on the underlying Whipple fault (cf. Frost, 1980, 1984; Gross and Hillemeyer, 1982). The principal problem with treating each plate as a self-contained, evolving unit is that cross-sections, through the Whipple upper plate (e.g., Figs. 10, 11) cannot be balanced, i.e., palinspastically restored to initial geologic configurations. Restoration of upper-plate geologic relations prior to upper-plate normal faulting clearly indicates that we are now missing rocks that were once contiguous with those still preserved. Such geometries require that most steep faults above the Whipple detachment formed prior to development of the presently observed Whipple fault across them. If it is assumed that these early, steeper normal faults developed above a coeval detachment fault of the Whipple system—an assumption supported by geometric and kinematic data—then the later truncation of these faults by a structurally higher detachment fault (the presently exposed Whipple fault) requires a transfer of some former upper-plate rocks to a lower-plate position. This is the phenomenon of excisement. It and the opposite phenomenon of incisement—the structural transfer of lower-plate rocks to an upper-plate position—have been important processes in the evo-

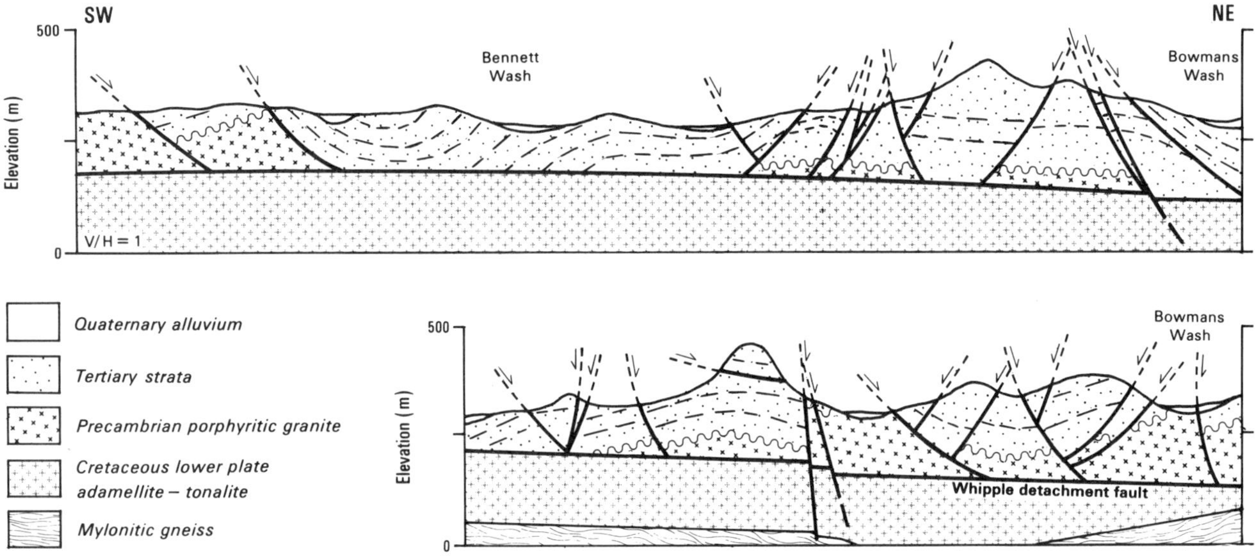

Figure 11. Geologic cross sections of the Whipple detachment fault, southern flank of Whipple Mountains (from Thurn, 1983). These parallel sections lie just southwest of the section of Figure 10. Upper-plate Tertiary strata are generally shallow-dipping to horizontal, broadly folded, and offset by both northeast- and southwest-dipping normal faults. Domino-style faulting did not occur in this area, and Tertiary strata have clearly not been rotated downward into contact with the Whipple fault. Cross sections above the Whipple fault cannot be balanced, suggesting strongly that the Whipple fault postdates upper-plate structure in this area. The Cretaceous lower-plate pluton shares a common mylonitic foliation with the mylonitic gneiss. Lower-plate chloritic breccias developed across the pluton are not illustrated in the sections.

lution of the northern lower Colorado River region (Lister and others, 1984; Dunn and others, 1986).

In Figure 12 we attempt to illustrate how warping of an initially more planar detachment fault during its operation can produce geometric complexities in the development of coeval upper-plate faults. The phenomenon of isostatically induced warping within the Colorado River extensional corridor has been discussed by Howard and others (1982a), and Spencer (1982, 1984, 1985). As an initially rather planar and uniformly dipping detachment fault becomes more and more curved as the lower plate warps upward, a time is reached when mechanical impedence against the continued functioning of the curved fault becomes so great that a new, more planar detachment develops as a higher splay off the preexisting fault (Fig. 12). As displacement (translation) is transferred to the younger fault, excised upper-plate rocks are carried away in the lower plate from their former upper-plate counterparts. Because of this transfer, cross sections drawn through the remaining, lessened portion of the upper plate can no longer be balanced. Figure 12 illustrates our belief that most upper-plate normal faults originally had listric geometries, and flattened into active underlying detachments. Their present more planar configurations are interpreted as the consequence of excisement of their flattened lower portions during development of higher, younger detachment splays.

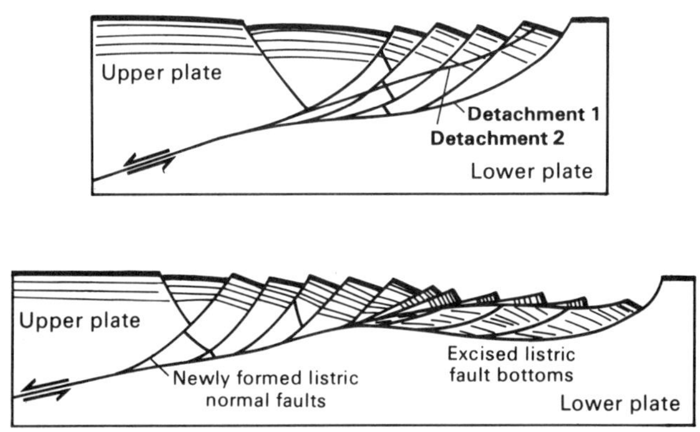

Figure 12. Hypothetic evolution of upper structural levels of an evolving detachment fault zone. Upper diagram shows a major fault splay (detachment 2) that has branched upward from the previously active detachment (1). In the lower diagram, this splay has become the active detachment, and excisement of the lower portion of the former upper plate has occurred. These excised rocks are now transferred to the lower plate of the younger, active fault. Previously active listric faults (phase 1) are abruptly terminated by the younger detachment. Younger listric normal faults now form with respect to the new detachment surface.

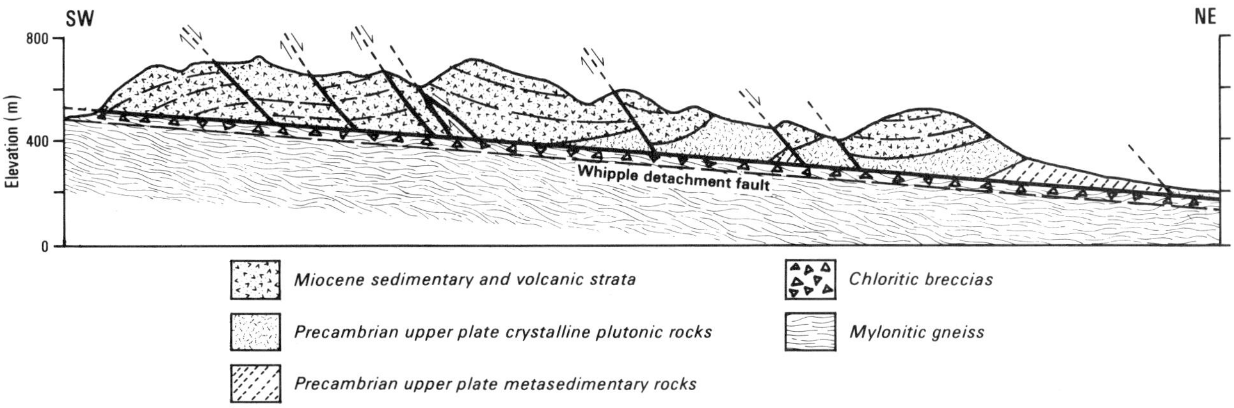

Figure 13. Geologic cross-section from Frost (1980) through the Whipple Wash gorge area, eastern Whipple Mountains, where the geometry of upper-plate faulting is well exposed. Numerous planar normal faults that terminate at, but do not cut the Whipple detachment fault (WDF) are shown (cf. Fig. 8b). No vertical exaggeration.

Domino faulting—the synchronous rotation of planar faults and crustal blocks between them—was described by Thompson (1960) and has been widely considered to be an important mechanism for continental extension (e.g., Wernicke and Burchfiel, 1982; Miller and others, 1983; Jackson and McKenzie, 1983). Two major geometric problems exist with domino mechanisms for upper-plate extension: (1) there is a space problem at the base of the blocks as they tilt, and it becomes progressively aggravated as the width of fault-bounded blocks increases; and (2) domino-faulting may allow fault blocks to rotate, but because abrupt changes in tilt angles between blocks would produce major fit problems, adjoining fault blocks cannot differ very much in their orientation or in the dip of strata contained within them. We do not observe major space problems in the basal part of the Whipple upper plate where it is cut by multiple, more-or-less planar faults, and we do see adjacent fault-bounded panels within which the dip of strata varies markedly from one panel to the other (cf. Fig. 9). Accordingly, we are skeptical of the importance of domino faulting in the lower Colorado River detachment terrane as the primary feature allowing extension of the upper plate, favoring instead the interpretation of primary listric faults offered above and developed further below.

We propose that a basic error made by previous workers in detachment terranes (e.g., in the lower Colorado River region by Gross and Hillemeyer, 1982; Howard and others, 1982a,b; and, until recently, ourselves, cf. Davis and others, 1980) is *the assumption that a given fault, e.g., the Whipple detachment fault, is everywhere the same tectonic surface of the same age.* Geologic observations tell us that this assumption is suspect. For example, portions of the Whipple detachment fault in the Whipple Mountains are underlain by impressively resistant ledges (typically <1 m) of black, flinty cataclasite or "microbreccia" (cf. Phillips, 1982), whereas in other areas the fault is underlain only by a zone several meters thick of poorly consolidated fault gouge. In some portions of the Whipple Mountains, Miocene strata dip vertically into the subhorizontal Whipple fault, but elsewhere only a few kilometers distant, strata of the same age rest subhorizontally (and tectonically) directly on the Whipple fault over large areas (cf. Fig. 11). How can these variable geometric and structural relations be accounted for?

A specific example, drawn from studies of the Whipple detachment fault, can be used to illustrate some of the geometric complexities that occur in the evolution of detachment zones. Figure 13 is a cross section by Frost (1980) of the Whipple detachment fault (WDF) as exposed in the deep canyon of Whipple Wash west of Lake Havasu. This cross section and those of Figures 10 and 11 are typical of cross sections drawn through this and other Cordilleran detachment complexes in that upper-plate rocks (here both Precambrian crystalline and Tertiary) cannot be restored to a balanced geometric configuration by simply reversing normal fault displacements along the upper-plate normal faults. Doing so results in unfilled spaces in the restored sections. These faults, as exposed in the canyon (Fig. 8b), do not flatten downward as they approach the Whipple detachment fault. Gross and Hillemeyer (1982) commented on the geometric difficulties of preparing balanced cross sections of extended upper-plate units and specifically discussed the geometry of faults in the Whipple Wash section (Fig. 13). Essentially they concluded that upper-plate extension is so complicated by successive generations of rotating planar faults, including microfaults, that balanced palinspastic reconstructions cannot be made. They stated, for example (1982, p. 263), that

> Normal faulting within the upper plate is much more extensive than has previously been recognized—to the extent that microfaults are penetratively developed. Such penetrative deformation appears to be essential in producing the rotation and extension in upper-plate rocks throughout the detachment terrane. Ironically, it seems that the abundance of faults is responsible for relatively few of them having been observed. The pres-

> ence of penetrative faulting (and thus great numbers of faults) means that offsets along individual faults need be very small, on the order of a few centimeters or less. . . .Very large faults certainly do exist, but the fault blocks which they bound . . . are not rigid blocks, rather they must be penetratively deformed via microfaults.

We do not agree with this analysis, partly because we are skeptical that penetrative microfaulting exists to the extent described, and partly because we do not believe that penetrative faulting in fault-bounded panels can geometrically compensate for the abrupt termination at the Whipple fault of upper-plate faults with dip-slip displacements of tens to hundreds of meters (Fig. 13). As an alternative we propose (Fig. 14) that the surface mapped as the Whipple fault in Whipple Wash is a second- (third? fourth?) generation splay of an earlier Whipple detachment fault, and not the original detachment itself. Figure 14a illustrates a phase of detachment hypothesized earlier, where Tertiary strata lay nonconformably atop upper-plate Precambrian basement rocks. Rotation of strata only in the northeastern part of the upper plate was accomplished along an early listric fault that flattened northeastward into the Whipple fault (WDF 1). Figure 14b illustrates the subsequent formation of an inferred low-angle, upper-plate splay—a low- to moderate(?)-dipping fault that branches upward from WDF 1, climbs across the upper-plate crystalline basement, and into its shallow-dipping Tertiary cover. In this section the hanging wall splay has become the active "Whipple" detachment fault (WDF 2). Basement rocks that were upper plate to WDF 1 have been excised by the development of the higher splay and transferred to the lower plate of WDF 2. The listric fault of phase 1 extension is truncated at depth by WDF 2. It (listric fault 1) now appears to be a planar hanging wall fault with the attendant geometric problems that such faults present at their abrupt intersection with a presumably coeval underlying detachment fault (actually, a younger, truncating structure). Once the upper-plate normal fault is recognized as an older structure, now kinematically dead, the problem of attempting to balance upper-plate strata on either side of it vanishes.

With the transfer of plate interaction from WDF 1 to WDF 2, lower plate chloritic breccias (phase 1), are "switched" (using railroad parlance) to move up beneath the hanging wall of the now-active higher Whipple splay (WDF 2). If the Whipple fault is routinely mapped only as the fault separating mylonitic gneisses and chloritic breccias from "upper-plate" nonmylonitic crystalline or Tertiary rocks, then WDF 2 would not be recognized as a younger phase of the master Whipple detachment updip from where the chloritic breccias depart from it. WDF 1 would still be recognized as the Whipple fault, but WDF 2 (with Tertiary strata in its footwall) would only be considered an "upper-plate" detachment fault offsetting Tertiary strata and its crystalline basement. Without such recognition, the Whipple fault is mistakenly interpreted as a single-generation structure across which sections cannot be balanced and variable dips in upper-plate strata cannot be accounted for. Although microbreccias exist beneath WDF 2 in Whipple Wash, other younger and/or higher

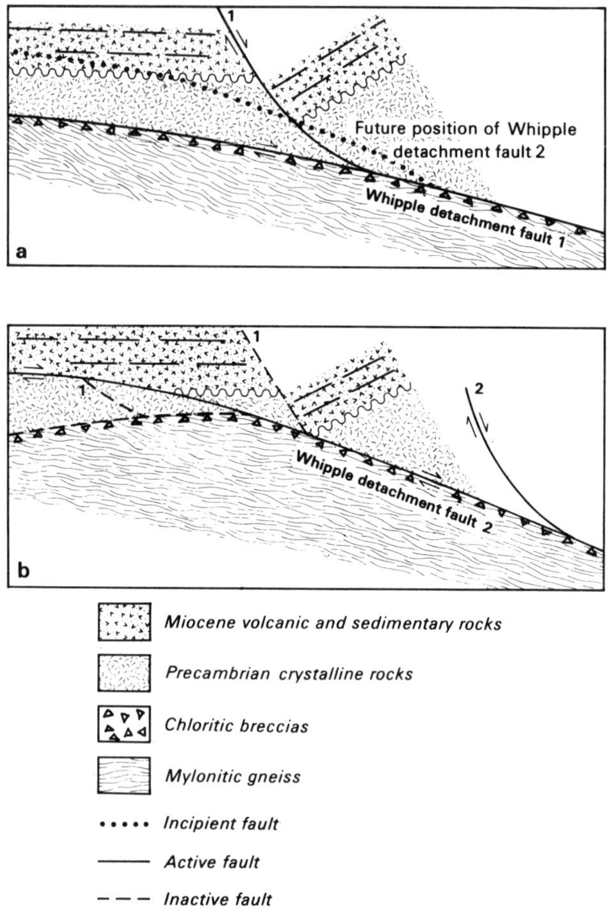

Figure 14. Diagrammatic representation of evolving geometry of Whipple detachment fault to explain geologic relations illustrated in Figure 11. a, Illustration of inferred geometry of Whipple fault during phase 1 of its development. WDF 1 lies at unknown depth beneath Tertiary-Precambrian crystalline nonconformity. Upper-plate listric fault 1 is responsible for rotating hanging wall strata; footwall strata are not rotated. The future trace of a higher detachment fault splay (WDF 2) that will cut upward across upper-plate rocks is shown by the heavy dotted line. b, Illustration of subsequent phase of extensional deformation. WDF 2 is now the active detachment fault. Crystalline rocks formerly in the upper plate of WDF 1 have been excised by the younger, higher WDF 2 splay and transferred to its lower plate. The upper-plate listric fault 1, active during phase 1, has been truncated and offset by WDF 2; it, like WDF 1, is now kinematically "dead." Lower-plate chloritic breccias (cb) are "switched" to follow up along and beneath WDF 2. A second phase upper-plate listric fault, 2, is shown on the right edge of section b. It, or other phase 2 faults, can theoretically rotate previously unrotated strata (e.g., left side of section) or re-rotate already tilted strata (e.g., right side of section).

splays elsewhere in the range may have formed at levels too shallow for microbreccias to develop along them, despite the fact that they, like their earlier phase parents, may be underlain by chloritic breccias and older mylonitic gneisses.

John (1987) saw similar evidence for the evolving geometry of detachment faults. She has suggested that early movement along the Chemehuevi fault (CDF 1 by analogy with Fig. 14) was followed by broad antiformal arching of the fault along a north-northwest trend. After arching, the Chemehuevi fault on the western flank of the arch is believed to have become inactive. However, on the eastern flank, John inferred continued or renewed movement along a postarching, upper-plate fault splay (which we would call CDF 2), which merges northeastward with the original detachment fault (CDF 1). Western exposures of the Chemehuevi fault are characterized by a wide zone of "breccia and cataclasite, lacking any throughgoing planar surface." In contrast, exposures of the presumably younger, shallower fault on the east flank of the arch display thinner breccias, gouge, and a sharp planar fault contact. The evolving geometry of faulting interpreted by John is closely similar to that hypothesized independently in Figure 14 for the Whipple fault.

Howard and others (1982b) and Howard and John (1987) have presented a different view of the Whipple detachment fault than that proposed by us above. They considered the Whipple fault (and the Chemehuevi fault that they correlate with it) to be the sole fault (in a geometric sense) of the detachment terrane at this latitude in the Colorado River extensional corridor. They hypothesized that those portions of the Whipple fault now exposed in easternmost parts of the Whipple Mountains originally lay at a paleodepth of 10 to 15 km, and that Tertiary strata that now dip into the fault originally lay many kilometers above it. In their view, progressive rotation (domino-style) of the upper crust (basement plus Tertiary cover) has occurred along multiple, closely spaced (1–2 km), once steep normal faults that transected the entire crust above the deep-seated Whipple detachment fault. Those upper-plate faults now dip gently, subparallel to the Whipple fault, and are said to bound upended (tilted) sections of the upper crust as thick as 12 or 13 km when measured perpendicular to the Tertiary-Precambrian nonconformity. We cannot agree with this geometric analysis as it applies to the Whipple fault and its upper-plate structures in the eastern Whipple Mountains for two reasons: (1) we consider the present Whipple fault in this area to be a very young detachment, perhaps one of the youngest such faults in the region, and not an early sole fault for the entire terrane above which all extension occurred; and (2) the present Whipple fault was a very shallow structure (<2 to 3 km) at the time it formed in the eastern Whipple Mountains.

Looking at western portions of Figure 13 or the cross sections of Figure 11, there is no basis for postulating that the present Whipple fault once lay 10 or more km below the subhorizontal Tertiary strata that now rest tectonically on it. How is the observed juxtaposition possible? Clearly, domino-block rotation of upper-plate normal faults cannot be called upon to accomplish the lowering of subhorizontal Tertiary strata by 10 or more km to a position of juxtaposition against a "deep-seated" Whipple fault. The Tertiary rocks have not been rotated significantly in some areas of the Whipple Mountains, and the thick section of crystalline rocks once beneath them and above an inferred 10-km-deep detachment fault is missing.

Again, these geometric relations are easily explained by the development of a shallow-dipping splay off an earlier generation Whipple fault. The splay cut upward at shallow angles (<20–25°) across the former upper plate, first through upper-plate basement and then into the thin (<2 to 3 km), faulted Tertiary section deposited atop it. We are convinced that some such splays reached the earth's surface during crustal extension, thus allowing lower plate rocks (e.g., chloritic breccias and mylonitized gneisses) to be eroded and to be preserved as detritus in nearby sedimentary basins. The presence of clasts of both Tertiary mylonitic gneisses, and, less commonly, chloritic breccias in Tertiary sedimentary rocks that are now allochthonous with respect to the present Whipple fault attest to multiple phases of detachment in the evolution of the Whipple core complex. John (1987) and S. Reynolds (personal communication, 1985) have also reported the occurrence of detachment fault-related chloritic cataclasites in allochthonous alluvial fan deposits.

A somewhat similar, but considerably more complicated scenario of detachment-related faulting must be postulated for the Sacramento Mountains, 60 km north-northwest of the Whipple Mountains (Fig. 2). Figure 15a is a diagrammatic southwest-northeast cross section through the southern part of the range, drawing from the study by McClelland (1982, 1984). Figure 15b is a true-scale cross section through McClelland's southern area to better illustrate how thin the detachment fault–bounded allochthons are in this range. As in the Whipple Mountains, mylonitic gneisses, a southwest-dipping mylonitic front, and chloritic breccias (retrograded and cataclasized mylonitic gneisses) underlie the major Sacramento detachment fault. Miocene strata and their nonmylonitized Precambrian basement overlie the fault, also as in the Whipple Mountains. However, significant differences exist between the upper-plate structures of the two ranges. As Figure 15 illustrates, three allochthons lie above the Sacramento Mountains detachment fault: (a) a lower allochthon consisting of Precambrian and Mesozoic(?) crystalline rocks; (2) a middle allochthon containing Precambrian basement and an unconformably overlying section of Miocene sedimentary and volcanic rocks; and (3) an upper allochthon consisting largely of Miocene fanglomerates. Field relations and geochronologic data indicate a consistent younger-over-older stacking of units within the three plates. The highest allochthon, somewhat surprisingly, contains southwest-tilted strata that dip considerably more steeply than do older strata in the underlying middle plate. Multiple generations of steep- to moderately-dipping normal faults cut and/or are truncated by detachment faults above the basal Sacremento detachment. As in the Whipple Mountains, the Sacramento fault appears to be the youngest major fault in the range (Fig. 15) because it cuts late, steep normal faults in its upper plate.

Spencer (1985) presented convincing data that in the north-

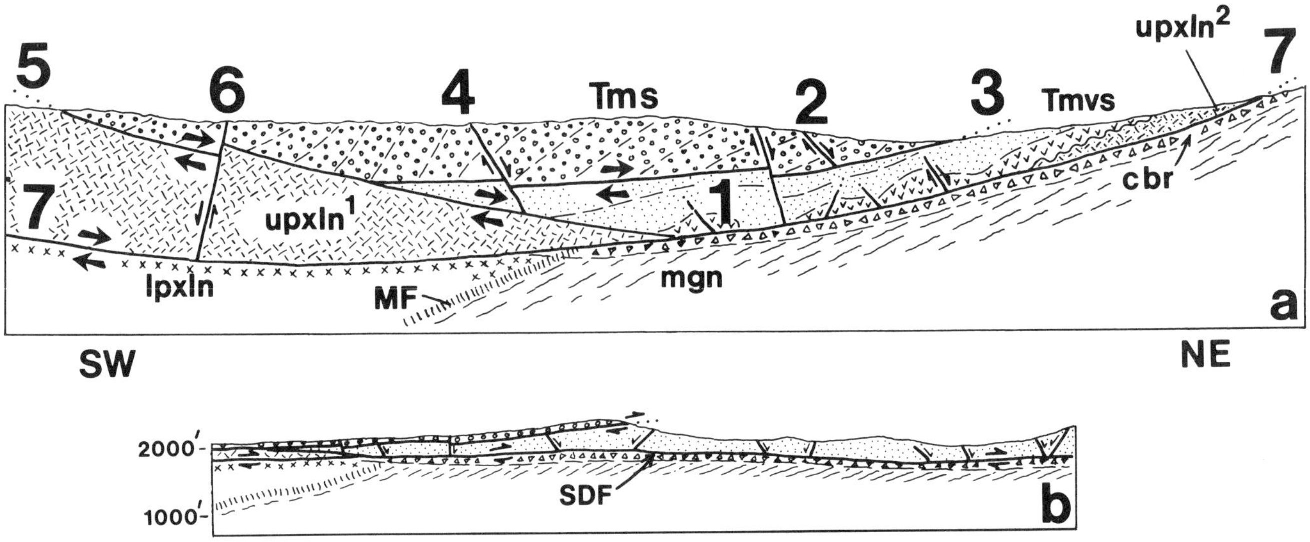

Figure 15. a, Diagrammatic southwest-northeast cross section through the southern Sacramento Mountains (redrawn from McClelland, 1984), illustrating geometric relations in this complexly deformed detachment terrane. Numbers designate sequence of formation (1 = oldest) of high- and low-angle (detachment) faults determined from field studies. Rock unit and structure abbreviations not used in previous figures include Tms = Tertiary Miocene sedimentary rocks (which include basalts in the northern Sacramento Mountains) and SDF = Sacramento detachment fault (section b). b, Southwest-northeast cross section through the southern Sacramento Mountains (McClelland, 1984). Figure illustrates how thin the detachment fault-bounded allochthons shown diagrammatically in section a are when drawn with horizontal: vertical scale = 1.

ern Sacramento Mountains (Fig. 2), as in the Whipple Mountains area (Davis, 1980), the basal detachment fault developed at a very shallow structural level:

> Flat-lying basalt at Flattop Mountains, yielding a K-Ar date of 14.6 ± 0.2 m.y. rests unconformably on tilted conglomerate . . . and is interpreted as post-dating detachment faulting and associated tilting. The contact at the base of the basalt of Flattop Mountain is about 250 to 300 meters above the basal detachment fault. Since the basalt immediately post-dates detachment faulting, the subhorizontal basal detachment fault appears to have been not much more than 300 meters below the ground surface at the time of latest fault movement.

From field relations determined by McClelland, we hypothesize a sequence of faulting events, both high and low angle, that generally explains geometric relations illustrated in Figure 15. Other scenarios can be proposed to explain McClelland's field relations, but we believe that all would require variations on our basic theme. That theme (Fig. 16) requires the formation during crustal extension of both excisement and incisement faults, the latter representing the development of detachment fault splays *below* preexisting active detachments. The consequence of incisement is that rocks formerly belonging to the lower plate of a controlling detachment fault (e.g., fault 5, Fig. 16c) are transferred to an upper-plate position with the development of a younger, structurally deeper splay (cf. Fig. 16c,d). The interpretation of evolving fault geometries in the Sacramento Mountains shown in Figure 16, rests on two assumptions: that moderately steep to steep normal faults of demonstrably different ages all flattened into active lower-angle detachment faults; and that, with the exception of faults developed during deposition of the basal stratigraphic sequence (phase 1 faulting, Fig. 16a), *all* faults in the Sacramento Mountains must have formed in the geologically brief period (<1 m.y., ca. 14 to 15 Ma) after deposition of now tilted basalt flows (14.6 Ma) in the uppermost fanglomerate sequence (Tms, Fig. 15) and before extrusion of the unconformably overlying Flattop basalts in the northern part of the range (14.6 Ma, Spencer, 1985).

Before concluding this discussion of the complexities of detachment faulting in the northern lower Colorado River region, two additional points are noteworthy. The first is that geologists may generally have underestimated both the magnitude and rate of displacement within detachment fault terranes. Reynolds and Spencer (1985) presented evidence that approximately 50 km of relative northeastward displacement of upper-plate rocks occurred on the Bullard (=Whipple?) detachment fault in west-central Arizona during middle to late Tertiary time. To the northwest in the Whipple Mountains, displacement of at least 40 km at a minimum rate of 0.8 cm/yr along the Whipple fault is strongly indicated by its post-20-Ma truncation of the Chambers

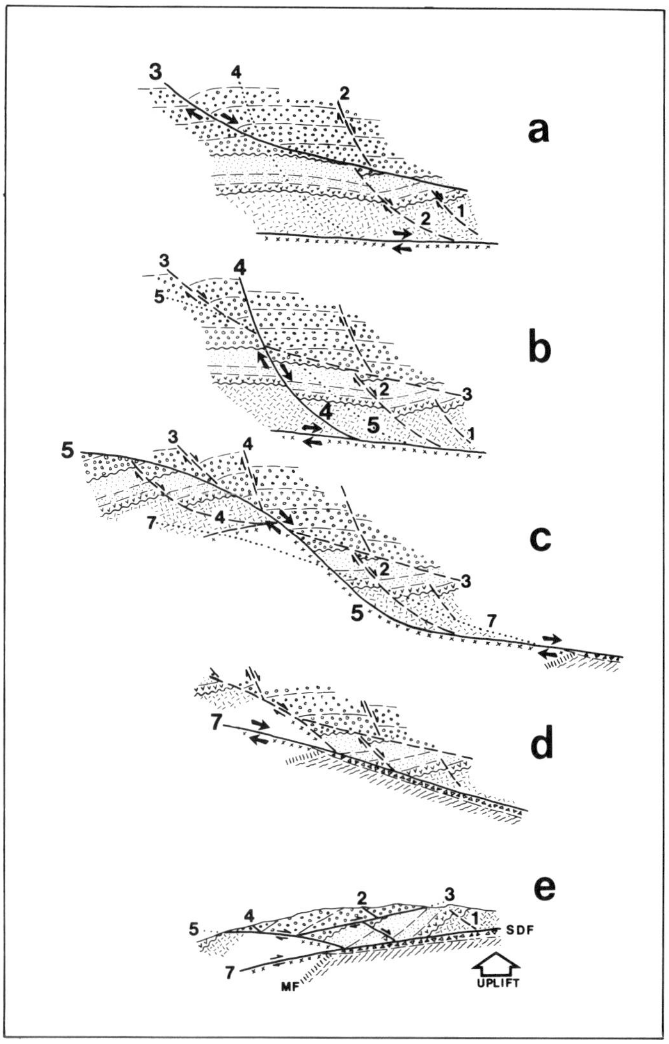

Figure 16. Possible scenario for development of multiple generations of high- and low-angle faults defined by McClelland (1984) and illustrated in Figure 15. Extension faults active in each section (a, b, etc.) are shown by solid lines. Faults previously active are shown by dashed lines. Faults that will become active in the underlying younger section are shown by dotted lines. All faults except generation 1 are interpreted as forming after deposition of uppermost unit (Tms, Fig. 15). All high-angle faults are interpreted as having once had listric geometries with respect to synchronously operative lower detachment faults. Fault 5 is an excisement splay, transferring former upper-plate rocks (b) to a lower-plate position (c). Fault 7, the present Sacramento Mountains detachment fault, has both excisement and incisement geometries (right- and left-hand sides of c, respectively). Compare (e) with present geologic relations of Figure 15.

Well dike swarm; much faster rates for the period 18 to 20 Ma are indicated by geologic and geochronologic data reviewed in this paper. Furthermore, since Tertiary stratigraphic evidence (Otton, 1982; Davis, 1986) and an Oligo-Miocene age for mylonitization (26 ± 5 Ma, this paper) indicate that crustal extension in the NLCR region predates displacement of the Chambers Well swarm, total lithospheric extension in the region may considerably exceed the 40+ km that occurred after 20 Ma at the latitude of the Whipple Mountains.

Finally, the relationship between Cordilleran thrust faults and younger detachment faults deserves a brief comment. Some workers have suggested that the mechanical problems surrounding the formation of primary, shallow-dipping detachment faults (especially the large angle between steep or vertical σ_1, and shallow-dipping faults) might be resolvable if the development of such faults was controlled by preexisting, shallow-dipping structural anisotropies. Thrust faults of Mesozoic or earliest Cenozoic age are the controlling structures most frequently alluded to (e.g., Allmendinger and others, 1983, for the Sevier Desert detachment, Fig. 1). Although the reactivation of Mesozoic thrust faults as Cenozoic extension faults has been well documented in the eastern Basin and Range area (e.g., Royse and others, 1975; Smith and Bruhn, 1984) we are convinced that at least some low-angle detachment faults must be generated as primary structures, i.e., that they transect crustal rocks without the presence of preexisting shallow-dipping anisotropies. Wernicke and others (1985) have reached a similar conclusion from their studies in the Mormon Mountains of southern Nevada. The Whipple and Sacramento detachment faults described in this chapter both cut upward (southwestward) at shallow angles across basement crystalline rocks and into shallow-dipping, unconformably overlying Miocene strata that clearly never experienced prior thrust faulting.

CONCLUSIONS

In this chapter we have set forth what we believe are several new perspectives on the origin of Cordilleran detachment faults during lithospheric extension. We have drawn heavily upon observations, mapping, and the structural interpretations of numerous workers in the magnificently exposed detachment terrane of the lower Colorado River region of southern Nevada, southeastern California, and western Arizona. Some of our major conclusions are summarized below.

Cordilleran detachment faults are best explained as evolving low-angle shear zones that probably root into lower upper crustal or midcrustal structural levels during continental extension. These detachment zones propagate upward across the overlying crust and either reach the surface directly as low-angle faults or terminate at shallow depths in pull-apart complexes of closely spaced normal faults. Along these evolving shear zones, lower-plate rocks are drawn rapidly upward and out from beneath brittlely extending upper-plate rocks. Footwall rocks, which formed at great depth but now lie in fault contact below broadly synchronous Tertiary supracrustal units, composed the active fault wall. This conclusion is supported by petrologic and geochronologic evidence from the Whipple Mountains that Oligo-Miocene footwall mylonitic gneisses, formed at depths >12 km and at temperatures as high as 535°C, had cooled from temperatures above 450°C to below 150°C between 20 and 18 Ma. Such rapid cooling is best explained by rapid uplift of the mylonitic gneisses in the footwall of the evolving Whipple detachment system.

Although mylonitic gneisses in the footwalls of some Cordilleran detachment faults show kinematic coordination with those faults in having the same direction and sense of shear, they are in some cases (e.g., Whipple Mountains) apparently formed at depth, either along down-dip portions of older detachment zones (Fig. 3) or within intracrustal zones of laminar flow that have no obvious connection to structural levels now exposed. Such mylonitic gneisses are later crosscut and carried upward in the footwalls of the cross-cutting detachment faults, where they become sheared, retrograded, and then brecciated at progressively higher, cooler, and more brittle structural levels. Mylonitic fronts, the tops of intracrustal shear zones characterized by predominantly crystal–plastic flow mechanisms, can also be transported upward in the active footwalls. As such they dip away from the capturing fault and return to considerable crustal depths away from the detachment fault (Fig. 3). This is a geometric necessity since only directly beneath and near the younger, capturing detachment are footwall rocks elevated to surface or near-surface levels. At some distance from the detachment fault the older, lower-plate mylonitic gneisses maintain their lower upper crustal or mid-crustal depths and are only transported laterally during crustal extension. Mylonitic detachment complexes form only when relative displacements along the evolving crustal shear zones are large enough to carry deep-seated mylonitic assemblages to shallow crustal levels. Here they become exposed through combinations of crustal warping, tectonic denudation, and erosion.

The angle now observed in outcrop between the detachment fault and the mylonitic front in its footwall (cf. Fig. 3, Fig. 17 of Davis and others, 1980) should record approximately the angle of initial discordance between deep, shallow-dipping mylonitic gneisses and the somewhat steeper, younger detachment fault zone that transected them and captured them in its rising footwall. For the Whipple Mountains, this angular discordance is approximately 10° to 25°, essentially the same range of angles at which higher levels of the Whipple detachment fault cut through upper-plate crystalline basement and into subhorizontal strata. These angular relationships at upper and lower ends of the evolving Whipple detachment fault system argue strongly for a low angle of dip during the time of its development. Detachment faults thus appear to be capable of transecting upper crustal rocks as primary, low-dipping shear zones uncontrolled by preexisting, shallow-dipping structural anisotropies, although the possible reactivation of older thrust faults elsewhere as detachment faults during crustal extension is not denied by this statement.

Contrary to popular belief, some of the major detachment faults exposed today in the Colorado River extensional corridor are not faults that formed at the start of extensional tectonics, but are rather only the youngest in a succession of major detachment structures. Detachment faults undergo various warpings at high angles to the direction of crustal extension during their development, probably due in large part to isostatically induced distortions of originally more planar faults (and possibly due to reverse-drag flexing above lower faults; Wernicke and others, 1985). Such warpings lead to the development of younger, more planar splays that either cut upward into former upper-plate rocks (excisement) or downward into former lower-plate rocks (incisement). Recognition of such geometric complexities offers fresh insights into deciphering the evolving strain patterns within major detachment zones. Studies in the Whipple Mountain region indicate that most upper-plate normal faults originally had listric geometries before losing their shallow-dipping lower segments as the consequence of excisement tectonics. The northeast-southwest–trending curviplanar geometry of major detachment faults in the Colorado River extensional corridor is probably an expression of primary corrugations (fluting) in the fault surfaces formed parallel to the direction of fault displacement. Some of these corrugations or flutes may have been influenced by preexisting structures now seen in footwall rocks.

Finally, the rate of displacement along some evolving detachment systems may be very rapid, perhaps greater than 1 cm/yr. Tertiary mylonitic gneisses formed within deep-seated intracrustal shear zones beneath the lower Colorado River region were carried upward very rapidly (<2 m.y.) to surface or near-surface levels in the footwalls of somewhat younger detachment faults. Geologists may have underestimated not only how rapidly intraplate extension along evolving detachment fault systems can occur, but the magnitude of extension produced by such systems.

ACKNOWLEDGMENTS

We are indebted to Lawford Anderson, Steve Reynolds, Keith Howard, Barbara John, Jon Spencer, George H. Davis, and Eric Frost, among many others, for their verbal and published contributions to the development of concepts presented here on the evolution of detachment terranes in the southwestern U.S. Cordillera. Field studies in the lower Colorado River region of California and Arizona by Davis, his colleague Lawford Anderson, and their students, have been supported by National Science Foundation grants EAR 77-09695 and GA-43309. Kinematic studies on mylonitic gneisses by Lister and Davis were supported by NSF grant EAR-82-1188 (with A. W. Snoke). Comments by Lawford Anderson, Richard Armstrong, Elizabeth Miller, Steve Reynolds, Robert Speed, Jon Spencer, and Brian Wernicke have considerably improved the original manuscript and are greatly appreciated. A very special debt of gratitude is owed to the late Terry Shackelford. Terry's doctoral dissertation (1976) in the Rawhide Mountains of Arizona (northeast of the Buckskin Mountains) began, in a splendid way, the University of Southern California's efforts to understand the origin of the Colorado River detachment terrane.

REFERENCES CITED

Allmendinger, R. W., and 7 others, 1983, Cenozoic and Mesozoic structure of the estern Basin and Range from COCORP seismic reflection data: Geology, v. 11, p. 532–536.

Anderson, J. L., 1981, Conditions of mylonitization in a Cordilleran metamorphic complex, Whipple Mountains, California, *in* Howard, K. A., Carr, M. D., and Miller, D. M., eds., Tectonic framework of the Mojave and Sonoran Deserts, California and Arizona: U.S. Geological Survey Open-File Report 81-503, p. 103.

——, 1983, Proterozoic anorogenic granite plutonism of North America: Geological Society of America Memoir 161, p. 133–154.

——, 1985, Contrasting depths of "core complex" mylonitization; Barometric evidence: Geological Society of America Abstracts with Programs, v. 17, p. 337.

——, 1987, Core complexes of the Mojave-Sonora Desert; Conditions of plutonism, mylonitization, and decompression, *in* Ernst, W. G., ed., Metamorphism and crustal evolution of the western United States; Rubey Vol. VII: Englewood-Cliffs, New Jersey, Prentice-Hall.

Anderson, J. L., and Frost, E. G., 1981, Petrologic, geochronologic, and structural evolution of the allochthonous crystalline terrane in the Copper Basin area, eastern Whipple Mountains, southeastern California: Unpublished technical report, 63 p.

Anderson, J. L., and Rowley, M. C., 1981, Synkinematic intrusion of two-mica and associated granitoids, Whipple Mountains, California: Canadian Mineralogist, v. 19, p. 83–101.

Anderson, R. E., 1971, Thin-skinned distension in Tertiary rocks of southeastern Nevada: Geological Society of America Bulletin, v. 82, p. 43–58.

Amstrong, R. L., 1972, Low-angle (denudation) faults, hinterland of the Sevier orogenic belt, eastern and western Utah: Geological Society of America Bulletin, v. 83, p. 1729–1754.

Brun, J.-P., and Choukroune, P., 1983, Normal faulting, block tilting, and décollement in a stretched crust: Tectonics, v. 2, p. 345–356.

Burchfiel, B. C., and Davis, G. A., 1972, Structural framework and evolution of the southern part of the Cordilleran orogen, western United States: American Journal of Science, v. 272, p. 97–118.

——, 1975, Nature and controls of Cordilleran orogenesis, western United States; Extensions of an earlier systhesis: American Journal of Science, v. 275-A, p. 363–396.

Carr, W. J., and Dickey, D. D., 1976, Cenozoic tectonics of eastern Mojave Desert: U.S. Geological Survey Professional Paper 1000, p. 75.

Coonrad, W. L., 1960, Geology and mineral resources of Township 6 North, Ranges 23 and 24 East, San Bernardino base and meridian, San Bernardino County, California: Report submitted to Southern Pacific Land Company, 23 p.

Crittenden, M. D., Jr., Coney, P. J., and Davis, G. H., eds., 1980, Cordilleran metamorphic core complexes: Geological Society of America Memoir 153, 490 p.

Davis, G. A., 1980, Problems of intraplate extensional tectonics western United States, *in* Continental tectonics: National Academy of Sciences, National Research Council, p. 89–95.

——, 1986a, Tectonic implications of variable southwestward tilts in Tertiary upper-plate strata of a Miocene detachment terrane, southeastern California and west-central Arizona: Geological Society of America Abstracts with Programs, v. 18, p. 98.

——, 1986b, Upward transport of mid-crustal mylonitic gneisses in the footwall of a Miocene detachment fault, Whipple Mountains, southeastern California: Geological Society of America Abstracts with Programs, v. 18, p. 98.

——, 1987, Rapid upward transport of mid-crustal mylonitic gneisses in the footwall of a Miocene detachment fault, Whipple Mountains, California: Geologisches Rundschau (in press).

Davis, G. A., and Burchfiel, B. C., 1973, Garlock fault; An intracontinental transform structure, southern California: Geological Society of America Bulletin, v. 84, p. 1407–1422.

Davis, G. A., Anderson, J. L., Frost, E. G., and Shackelford, T. L., 1979, Regional Miocene detachment faulting and early Tertiary(?) mylonitization, Whipple-Buckskin-Rawhide Mountains, southeastern California and western Arizona, *in* Abbott, P. L., ed., Geologic excursions in the southern California area: San Diego State University, p. 75–108.

——, 1980, Mylonitization and detachment faulting in the Whipple-Buckskin-Rawhide Mountains terrane, southeastern California and western Arizona: Geological Society of America Memoir 153, p. 79–129.

Davis, G. A., Anderson, J. L., Martin, D. L., Krummenacher, D., Frost, E. G., and Armstrong, R. L., 1982, Geologic and geochronologic relations in the lower plate of the Whipple detachment fault, Whipple Mountains, southeastern California; A progress report, *in* Frost, E. G., and Martin, D. L., eds., Mesozoic-Cenozoic tectonic evolution of the Colorado River region, California, Arizona, and Nevada: San Diego, California, Cordilleran Publishers, p. 408–432.

Davis, G. A., Lister, G. S., and Reynolds, S. J., 1983, Interpretation of Cordilleran core complexes as evolving crustal shear zones in an extending orogen: Geological Society of America Abstracts with Programs, v. 15, no. 5, p. 311.

——, 1986, Structural evolution of the Whipple and South Mountains shear zones, southwestern United States: Geology, v. 14, p. 7–10.

Davis, G. H., 1980, Structural characteristics of metamorphic core complexes, southern Arizona: Geological Society of America Memoir 153, p. 35–77.

——, 1983, Shear-zone model for the origin of metamorphic core complexes: Geology, v. 11, p. 342–347.

Davis, G. H., and Coney, P. J., 1979, Geologic development of the Cordilleran metamorphic core complexes: Geology, v. 7, p. 120–124.

Davis, S., 1984, Structural geology of the Highland Spring Range, southern Nevada [M.S. thesis]: Los Angeles, University of Southern California, 125 p.

Dibblee, T. W., 1971, A great middle Tertiary buttress unconformity in the Newberry Mountains, Mojave Desert, California, and its paleogeologic implications: Geological Society of America Abstracts with Programs, v. 3, no. 3, p. 110.

Dokka, R. K., and Baksi, A. K., 1986, Structure and sedimentation along the lateral boundaries of a detachment fault terrane, central Mojave extensional complex, California: Geological Society of America Abstracts with Programs, v. 18, p. 102.

Dokka, R. K., and Lingrey, S. H., 1979, Fission track evidence for a Miocene cooling event, Whipple Mountains, southeastern California, *in* Armentrout, J. M., Cole, M. R., and TerBest, H., Jr., eds., Cenozoic paleogeography of the western United States: Los Angeles, Pacific Section, Society of Economic Paleontologists and Mineralogists, p. 141–146.

Dokka, R. K., and Woodburne, M. D., 1986, Mid-Tertiary extensional tectonics and sedimentation, central Mojave Desert, California: Louisiana State University Publications in Geology and Geophysics, Tectonics and Sedimentation, no. 1, 55 p.

Dunn, J. F., Yin, A., and Davis, G. A., 1986, Evidence for multiple detachment faulting in the northeastern Whipple Mountains area, San Bernardino County, California: Geological Society of America Abstracts with Programs, v. 18, p. 103.

Engel, A.E.J., and Schultejann, P. A., 1984, Late Mesozoic and Cenozoic tectonic history of south-central California: Tectonics, v. 3, p. 659–675.

Frost, E. G., 1980, Whipple Wash Pump Storage Project, California; Appraisal design data, structural geology, and water-holding capability of rocks in the Whipple Wash area, San Bernardino County, California: Unpublished report submitted to U.S. Water and Power Resources Service, Boulder, City, Nevada, 88 p.

——, 1984, Structural geology of a portion of the Whipple-Buckskin detachment terrane, southeastern California and western Arizona [Ph.D. thesis]: Los Angeles, University of Southern California, 333 p.

Frost, E. G., and Martin, D. L., eds., 1982, Mesozoic-Cenozoic tectonic evolution of the Colorado River region, California, Arizona, and Nevada: San Diego, California, Cordilleran Publishers, 608 p.

Gans, P. B., and Miller, E. L., 1983, Style of Mid-Tertiary extension in east-central Nevada, *in* Guidebook, Part 1, Geological Society of America Rocky Mountain and Codilleran Sections Meeting: Utah Geological and Mining Survey Special Studies, v. 59, p. 107–160.

Garner, W. E., Frost, E. G., Tanges, S. E., and Germinario, M. P., 1982, Mid-Tertiary detachment faulting and mineralization in the Trigo Mountains, Uma County, Arizona, *in* Frost, E. G., and Martin, D. L., eds., Mesozoic-Cenozoic tectonic evolution of the Colorado River region, California, Arizona, and Nevada: San Diego, California, Cordilleran Publishers, p. 158–172.

Glazner, A. F., Nielson, J. E., Howard, K. A., and Miller, D. M., 1986, Correlation of the Peach Springs Tuff, a large-volume Miocene ignimbrite sheet in California and Arizona: Geology, v. 14, p. 840–843.

Gross, W. W., and Hillemeyer, F. L., 1982, Geometric analysis of upper-plate fault patterns in the Whipple-Buckskin detachment terrane, *in* Frost, E. G., and Martin, D. L., eds., Mesozoic-Cenozoic tectonic evolution of the Colorado River region, California, Arizona, and Nevada: San Diego, California, Cordilleran Publishers, p. 256–266.

Hamilton, W., 1982, Structural evolution of the Big Maria Mountains, northeastern Riverside County, southeastern California, *in* Frost, E. G., and Martin, D. L., eds., Mesozoic-Cenozoic tectonic evolution of the Colorado River region, California, Arizona, and Nevada: San Diego, California, Cordilleran Publishers, p. 1–28.

Hamilton, W., and Myers, W. B., 1966, Cenozoic tectonics of the western United States: Reviews of Geophysics, v. 4, p. 509–549.

Hewitt, D. F., 1956, Geology and mineral resources of the Invanpah Quadrangle, California and Nevada: U.S. Geological Survey Professional Paper 275, 172 p.

Howard, K. A., and John B. E., 1987, Crustal extension along a rooted system of imbricate low-angle faults; Colorado River extensional corridor, California and Arizona: Geological Society of London Special Publications, Continental extensional tectonics (in press).

Howard, K. A., Stone, P., Pernokas, M. A., and Marrin, R. F., 1982a, Geologic and geochronologic reconnaissance of the Turtle Mountain area, California; West border of the Whipple Mountains detachment terrane, *in* Frost, E. G., and Martin, D. L., eds., Mesozoic-Cenozoic tectonic evolution of the Colorado River region, California, Arizona, and Nevada: San Diego, California, Cordilleran Publishers, p. 341–355.

Howard, K. A., Goodge, J. W., and John B. E., 1982b, Detached crystalline rocks of the Mohave, Buck, and Bill Williams Mountains, *in* Frost, E. G., and Martin, D. L., eds., Mesozoic-Cenozoic tectonic evolution of the Colorado River region, California, Arizona, and Nevada: San Diego, California, Cordilleran Publishers, p. 377–392.

Hunt, C. B., and Mabey, D. R., 1966, General geology of Death Valley, California; Stratigraphy and structure: U.S. Geological Survey Professional Paper 494-A, p. A1–A165.

Jackson, B. E., 1982, Geologic framework of the Chemehuevi Mountains, southeastern California, *in* Frost, E. G., and Martin, D. L., eds., Mesozoic-Cenozoic tectonic evolution of the Colorado River region, California, Arizona, and Nevada: San Diego, California, Cordilleran Publishers, p. 317–325.

——, 1987, Geometry and evolution of a mid-crustal extensional fault system, Chemehuevi Mountains, southeastern California: Geological Society of London Special Publication, Continental extensional tectonics (in press).

Kemnitzer, L. E., 1937, Structural studies in the Whipple Mountains, southeastern California [Ph.D. thesis]: Pasadena, California Institute of Technology, 150 p.

Lee, D. E., Marvin, R. F., Stern, T. W., and Peterman, Z. E., 1970, Modification of potassium-argon ages by Tertiary thrusting in the Snake Range, White Pine County, Nevada: U.S. Geological Survey Professional Paper 700-D, p. 92–102.

Lee, J., Miller, E. L., and Sutter, J. F., 1987, Ductile strain and metamorphism in an extensional tectonic setting; A case study from the northern Snake Range, Nevada, U.S.A.: Geological Society of London Special Publication, Continental extensional tectonics (in press).

Lister, G. S., and Davis, G. A., 1983, Development of mylonitic rocks in an extending orogen: Geological Society of America Abstracts with Programs, v. 15, no. 5, p. 310.

——, 1987, Models for the formation of metamorphic core complexes or mylonitic detachment terranes (in review).

Lister, G. S., and Snoke, A. W., 1984, S-C mylonites: Journal of Structural Geology, v. 6, p. 617–638.

Lister, G. S., Davis, G. A., McClelland, W. C., and Marcott, D. T., 1984, Complexities in the evolution of low-angle crustal shear zones during continental extension: Geological Society of America Abstracts with Programs, v. 16, no. 6, p. 577.

Loring, A. K., 1972, Temporal and spatial distribution of basin-range faulting in Nevada and Utah [M.S. thesis]: Los Angeles, University of Southern California, 163 p.

——, 1974, Distribution in time and space of late Phanerozoic normal faulting in Nevada and Utah: Utah Geology, v. 3, p. 97–109.

Luccita, I., and Suneson, N., 1981, Observations and speculations regarding the relations and origins of mylonitic gneiss and associated detachment faults near the Colorado Plateau boundary in western Arizona, *in* Howard, K. A., Carr, M. D., and Miller, D. M., eds., Tectonic framework of the Mojave and Sonoran Deserts, California and Arizona: U.S. Geological Survey Open-File Report 81-503, p. 53–55.

McClelland, W. C., 1982, Structural geology of the central Sacramento Mountains, San Bernardino County, California, *in* Frost, E. G., and Martin, D. L., eds., Mesozoic-Cenozoic tectonic evolution of the Colorado River region, California, Arizona, and Nevada: San Diego, California, Cordilleran Publishers, p. 401–407.

——, 1984, Low-angle Cenozoic faulting in the central Sacramento Mountains, San Bernardino County, California [M.S. thesis]: Los Angeles, University of Southern California, 94 p.

McDonald, R. E., 1976, Tertiary tectonics and sedimentary rocks along the transition; Basin and Range province to plateau and thrust belt province, Utah, *in* Hill, G. J., ed., Geology of the Cordilleran hingeline: Denver, Colorado, Rocky Mountain Association of Geologists, p. 218–317.

Miller, D. M., Howard, K. A., and Anderson, J. L., 1981, Mylonitic gneiss related to emplacement of a Cretaceous batholith, Iron Mountains, southern California, *in* Howard, K. A., Carr, M. D., and Miller, D. M., eds., Tectonic framework of the Mojave and Sonora Deserts, California and Arizona: U.S. Geological Survey Open-File Report 81-0503, p. 73–75.

Miller, E. L., and Gans P. B., 1984, Continental transects C1 and C2; Cenozoic extension of the northern Basin-Range province: Geological Society of America Abstracts with Programs, v. 16, no. 6, p. 596.

Miller, E. L., Gans, P. B., and Garing, J., 1983, The Snake Range décollement; An exhumed mid-Tertiary ductile-brittle transition: Tectonics, v. 2, p. 239–264.

Misch, P., 1960, Regional structural reconnaissance in central-northeast Nevada and some adjacent areas; Observations and interpretations, *in* Geology of east-central Nevada: Intermountain Association of Petroleum Geologists 11th Annual Field Conference Guidebook, p. 17–42.

——, 1971, Geotectonic implications of Mesozoic décollement thrusting in parts of eastern Great Basin: Geological Society of America Abstracts with Programs, v. 3, p. 164–166.

Nakata, J. K., 1982, Preliminary report on diking events in the Mohave Mountains, Arizona, *in* Frost, E. G., and Martin, D. L., eds., Mesozoic-Cenozoic tectonic evolution of the Colorado River region: San Diego, California, Cordilleran Publishers, p. 85–90.

Nelson, R. B., 1966, Structural development of northernmost Snake Range, Kern Mountains, and Deep Creek Range, Nevada and Utah: American Association of Petroleum Geologists Bulletin, v. 50, p. 921–951.

——, 1969, Relation and history of structures in sedimentary succession with deeper metamorphic structures, eastern Great Basin: American Association of Petroleum Geologists Bulletin, v. 53, p. 307–339.

Nielson, J. E., and Glazner, A. F., 1986, Miocene stratigraphy and structure; Colorado Plateau to the central Mojave Desert, *in* Field trip number 5,

Miocene stratigraphy and structure; Colorado Plateau to the central Mojave Desert: Los Angeles, California State University, Department of Geology, 1986, Geological Society of America Cordilleran Section Guidebooks, p. 1–6.

Noble, L. F., 1941, Structural features of the Virgin Spring area, Death Valley California: Geological Society of America Bulletin, v. 52, p. 941–1000.

Okaya, D. A., and Frost, E. G., 1986, Crustal structure of the Whipple–Turtle–Old Woman Mountains region based on CALCRUST and industry seismic profiles: Geological Society of America Abstracts with Programs, v. 18, p. 166.

Osborne, G. M., 1981, The structural geology of the Squaw Peak area of the Buckskin Mountains, Yuma County, Arizona [M.S. thesis]: Los Angeles, University of Southern California, 164 p.

Otton, J. K., 1982, Tertiary extensional tectonics and associated volcanism in west-central Arizona, *in* Frost, E. G., and Martin, D. L., eds., Mesozoic-Cenozoic tectonic evolution of the Colorado River region, California, Arizona, and Nevada: San Diego, California, Cordilleran Publishers, p. 143–157.

Phillips, J.-C., 1982, Character and origin of cataclasite developed along the low-angle Whipple detachment fault, Whipple Mountains, California, *in* Frost, E. G., and Martin, D. L., eds., Mesozoic-Cenozoic tectonic evolution of the Colorado River region, California, Arizona, and Nevada: San Diego, California, Cordilleran Publishers, p. 109–116.

Proffett, J. M., Jr., 1977, Cenozoic geology of the Yerington District, Nevada, and implications for the nature and origin of Basin and Range faulting: Geological Society of America Bulletin, v. 88, p. 247–266.

Proffett, J. M., Jr., and Dilles, J. H., 1984, Geologic map of the Yerington District, Nevada: Nevada Bureau of Mines and Geology Map 77, scale 1:24,000.

Ransome, F. L., 1931, Geological reconnaissance of the revised Parker route through the Whipple Mountains: Unpublished technical report to the Metropolitan Water District of Southern California, 10 p.

Rehrig, W. A., and Reynolds, S. J., 1980, Geologic and geochronologic reconnaissance of a northwest-trending zone of metamorphic complexes in southern Arizona: Geological Society of America Memoir 153, p. 131–158.

Reynolds, S. J., 1982, Geology and geochronology of the South Mountains, central Arizona [Ph.D. thesis]: Tucson, University of Arizona, 220 p.

—— , 1985, Geology of the South Mountains, central Arizona: Arizona Bureau of Geology and Mineral Technology Bulletin 195, 61 p.

Reynolds, S. J., and Rehrig, W. A., 1980, Mid-Tertiary plutonism and mylonitization, South Mountains, central Arizona: Geological Society of America Memoir 153, p. 159–176.

Reynolds, S. J., and Spencer, J. E., 1985, Evidence for large-scale transport on the Bullard detachment fault, west-central Arizona: Geology, v. 13, p. 353–356.

Royse, F., Warner, M. A., and Reese, D. L., 1975, Thrust belt structural geometry and related stratigraphic problems, Wyoming–Idaho–northern Utah, *in* Bolyard, D. W., ed., Deep drilling frontiers in the central Rocky Mountains: Denver, Rocky Mountain Association of Geologists, p. 41–54.

Shackelford, T. J., 1976, Structural geology of the Rawhide Mountains, Mohave County, Arizona [Ph.D. thesis]: Los Angeles, University of Southern California, 175 p.

Sibson, R. H., 1977, Fault rocks and fault mechanisms: Journal of the Geological Society of London, v. 133, p. 191–213.

Smith, R. B., and Bruhn, R. L., 1984, Intraplate extensional tectonics of the eastern Basin-Range; Inferences on structural styles from seismic reflection data, regional tectonics, and thermal-mechanical models of brittle-ductile deformation: Journal of Geophysical Research, v. 89, p. 5733–5762.

Snelson, S., 1957, The geology of the northern Ruby Mountains, East Humboldt Range, Elko County, Nevada [Ph.D. thesis]: Seattle, University of Washington, 214 p.

Spencer, J. E., 1982, Origin of folds of Tertiary low-angle fault surfaces, southeastern California and southwestern Arizona, *in* Frost, E. G., and Martin, D. L., eds., Mesozoic-Cenozoic tectonic evolution of the Colorado River region, California, Arizona, and Nevada: San Diego, California, Cordilleran Publishers, p. 123–143.

—— , 1984, Role of tectonic denudation in uplift and warping of low-angle normal faults: Geology, v. 12, p. 95–98.

—— , 1985, Miocene low-angle normal faulting and dike emplacement, Homer Mountain and surrounding areas, southeastern California and southernmost Nevada: Geological Society of America Bulletin, v. 96, p. 1140–1155.

Spencer, J. E., and Turner, R. D., 1982, Dike swarms and low-angle faults, Homer Mountain and the northwestern Sacramento Mountains, southeastern California, *in* Frost, E. G., and Martin, D. L., eds., Mesozoic-Cenozoic tectonic evolution of the Colorado River region, California, Arizona, and Nevada: San Diego, California, Cordilleran Publishers, p. 97–108.

Stewart, J. H., 1971, Basin and Range structure; A system of horsts and grabens produced by deep-seated extension: Geological Society of America Bulletin, v. 82, p. 1019–1044.

—— , 1978, Basin-Range structure in western North America; A review: Geological Society of America Memoir 152, p. 1–31.

Terry, A. H., 1972, The geology of the Whipple Mountains thrust fault, southeastern California [M.S. thesis]: California State University at San Diego, 90 p.

Thompson, G. A., 1959, Gravity measurements between Hazen and Austin, Nevada; A study of Basin-Range structure: Journal of Geophysical Research, v. 64, p. 217–229.

Thorman, C. H., 1970, Metamorphosed and non-metamorphosed Paleozoic rocks in the Wood Hills and Pequop Mountains, northeast Nevada: Geological Society of America Bulletin, v. 81, p. 2417–2448.

Thurn, L. C., 1983, Geology of a portion of the southeastern Whipple Mountains, San Bernardino County, California [M.S. thesis]: Los Angeles, University of Southern California, 186 p.

Wallace, R. D., and English, D. J., 1982, Evaluation of possible detachment faulting west of the San Andreas, southern Santa Rosa Mountains, California, *in* Frost, E. G., and Martin, D. L., eds., Mesozoic-Cenozoic tectonic evolution of the Colorado River region, California, Arizona, and Nevada: San Diego, California, Cordilleran Publishers, p. 502–510.

Wernicke, B., 1981, Low-angle normal faults in the Basin and Range province; Nappe tectonics in an extending orogen: Nature, v. 291, p. 645–648.

—— , 1985, Uniform-sense normal simple shear of the continental lithosphere: Canadian Journal of Earth Sciences, v. 22, p. 108–125.

Wernicke, B., and Burchfiel, B. C., 1982, Modes of extensional tectonics: Journal of Structural Geology, v. 4, p. 105–115.

Wernicke, B., Walker, J. D., and Beaufait, M. S., 1985, Structural discordance between Neogene detachments and frontal Sevier thrusts, central Mormon Mountains, southern Nevada: Tectonics, v. 4, p. 213–246.

Wilkins, J., Jr., and Heidrick, T. L., 1982, Base and precious metal mineralization related to low-angle tectonic features in the Whipple Mountains, California and Buckskin Mountains, Arizona, *in* Frost, E. G., and Martin, D. L., eds., Mesozoic-Cenozoic tectonic evolution of the Colorado River region, California, Arizona, and Nevada: San Diego, California, Cordilleran Publishers, p. 182–203.

Wilson, E. D., 1962, A resume of the geology of Arizona: Arizona Bureau of Mines Bulletin 171, 140 p.

Woodward, R. J., 1980, Structural geology of the Swansea area, Buckskin Mountains, Yuma County, Arizona [M.S. thesis]: Los Angeles, University of Southern California, 130 p.

Woodward, R. J., and Osborne, G. M., 1980, Low-angle detachment faulting and multiple deformation of the central Buckskin Mountains, Yuma County, Arizona: Geological Society of America Abstracts with programs, v. 12, p. 245.

Wright, J. E., Anderson, J. L., and Davis, G. A., 1986, Timing of plutonism, mylonitization, and decompression in a metamorphic core complex, Whipple Mountains, California: Geological Society of America Abstracts with Programs, v. 18, p. 201.

Manuscript Accepted by the Society June 1, 1987

Printed in U.S.A.

Geological Society of America
Special Paper 218
1988

Mechanics, growth, and erosion of mountain belts

F. A. Dahlen and John Suppe
Department of Geological and Geophysical Sciences, Princeton University, Princeton, New Jersey 08544

ABSTRACT

Active fold-and-thrust belts and accretionary wedges along compressive plate boundaries are analogous to the wedges of soil or snow that form in front of moving bulldozers. Such wedges deform until they attain a critical taper, which corresponds to an internal state of stress that is everywhere on the verge of Coulomb failure. The critical taper depends on the wedge cohesion S_0 as well as the internal and basal coefficients of friction μ and μ_b and the internal and basal Hubbert-Rubey pore-fluid pressure ratios λ and λ_b; an exact relation can be obtained if the cohesion increases linearly with depth, i.e., $S_0 = Kz$. Typical laboratory rock-strength parameters in the range $\mu = \mu_b = 0.6$ to 0.85 and $S_0 = 5$ to 18 MPa are consistent with the known taper, thrust-fault dips, and pore-fluid pressures in the western Taiwan fold-and-thrust belt. Much lower rock strengths in the range $\mu = \mu_b \approx 0.2$ and $S_0 \approx 0$, however, can also satisfy the Taiwan data.

As fresh material enters the toe of a fold-and-thrust belt, accretionary wedge, or bulldozer wedge, the wedge grows self-similarly, maintaining its critical taper. A submarine or other noneroding wedge widens with time like $t^{1/2}$, whereas an eroding wedge attains a steady-state width when the accretionary influx is balanced by erosion. The rate of erosion exerts a significant control on the deformation and metamorphic histories of rocks incorporated into mountain belts. To illustrate this, we develop a simple kinematical model of self-similar wedge growth, and use it to infer the mean trajectories and residence times of rocks in the rapidly eroding fold-and-thrust belt in central Taiwan, which develops a steady-state width of about 90 km. A typical rock resides in the Taiwan wedge for 2 to 3 m.y. before being uplifted and eroded, and it experiences strain rates in the range $\dot{\epsilon} = 10^{-13}$ to 10^{-14} s^{-1} and finite strains in the range $S = 1$ to 10. Qualitatively, the model predicts that rocks transported farther into the wedge have been buried deeper and thus have been subjected to greater pressures and temperatures, in general agreement with the observed gradient of metamorphism in Taiwan and other mountain belts. In contrast with rapidly eroding mountain belts like Taiwan, more slowly eroding belts can grow to widths in excess of 300 to 500 km and have mean residence times greater than 50 m.y., which may be too long to reach true steady state before conditions change.

INTRODUCTION

The critical taper of a wedge that is failing under the influence of horizontal compression has been recognized in recent years to be a useful concept for understanding the large-scale mechanics of fold-and-thrust belts and accretionary wedges. The concept is illustrated with the small bulldozer shown in Figure 1. The soil or snow deforms in front of the bulldozer until a wedge with a constant critical surface slope or taper is developed. Deformation then ceases and if no fresh material is encountered at the toe, the wedge slides stably as it is pushed. The wedge will grow if more soil or snow is encountered but will always maintain its critical taper, which is governed by the relative magnitudes of the sliding resistance along the base and the strength of the wedge material—increasing the sliding resistance increases the critical taper, whereas increasing the strength within the

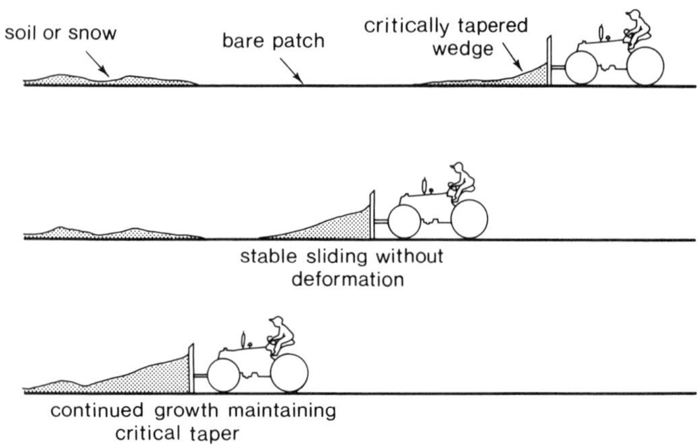

Figure 1. The wedge of soil or snow in front of a moving bulldozer grows self-similarly as freshly plowed material is encountered at the toe. The critical taper is the shape that is just strong enough to be slid over a bare patch without steepening or deforming.

wedge decreases it. In this chapter we briefly review the quantitative theory that relates the critical taper to the wedge material properties, as well as the results of applying that theory to fold-and-thrust belts and accretionary wedges. We then present a simplified model of the kinematics of wedge deformation, including the effects of erosion.

Erosion significantly modifies the kinematics of compressive wedges because an eroding wedge must continually deform throughout to maintain its critical taper. In contrast, submarine or other noneroding wedges undergo only the deformation required to accommodate the influx of newly accreted material, so their deformation is less intense and is more concentrated toward the toe. An eroding mountain belt can attain a steady-state width when the accretionary influx is balanced by the erosive outflux, whereas a noneroding wedge continues to grow self-similarly as long as fresh material is being accreted. As a consequence, the strain, stress, and temperature paths—and therefore the metamorphic histories—of rocks incorporated into eroding and noneroding wedges are significantly different.

REVIEW OF WEDGE MECHANICS

Material Behavior

The key physical assumption made in the critical taper model is that the stress within the wedge is everywhere equal to the rock strength, inasmuch as fold-and-thrust belts and accretionary wedges attain their shape by deformation. The taper is "critical" in the sense that the compressive stress within a thinner (subcritical) wedge will exceed the rock strength, so the wedge will fail and deform by thickening until the critical taper is achieved, whereas a thicker (supercritical) wedge will be strong enough to slide stably over its basal décollement without internal failure or deformation. Thus, a failure law must be specified both within the wedge and along the basal décollement in order to solve for the state of stress and the critical shape or taper. Quantitative wedge-deformation models could be based on a wide variety of possible stress-limiting rheologies, and critical wedge analyses have been published for perfectly plastic and viscous material (Chapple, 1978; Stockmal, 1983; Emerman and Turcotte, 1983) as well as for brittle materials (Davis and others, 1983; Dahlen and others, 1984; Dahlen, 1984; Zhao and others, 1986). A valid theory for application to the earth requires the correct choice of the stress-limiting material behavior.

Stress within the lithosphere is generally thought to be limited by brittle fracture and frictional sliding at low pressures and by thermally activated processes, especially dislocation climb, at high temperatures (Brace and Kohlstedt, 1980), although other less well-understood mechanisms, such as pressure solution, could conceivably be stress-limiting under certain circumstances. Figure 2 illustrates the dependence of typical rock strengths on depth if brittle friction and high-temperature creep are the two limiting mechanisms. Brittle behavior is seen to be stress-limiting in at least the upper 10 to 15 km of the crust for most earth materials except evaporites, which flow plastically at substantially shallower depths. Use of the Coulomb failure law to describe the material behavior within a fold-and-thrust belt should therefore be appropriate under most upper crustal conditions. We confine this review to Coulomb wedges exhibiting frictional behavior on their basal décollement—the important exception of wedges with a salt décollement was discussed by Davis and Engelder (1985).

Coulomb behavior, including both brittle fracture and friction, is strongly pressure- and fluid-pressure–dependent but is largely independent of temperature and strain rate (Paterson, 1978). The shear strength $|\tau|$ in the well-known Coulomb equation is given by:

$$|\tau| = S_0 + \mu\,(\sigma_n - P_f), \tag{1}$$

where S_0 is the cohesive strength and μ is the coefficient of either friction or internal friction. The quantity P_f is the pore-fluid pressure and $\sigma_n - P_f$ is the effective normal stress (Hubbert and Rubey, 1959) with compression being regarded as positive. Available rock mechanics data suggest that μ for either friction or fracture is typically in the range 0.6 to 0.85 for essentially all rocks. Cohesion S_0 is negligible in the case of friction but varies in fracture experiments over a very large range, 1 to 150 MPa, with a strong dependence on porosity, cementation, mineralogy, microcrack density, and other factors (Byerlee, 1978; Mogi, 1966; Hoshino and others, 1972). It is evident from Equation 1 that cohesion is important only at shallow depths or where the pore-fluid pressure is high; therefore a noncohesive Coulomb wedge theory is an adequate approximation for some geologic applications, particularly away from the toe of the wedge.

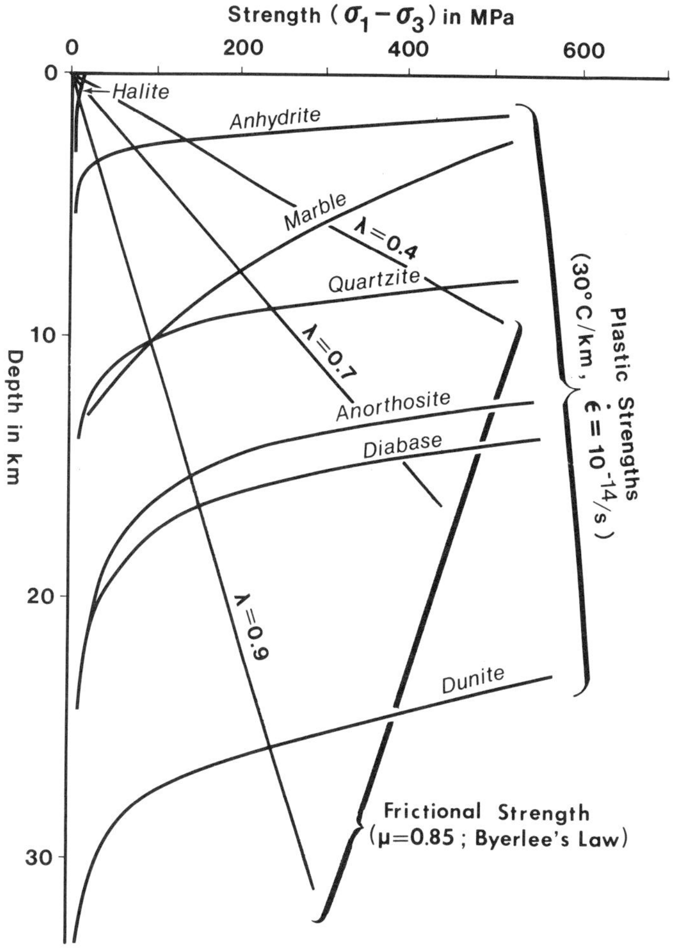

Figure 2. Brittle and plastic rock strengths as a function of depth; the latter are calculated for a typical geologic strain rate and geothermal gradient. Brittle friction is largely independent of rock type, whereas plastic strengths vary considerably. This leads to a corresponding variation in depth to the brittle-plastic transition: 0–5 km for evaporites, 10–15 km for carbonate- and quartz-rich rocks, 15–20 km for quartz-free feldspar-bearing rocks, and 25–30 km for olivine-rich rocks. (Based on compilation of Suppe, 1985.)

Critical-Taper Theory

There are several versions of the mathematical relationship between the critical taper and the Coulomb material properties, which differ in detail. Davis and others (1983) presented an approximate theory for a noncohesive wedge using methods and notation intended to be familiar to a general geological audience. Subsequently an exact solution for a noncohesive wedge was discovered (Dahlen, 1984), and, somewhat embarrassingly, the final form of the exact results proved to be even simpler than the original approximate theory. Lehner (1986) showed that the exact formula can also be obtained by a graphical method employing Terzaghi's (1943) pole construction on the Mohr diagram. It is not possible to obtain an exact solution for the case of a wedge having a uniform cohesion S_a, and an approximate treatment incorporating such a constant cohesion was developed by Dahlen and others (1984). Finally, Zhao and others (1986) extended Dahlen's exact cohesive results to a wedge having a linear increase in cohesion with depth,

$$S_0 = Kz. \tag{2}$$

This is the most general circumstance for which an exact analytic treatment is possible and we summarize the results here. We do not repeat the derivation, which consists of solving the static equilibrium equations subject to the specification that the stress be equal to the Coulomb strength (1). A systematic derivation, which establishes the uniqueness of the solution, is given by Barcilon (1987).

As shown in Figure 3, we consider a perfectly triangular wedge with surface slope α and basal dip β, and the axes x, z are chosen to be aligned with the upper surface of the wedge. Denote the wedge density by ρ, the overlying water density by ρ_w, the acceleration of gravity by g, and define the modified Hubbert-Rubey pore-fluid pressure ratio λ by:

$$\lambda = \frac{P_f - \rho_w g D}{\sigma_z - \rho_w g D}, \tag{3}$$

where D is the water depth. For a subaerial wedge the density ρ_w is that of air, which is negligible. The basal décollement is assumed to be governed by a coefficient of sliding friction μ_b and a Hubbert-Rubey pore-fluid pressure ratio λ_b that may differ from the parameters μ and λ within the wedge. We restrict attention to a wedge with uniform properties by requiring that ρ, ρ_w, μ, μ_b, λ, λ_b, g, and K all be constant. The critical taper of such a wedge is then given implicitly by:

$$\alpha + \beta = \psi_b - \psi_0 \tag{4}$$

$$\psi_0 = \frac{1}{2}\arcsin\left[(1 + 1/\mu^2)^{1/2}\sin\alpha'\right] - \frac{1}{2}\alpha' \tag{5}$$

$$\psi_b = \frac{1}{2}\arcsin\left[(1 + 1/\mu_b'^2)^{-1/2}\left(\frac{(K/\rho g\mu)\cos 2\psi_0 + (1-\lambda)(1 + 1/\mu^2)^{1/2}\cos\alpha}{(K/\rho g\mu) + (1-\lambda)\cos\alpha}\right)\right] - \frac{1}{2}\arctan\mu_b', \tag{6}$$

where the two auxiliary variables α' and μ'_b are defined by

$$\alpha' = \arctan\left[\frac{(1 - \rho_w/\rho)\sin\alpha}{(K/\rho g\mu) + (1-\lambda)\cos\alpha}\right] \tag{7}$$

$$\mu'_b = \arctan\left[\frac{(1-\lambda_b)}{(1-\lambda)}\mu_b\right] \tag{8}$$

The orientation of the state of stress is the same everywhere within the wedge, and the quantities ψ_0 and ψ_b are the acute angles between the axis of principal compressive stress and the top and bottom, so that the critical taper Equation 4 expresses an elementary relationship between two internal angles and the opposite external angle of a triangle (see Fig. 3).

It is noteworthy that the critical taper is independent of

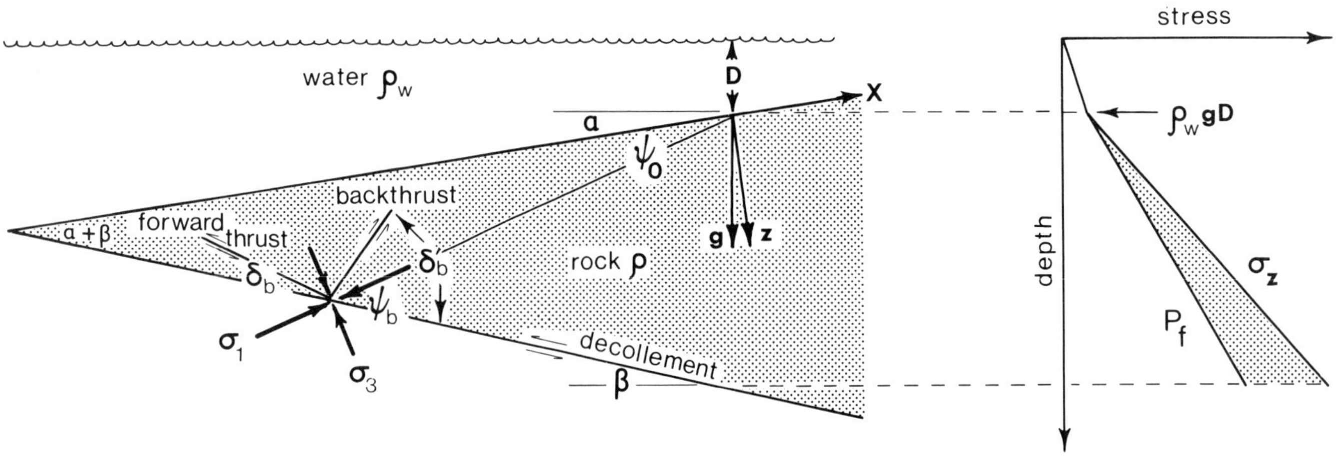

Figure 3. Cross-sectional sketch of a critically tapered submarine wedge showing the Cartesian coordinates x, z, and the angles α, β, ψ_o, ψ_b, δ_b, and δ_b' that define the taper and orientation of principal stresses and thrust faults within the wedge. Effective stress σ_z–P_f is shown schematically by the shaded area on the right. Hubbert-Rubey pore-fluid pressure ratio is the ratio of the (constant) gradients of P_f and σ_z.

horizontal distance x, inasmuch as Equations 4 through 8 involve only angles and the dimensionless quantities μ, μ_b, λ, λ_b, ρ_w/ρ, and $K/\rho g$. A cohesive Coulomb wedge with $S_0 = Kz$, like a noncohesive wedge ($K = 0$), is thus self-similar in the sense that a magnified version of any portion of it near the toe is indistinguishable from the wedge as a whole.

Shear failure within a Coulomb material occurs along conjugate shear planes oriented at angles $\pm$ 1/2 arctan $(1/\mu)$ with respect to the axis of principal compressive stress (Jaeger and Cook, 1969). Because of the self-similarity these slip surfaces have the same orientation everywhere within the wedge. Forward verging thrusts step up from the basal décollement at an angle

$$\delta_b = \frac{1}{2} \arctan (1/\mu) - \psi_b, \tag{9}$$

whereas backthrusts step up at the steeper angle (see Fig. 3)

$$\delta'_b = \frac{1}{2} \arctan (1/\mu) + \psi_b. \tag{10}$$

Forward thrusts are more commonly observed in most fold-and-thrust belts and accretionary wedges not underlain by salt, and their attitude δ_b together with the taper $\alpha + \beta$ are the two main predictions of the wedge theory that may be compared with observations.

The failure stress $|\tau|$ on the thrusts at the point where they step up into the wedge is given by:

$$|\tau| = (\rho - \rho_w)\, gz \sin\alpha \left[\frac{(1 + \mu^2)^{-1/2}}{\sin 2\psi_0} \right], \tag{11}$$

whereas the frictional resistance $\tau_b = \mu'_b (\sigma_n - \mathrm{P_f})$ on the basal décollement at the same point is:

$$\tau_b = (\rho - \rho_w)\, gz \sin\alpha \left[\frac{\sin 2\psi_b}{\sin 2\psi_0} \right]. \tag{12}$$

The quantity

$$\chi = \frac{\tau_b}{|\tau|} = (1 + \mu^2)^{1/2} \sin 2\psi_b \tag{13}$$

is a depth-independent measure of the ratio of décollement strength to wedge strength, which must be in the range $0 \leqq \chi \leqq 1$ for every fold-and-thrust belt or accretionary wedge.

Applications of Wedge Theory

Quantitative analysis of the mechanics of compressive mountain belts has been severely hampered by an almost complete lack of deep in situ measurements of the regional stress field and other key physical quantities. Therefore we must use indirect measurements and simplifying assumptions to test or apply theories such as the critical taper theory. For example, since regional stress and strength measurements are unavailable, we must make use of laboratory measurements of rock strength, in spite of the disparity between the small scale of laboratory specimens and the large scale on which the critical wedge theory is applicable. An alternate approach that is slightly more direct may be to use empirical velocity-porosity-strength relations to estimate the variation of cohesion or other strength parameters within a wedge from well-constrained seismic velocity data, where it is available.

The noncohesive Coulomb wedge theory was first applied to laboratory wedges of homogeneous dry sand sliding over a Mylar décollement (Davis and others, 1983; Dahlen and others, 1984). The quantities measured were the internal friction of dry sand $\mu = 0.6$, the friction of sand on Mylar $\mu_b = 0.3$, the imposed

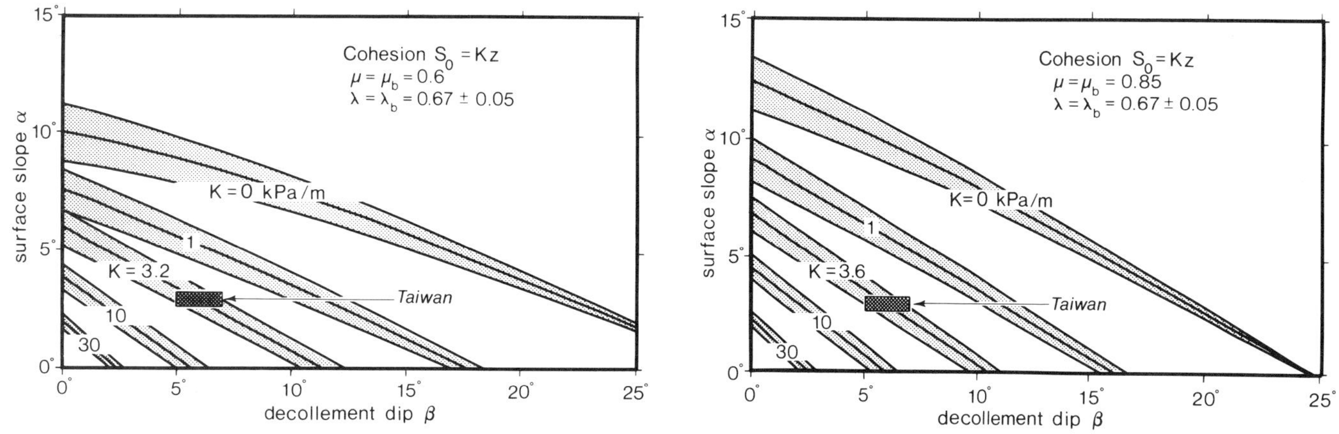

Figure 4. Critical surface slope α versus décollement dip β for Taiwan fluid-pressure conditions. The best-fitting cohesion gradient is $K = 3.2$ kPa/m if $\mu = \mu_b = 0.6$ or $K = 3.6$ kPa/m if $\mu = \mu_b = 0.85$. A noncohesive ($K = 0$) wedge would exhibit a significantly steeper surface slope—three to four times as great.

inclination of the décollement $0° \leqq \beta \leqq 6°$, and the resulting critical surface slopes α and thrust fault inclinations δ_b and δ_b'. The observed critical tapers and step-up angles agreed with the theoretical predictions. This result, in light of the nondimensionality or scale independence of the critical-taper theory, encouraged us to apply it to the obviously more complex large-scale geologic phenomena of fold-and-thrust belts and accretionary wedges.

The first geologic application of the Coulomb wedge theory was in western Taiwan, where the toe of the fold-and-thrust belt is actively deforming, as shown by the presence of seismicity (Wu and others, 1979) and geomorphologic studies (Suppe, 1981). We are able to measure the critical surface slope ($\alpha = 2.9° \pm 0.3°$) from topographic maps, and the dip of the décollement ($\beta = 6° \pm 1°$) and the forward verging step-up angle ($\delta_b = 13.3° \pm 2.4°$) from surface mapping and drilling and seismic reflection data (Davis and others, 1983; Dahlen and others, 1984). The density (ρ = 2,400 to 2,700 kg/m^3) is known from surface sampling, and the Hubbert-Rubey fluid-pressure ratio ($\lambda = \lambda_b = 0.67 \pm 0.05$) can be estimated from formation tests and geophysical logging of wells that penetrate the wedge near the deformation front (Suppe and Wittke, 1977; Davis and others, 1983). The critical-wedge model can be used to determine the range of wedge rock strength and décollement friction values consistent with the Taiwan data.

As an illustration, we consider the model described above, although the prescription that $S_0 = Kz$ is not especially appropriate for such a rapidly eroding and well-lithified fold-and-thrust belt as that in western Taiwan. Even though we have simplified the problem as much as possible by assuming that the wedge properties are all uniform, there are still three unknown parameters, μ, μ_b, and the cohesion gradient K. To reduce this number to a more manageable level, we assume that the basal and internal friction coefficients are equal, i.e., $\mu = \mu_b$, as indicated generally by laboratory data (see Zhao and others, 1986). Figure 4 shows the theoretical dependence of the critical surface slope α on K and the décollement dip β for $\mu = \mu_b = 0.6$ and $\mu = \mu_b = 0.85$. The shaded bands denote the range associated with the uncertainty in $\lambda = \lambda_b$, and the box denotes the geometry of Taiwan. A cohesion gradient in the range K = 3.2 to 3.6 kPa/m is seen to be consistent with the observed conditions. Figure 5 shows that a slightly larger but overlapping range K = 1 to 3.6 kPa/m is also consistent with the observed finite step-up angle δ_b if $\mu = \mu_b$ is in the laboratory range 0.6 to 0.85. The average cohesion of a column of rock 10 km thick is 5 to 18 MPa if K = 1 to 3.6 kPa/m, in good agreement with typical laboratory measurements of the fracture strengths of low-porosity shales and sandstones (Hoshino and others, 1972). This good agreement suggests that stress in the Taiwan wedge is limited by fracture, although a more complete analysis (Dahlen and others, 1984) shows that friction along nonoptimally oriented faults could also contribute substantially to the wedge strength.

The above results support but do not prove the contention that normal laboratory fracture and frictional strengths generally govern large-scale brittle deformation in compressive mountain belts in the upper crust. Figure 6, which shows the locus of values of K and $\mu = \mu_b$ consistent with both the observed taper and step-up angle, illustrates our inability to place firm constraints on the large-scale strength of the rocks in Taiwan. When all the observational errors in α, β, δ_b, and $\lambda = \lambda_b$ are taken into account, any value in the cleaver-shaped region is consistent with the data. In particular the strength may be much lower than typical laboratory estimates—as low as $\mu = \mu_b \approx 0.2$ and $S_0 \approx 0$.

The ratio $\chi = \tau_b / |\tau|$ of décollement strength to wedge strength is better constrained than the strength itself. This is illustrated in Figure 7 by a plot of χ versus $\mu = \mu_b$ corresponding to

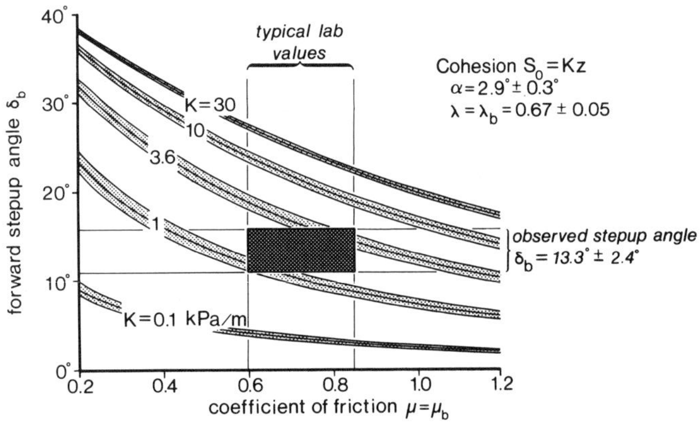

Figure 5. Dependence of forward thrust step-up angle δ_b on K and $\mu = \mu_b$ for Taiwan fluid-pressure conditions. A cohesion gradient K between 1 and 3.6 kPa/m is consistent with the observed range of thrust orientations, provided $\mu = \mu_b$ is in the range 0.6 to 0.85 of laboratory estimates.

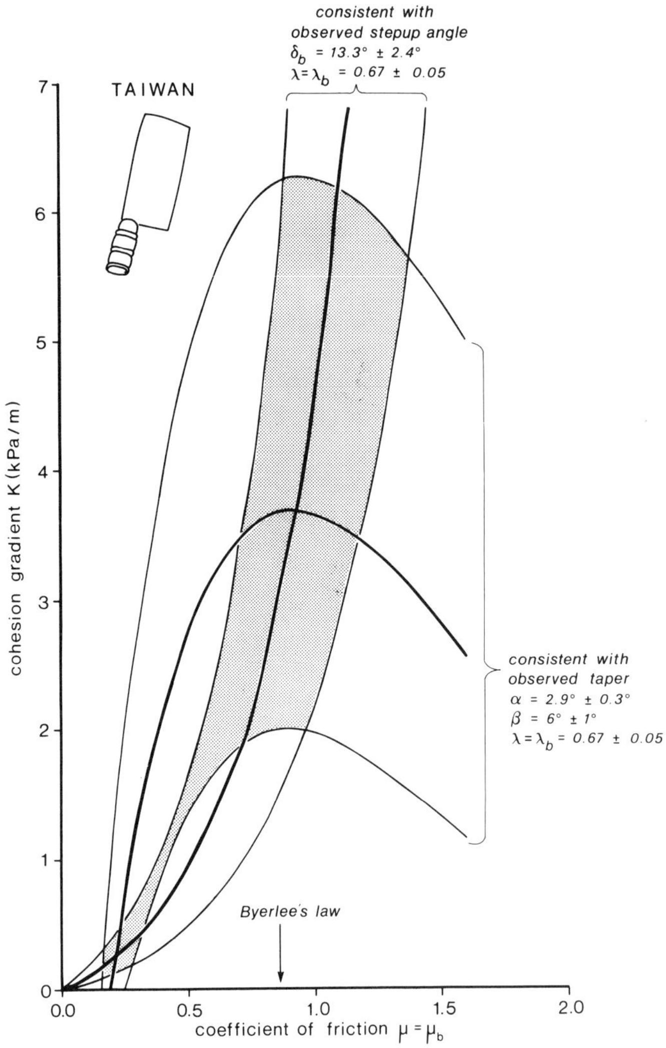

Figure 6. Values of K and $\mu = \mu_b$ consistent with the range of observations in Taiwan lie within the shaded region. The heavy curves, which represent the loci of values consistent with the best estimates of the Taiwan parameters, intersect at two points—one is in reasonable agreement with Byerlee's law ($\mu = \mu_b \approx 0.85$), whereas the other corresponds to much lower rock strengths ($\mu = \mu_b \approx 0.2$ with negligible cohesion).

the cleaver-shaped region in Figure 6. If the strength is high ($\mu = \mu_b \geqq 0.6$), then χ is between 0.4 and 0.6, whereas if it is low ($\mu = \mu_b \approx 0.2$), then χ may be as great as 0.8. In either case, χ must definitely be less than 1, which is an indication that the wedge is not so pervasively fractured that frictional sliding is possible on surfaces of optimum orientation everywhere. If it were, it would have a much steeper critical surface slope (α = 8° to 10° if $\mu = \mu_b$ = 0.6 to 0.85), in addition to listric (δ_b = 0°) thrust faults stepping up from its base.

The sensitivity of wedge shape to variations in both strength and fluid pressure is greater in wedges that are more highly overpressured. Therefore, high fluid-pressure thrust belts are particularly important in testing the critical wedge concept. Davis and others (1983) and Dahlen (1984) used the noncohesive Coulomb theory to infer the fluid pressure in four narrowly tapered accretionary wedges with known geometries: the Barbados wedge ($\alpha + \beta \approx 3°$), the Makran wedge in the Gulf of Oman ($\alpha + \beta \approx 4°$), the eastern Aleutian wedge in the Gulf of Alaska ($\alpha + \beta \approx 7°$), and the wedge off the Oregon-Washington coast ($\alpha + \beta \approx 8°$). The inferred fluid-pressure ratios $\lambda = \lambda_b$ exceeded 0.8 in Oregon and the Aleutians and were nearly lithostatic in Barbados and the Makran, in good agreement with nearby fragmentary well data and the observed occurrence of mud volcanic activity in the latter two instances.

The Barbados accretionary wedge was studied in more detail by Zhao and others (1986), who focused on the origin of the convex surface slope commonly observed in cross section near the toe of many overpressured submarine wedges. This convex shape is inconsistent with a homogeneous Coulomb-wedge model, which predicts either a constant surface slope α in the case $S_0 = Kz$ or a slightly concave upper surface in the case of constant S_0. At Barbados the wedge geometry is well known from excellent seismic and seabeam surveys, and direct evidence for very high fluid pressures ($\lambda = \lambda_b \approx 0.95$) was obtained during DSDP drilling at one site in the toe of the wedge. A line drawing of Westbrook and others' (1982) seismic profile across the Barbados wedge at 16°N is shown at the top of Figure 8. A clearly visible shallow-dipping (β = 2°) planar décollement extends under the wedge for at least 70 km, and is underlain for at least the first 30 km by undisrupted subducted sediments. The surface slope is markedly convex, decreasing from $\alpha \approx 3°$ at the deformation front to $\alpha \approx 0°$ at a distance of 30 km arcward.

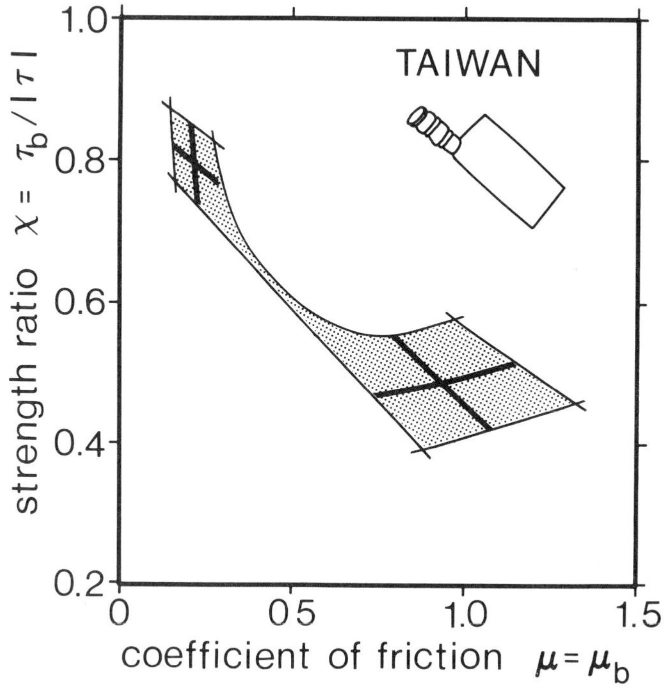

Figure 7. Values of the décollement/wedge strength ratio χ consistent with the Taiwan data. The region shown here is the transformation of the region in Figure 6 from K versus $\mu = \mu_b$ space to χ versus $\mu = \mu_b$ space. The relative orientations of the two regions are indicated by the small Chinese cleavers.

The most probable cause of this widely observed convexity is the increase in rock strength accompanying lithification during sediment accretion—the entering sediments are typically weak and unconsolidated with porosities as high as 70 percent. Zhao and others (1986) argued that the principal effect of compaction should be to increase the cohesive strength of the sediments within the wedge, and used the critical Coulomb model with constant friction and fluid-pressure ratio to infer the distribution of $S_0 = Kz$ needed to account for the observed shape. The inferred contours of cohesion assuming $\mu = \mu_b = 0.85$ and $\lambda = \lambda_b = 0.95$ are shown in Figure 8, together with the corresponding distributions of porosity and P-wave velocity calculated from empirical cohesion-porosity-velocity relations. The inferred porosity decreases from its entering value of 70 percent to approximately 30 percent near the base of the wedge 3 km below the sea floor and 60 km arcward from the toe, whereas the cohesion increases from near zero to about 14 MPa (the gradient K increases from near zero to 5 kPa/m). The velocity of 1.53 km/s near the toe and at the top of the wedge is essentially that of sea water, and it increases to about 2.4 km/s as the sediments compact. These inferred properties are consistent with the available porosity, velocity, and sediment-strength data from DSDP drilling in the uppermost 500 m of the Barbados wedge, although additional fluid-pressure data and better deep velocity estimates from wide-angle seismic surveys would enable the model to be tested more critically. Karig (1986a), in a study of the accretionary wedge along the Nankai Trough, has attributed its convex shape to an increase in internal friction μ rather than S_0; this difference in opinion points up the need for further laboratory testing of fluid-saturated weak rocks in the brittle regime.

In summary, the Coulomb wedge model has been successfully applied to laboratory experiments and to several active fold-and-thrust belts and accretionary wedges, with good agreement between observed wedge properties, theory, and rock mechanics data. Therefore, in spite of the remarkable lack of deep in situ measurements of key parameters such as stress and strength in active wedges, we have a good theoretical and observational basis for applying the results of the wedge theory to compressive mountain belts in general.

KINEMATICS OF WEDGE GROWTH AND DEFORMATION

Self-Similar Growth

Although the state of stress within a critical wedge is determined by means of a purely static analysis of the balance of forces, there is an important dynamical principle implicit in the model, which is that an actively accreting wedge grows self-similarly, maintaining its critical taper. This self-similar nature of the growth of a critical wedge—when combined with mass balance—places a significant general constraint on the overall kinematics and deformation of fold-and-thrust belts and accretionary wedges, as we now illustrate.

We begin by considering the growth of a small bulldozer wedge, shown in Figure 9. Suppose the bulldozer is moving forward at a constant velocity V, scraping up a layer of soil or snow of constant thickness H. The cross-sectional compressive flux into the toe is then:

$$\dot{A} = (1 - \Delta\rho/\rho)HV, \tag{14}$$

where $\dot{A}$ denotes differentiation with respect to time, and $\Delta\rho$ is the increase in density due to compaction at the time of accretion. By conservation of mass, if there is no subsequent change in density ρ, this compressive flux must be equal to the rate of change of the cross-sectional area of the noneroding wedge, i.e.,

$$\frac{d}{dt}\left[\frac{1}{2}\tan(\alpha + \beta)\, W^2\right] = \dot{A}, \tag{15}$$

where W is the wedge width. Self-similarity implies that the taper $\alpha + \beta$ is constant, so that equation (15) reduces to

$$W\dot{W} = \frac{\dot{A}}{\tan(\alpha + \beta)}, \tag{16}$$

which has the solution

$$W = \left[W_0^2 + \left(\frac{2\dot{A}}{\tan(\alpha + \beta)} \right) t \right]^{1/2}, \tag{17}$$

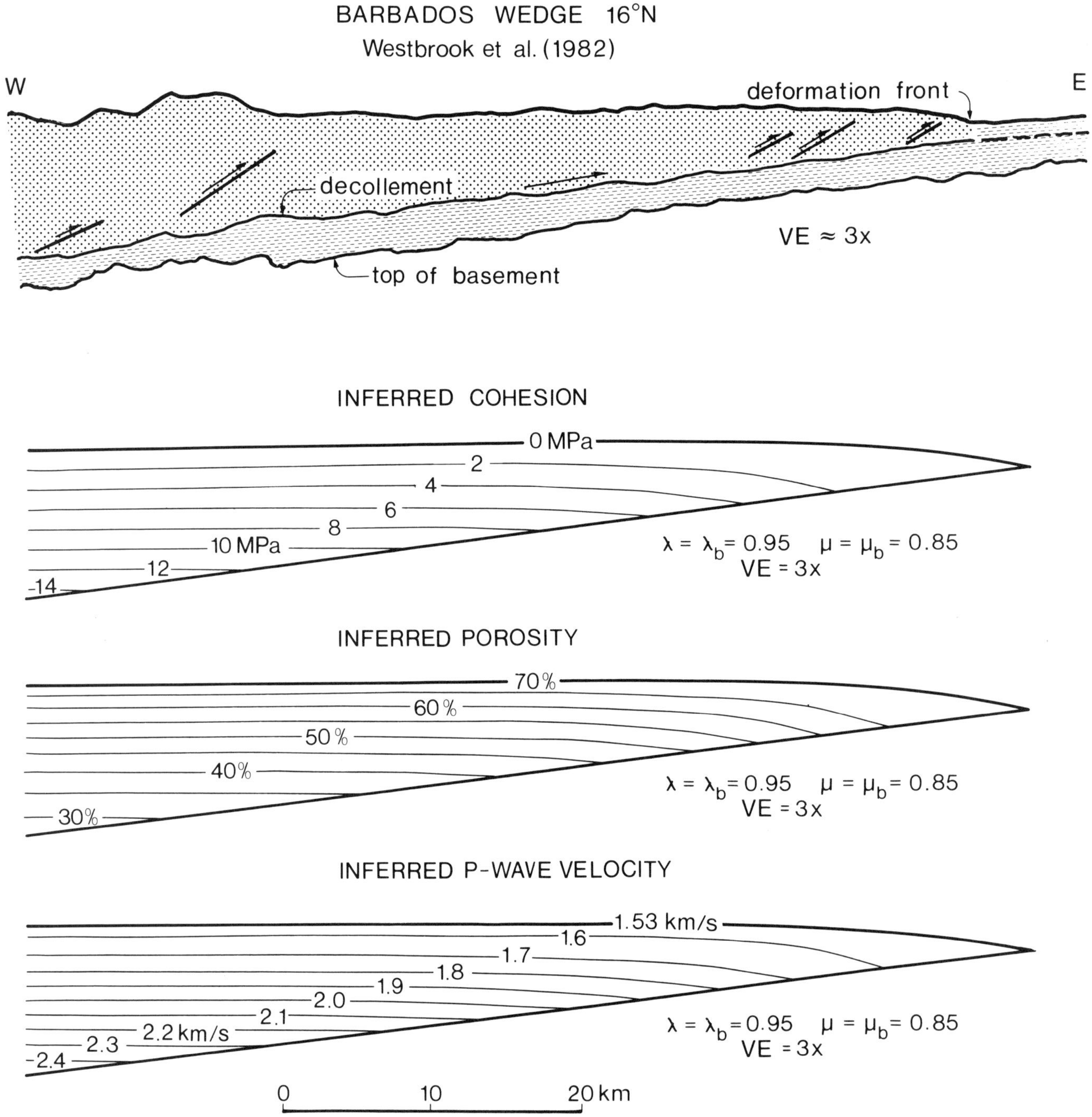

Figure 8. Cross section of the Barbados accretionary wedge at 16°N, showing inferred contours of cohesion, porosity, and seismic velocity required to explain the observed convexity if $\lambda = \lambda_b = 0.95$ and $\mu = \mu_b = 0.85$. The most recently accreted sediments in the first 10 km arcward of the deformation front have very high porosity (60–70%) and low cohesion ($K = 0$ to 10 kPa/m), so they deform more readily than the stronger sediments farther back, giving rise to the observed convexity. The inferred properties depend strongly on the fluid-pressure ratio $\lambda = \lambda_b$, and are relatively insensitive to the assumed value of friction $\mu = \mu_b$.

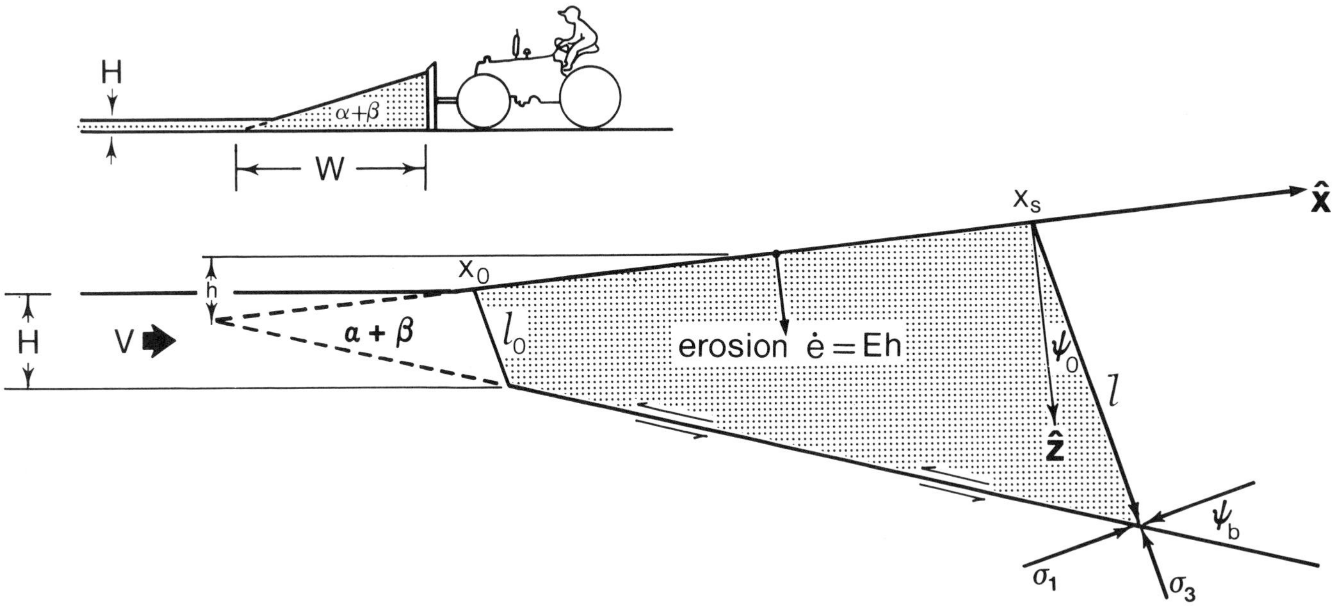

Figure 9. Cross-sectional sketch of a critically tapered wedge, illustrating the parameters employed in deriving the pure-shear kinematical model. The origin of coordinates x, z is at the submerged vertex where the dashed lines intersect. An undeformed, infinitesimally thin column of length l_0 is accreted at x_0. It stretches uniformly in the σ_3 direction and thins uniformly in the σ_1 direction as it moves toward the rear, so that its length at x_s is l. Erosion $\dot{e}$ is measured perpendicular to the regional slope of the mountain belt, but is proportional to the vertical height h above the vertex. Inset cartoon illustrates the self-similar growth of a bulldozer wedge.

where W_0 is the initial width at time $t = 0$. Both the width and the height of a bulldozer wedge thus grow with time like $t^{1/2}$, i.e., less rapidly as time goes on.

A similar $t^{1/2}$ behavior should very roughly govern the growth of accretionary wedges and the migration of trenches away from island arcs. Sediment thicknesses entering present-day mountain belts and accretionary wedges are in the range $H = 1$ to 10 km, and the density increase due to compaction varies from nearly zero up to $\Delta\rho/\rho = 40\%$, depending on the initial porosity and state of lithification. Plate convergence velocities at compressive margins are in the range $V = 10$ to 100 km/m.y., so compressive fluxes can vary by more than two orders of magnitude from less than 10 km^2/m.y. to as much as 1,000 km^2/m.y. Barbados is an example of a slowly growing wedge. The compressive influx in the north, where it is unaffected by the Orinoco delta, is only $\dot{A} \approx 20$ km^2/m.y., and the average taper $\alpha + \beta$ is about 3°. Equation (17) thus predicts that it should have grown to half its present width of 300 km in about 30 m.y. and attained its full width in about 120 m.y. Conversely, Taiwan is an example of a mountain belt with a relatively high compressive influx, $\dot{A} \approx 500$ km^2/m.y. A typical entering rock resides in the Taiwan wedge only for 2 to 3 m.y. before it is uplifted and eroded off the top, as discussed in more detail below.

The $t^{1/2}$ law of self-similar growth contradicts the more nearly linear relationship between forearc width and age found by Dickinson (1973) in a worldwide data compilation. This may be explained by the tendency for large wedges to level off, i.e., exhibit a trench-slope break, due to the decrease in strength once their bases protrude below the brittle-plastic transition (Davis and others, 1983). In addition, older accretionary wedges in the compilation may be biased toward continental margins where compressive fluxes $\dot{A}$ are higher.

Erosion

Mass balance and wedge kinematics are also strongly affected by erosion and the uplift that accompanies it, and erosion rates in tectonically active regions show almost as much as variation as accretion rates. For example, in Taiwan streamload data indicate a mean erosion rate of 5 to 6 km/m.y. over the last 30 yr (Li, 1976), radiocarbon dating of uplifted terraces indicates a rate of about 5 km/m.y. over the last 10,000 yr (Peng and others, 1977), and fission-track data indicate a rate of 3 to 10 km/m.y. over the last 1 m.y. (Liu, 1982). Radiometric, streamload, and geodetic data from the Himalayas indicate average erosion rates of 0.3 to 0.9 km/m.y. Locally, however, the rates can be much higher: as much as 1.4 to 9 km/m.y. for the Nanga Parbat massif (Mehta, 1980; Zeitler and others, 1982a,b). Similar regionally variable rates in the range 0.5 to 10 km/m.y. are indicated by contemporary uplift and streamload data from the South Island of New Zealand (Wellman, 1979; Adams, 1980). Late Cenozoic erosion rates in the western Alps are estimated to have been 0.4 to 1 km/m.y., based on modeling of heat-flow data and radiometric cooling dates (Clark and Jäger, 1969; Wagner and others, 1977). Finally, early Cenozoic erosion rates of about 2 km/m.y. have been inferred for the Coast Mountains of British Columbia from an analysis of metamorphic assemblages, fluid inclusions, and geochronology (Hollister, 1982). These examples show that erosion and uplift rates in active mountain belts vary from at least as low as 0.1 to 0.5 km/m.y. to as high as 5 to 10 km/m.y. It is interesting to note that modern spatial geodesy techniques, which have a vertical accuracy of ±2 cm (Bock and others, 1985), should be capable of directly observing uplift rates of 1 km/m.y. after only a few decades of observation.

The cross-sectional erosive flux off the top of a mountain belt of width W is $< \dot{e} > W$ where $< \dot{e} >$ is the mean erosion rate. A subaerial mountain belt will be in steady state with the compressive influx balanced by erosion if $W = \dot{A} / < \dot{e} >$. Noneroding submarine accretionary wedges such as Barbados cannot attain a steady state but instead grow, essentially without limit, until the tectonic regime changes. High erosion rates ($< \dot{e} > = 5$ to 10 km/m.y.) give rise to narrow steady-state mountain belts with widths in the range $W = 50$ to 200 km, as in Taiwan (Suppe, 1981) and New Zealand. Both the time required to attain steady state and the average residence time of material in these narrow, highly erosive mountain belts is only 1 to 10 m.y. In contrast, wide ($W > 500$ km), more slowly eroding ($< \dot{e} > = 0.5$ km/m.y.) mountain belts such as the central Andean Altiplano have growth and average residence times greater than 50 m.y., which may be too long for a steady state to be established before conditions change. We see from these simple considerations that erosion rate exerts a dominant control over the rate of material throughput and therefore the tectonic histories of mountain belts. It is also worth noting that erosion and accretion rates may frequently be coupled, since erosional products that are deposited in foredeeps immediately affect $\dot{A}$. In thinking of the role of erosion in ancient orogenic belts, we see that the Proterozoic Grenville belt is more than 600 km wide and appears to have cooled over about 200 m.y. (Berger and York, 1981), which suggests slow syntectonic as well as posttectonic erosion. In contrast, the Ordovician Taconic belt in the Appalachians must have been relatively erosive because it attained a width of only 100 to 200 km (Rodgers, 1971), and the interval between sedimentation and metamorphic cooling dates in slates and phyllites is only 10 to 20 m.y. (Stanley and Ratcliffe, 1985). However, the relationship between cooling dates and deformation can also exhibit complexities that are not directly related to wedge mechanics; for example, the more metamorphic interior of the Taconic belt in western New England underwent a slow postorogenic cooling followed by rapid uplift and exposure in the Devonian Acadian orogeny (Sutter and others, 1985; Stanley and Ratcliffe, 1985).

Several diverse phenomena, such as mineral stability and climate, control the regional erosion rates of mountain belts and thus their kinematic and tectonic histories. With regard to climate, examination of the present globe indicates that wide ($W >$ 500 km), active compressive mountain belts lie almost exclusively within semiarid zones by virtue of either latitudinal or orographic effects. The role of climate in the growth of mountain belts is particularly striking in the Andes because of their large latitudinal range; in the dry zone between 10° and 30°S latitude, the mountains have grown to an average width of more than 500 km, whereas in the more humid zones to the north and south their average width is about 250 km.

Wedge mechanics also exerts an important control on erosion through its effect on the regional surface slope. Once a wedge thickens beyond about 10 to 15 km, it commonly experiences a drop in surface slope caused by the drop in strength across the brittle-plastic transition, as noted earlier. When the foreland width of a mountain belt exceeds about 150 km the hinterland becomes nearly flat and essentially nonerosive, as in Tibet and the Altiplano. A salt décollement also produces a very low surface slope, which apparently contributes to the low erosive flux of the Zagros Mountains in Iran.

A Pure-Shear Kinematic Model

In order to explore further the role of self-similar wedge growth and erosion on orogenic history, we must consider rock trajectories within the wedge and the associated stress, strain and temperature histories. We restrict attention to perfectly triangular wedges in which the principal stresses σ_1 and σ_3 have the same alignment everywhere, and assume that a column of rock aligned with those stresses near the toe of the wedge retains its alignment and simply stretches uniformly in the σ_3 direction and shrinks uniformly in the σ_1 direction, conserving its cross-sectional area as it moves toward the back. The wedge through which the columns move has a constant geometry as long as the material properties do not change; this is the constraint of self-similarity. Such a kinematic model has been considered before, for the case where there is no erosion, by Karig (1986b).

The assumption that a column conserves its cross-sectional area so that there is no change in density anywhere within the wedge implies either that compaction is negligible or that it occurs at the instant of accretion in the immediate vicinity of the toe. This is a reasonable simplifying assumption, particularly for the rapidly eroding fold-and-thrust belts that will be our main concern here, since the thick sediments entering such belts generally have rather low mean porosity compared with the highly overpressured pelagic sediments entering submarine accretionary wedges such as Barbados. The assumption that a column deforms in pure shear without rotation or disruption is less realistic, since it does not take into account the presence of discrete, large-scale, forward-verging thrust sheets in fold-and-thrust belts. Like the critical wedge model itself, such a highly idealized kinematic description is intended to be valid only in an averaged or homogenized sense, and one of its principal virtues is its explicit dependence on a small number of parameters.

Let H be the thickness of the incoming sedimentary sequence at the toe of the wedge and let V be the horizontal convergence velocity of the subducting and overriding plates, so that the accretionary influx $\dot{A}$ in km^2/m.y. is given by Equation 14 as for the bulldozer. Suppose deformation commences at a distance x_0 from the vertex, measured along the buried extension of the top of the wedge as shown in Figure 9, so that the initial length of the entering column is:

$$l_o = x_0 \frac{\sin(\alpha + \beta)}{\cos\psi_b}. \tag{18}$$

The erosion rate $\dot{e}$ perpendicular to the surface is assumed to be proportional to the height h above the vertex, i.e.,

$$\dot{e} = Eh, \tag{19}$$

where E is a constant with units of (m.y.)$^{-1}$. Such a linear dependence of erosion rate on relief has been justified in several studies of erosion under different climatic and vegetative conditions (Ruxton and McDougall, 1967; Ahnert, 1970), and has been used by England and Richardson (1977) and others to model the effects of erosion on the thermal evolution of large thrust sheets. The total erosive flux in km^2/m.y. off the top of a wedge from the onset of deformation at x_0 out to a distance x_s measured along its surface is

$$\int_{x_0}^{x_s} \dot{e}\, dx = \frac{1}{2} E \sin\alpha\, (x_s^2 - x_0^2). \tag{20}$$

The motion $x_s(t)$ of a rock column toward the back of the wedge can be determined by balancing the fluxes entering and leaving the shaded region shown in Figure 9. The flux due to motion of the elongated column of length l out the back is:

$$l\, \dot{x}_s \cos\psi_0 = x_s\, \dot{x}_s \frac{\sin(\alpha + \beta)\cos\psi_0}{\cos\psi_b}, \tag{21}$$

so the flux balance equation is:

$$\frac{1}{2} E \sin\alpha\, (x_s^2 - x_0^2) + x_s\, \dot{x}_s \frac{\sin(\alpha + \beta)\cos\psi_0}{\cos\psi_b} = \dot{A}. \tag{22}$$

Like Equation 16, this can be solved by separation of variables with the result:

$$x_s(t) = (W + x_0)\left[1 - [1 - (x_0/(W + x_0))^2] \exp(-\eta t)\right]^{½}, \tag{23}$$

where $t = 0$ coincides with the entry of a rock column into the wedge so that $x_s(0) = x_0$ and:

$$W = \left[\frac{2\dot{A}}{E \sin\alpha} + x_0^2\right]^{½} - x_0 \tag{24}$$

$$\eta = \frac{E \sin\alpha \cos\psi_b}{\sin(\alpha + \beta)\cos\psi_0}. \tag{25}$$

In the limit $t \to \infty$ the position of the column approaches $x_0 + W$ so that W is the theoretical steady-state width of an eroding, critically tapered mountain belt. In the absence of erosion, $W \to \infty$ and equation (23) reduces to:

$$x_s(t) = \left[x_0^2 + \left(\frac{2\dot{A}\cos\psi_b}{\sin(\alpha + \beta)\cos\psi_0}\right) t\right]^{½}, \tag{26}$$

in which case rocks appear to "diffuse" into the wedge like $t^{½}$, i.e., in the same manner that such a noneroding wedge grows.

The instantaneous strain rate of a rock column situated at x_s at time t is:

$$\dot{\epsilon} = \dot{\epsilon}_3 = -\dot{\epsilon}_1 = \dot{l}/l, \tag{27}$$

where $\dot{\epsilon}_1$ and $\dot{\epsilon}_3$ are aligned with σ_1 and σ_3. The rate of change of length $\dot{l}$ is given by:

$$\dot{l} = \dot{x}_s \frac{\sin(\alpha + \beta)}{\cos\psi_b} + \frac{\dot{e}}{\cos\psi_0}, \tag{28}$$

where the first term is due to the backward motion of the column and the second is due to erosion. Erosion is constantly seeking to bevel off the wedge and render its taper subcritical, and the rock columns must stretch to counteract this tendency. The strain rate expressed as a function of a column's position x_s reduces to:

$$\dot{\epsilon}(x_s) = \frac{1}{2}\eta\left[1 + \left(\frac{W + x_0}{x_s}\right)^2\right], \tag{29}$$

whereas if expressed as a function of time t, it is:

$$\dot{\epsilon}(t) = \frac{1}{2}\eta\left[\frac{2 - \left[1 - (x_0/(W + x_0))^2\right]\exp(-\eta t)}{1 - \left[1 - (x_0/(W + x_0))^2\right]\exp(-\eta t)}\right]. \tag{30}$$

As $t \to \infty$ and the rock column nears the rear of the mountain belt at $x_0 + W$, the strain rate $\dot{\epsilon} \to \eta$, which is the strain rate of a stationary column due to erosion alone. In the absence of erosion, equations (29) and (30) reduce to:

$$\dot{\epsilon} = \left[\frac{\dot{A}\cos\psi_b}{\sin(\alpha + \beta)\cos\psi_0}\right] x_s^{-2} = \left[\left(\frac{\sin(\alpha + \beta)\cos\psi_0}{\dot{A}\cos\psi_b}\right) x_0^2 + 2t\right]^{-1}. \tag{31}$$

The dependence of the nonerosive strain rate on x_s^{-2} and t^{-1} illustrates the extent to which the deformation of a submarine wedge is concentrated near the toe.

The cumulative or finite longitudinal strain S of a rock column at time t, measured relative to its initial length l_0, is given by:

$$S = 1 + l_0^{-1} \int_0^t \dot{l}\, dt, \tag{32}$$

which can be integrated to give the result:

$$S = 1 - \left[\frac{x_s - x_0}{x_0}\right] + \left[\frac{W + x_0}{x_0}\right] \log\left[\frac{W(x_0 + W + x_s)}{2x_0 + W)(x_0 + W - x_s)}\right]. \tag{33}$$

In the absence of erosion ($W \to \infty$) this finite strain reduces to:

$$S = x_s/x_0 = l/l_0, \tag{34}$$

which is just the geometric change in length required to maintain the critical taper as the column moves back in the wedge.

Rock trajectories through the deforming wedge can also be determined from the above kinematic model. An individual rock is characterized by its initial depth $0 \leqq d \leqq H$ when it enters the toe of the wedge. The uneroded fraction of an infinitesimally thin rock column at any time t is just the ratio of its cross-sectional area at that time to its initial area, i.e.,

$$l\dot{x}_s/l_0\dot{x}_0 = x_s\,\dot{x}_s/x_0\dot{x}_0 = \exp(-\eta t). \tag{35}$$

Since the instantaneous strain in this approximate model is distributed uniformly within a column, a rock that enters the wedge at depth d will be situated a distance:

$$\zeta(t) = \frac{\sin(\alpha + \beta)}{\cos\psi_b}\left[1 + (d/H - 1)\exp(-\eta t)\right] x_s(t) \tag{36}$$

down the column at time t. Rock trajectories within the wedge are thus given by:

$$x(t) = x_s(t) + \zeta(t)\sin\psi_0 \tag{37}$$

$$z(t) = \zeta(t)\cos\psi_0, \tag{38}$$

where $x_s(t)$ is given by Equation 23. The time t at which a rock outcrops and is eroded off the top is simply related to its entering depth d by:

$$t = \frac{1}{\eta}\log(1 - d/H). \tag{39}$$

The rate of change of depth or outward velocity of a rock at the instant it outcrops is:

$$\dot{z} = \dot{\zeta}\cos\psi_0 = -E\sin\alpha\, x_s = -\dot{e}\,, \tag{40}$$

which is simply an indication that the uplift rate of a steady-state mountain belt is equal and opposite to the erosion rate as expected.

All the above results can be simplified if the sediments entering the wedge are very thin, so that the distance x_0 between the vertex and the onset of accretion can be ignored. Most notably, the strain rate in a noneroding wedge reduces in that case to $\dot{\epsilon} = 1/2t$, independent of the wedge geometry.

Application to Taiwan

Taiwan is the site of an oblique ongoing collision between the Luzon volcanic arc at the western edge of the Philippine Sea plate and the stable Chinese continental margin (Suppe, 1987). The accretionary wedge south of Taiwan expands in size northward as the thick margin sediments are encountered, rising above sea level to become the Central Mountains of Taiwan. In essence, to move northward along the mountain belt by 100 km is to move forward in time by about 1 m.y. because of the oblique nature of the collision (Suppe, 1981, 1984). The mountains grow steadily wider and higher until at about 100 km north of the southern tip of Taiwan they attain a constant width, which prevails until the polarity of subduction flips about 200 km farther to the north. The constant-width segment of the mountain belt in central Taiwan appears to be in steady state, and we use it to illustrate our kinematic model.

We chose wedge angle parameters $\alpha = 2.9°$, $\beta = 6°$, $\psi_0 = 2.6°$, and $\psi_b = 11.5°$ that are consistent with Byerlee's law $\mu = 0.85$, and the observed step-up angle $\delta_b = 13.3°$. The plate convergence velocity is $V \approx 70$ mm/yr (Seno, 1977; Ranken and others, 1984), and the thickness of the already well-lithified incoming sediments at the toe is $H \approx 7$ km (Suppe, 1981), so the accretionary influx, as noted earlier, is:

$$\dot{A} \approx 500 \text{ km}^2/\text{m.y.} \tag{41}$$

The distance of the deformation front from the submerged vertex we use as the origin of our coordinates is:

$$x_0 = \frac{H\cos\psi_b}{\sin(\alpha + \beta)\cos(\psi_0 + \alpha)} \approx 45 \text{ km}, \tag{42}$$

which is too large to be ignored.

Figure 10 shows the theoretical dependence of the steady-state width W on the erosion-rate constant E for the above choice of parameters. The shaded band denotes the observed width of the steady-state Central Mountains in Taiwan, which is (Suppe, 1981):

$$W = 87 \pm 4 \text{ km}. \tag{43}$$

From this we infer the value:

$$E = 1.27 \text{ (m.y.)}^{-1}. \tag{44}$$

The corresponding theoretical erosion rate varies from 2.9 km/m.y. at the deformation front to 8.4 km/m.y. at the crest of the mountains, with an average rate across the mountain belt given by:

$$<\dot{e}> = \frac{1}{W}\int_{x_0}^{x_0 + W} \dot{e}\,dx = \frac{1}{2}E\sin\alpha\,(W + 2x_0) = 5.7 \text{ km/m.y.} \tag{45}$$

The good agreement with the measured erosion and uplift rates quoted earlier is an added confirmation that the steady-state model with accretion balanced by erosion is appropriate in central Taiwan.

The predicted horizontal motion $x_s - x_0$ of rock columns through the Taiwan wedge is shown in Figure 11 for $E = 0$, 0.5 and 3.0 (m.y.)$^{-1}$, as well as for the steady-state value $E = 1.27$ (m.y.)$^{-1}$. It is seen that columns are transported across almost the entire width W in about 3 m.y. The rate of horizontal transport is slowed if the erosion rate increases, since more of the incoming

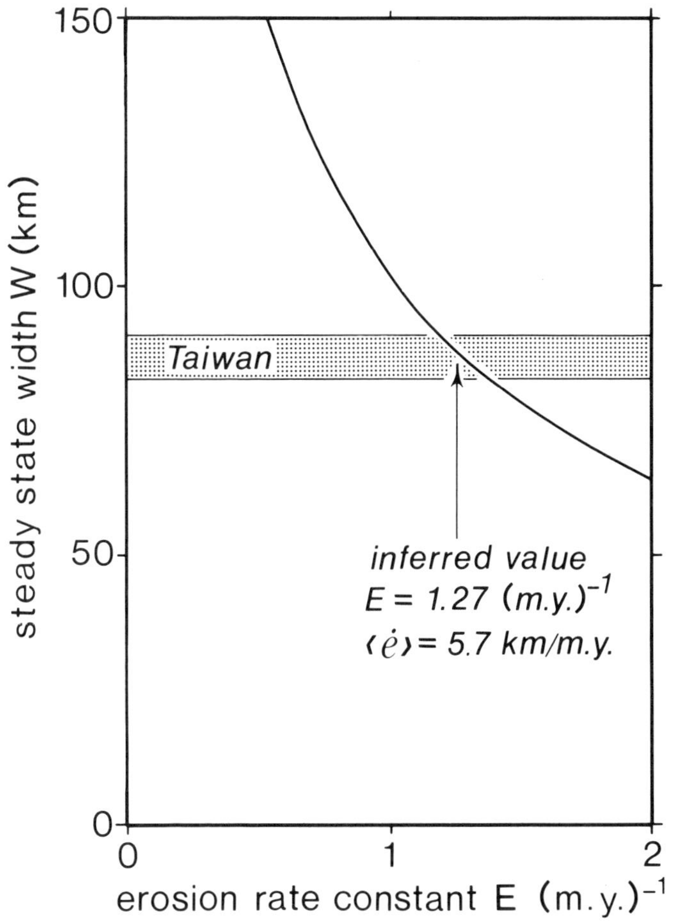

Figure 10. Dependence of the theoretical steady-state width W on the erosion rate constant E for $\alpha = 2.9°$ and $x_0 = 45$ km. The inferred value $E = 1.27$ (m.y.)$^{-1}$ implies $< \dot{e} > = 5.7$ km/m.y., in good agreement with observed uplift and erosion rates in Taiwan from geomorphological and fission-track data.

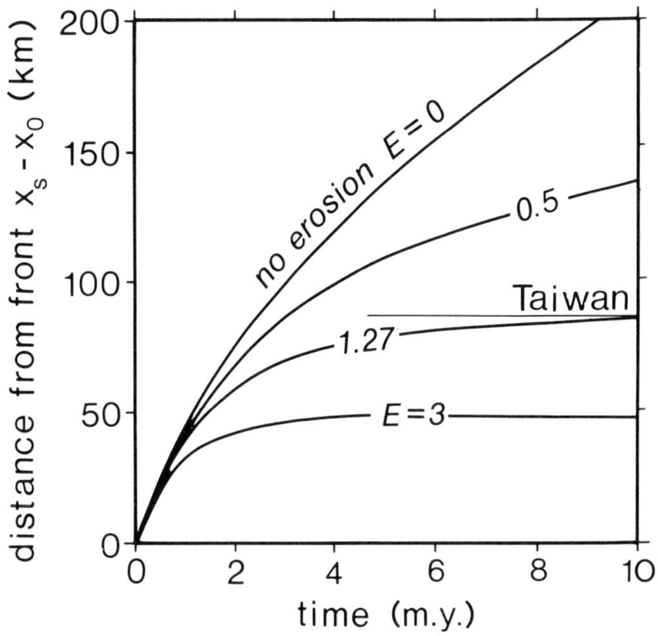

Figure 11. Theoretical horizontal transport of rocks through the Taiwan wedge for various values of the erosion rate constant E. The transport is fastest near the toe where $x_s - x_0$ varies like $t^{1/2}$. If $E > 0$, a steady state is possible, in which case rocks approach the rear at $x_s - x_0 = W$ exponentially slowly. For the preferred value $E = 1.27$ m.y., rocks are transported halfway across the mountain belt in about 1 m.y.

material must then be used to replace that lost by erosion in order to maintain the critical taper. The theoretical strain rate $\dot{\epsilon}$ versus time is shown in Figure 12, and the corresponding finite strain S versus $x_s - x_0$ is shown in Figure 13, both for the same choices of E, including $E = 1.27$ (m.y.)$^{-1}$. The strain rate is greatest near the deformation front as expected, and is in the range 10^{-13} to 10^{-14} s^{-1} throughout the entire mountain belt. The corresponding strain S diverges as $x_s - x_0 \rightarrow W$ and $\dot{\epsilon} \rightarrow \eta$, but it is in the range $S = 1$ to 10 throughout most of the mountain belt. Similar values of both strain rate and finite strain have been inferred for other orogenic belts from studies of naturally deformed rock samples (Pfiffner and Ramsay, 1982). A direct comparison with such studies may not be too meaningful, however, since $\dot{\epsilon}$ and S are measures of the total strain due to both large-scale and small-scale processes, and much of the regional horizontal shortening in mountain belts is the result of brittle imbricate thrusting on major discrete faults. The important effect of erosion on throughput and deformation is apparent in Figures 11 through 13; rocks move more slowly through rapidly eroding mountain belts, but upon reaching the core they have undergone total strains S that are three to five times those in a noneroding mountain belt of the same size.

The trajectories or flowlines followed by rocks that enter the model Taiwan wedge at various depths $0 \leqq d \leqq 7$ km are shown in Figure 14 together with the slip surfaces on which the failure criterion (1) is satisfied. Superficially, the flowlines resemble those in the viscous corner flow models of Cowan and Silling (1978), Cloos (1982), and Wang and Shi (1984), even though the assumptions underlying the two models are very different. In corner flow models the flow is driven upward by a rigid "backstop" at the back of the wedge (Pavlis and Bruhn, 1983), whereas in the present model the uplift occurs naturally in response to

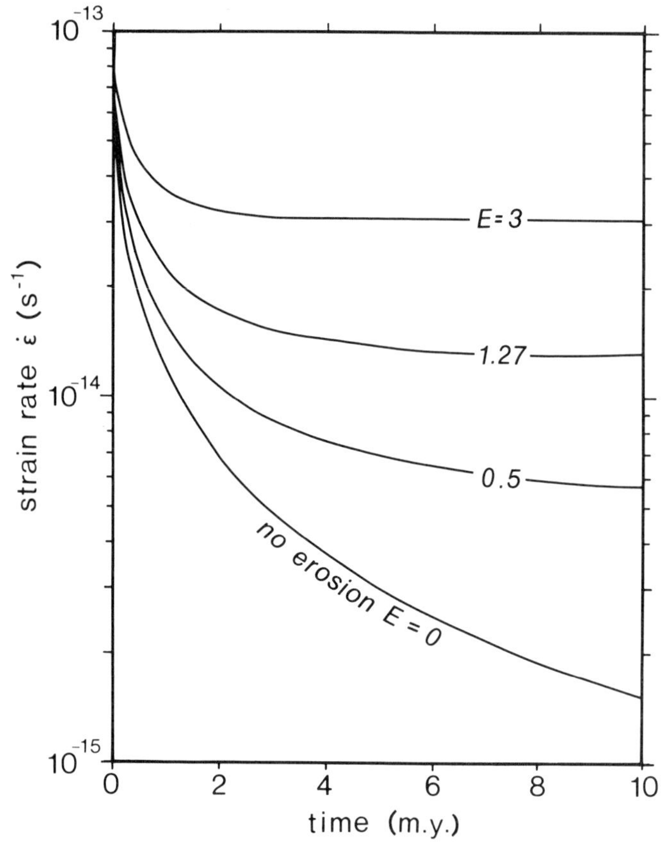

Figure 12. Theoretical strain rate $\dot{e}$ of rock columns in the Taiwan wedge for various values of the erosion rate constant E. For $E > 0$, the strain rate approaches a constant value as $t \rightarrow \infty$, which is the strain rate η due to erosion alone.

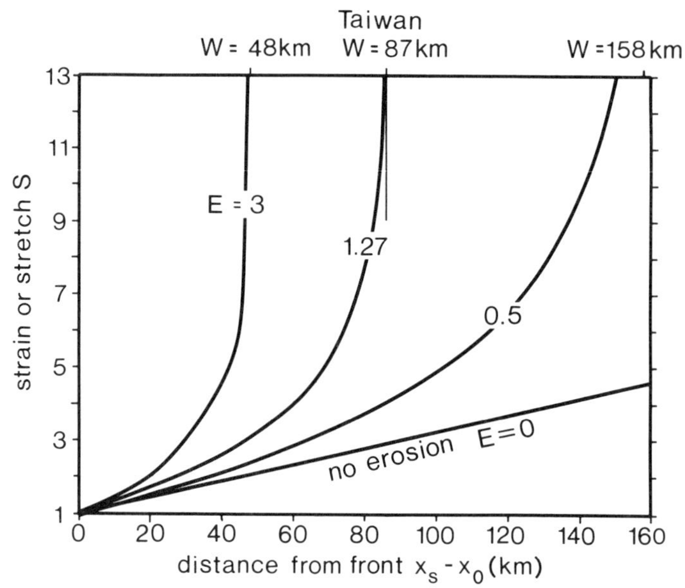

Figure 13. Theoretical finite strain or "stretch" of rock columns in the Taiwan wedge for various values of the erosion rate constant E. The strain S depends linearly on $x_s - x_0$ if $E = 0$ because of the wedge's perfectly triangular geometry.

erosion. In addition, the present model is not based on an unrealistic viscous rheology, but rather is intended to represent the purely brittle deformation of a critically tapered wedge in the upper crust.

The deformation and transport shown in the bottom of Figure 14 can be envisaged as occurring by slip along the continuously distributed network of conjugate thrust faults shown at the top. The acute angle between the conjugate slip surfaces is arctan $(1/\mu) = 49.6°$, assuming $\mu = 0.85$, so that the backthrusts step up at an angle $\delta'_b = 36.3°$ that is almost three times greater than the forward stepup angle $\delta_b = 13.3°$. To avoid space problems and satisfy the condition of tangential slip along the basal décollement, the relative displacement on the two sets of thrusts there should theoretically be in the ratio:

$$\frac{\Delta\mu_{forward}}{\Delta\mu_{back}} = \frac{\sin\delta'_b}{\sin\delta_b} = 2.6. \tag{46}$$

The observed predominance of forward thrusting along major faults is actually greater than this, presumably because of stratigraphically controlled anisotropy in the fracture strength.

Both the attitude of the slip surfaces and the shapes and relative positions of the trajectories depend only upon the wedge geometry α, β, ψ_b, and ψ_0, and they are independent of the accretionary and erosion rate parameters $\dot{A}$ and E; this is a direct consequence of the property of self-similarity. All that the flux rates control is the overall scale or width W and the rate at which rocks deform and move along the trajectories.

Metamorphic Conditions

Kinematic models of wedge growth, such as the one just presented and applied to Taiwan, imply specific metamorphic histories for rocks traversing the particle trajectories. Typical dehydration reactions will depend on the temperature T, the porewater pressure $P_f = P_{H_2O}$, and the thermodynamic solid pressure P_s. Estimates of the temperature variations are beyond our present scope, but knowledge of the stress and fluid-pressure variations experienced by rocks follows directly from knowledge of their trajectories, since the state of effective stress is known completely within a deforming wedge from the critical taper

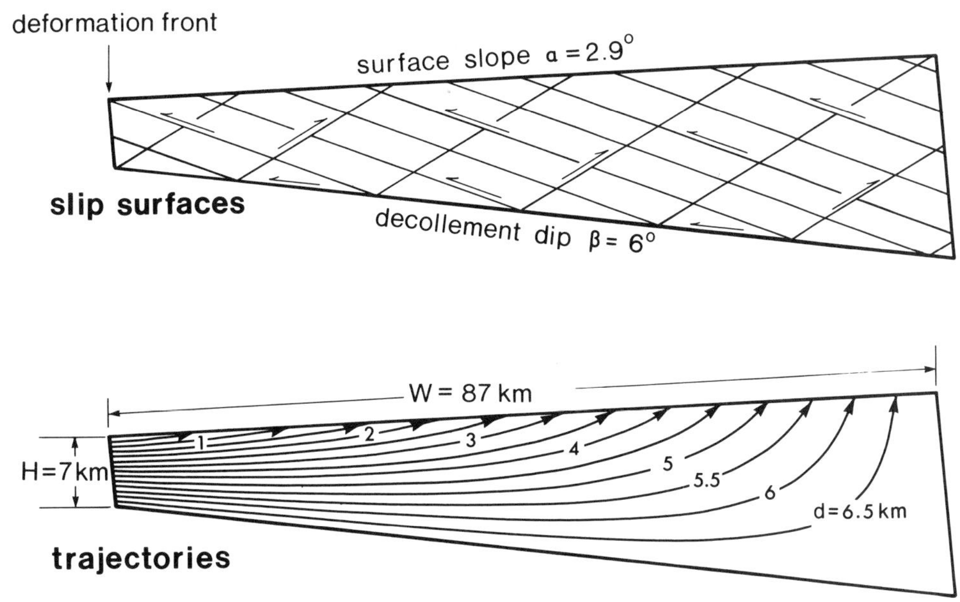

Figure 14. Theoretical slip surfaces and rock trajectories in the critically tapered Taiwan wedge, assuming $\mu = 0.85$.

theory. In a cohesive wedge having $S_0 = Kz$, both P_{H_2O} and P_s depend only on the rock depth z and the overlying water depth D; for example, the pore-water pressure in a submarine wedge is given by:

$$P_{H_2O} = \rho_w g D + \lambda \rho g \cos\alpha \; z. \tag{47}$$

Likewise, the thermodynamic solid pressure is simply the mean stress:

$$P_s = \frac{\sigma_1 + \sigma_3}{2} = \rho_w g D + \rho g \cos\alpha \; z + \left[\frac{Kz + \mu (1 - \lambda) \rho g \cos\alpha \; z}{(1 + \mu^2)^{1/2} \sec 2\psi_0 - \mu} \right], \tag{48}$$

where it has been assumed for simplicity that $\sigma_2 = P_s$, since the along-strike component of the stress is not otherwise constrained by our two-dimensional wedge model.

The derivative dP_{H_2O}/dP_s is important in metamorphism because it controls the slopes dP_s/dT of divariant dehydration reaction boundaries via the generalized Clapeyron relation (Thompson, 1955):

$$\left[\frac{dP_s}{dT} \right]_{\Delta G = 0} = \frac{\Delta S}{\Delta V_s + \Delta V_{H_2O} \, (dP_{H_2O}/dP_s)}, \tag{49}$$

where ΔS, ΔG, ΔV, and ΔV_{H_2O} are the changes in entropy, Gibbs free energy, and molar volumes of the solids and fluid. In a subaerial wedge ($\rho_w = 0$), the derivative dP_{H_2O}/dP_s is simply the constant ratio P_{H_2O}/P_s; in fact:

$$\frac{dP_{H_2O}}{dP_s} = \frac{P_{H_2O}}{P_s} = \frac{\lambda}{\left[1 + \frac{(K/\rho g \cos\alpha) + \mu (1 - \lambda)}{(1 + \mu^2)^{1/2} \sec 2\psi_0 - \mu} \right]}. \tag{50}$$

This constant pressure ratio P_{H_2O}/P_s and hence the location of metamorphic facies boundaries depends significantly on the wedge strength parameters K and μ in addition to the pore-fluid pressure ratio λ. Only if the strength is negligible and the stress-state isotropic does Equation 50 reduce to $P_{H_2O} = \lambda P_s$. More generally, since there must be some deviatoric stress to support the topography in orogenic belts, we expect $P_{H_2O} < \lambda P_s$. Figure 15 shows the possible values of the pressure ratio P_{H_2O}/P_s in Taiwan corresponding to the cleaver-shaped region of possible rock-strength parameters in Figure 6. If the coefficient of friction is as low as $\mu = \mu_b \approx 0.2$ then P_{H_2O}/P_s is about 0.6, i.e., only slightly less than $\lambda = \lambda_b = 0.67$. On the other hand, if $\mu = \mu_b = 0.6$ to 0.85, then $P_{H_2O}/P_s = 0.3$ to 0.4. Thus, if normal laboratory rock strengths are indicative of regional rock strengths in compressive mountain belts, then the significant deviatoric stresses that are implied can significantly affect the conditions of metamorphism. Since the change ΔV_s in Equation 49 is normally small compared to ΔV_{H_2O}, the reaction boundary slopes dP_s/dT can be steepened considerably: in Taiwan, dP_s/dT is twice what it would be if $P_{H_2O} = \lambda P_s$ and three times what it would be if $P_{H_2O} = P_s$, as is sometimes assumed. The implication of the steepened reaction boundaries is that rocks can attain higher grades of metamorphism at lower temperatures than if the stress were isotropic.

The metamorphic histories implied by the particle trajectories shown in Figure 14 can be compared, at least qualitatively, with the observed metamorphic grade of rocks exposed at the surface in Taiwan. An increase in maximum burial away from the toe of the wedge is predicted, which agrees with the normal

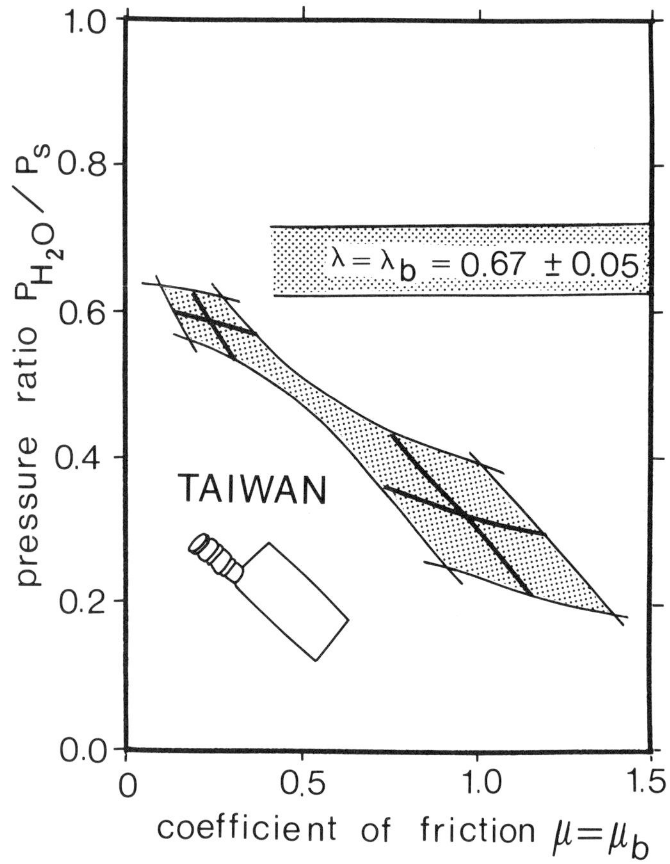

Figure 15. Values of the pressure ratio P_{H_2O}/P_s consistent with the structural and fluid-pressure data in Taiwan. The distorted cleaver-shaped region is the transformation of the region in Figure 6 from K versus $\mu = \mu_b$ space to P_{H_2O}/P_s versus $\mu = \mu_b$ space, with the relative orientations indicated by the small Chinese cleavers.

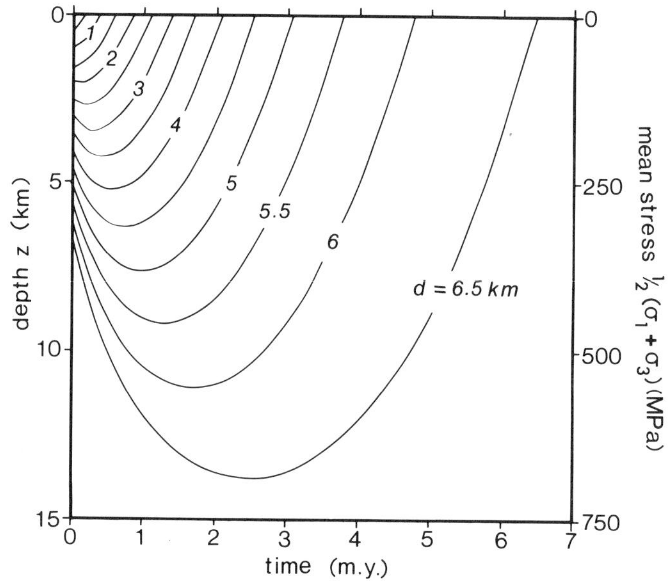

Figure 16. Change in depth z versus time and corresponding change in the mean stress $P_s = \frac{1}{2}(\sigma_1 + \sigma_3)$ for rocks of various entering depths d.

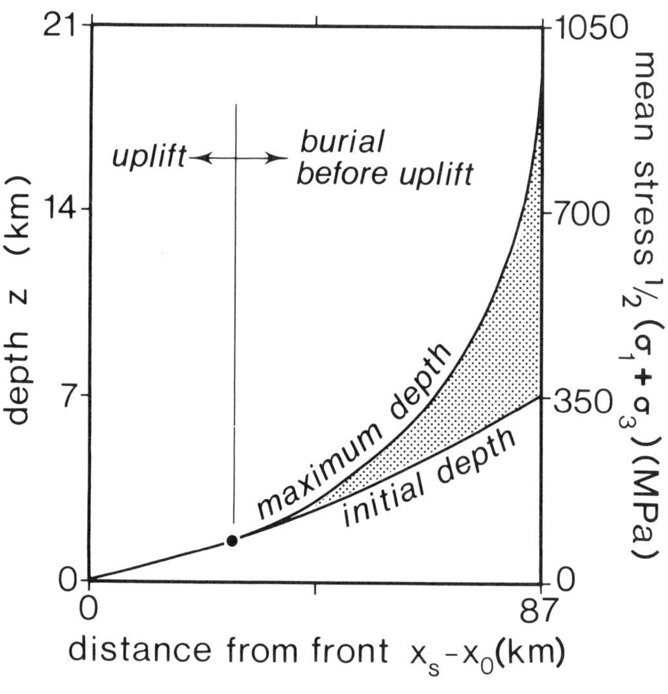

Figure 17. Comparison of maximum depths attained by rocks in the model Taiwan wedge with their entering depths, as a function of their eventual outcrop distance $x_s - x_0$ from the toe. Shading denotes the increase in mean stress $P_s = \frac{1}{2}(\sigma_1 + \sigma_3)$ experienced by the lowermost 5 km of incoming section.

increase in metamorphic grade into mountain belts. The change in depth $z(t)$ of rocks entering the Taiwan wedge is shown in Figure 16, assuming $\dot{A} \approx 500$ km^2/ m.y. and $E = 1.27$ (m.y.)$^{-1}$. It is seen that the uppermost 5 km of the incoming section has been removed by erosion after about 3 m.y. and the uppermost 6.5 km has been eroded after about 6 m.y. Figure 17 shows the entering depth and the maximum depth attained by rocks that outcrop at various distances x_s to x_0 from the toe. As is evident also in Figure 16, the uppermost 2 km of rock is simply uplifted before being eroded, whereas the lowermost 5 km is transported to deeper depths before being uplifted and exposed. The right-hand scales in Figures 16 and 17 indicate the associated changes in mean stress P_s, assuming $\rho = 2{,}700$ kg/m^3, $\mu = 0.85$, and $K = 3.6$ kPa/m. The gradient in mean stress within such a strong model wedge is 50 kPa/m, or roughly twice the lithostatic gradient, reflecting the strong state of horizontal compression. The shading in Figure 17 depicts the increase in mean stress associated with the tectonic burial of the rocks in the lowermost section.

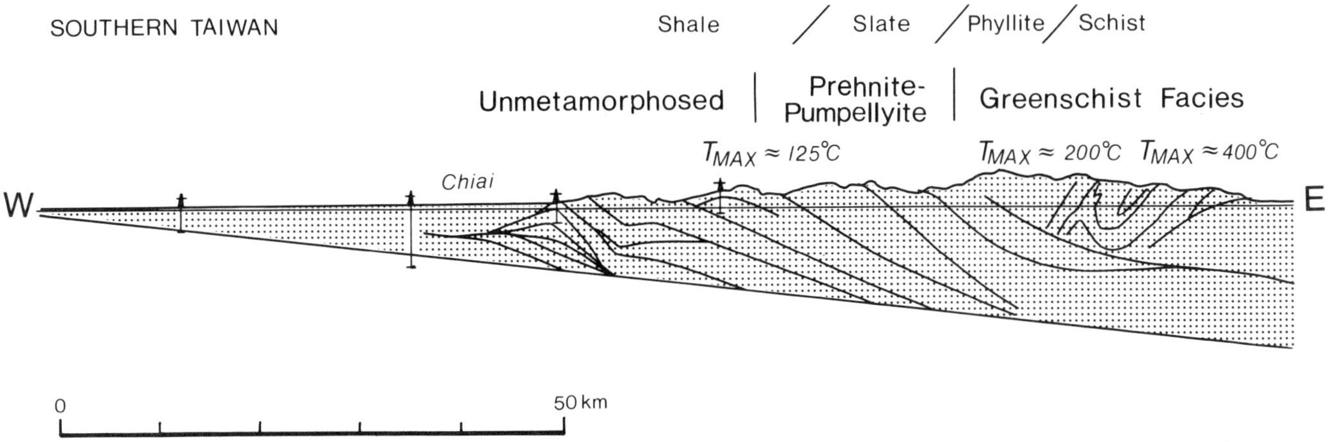

Figure 18. Metamorphic grade of surface rocks along a cross section of southern Taiwan (compiled from Davis and others, 1983; Chen and others, 1983; Warneke and Ernst, 1984).

When these predictions are compared with the cross section of southern Taiwan shown in Figure 18, we see a good qualitative agreement with the surface rocks, which change progressively from shale to slate to phyllite to schist, with a corresponding progressive increase in the maximum temperature and mean stress experienced. A more quantitative comparison cannot be made without a model for the temperature structure of an eroding fold-and-thrust belt, and that will be the subject of a future study.

ACKNOWLEDGMENTS

This chapter is dedicated to Professor John Rodgers, to whom Suppe is thankful for his interest, guidance, and friendship over the last 20 years. It was particularly enjoyable to take a geologic excursion through Taiwan with Rodgers in 1978. We acknowledge Dan Davis and Wu-Ling Zhao for their contributions to the theory of wedge mechanics reviewed in this paper; we also thank Dan Karig and Jason Morgan for their speedy and constructive reviews. Suppe is grateful for the support and hospitality of the Chinese Petroleum Corporation, the National Taiwan University, the Central Geological Survey, and the Mining Research and Service Organization while working in Taiwan. This work was supported by National Science Foundation grants EAR83-06157 and EAR84-07513.

REFERENCES CITED

Adams, J., 1980, Contemporary uplift and erosion of the Southern Alps, New Zealand: Geological Society of America Bulletin, v. 91, p. 1–114.

Ahnert, F., 1970, Functional relationships between dunudation, relief, and uplift in large mid-latitude drainage basins: American Journal of Science, v. 268, p. 243–263.

Barcilon, V., 1987, A note on "Noncohesive critical Coulomb wedges; An exact solution" by F. A. Dahlen: Journal of Geophysical Research, v. 92, p. 3681–3682.

Berger, G. W., and York, D., 1981, Geothermometry from $^{40}Ar/^{39}Ar$ dating experiments: Geochimica and Cosmochimica Acta, v. 45, p. 795–811.

Bock, Y., Abbot, R. I., Counselman, C. C., Gourevitch, S. A., and King, R. W., 1985, Establishment of three-dimensional geodetic control by interferometry with the Global Positioning System: Journal of Geophysical Research, v. 90, p. 7689–7703.

Brace, W. F., and Kohstedt, D. L., 1980, Limits on lithospheric stress imposed by laboratory experiments: Journal of Geophysical Research, v. 85, p. 6248–6252.

Byerlee, J., 1978, Friction of rocks: Pure and Applied Geophysics, v. 116, p. 615–626.

Chapple, W. M., 1978, Mechanics of thin-skinned fold-and-thrust belts: Geological Society of America Bulletin, v. 89, p. 1189–1198.

Chen, C. H., Chu, H. T., Liou, J. G., and Ernst, J. G., 1983, Explanatory notes for the metamorphic facies map of Taiwan: Central Geological Survey Special Publication, no. 2, 32 p.

Clark, S. P., and Jäger, E., 1969, Denudation rate in the Alps from geochronologic and heat flow data: American Journal of Science, v. 267, p. 1143–1160.

Cloos, M., 1982, Flow mélanges: numerical modelling and geologic constraints on their origin in the Franciscan subduction complex, California: Geological Society of America Bulletin, v. 93, p. 330–345.

Cowan, D. S., and Silling, R. M., 1978, A dynamic scaled model of accretion at trenches and its implications for the tectonic evolution of subduction complexes: Journal of Geophysical Research, v. 83, p. 5389–5396.

Dahlen, F. A., 1984, Noncohesive critical Coulomb wedges; An exact solution: Journal of Geophysical Research, v. 89, p. 10125–10133.

Dahlen, F. A., Suppe, J., and Davis, D., 1984, Mechanics of fold and thrust belts and accretionary wedges; Cohesive Coulomb theory: Journal of Geophysical Research, v. 89, p. 10087–10101.

Davis, D., and Engelder, T., 1985, The role of salt in fold-and-thrust belts: Tectonophysics, v. 119, p. 67–88.

Davis, D., Suppe, J., and Dahlen, F. A., 1983, Mechanics of fold-and-thrust belts and accretionary wedges: Journal of Geophysical Research, v. 88, p. 1153–1172.

Dickinson, W. R., 1973, Widths of modern arc-trench gaps proportional to past duration of igneous activity in associated magmatic arcs: Journal of Geo-

physical Research, v. 78, p. 3376–3389.
Ernerman, S. H., and Turcotte, D. L., 1983, A fluid model for the shape of accretionary wedges: Earth and Planetary Science Letters, v. 63, p. 379–384.
England, P. C., and Richardson, S. W., 1977, The influence of erosion upon the mineral facies of rocks from different metamorphic environments: Geological Society of London Journal, v. 134, p. 201–213.
Hollister, L. S., 1982, Metamorphic evidence for rapid (2 mm/yr) uplift of a portion of the Central Gneiss Complex, Coast Mountains, B.C.: Canadian Mineralogist, v. 20, p. 319–332.
Hoshino, K., Koide, H., Inami, K., Iwamura, S., and Mitsui, S., 1972, Mechanical properties of Japanese Tertiary sedimentary rocks under high confining pressures: Geological Survey of Japan Report 244, 200 p.
Hubbert, M. K., and Rubey, W. W., 1959, Role of fluid pressure in mechanics of overthrust faulting; 1. Mechanics of fluid-filled solids and its application to overthrust faulting: Geological Society of America Bulletin, v. 70, p. 115–166.
Jaeger, J. C., and Cook, N.G.W., 1969, Fundamentals of rock mechanics: London, Methuen, p. 87–91.
Karig, D. E., 1986a, Physical properties and mechanical state of accreted sediments in the Nankai Trough, S. W. Japan Arc, *in* Moore, J. C., ed., Structural fabric in Deep Sea Drilling Project cores from Forearcs, Geological Society of America Memoir 166, p. 117–134.
—— , 1986b, Kinematics and mechanics of deformation across some accreting forearcs, *in* Nasu, N., Kobayashi, K., Uyeda, S., Kushiro, I., and Kagami, H., eds., Formation of active ocean margins: Tokyo, Terra Scientific Publishing Co., p. 155–177.
Lehner, F. K., 1986, Comments on "Noncohesive critical Coulomb wedges; An exact solution" by F. A. Dahlen: Journal of Geophysical Research, v. 91, p. 793–796.
Li, Y. H., 1976, Denudation of Taiwan island since the Pliocene epoch: Geology, v. 4, p. 105–107.
Liu, T.-K., 1982, Tectonic implications of fission-track ages from the Central Range, Taiwan: Geological Society of China Proceedings, no. 25, p. 22–37.
Mehta, P. K., 1980, Tectonic significance of the young mineral dates and the rates of cooling and uplift in the Himalayas: Tectonophysics, v. 62, p. 205–217.
Mogi, K., 1966, Pressure dependence of rock strength and transition from brittle fracture to ductile flow: Earthquake Research Institute Bulletin, Tokyo University, v. 44, p. 215–232.
Paterson, M. S., 1978, Experimental rock deformation; The brittle field: New York, Springer-Verlag, 254 p.
Pavlis, T. L., and Bruhn, R. L., 1983, Deep-seated flow as a mechanism for the uplift of broad forearc ridges and its role in the exposure of high *P/T* metamorphic terranes: Tectonics, v. 2, p. 473–497.
Peng, T.-H., Li, Y.-H. and Wu, F. T., 1977, Tectonic uplift rates of the Taiwan island since the early Holocene: Geological Society of China Memoirs, no. 2, p. 57–69.
Pfiffner, O. A., and Ramsay, J. G., 1982, Constraints on geological strain rates; Arguments from finite strain states of naturally deformed rocks: Journal of Geophysical Research, v. 87, p. 311–321.
Ranken, B., Cardwell, R. K., and Karig, D. E., 1984, Kinematics of the Philippine Sea Plate: Tectonics, v. 3, p. 555–575.
Rodgers, J., 1971, The Taconic orogeny: Geological Society of America Bulletin, v. 82, p. 1141–1178.
Ruxton, B. P., and McDougall, I., 1967, Denudation rates in northeast Papua from potassium-argon dating of lavas: American Journal of Science, v. 265, p. 545–561.
Seno, T., 1977, The instantaneous rotation vector of the Philippine Sea plate relative to the Eurasion plate: Tectonophysics, v. 42, p. 209–226.
Stanley, R. S., and Ratcliffe, N. M., 1985, Tectonic synthesis of the Taconian orogeny in western New England: Geological Society of America Bulletin, v. 96, p. 1227–1250.
Stockmal, G. S., 1983, Modeling of large-scale accretionary wedge formation: Journal of Geophysical Research, v. 88, p. 8271–8287.
Suppe, J., 1981, Mechanics of mountain building and metamorphism in Taiwan: Geological Society of China Memoir, no. 4, p. 67–89.
—— , 1984, Kinematics of arc-continent collision, flipping of subduction, and back-arc spreading near Taiwan: Geological Society of China Memoir, no. 6, p. 21–33.
—— , 1985, Principles of structural geology: Englewood Cliffs, New Jersey, Prentice-Hall, 537 p.
—— , 1987, The active Taiwan mountain belt, *in* Schaer, J. P., and Rodgers, J., eds., The anatomy of mountain ranges: Princeton, New Jersey, Princeton University Press, p. 277–293.
Suppe, J., and Wittke, 1977, Abnormal pore-fluid pressures in relation to stratigraphy and structure in the active fold-and-thrust belt of northwestern Taiwan: Petroleum Geology of Taiwan, v. 14, p. 11–24.
Sutter, J. F., Ratcliffe, N. M., and Mukasa, S. B., 1985, $^{40}Ar/^{39}Ar$ and K-Ar data bearing on the metamorphic and tectonic history of western New England: Geological Society of America Bulletin, v. 96, p. 123–136.
Terzaghi, K., 1943, Theoretical soil mechanics: New York, John Wiley, 510 p.
Thompson, J. B., Jr., 1955, The thermodynamic basis for the mineral facies concept: American Journal of Science, v. 253, p. 65–103.
Wagner, G. A., Reimer, G. M., and Jäger, E., 1977, Cooling ages derived by apatite fission-track, mica Rb-Sr and K-Ar dating; The uplift and cooling history of the Central Alps: Istituto di Geologia e Mineralogia dell'Universita di Padova Memoire, v. 30, p. 1–27.
Wang, C.-Y., and Shi, Y.-L., 1984, On the thermal structure of subduction complexes; A preliminary study: Journal of Geophysical Research, v. 89, p. 7709–7718.
Warneke, L. A., and Ernst, W. G., 1984, Progressive Cenozoic metamorphism of rocks cropping out along the southern east-west cross-island highway, Taiwan: Geological Society of China Memoir, no. 6, p. 105–132.
Wellman, H. W., 1979, An uplift map for the South Island of New Zealand, and a model for uplift of the Southern Alps: Royal Society of New Zealand Bulletin, no. 18, p. 13–20.
Westbrook, G. K., Smith, M. J., Peacock, J. H., and Poulter, M. J., 1982, Extensive underthrusting of undeformed sediment beneath the accretionary complex of the Lesser Antilles subduction zone: Nature, v. 300, p. 625–628.
Wu, F. T., Yeh, Y. H., and Tsai, Y. B., 1979, Seismicity in the Tsengwen reservoir area: Seismological Society of America Bulletin, v. 69, p. 1783–1796.
Zhao, W.-L., Davis, D., Dahlen, F. A., and Suppe, J., 1986, The origin of convex accretionary wedges; Evidence from Barbados: Journal of Geophysical Research, v. 91, p. 10246–10258.
Zeitler, P. K., Johnson, N. M., Naeser, C. W., and Tahirkheli, R.A.K., 1982a, Fission-track evidence for Quaternary uplift of the Nanga Parbat region, Pakistan: Nature, v. 298, p. 255–257.
Zeitler, P. K., Tahirkheli, R.A.K., Naeser, C. W., and Johnson, N. M., 1982b, Unroofing history of a suture zone in the Himalaya of Pakistan by means of fission-track annealing ages: Earth and Planetary Science Letters, v. 57, p. 227–240.

MANUSCRIPT ACCEPTED BY THE SOCIETY JUNE 1, 1987

Geological Society of America
Special Paper 218
1988

Some simple physical aspects of the support, structure, and evolution of mountain belts

Peter Molnar and Hélène Lyon-Caen
Department of Earth, Atmospheric, and Planetary Sciences, Massachusetts Institute of Technology, Cambridge, Massachusetts 01239

ABSTRACT

We review separately aspects of two types of forces that resist mountain building and therefore that affect the support, deep structure, and evolution of mountain ranges, using observations from the large-scale tectonics of Asia and the Andes to illustrate them. The first such force might be termed mechanical strength. In its simplest description, the lithosphere is flexed as an elastic plate under the weight of a mountain range thrust on top of the lithosphere. The gross shapes of foredeep basins and gravity anomalies over them show that the simple analyses, in terms of elastic plates over inviscid fluids, are reasonable first approximations. At the same time, the weights of some mountain ranges are inadequate to depress lithospheric plates to the depths of neighboring foredeep basins, and the weights of others would create deeper foredeep basins than observed if additional forces were not present. Thus, the strength of the lithosphere alone does not support all ranges. The strength of the lithosphere, however, does affect the geometry of major, deeply rooted thrust faults, which seem to behave as crustal-scale ramp-overthrust faults. Seismic activity seems to be concentrated on the steeper, deeper sections, whereas slip on the flatter planes parallel to the underthrusting basement seems to occur aseismically. The second force that resists mountain building is gravity; the forces that drive two plates together and that cause crustal thickening must do work against gravity. More gravitational potential energy is stored in a column of mass that includes a high range and thick crustal root than is in a column that is in isostatic equilibrium with the mountain belt but with thinner crust. Because of the increasing amount of work that must be done against gravity acting on an increasingly higher range, the range should reach a maximum mean elevation related to the force at which the plates are pushed together. In this sense, the mean elevations of high plateaus serve as crude pressure gauges for the average compressive stresses pushing on the margins of the plateaus. When the maximum elevations are reached, crustal shortening need not cease; convergence can continue as the range builds outward, growing laterally into a high plateau. Moreover, changes, possibly small ones, in the driving forces can lead to situations in which the crests of high ranges or high plateaus can undergo crustal extension while crustal shortening continues on the flanks of the range. From the simple analogy with the pressure gauge, we can use the simultaneous occurrence of crustal extension at high altitudes with crustal shortening on the flanks of the range to place crude limits on the average strength of the lithosphere at mountain ranges. Finally, we discuss aspects of the tectonic evolution of western North America that appear to be analogous with aspects of the active tectonics of the Himalaya, the Tibetan plateau, and the Tien Shan in Asia and the Andes.

PREFACE

Many years ago, the first author of this chapter (PM) heard John Rodgers lecture at Boston University on the mountain belts of Soviet Asia. Rodgers stated that he had begun his study with the supposition that all mountain belts are basically similar, and that his favorite one—the southern Appalachians—would serve as the model for the structure and evolution of these Asian mountain belts. Rodgers' talk, however, concluded with a resigned acceptance that essentially all mountain ranges evolve differently from one another and that no mountain belt should serve as the model for others.

On that occasion, the one of us that heard the talk thought to himself that those conclusions might, in part, be a consequence of the difficulties that geologists have in seeing more deeply than the upper few kilometers of the earth, whether they be 2km of subaerially deposited sedimentary rocks or 2 km of deeply exhumed, high-grade crystalline rocks. This individual pursued the idea that when one looked below the top few kilometers in order to study the processes controlling the large-scale aspects of mountain building, the structure and tectonics of different belts might appear more similar to one another than they do at the surface. Indeed this proved to be the case; at that time our ignorance was sufficiently vast that only a few clear differences in the deep structure and active tectonics of the various active belts could be recognized. Ascribing them to different plate tectonic scenarios was easy, but accounting for these few differences in terms of well-understood physical processes and properties was largely speculation.

As our knowledge has grown, however, some such differences in active belts can not only be recognized, but they can also be described and interpreted quantitatively, or at least in terms of simple physics and mathematics.* Thus, we are now, in 1985, trying to go beyond the understanding John Rodgers had years ago by also exploiting differences among various active mountain belts. Active belts can differ from one another because of different physical processes and material properties and because of different stages in their evolutions. Accordingly, current understanding of how various belts are supported and how deformation occurs not only can offer clues about how these belts have evolved, but also can provide "snapshots" of structures and processes analogous to those that existed and occurred in older ranges at different stages in their evolutions.

This chapter is a brief summary of some of the preliminary steps taken toward placing constraints on the deep structure and active tectonics of some mountain ranges and brief discussions of very simple aspects of some of the different physical processes and properties that affect orogenic belts. Rather than being a general review, this paper focuses on our collaborative work and on work by Molnar and Paul Tapponnier, now at the Université de Paris. As such, it attempts to develop some thoughts that underlie our published work but that have heretofore been discussed only briefly. Moreover, in certain respects this work is postdated by some studies, such as those of Dahlen (1984), Dahlen and others (1984), Davis and others (1983), England and McKenzie (1982), England and Houseman (1985, 1986), Houseman and England (1986), Fleitout and Froidevaux (1982, 1983), and Houseman and others (1981), which have already, independently, gone beyond some of the concepts examined here. Nevertheless, this chapter describes—in more basic terms than those used in the papers cited above—how some of the various physical processes, or how differences in some of the relevant parameters, can affect the structure or evolution of mountain belts.

INTRODUCTION

Virtually all major mountain ranges in the world are a consequence of crustal shortening. Because most of the constituent rocks undergo neglible or no change in volume, crustal shortening requires crustal thickening, at least within linear mountain belts. Isostatic compensation of the thickened crust then buoys up a mountain range. Consequently, to form mountain ranges, in general, horizontal forces must be applied to masses of crust and mantle, to lithospheric plates, to drive them together and to cause crustal shortening and crustal thickening. In these very broad respects, most mountain ranges should be similar to one another.

One can imagine, however, that differences among the structures and evolutions of mountain belts arise for several reasons, such as different strengths of materials, which might result from differences in composition, temperature, or strain rates; differing amounts and causes of regional isostatic compensation; and different mechanisms by which forces are applied to the converging plates. Moreover, all of these possible controlling factors can change with time and can simply vary along or across a given orogenic belt. Thus, a little imagination can uncover numerous possibilities for a wide variety of structural styles and mechanisms for support of mountain ranges, as well as for variations in these styles and support mechanisms during the evolution of a belt. Correspondingly, we probably will gain more by recognizing that such differences are likely than by trying to discover a simple formula that describes the evolution of all orogenic belts. In this chapter, we attempt to isolate some of the simple physical aspects of mountain building that allow quantification of some of the differences in various belts.

We begin with a discussion of the support of mountain ranges that formed by underthrusting of continental lithosphere beneath such ranges. It concerns how an elastic plate can be used, with limitations, to analyze the structure of the belt and how differences in the support of different ranges can be quantified. Then one of the geometric consequences of such underthrusting—the "ramping" of crystalline terrains onto the strong plate, and its implications for uplift rates and for the exposed geology in cross sections of some belts—is considered. Ignoring the strength

*The word "quantitative" is reserved for those processes, properties, or measurements to which meaningful and useful uncertainties can be assigned.

of the lithosphere, we then consider the gravitational potential energy created by the formation of mountain belts, and which later can be a source of energy expended in driving the neighboring regions apart. In this context we note how the elevation of a high plateau can be used as crude measures of both the horizontal forces applied to the margins of the plateau and the average regional compressive stress. We use this concept to argue that the tectonic style in a particular region depends less on variations of a "regional horizontal stress field" than on variations in the topography, crustal thickness, and the vertical normal stress. We then discuss how variations in the style of deformation across a mountain range or plateau can be used to place a constraint on the average strength of the lithosphere. Finally, we end by using these concepts to describe how the tectonic evolution of a mountain belt is likely to change with time and how different portions are likely to evolve differently from one another.

FOREDEEP BASINS AND FLEXURAL SUPPORT OF MOUNTAIN RANGES

Most active mountain ranges are flanked by neighboring basins. Some ranges, such as the Himalaya, abut against foredeep basins over which the ranges are being thrust with a well-defined polarity (Fig. 1). In others, such as the Venezuelan Andes or the Tien Shan, basins lie adjacent to and underthrust both sides of the ranges (Fig. 2). Finally, while subduction of oceanic lithosphere occurs at a rate of as much as 100 mm/yr beneath some continental margins, such as beneath the Andes, both a foredeep basin and its underlying basement on the continent side of the range underthrust the range with a sense opposite to that of the subduction zone and at a much lower rate (Fig. 3). The depths of these basins increase gradually, and in most cases quite smoothly, toward the mountain ranges so that the maximum depths are very close to the ranges. In turn, the basins are generally filled with material derived from the adjacent ranges. Thus, in both space and time, the association of such foredeep basins with the neighboring ranges is inescapable and virtually ensures an origin that couples them.

For us, gravity anomalies over mountain ranges and foredeep basins not only provide an important clue for relating the two features but also allow variations in both the deep structure and the support of different belts to be quantified. Gravity anomalies generally show that mountain belts and adjacent basins are not in local isostatic equilibrium. Foredeep basins are characterized by negative isostatic anomalies of 100 mgal or more and therefore by large deficits in mass (Fig. 4). In contrast, isostatic anomalies over the mountain ranges are generally positive and often exceed 100 mgal; thus they are underlain by excess mass. The juxtaposition of excess mass beneath the mountain ranges next to basins, which are deficient in mass and over which the mountains are being thrust, can be explained by various mechanisms of regional compensation in which a strong layer underlies both the basin and at least part of the range. This layer distributes the weight of the mountains over a broad area that includes the basin (e.g., Gunn, 1943; Karner and Watts, 1983; Lyon-Caen and Molnar, 1983). The gravity anomalies and the shape of the basin can then be used to place constraints on parameters that describe the strength of the underlying strong plate and its response to loads placed on it.

The model most commonly used for analyzing regional compensation employs an elastic plate overlying an inviscid fluid (e.g., Gunn, 1943; Hanks, 1971; Turcotte and Schubert, 1982; Walcott, 1970). The strength of the plate is defined by one parameter, the flexural rigidity:

$$D = ET^3/12\ (1 - \gamma^2),$$

where E is Young's modulus, T is the thickness of such an elastic plate, and γ equals Poisson's ratio (≈0.25). Both the shape of the flexed plate and the gravity anomalies depend on the flexural rigidity, on density contrasts among the crust, mantle, and material above the plate (which can be crust, sediment, water, or air), and on the amount and distribution of the load that flexes the plate down. One reason for using such a model is that few parameters need be specified, many fewer than are usually specified arbitrarily in block models used to fit gravity anomalies.

The shape of a flexed elastic plate immersed between two fluids of different densities $\Delta\rho$ is proportional to:

$$\cos x/\lambda\ e^{-x}/\lambda,$$

where $\lambda = \sqrt[4]{4D/\Delta\rho g}$ and g is gravity. Thus the characteristic wavelength λ is proportional to the 3/4 power of the thickness of the plate (Fig. 5). Thin plates will give rise to disturbances with short wavelengths, and we might expect thin plates to be associated with narrow foredeep basins and thick plates with wide basins (Fig. 5). For the simple case of a semi-infinite plate with a line load (force per unit length) F_o on the end, the maximum deflection is $2F_o/\Delta\rho g\lambda$. Clearly, the deflection is more substantial for larger loads than for small ones, but it is also greater for thin plates than for thick ones (Fig. 5).

Usually, we begin by assuming that the weight of the mountain range provides the load on such a plate, but the plate can be bent by other mechanisms. For instance, a slab of oceanic lithosphere could provide an additional load that creates a deeper basin than would be expected from the weight of the mountains alone (Fig. 6; Royden and Karner, 1984). Other buried loads are also possible (Karner and Watts, 1983). In addition, a bending moment can be applied to the end of the plate, and this bending moment can flex the plate up or down, depending on its sign (Fig. 6). We suspect that the Indian crust may have been detached from the mantle portion of Indian lithosphere to form the Himalaya. Gravity acting on this slab of mantle lithosphere may apply a bending moment to the intact part of the Indian plate that not only flexes it up but also contributes to support the Himalaya (Lyon-Caen and Molnar, 1983, 1985).

More realistic models that parameterize the rheology of the plate and the underlying material by yield criteria, temperature

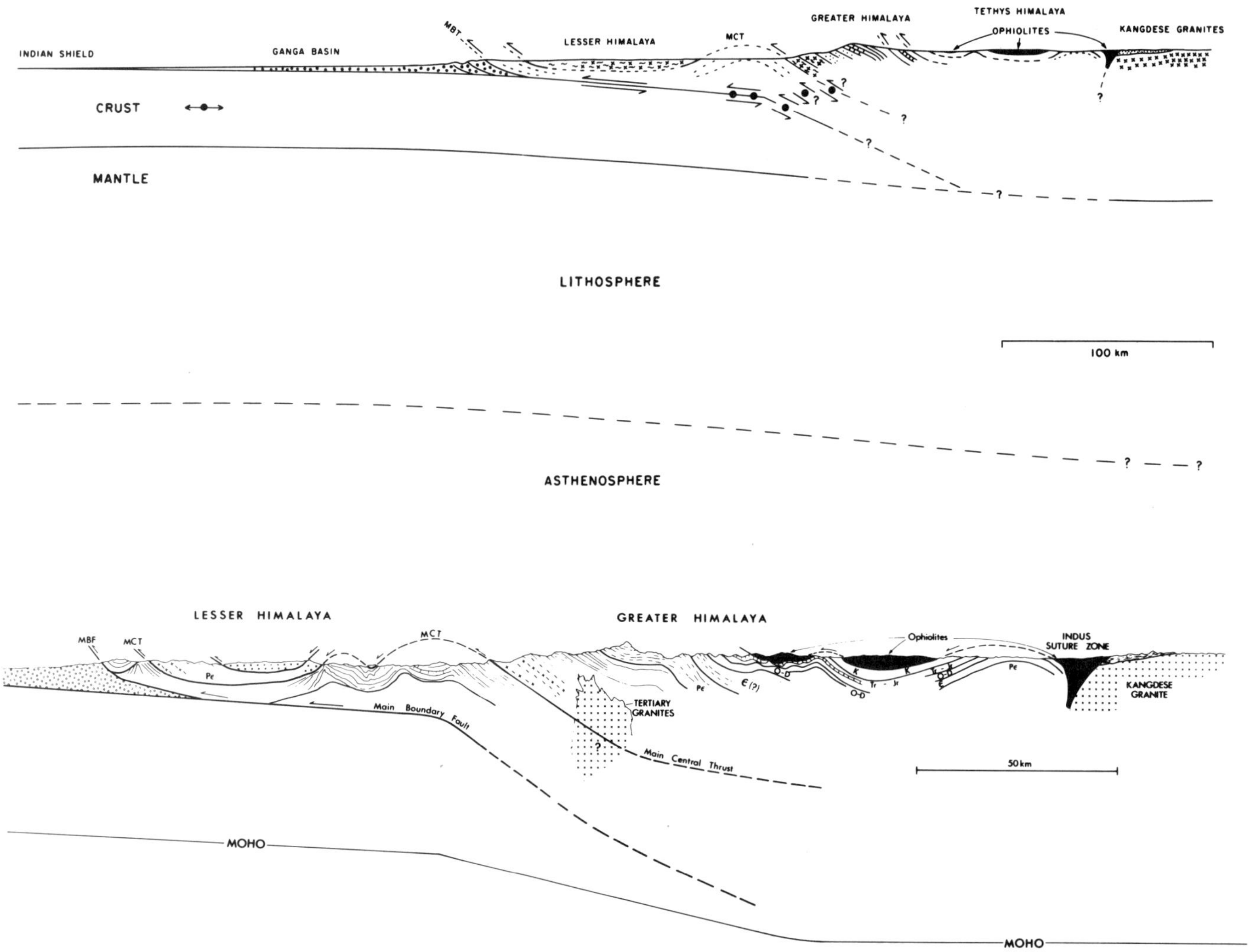

Figure 1. Schematic cross sections across the Himalaya, from south (left) to north (right), adapted from Gansser's (1964) cross section through the Kumaon Himalaya. Top cross section shows the Ganga basin to be the result of a downward flexure of the Indian shield, which underthrusts beneath the Himalaya. Black dots indicate hypocenters; arrows show sense of slip on thrust faults. For one event beneath the Ganga basin, arrows indicate the orientations of extension; this event presumably results from stretching of the upper part of the flexed Indian plate. Lower cross section shows an expanded view of the Himalaya (from Molnar and others, 1983). Top 3 km are taken from Gansser's (1964) cross section across the Kumaon Himalaya. The lower parts are redrawn, in part, using results from an analysis of gravity anomalies by Lyon-Caen and Molnar (1983) from the Mount Everest region; that study showed that the Moho dips gently at a few degrees beneath the Lesser Himalaya. As Seeber and others (1981) inferred, and as fault-plane solutions and depths of foci of earthquakes imply, the Indian plate is shown as a coherent plate beneath the Lesser Himalaya. The gravity anomalies from the Mount Everest region, however, require that the dip of the Moho to steepen to about 15° beneath the Greater Himalaya in that area, presumably because the Indian plate is weaker and can bend more there than it can beneath the Lesser Himalaya (Lyon-Caen and Molnar, 1983). The Indian plate is shown here with a sliver of crust having been detached from the rest of the lithosphere when slip began on the Main Boundary Fault (MBF). After some 100 to 150 km of slip on that fault, the Lesser Himalaya has ramped up onto the Indian plate, causing the steepening of the now-presumed–inactive Main Central Thrust (MCT) and rapid erosion of the portion that has already ramped on to the coherent portion of the Indian plate. In this scheme, developed in more detail by Lyon-Caen and Molnar (1983), uplift and erosion of the Greater Himalaya where the MCT crops out would be very recent (less than 10 m.y.—and maybe only a couple of million years). The possibility that granites in the Greater Himalaya are derived from below the MCT is shown as questionable; this idea is discussed in Molnar and others (1983), but has not been inferred from geologic mapping.

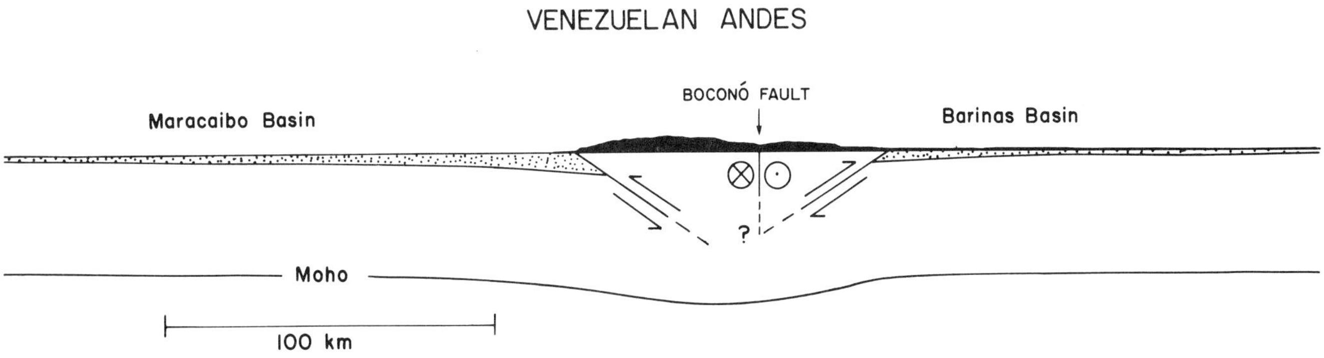

Figure 2. Simplified northwest (left)–southeast (right) cross section across the Venezuelan Andes. Both the Maracaibo and Barinas basins seem to be the result of flexures of their underlying basements, both of which underthrust the Venezuelan Andes. The greater depth of the Maracaibo basin might be a consequence of a lower flexural rigidity of the plate beneath it than that of the Guyana shield beneath the Barinas basin. Right-lateral strike slip on the Boconó fault probably accommodates some of the convergence, which may not be perpendicular to the range.

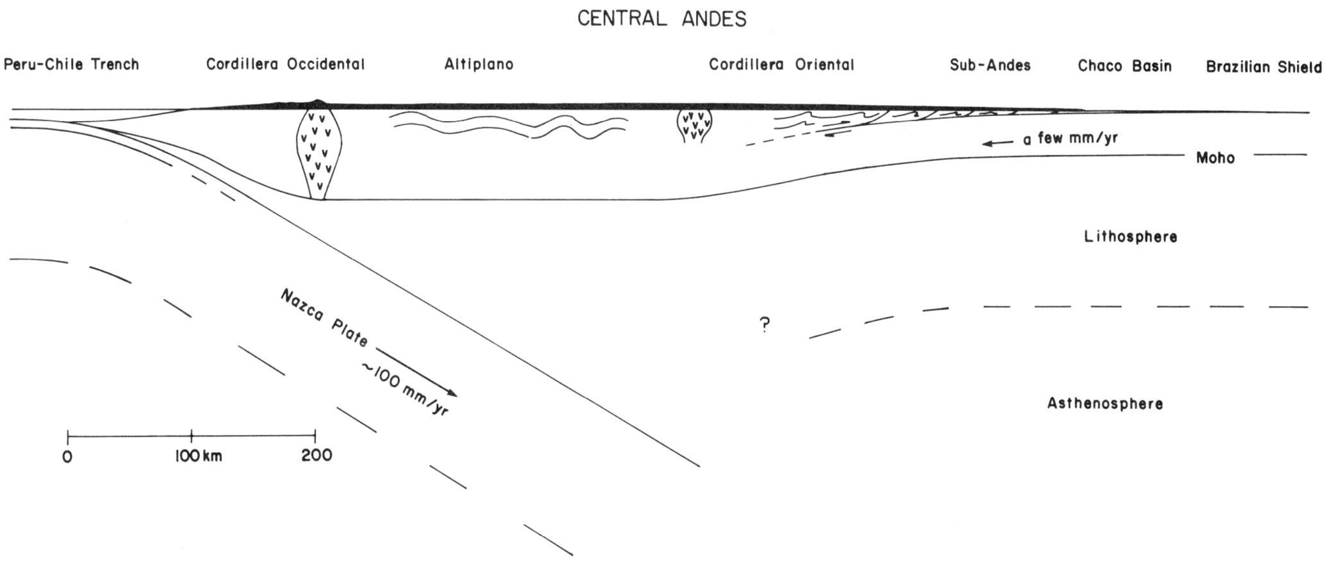

Figure 3. Simplified east (right)–west (left) cross section across Atliplano through Bolivia. Rapid eastward subduction of the Nazca plate is associated with offshore seismicity and with volcanism in the Cordillera Occidental. Simultaneously, much slower underthrusting of the Brazilian shield beneath the Andes has caused folding of the Paleozoic, Mesozoic, and Cenozoic sedimentary cover. Minor, probably anatectic, intrusions crop out in parts of the Cordillera Oriental. Between the Cordilleras Oriental and Occidental is the Altiplano, a broad flat plain buried in upper Cenozoic sediment through which older, folded sedimentary rocks crop out.

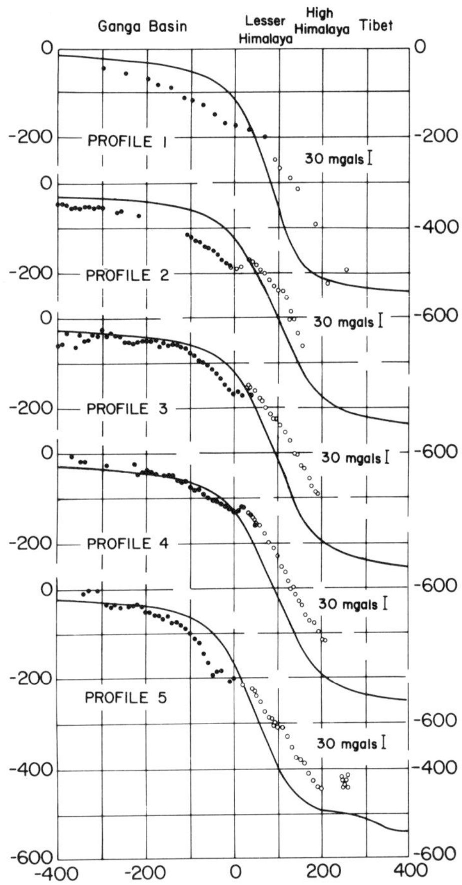

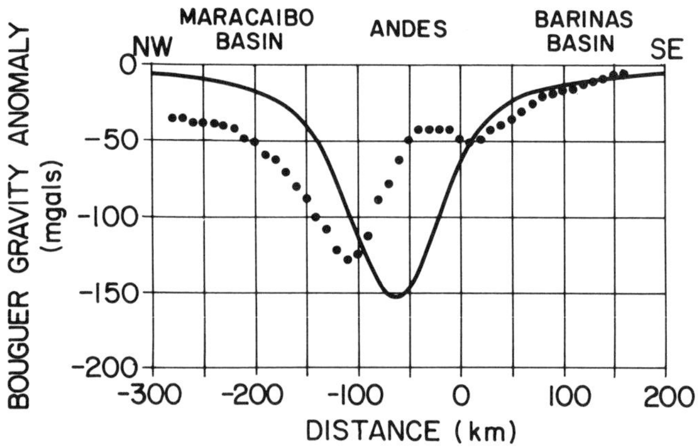

Figure 4. Comparisons of measured Bouguer gravity anomalies and those calculated assuming local isostatic compensation. (a) Five profiles across the Himalaya (from Lyon-Caen and Molnar, 1985). Closed circles represent points taken from the listing compiled by the Defense Mapping Agency; open circles represent points from Kono (1974) for profile 1 and from Das and others (1979) for the other profiles. Note the more negative observed values over the Ganga basin, which reflect a deficit of mass there, and the less negative observed values over the Lesser Himalaya, which reflect an excess mass there. (b) One profile across the Venezuelan Andes. Data are from Bowin (1976), Folinsbee (1972), and Bonini and others (1977), and calculations are by R. Benites and others (in preparation). As for the Himalaya, note the more negative observed values over the basins and the less negative values over the mountains. Again, the flexed plates beneath the basins are filled with low-density sediments, while the strength of the plates redistributes the support of the mountains.

dependent, nonlinear creep laws, and assumed geotherms show that flow laws deduced from laboratory experiments of ductile creep can be applied to real geophysical problems (e.g., Goetze and Evans, 1979). Indeed, profiles of topography across oceanic trenches and outer rises cannot be matched well by the flexure of elastic plates of constant thickness; the inclusion of a finite yield stress, at the least, seems to be required to match the relatively steep slopes near the bottoms of the trenches (Bodine and Watts, 1979; Chapple and Forsyth, 1979; McAdoo and others, 1978; McNutt and Menard, 1982). To the best of our knowledge, however, neither gravity measurements nor cross-sectional profiles of the basement across foreland basins are sufficiently precisely known to require rheologies more sophisticated than elastic plates. Thus, although a better model for analyzing regional compensation of mountain ranges than an elastic plate over an inviscid fluid is surely forthcoming, it is not yet obvious how more can be learned about mountain building by using such "more realistic" models.

Analyses of gravity anomalies and profiles of foredeep basins yield constraints both on flexural rigidities of equivalent elastic plates and on the forces that deflect them. Consequently, such analyses can be considered useful and worthwhile in the sense that data fit the calculations and that, in general, plausible flexural rigidities are deduced. Moreover, some insight is gained about what forces flex the plate down and about how the ranges are supported (see Benites and others, in preparation; Karner and Watts, 1983; Lyon-Caen and Molnar, 1983, 1984, 1985; Lyon-Caen and others, 1985; Royden and Karner, 1984; Sheffels and McNutt, 1986).

It can be justifiably argued, however, that such models are inadequate for two reasons:

First, the variations in flexural rigidities are larger than what would be expected if they were largely a measure of different geotherms. Since the flexural rigidity is proportional to the cube of the equivalent elastic plate thickness, if the equivalent elastic plate were defined as that part of the lithosphere colder than some temperature, then it might be expected that the flexural rigidity would be approximately inversely proportional to the cube root of the geothermal gradient. Differences of 10^3 times in estimated flexural rigidities for plates beneath India (Karner and Watts, 1983; Lyon-Caen and Molnar, 1983, 1985) and the Adriatic Sea (Royden and Karner, 1984) would imply differences in geothermal gradients of 10 times. Similarly, variations of a factor of 10 in the flexural rigidity within the Indian plate along the Himalaya

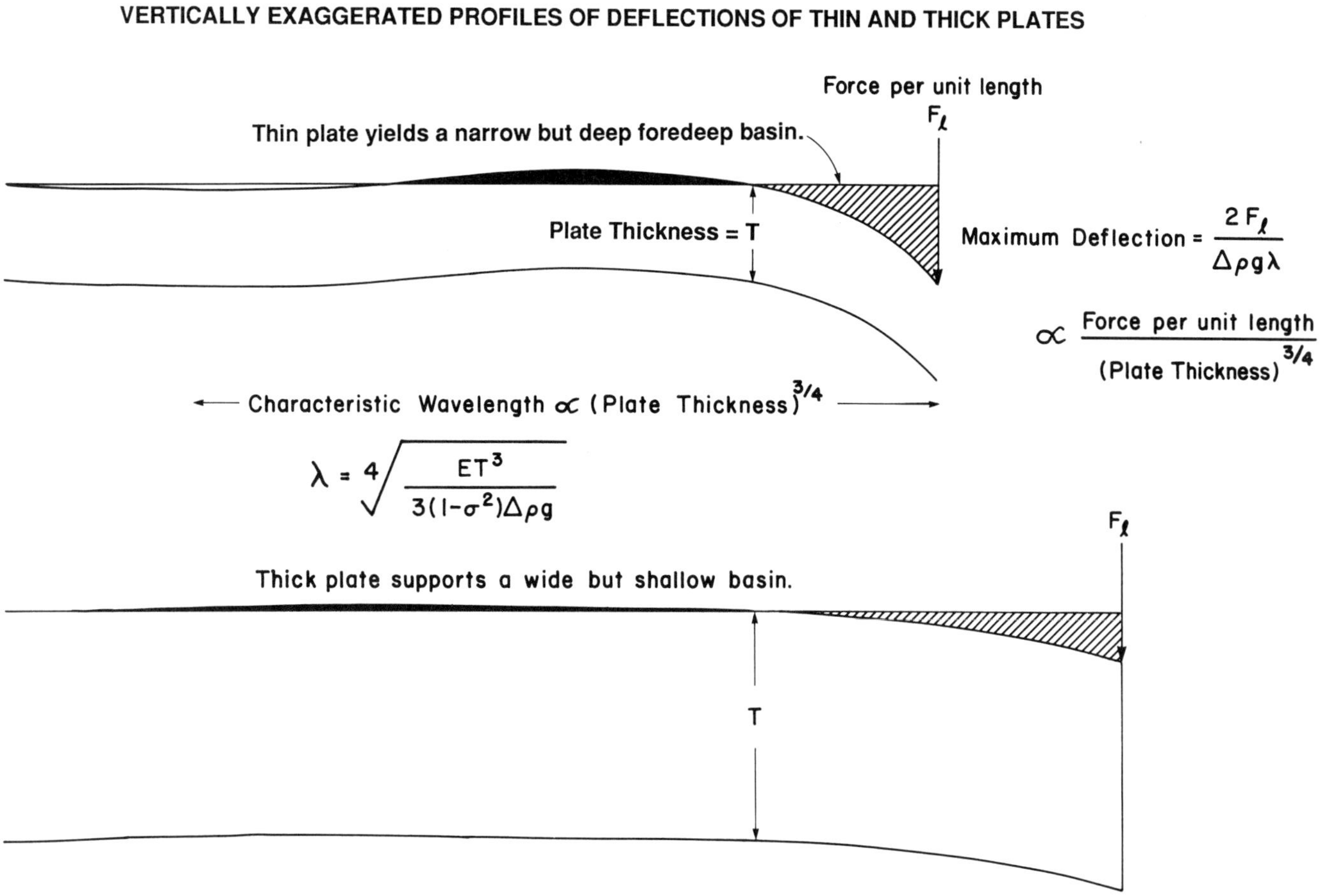

Figure 5. Schematic cross sections of semi-infinite plates flexed by a line load (force per unit length) at one end of the plate. The deflection $w(x)$ as a function of horozintal position x (increasing to the left) is given by:

$$w(x) = \frac{2\,F_l}{\Delta\rho g\lambda}\,e^{-x/\lambda}\cos x/\lambda\,,$$

where F_l is the load, $\Delta\rho$ is the difference in density between the underlying and overlying fluid, g is gravity, and λ is the flexural parameter defined in the figure. T is the thickness of the equivalent elastic plate, and E is Young's modulus. Note that a thick, and therefore strong, plate distributes the load over broader region than a thin plate. Note also the vertical exaggeration of the order of 10 times.

(Lyon-Caen and Molnar, 1985) would imply geothermal gradients differing by more than two times. These variations in geothermal gradients seem to us to be unlikely, and therefore they suggest that the simple elastic plate model may not be adequate, or at least that the flexural rigidity depends on temperature in a more complicated way than simply the cube root of the gradient (e.g., Lyon-Caen and Molnar, 1985).

Second, the forces needed to bend the plates down bear little relation to the cross-sectional areas, or the masses per unit length, of the different mountain ranges. In some belts, such as the Himalaya (Lyon-Caen and Molnar, 1983, 1985) or the Venezuelan Andes (Benites and others, in preparation), additional upward forces acting on the flexed elastic plates are necessary to support the ranges. In others, such as the Alps, Appennines, and Carpathians, the weights of mountain ranges appear to be too small to flex the plates down to create foredeep basins as deep as those observed; some other force must help pull them down (Fig. 6; Karner and Watts, 1983; Royden and Karner, 1984). The deduction that additional forces are needed does not invalidate the simple flexural models, but it does imply that some unknown (or poorly defined) process must be introduced, almost as a *deus ex machina,* to explain the observed average elevations of the ranges and the measured depths of basement in the basins.

Thus, regardless of whether it is reasonable to treat the lithosphere underthrusting a mountain range as elastic, the strength of the lithosphere may not depend simply on the geo-

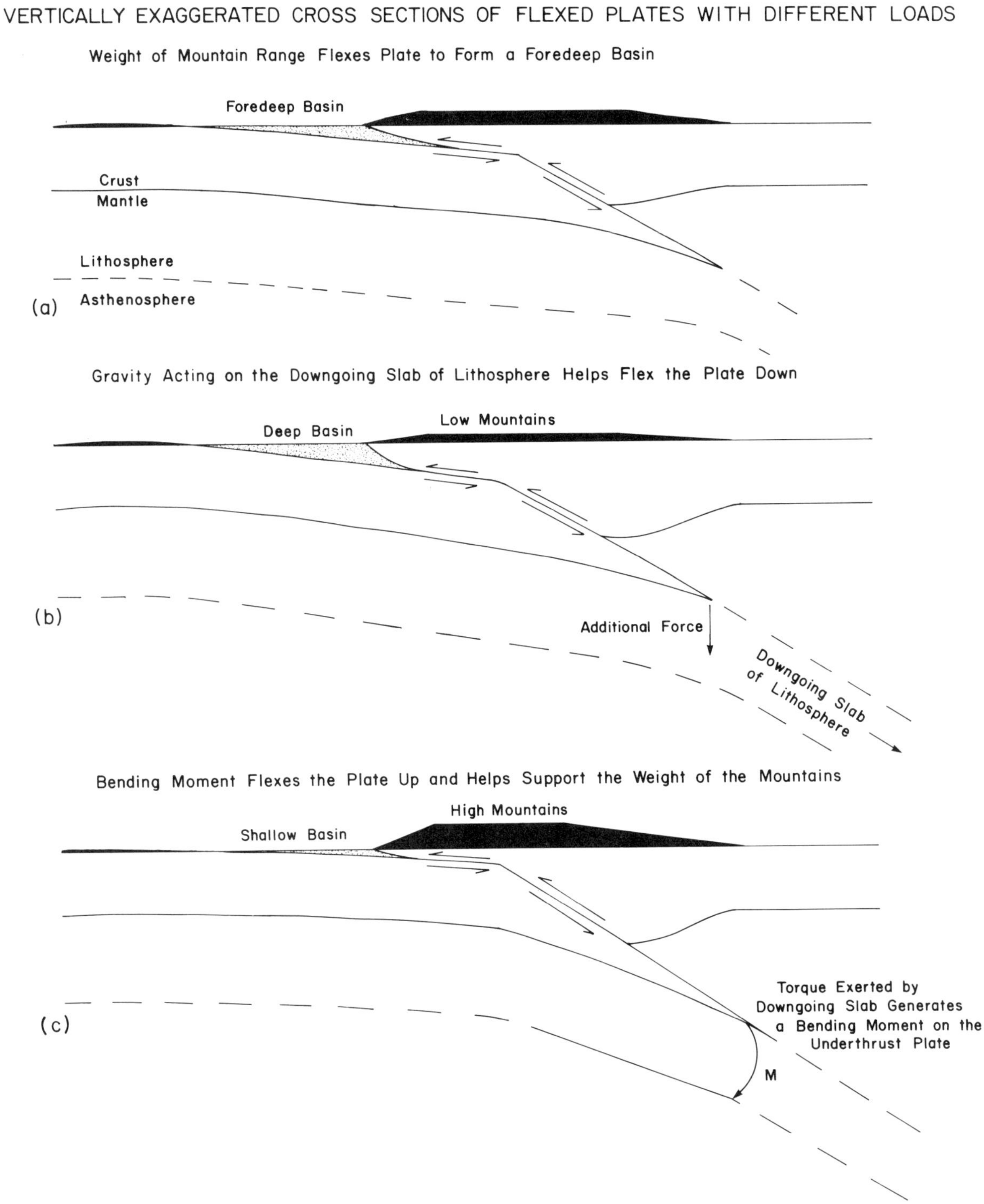

Figure 6. Response of underthrust plates to different forces. (a) The weight of the mountains exerts a force on the underthrust lithosphere that causes a foredeep basin. In the absence of other forces, the depth of the basin and the average height of the mountains should be proportional, with the proportionality constant, depending on the flexural parameter (see Fig. 5). (b) A downgoing slab of oceanic lithosphere (or continental lithosphere) can contribute an additional force that pulls the plate down. In such a case, the basin should be deeper (or the mountains lower) than when no buried load exists. (c) The distribution of mass in the downgoing slab and the mechanism by which it exerts a force on the rest of the lithosphere can have an additional affect on depth of the basin and height of the mountains. If the downgoing slab exerts a bending moment on the surficial lithosphere, it can flex the lithosphere up. In such a case, the basin will be shallower and the mountains higher than for the case in (a).

therm, and it appears that forces other than gravity acting on the mountain ranges and the buoyant response of the asthenosphere are necessary both to support such ranges and in some cases to cause foredeep basins. Nevertheless, in a qualitative sense, analyses of mountain ranges and foredeep basins in terms of flexed plates call attention to variations in the strengths of the underthrust lithosphere and in the forces acting on such plates. Even if we do not understand fully how various ranges are supported, we should expect higher ranges over stronger plates. The flexural rigidity of India beneath the Himalaya is indeed larger than that of the European platform beneath the Alps, which reach maximum elevations approximately half of those of the Himalaya. We might also expect deep basins to form where the underthrust continental lithosphere is still attached to a cold slab of subducted oceanic lithosphere (Fig. 6). The occurrence of intermediate-depth earthquakes beneath the Carpathian and Calabrian arcs suggests that oceanic lithosphere was subducted beneath these arcs in the last 10 to 15 m.y.; gravity acting on the excess mass in these slabs of oceanic lithosphere may provide the force responsible for the deep basins adjacent to the Carpathian and Appennine chains (Royden and Karner, 1984). Thus, despite some failings in the simple elastic plate models, they allow us to recognize and quantify differences among ranges that may have formed by different processes. Specifically, although the Himalaya, Alps, Appennines, and Carpathians all owe their origins to the overthrusting of the belt toward and over neighboring foredeep basins, both the deep structures and the forces supporting these ranges differ from one another in ways that can be quantified, if only crudely.

GEOMETRY OF UNDERTHRUST PLATES AND THRUST ZONES

Oceanic lithosphere bends down and plunges beneath island arcs so that virtually all oceanic lithosphere is subducted coherently. Fault plane solutions of shallow-focus earthquakes (depths of less than 70 km) show thrust faulting of island arcs over the oceanic plates on planes dipping at angles of only a few degrees to (rarely) more than 20°. The same is not observed for intracontinental deformation in most regions. The only major exception appears to be the Himalaya, where fault plane solutions and focal depths of some earthquakes imply coherent underthrusting of the Indian plate beneath at least the Lesser Himalaya (Baranowski and others, 1984; Molnar and Chen, 1982; Ni and Barazangi, 1984), and where slip during great earthquakes probably occurs on very gently dipping planes (Seeber and others, 1981).

The active deformation of most intracontinental mountain ranges is much more complex than that at island arcs. In most belts, and for some earthquakes in the Himalaya as well, fault plane solutions show thrust faulting on planes with the rather steep dips of 30° to 60°, for instance on the east side of the Andes (e.g., Chinn and Isacks, 1983; Stauder, 1975; Suárez and others, 1983), in the Zagros (e.g., Jackson and Fitch, 1981; Jackson and McKenzie, 1984; Kadinsky-Cade and Barazangi, 1982), or in the Tien Shan (e.g., Ni, 1978; Tapponnier and Molnar, 1979). Focal depths of these earthquakes are usually greater than 10 km and therefore assure that most such events have occurred in the basement, not the sedimentary cover. Thus, whereas the surface geology of many orogenic belts reveals widespread décollement, thin sedimentary nappes, and sometimes thin crystalline nappes transported on gently dipping thrust faults, the evidence from earthquakes in active belts gives little direct evidence for these processes (except in the Himalaya); see Molnar and Chen (1982) for a longer summary. At first glance, these fault plane solutions seem to reveal a style of deformation not often revealed by geologic work at the surface, if nevertheless inferred for the Rocky Mountains of Colorado, Utah, and Wyoming, areas not characterized by thin-skin deformation (e.g., Berg, 1962; Gries, 1983; Smithson and others, 1979).

The focal depths and fault plane solutions of shallow earthquakes on the east side of the Andes and those with relatively steep nodal planes in the Himalaya may offer a clue to how these two different styles of deformation are related. Shallow earthquakes in the Andes are concentrated on the eastern margin, beneath the sub-Andes (Fig. 7). The sub-Andes are characterized by late Tertiary and Quaternary folding and thrust faulting (e.g., Dalmayrac and others, 1980; Martinez, 1981; Mégard, 1978, 1984). Anticlines expose both pre-Tertiary and Tertiary sedimentary rocks; in some areas, detachment of sedimentary rocks from the basement is required by subsurface data (Allmendinger and others, 1983; Jordan and others, 1983; Mingramm and others, 1979). Fault plane solutions of earthquakes beneath the sub-Andes reveal thrust faulting on planes dipping 30° to 50° west (Chinn and Isacks, 1983; Stauder, 1975; Suárez and others, 1983), and most focal depths are between 15 and 25 km (Fig. 7). Epicenters of those earthquakes studied lie close to the western edge of the sub-Andes, generally not beneath the high part of the range, but also not near the active eastern edge of the sub-Andes (Fig. 8). Thus, these earthquakes, which occur deep within the basement, seem to occur as the range is thrust up onto the shield to the east along relatively steep planes (e.g., Suárez and others, 1983). As this fault zone reaches the surface, it probably then flattens, and slip within the sedimentary cover occurs aseismically (Fig. 9), an inference also drawn by Mégard (1984). Insofar as the present deformation of the sub-Andes and Cordillera Oriental is analogous with that occurring in the Canadian Rockies 100 m.y. ago, the sub-Andes would correspond to the region east of the Purcell anticlinorium, and the earthquakes would occur on the steep faults that Price (1981) draws dipping west beneath the western edge of that anticlinorium (Figs. 9 and 10) (see also Mattauer and others, 1983).

Fault plane solutions of some earthquakes in the Himalaya indicate thrust faulting on planes dipping about 25° northeast (e.g., Baranowski and others, 1984; Ni and Barazangi, 1984). They occur 80 to 110 km northeast of the front of the range and at depths of 15 to 20 km; they may indicate a steepening of the fault plane along which India underthrusts the Himalaya (Fig. 1b). Slip on such a fault would cause the area near the Greater

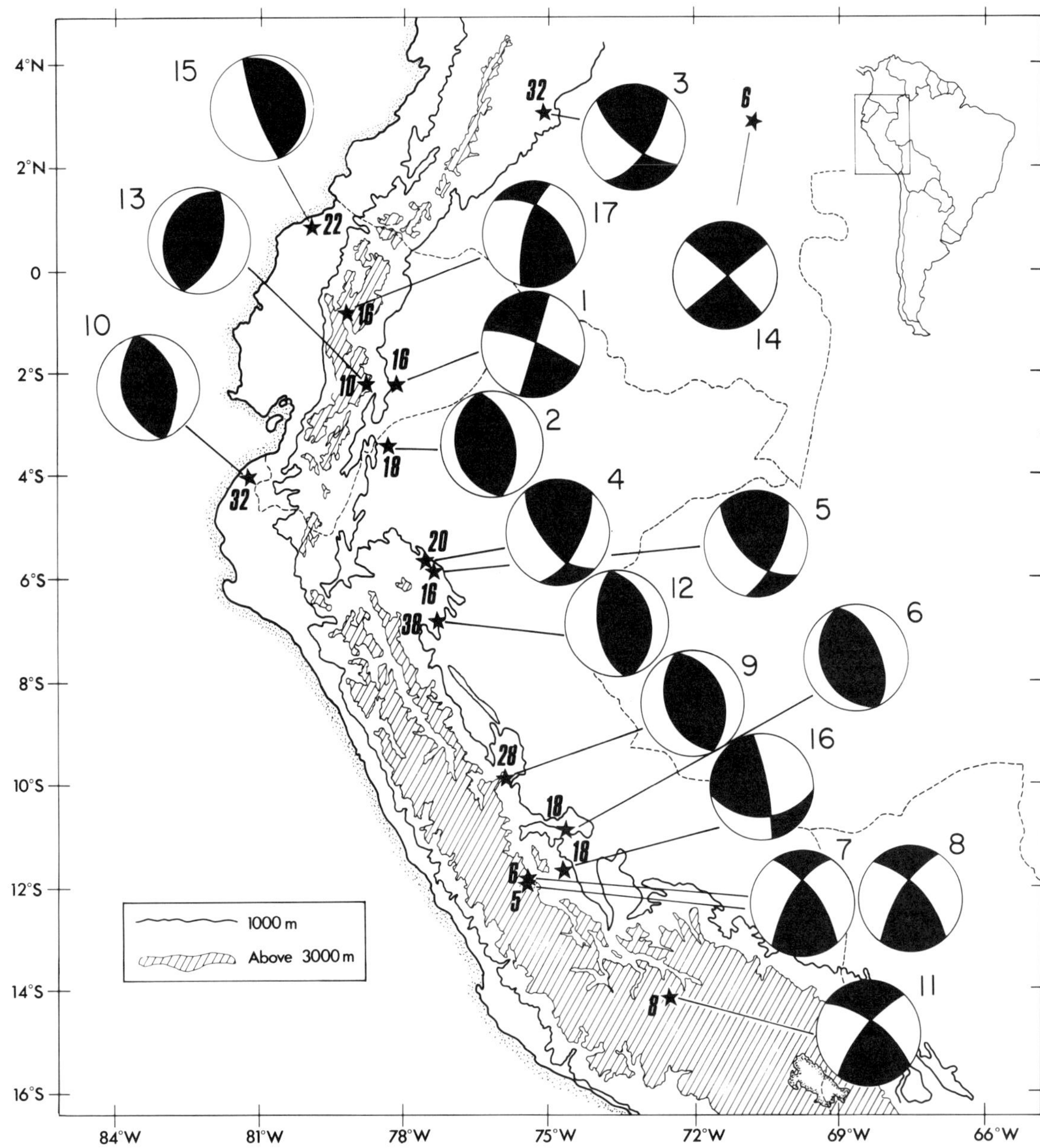

Figure 7. Map of the Peruvian Andes, summarizing fault plane solutions and focal depths (from Suárez and others, 1983). Shaded areas in lower hemispheric projections of focal sphere indicate quadrants with compressional first motion. Epicenters are plotted as stars; numbers beside them indicate estimated focal depths in kilometers. Numbers next to balloons correspond to those in Suárez and others (1983). Note that most of these earthquakes occurred where elevations are about 1,000 m. Thus they did not occur on the margin of the sub-Andes, and therefore on the eastern edge of the range, but rather west of the active fold and thrust belt. Note also that the events occurred in the basement.

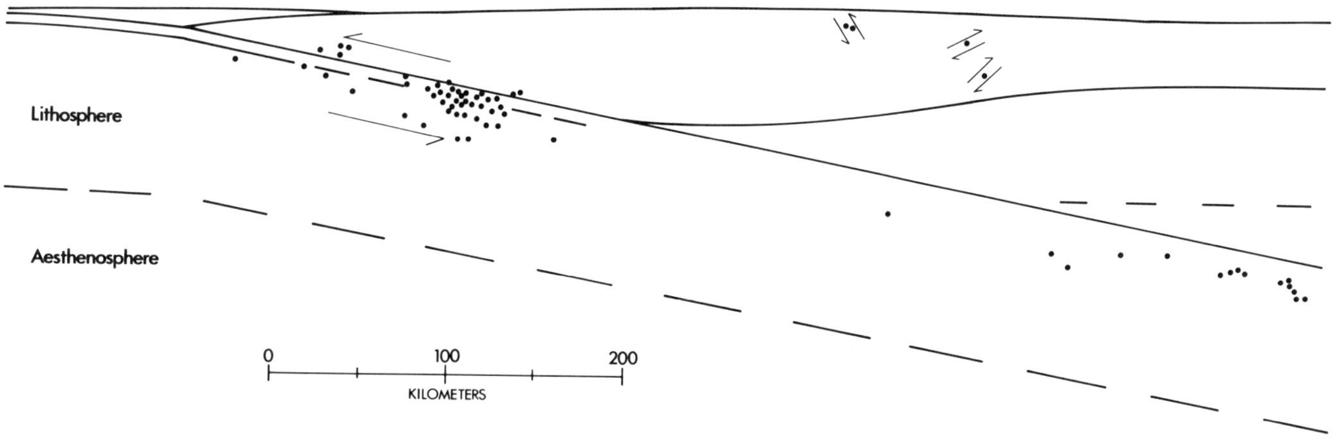

Figure 8. Cross section for seismicity through central Peru (from Suárez and others, 1983). The dip of the slab is inferred from the hypocenters of earthquakes selected by Barazangi and Isacks (1979) in their cross section C–C[1]. West-dipping nodal planes of two of the earthquakes studied here (6 and 16) from sub-Andes are projected onto the cross section (see Fig. 7). The fault plane of the Pariahuanca earthquakes (7 and 8) in the high Andes in known to dip east, as observed in the field (Philip and Mégard, 1977).

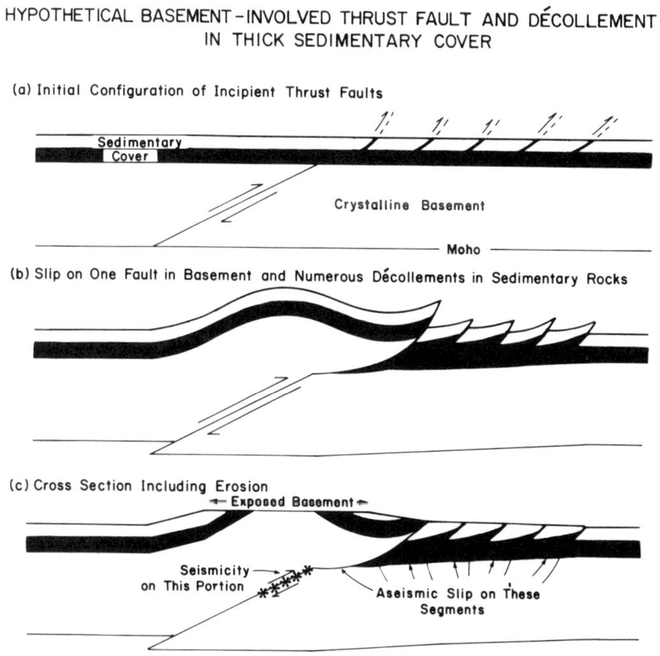

Figure 9. Hypothetical scenario relating steep faulting in basement and décollement on gently dipping thrust faults in the sedimentary cover: (a) Future thrust fault in the crystalline basement and several future detachment surfaces in the overlying sedimentary cover. (b) Balanced cross section after slip on the main fault. (c) Section b after erosion, with indication of where earthquakes might be expected.

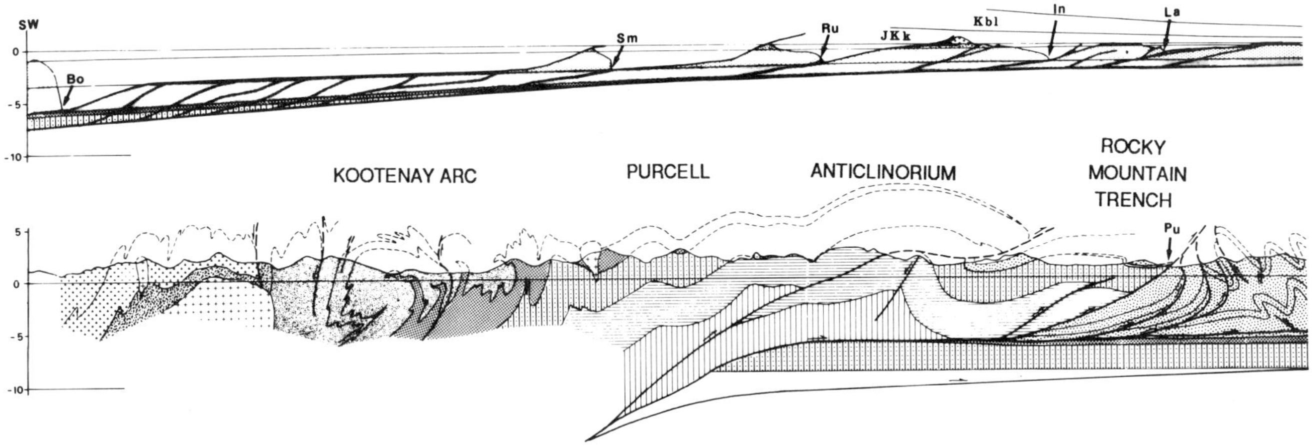

Figure 10. Structural cross section through the Cordilleran foreland thrust and fold belt in southern Canada (from Price, 1981). Symbols identifying equivalent points on the hanging-wall side of the more important faults in each structure section are listed. Bo = Bourgeau; BR = Bull River; BT = Burnt Timber; Bz = Brazeau; Co = Coleman; HL = Hall Lake; In = Inglismaldie; La = Lac des Arcs; Le = Lewis; Li = Livingstone; Mc = McConnell; Mi = Mill Creek; OB = Old Baldy; RMT = Rocky Mountain Trench; Ru

Himalaya to ramp up onto the intact part of the Indian plate (e.g., Lyon-Caen and Molnar, 1983; Molnar, 1984). The vertical component of slip would be much greater on this steep part of the fault than where India slides beneath the Lesser Himalaya along a very gentle plane (Fig. 11). Thus both the abruptly rising wall of mountains that characterizes the southern slope of the Greater Himalaya and the exposures of the high grade of metamorphic rocks there may be a consequence of localized rapid uplift on this steeper segment of the underlying Main Boundary Fault.

Thus, we suspect that, in most areas, seismicity is concentrated on the deeper and steeper portions of major thrust faults that curve and connect with shallower, more gently dipping segments. Molasse deposits are detached from their basement along these gentle segments, and already detached sedimentary rocks and crystalline nappes are carried along them over foreland basins (Fig. 9). Presumably, slip occurs aseismically, or nearly so, on the gently dipping segments of these faults (except in the Himalaya).

The absence (or limited extent) of such nappes in the Rocky Mountains of Colorado, Utah, and Wyoming might be due to a combination of the following factors: (1) a small amount of displacement on the reverse faults, and hence a small load on the underthrust plate; (2) the presence of strong lithosphere that might have prohibited deep foreland basins from forming; and (3) the absence of a thick sequence of sedimentary rock within which detachment could occur (Fig. 12). For such types of regions, the steep reverse faults revealed by fault plane solutions might break the earth's surface instead of flattening in the sedimentary cover. Reverse faults do reach the surface both on the edges of and within the Tien Shan, which appears to be a modern analogue of the Colorado-Wyoming Rocky Mountains (Molnar and Chen, 1982; Tapponnier and Molnar, 1979). Norin (1941) mapped steep thrust faults on the southern edge of the range, many of which are clearly expressed on Landsat imagery (Tapponnier and Molnar, 1979). Surface faulting with a reverse component was also associated with the 1911 Kebin earthquake (Bogdanovich and others, 1914), and possibly with the 1906 Malasi earthquake (Bei and Fang, 1981; see also Molnar and Deng, 1984). We have not yet studied the Tien Shan sufficiently to know if any of the possible reasons suggested above for why nappes are not prevalent in the Colorado-Wyoming Rockies might also apply to the Tien Shan. In any case, however, the existence of the steep faults traced on a deep reflection profile beneath the Wind River Range (Smithson and others, 1979) came as no surprise in light of the fault plane solutions of earthquakes in the Tien Shan.

POTENTIAL ENERGY CREATED BY MOUNTAIN BUILDING AND CRUSTAL THICKENING

When crustal shortening occurs, the forces driving the material together and pushing one block of crust over another must do work against resisting stresses. Because crustal shortening also causes crustal thickening, these forces must also do work against gravity. Thus the formation of both mountains and crustal roots is associated with an increase in gravitational potential energy, and part of the work done by the forces that drive the plates together creates that potential energy—the next focus of this chapter.

Potential energy is defined as the line integral of a force causing displacement integrated along the component of displacement:

$$\int \vec{F} \cdot \vec{ds} .$$

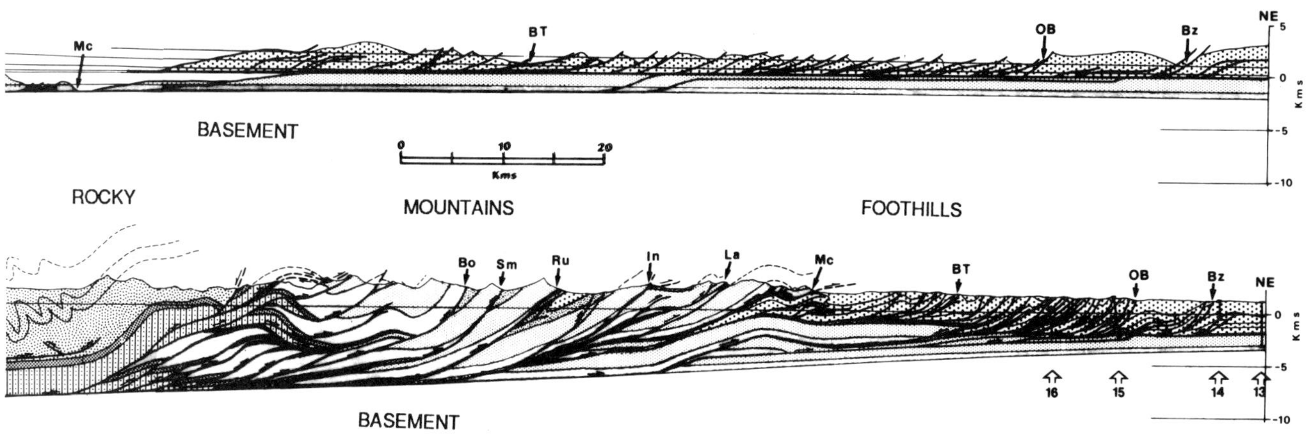

= Rundle; Sm = Sulphur Mountain; Wa = Watson; Wh = Whaleback. Locations of wells drilled for oil and gas and from which logs were used in the preparation of the structure sections are shown by heavy lines and identified by number. When active, seismicity may have been most significant on the deepest and most steeply dipping fault beneath the western edge of the Purcell anticlinorium.

Consider the potential energy stored in a column of material of unit cross-sectional area, which is the potential energy per unit surface area of the earth. The vertical force per unit area is the product of gravity, density, and thickness of the overlying layer. Work is done against such a force in the vertical direction. Consequently, the potential energy per unit area of a column of material above a given depth can be calculated as the integral of the lithostatic pressure from the earth's surface to that depth. In lowlands, lithostatic pressure $P_L(z)$ is given by:

$$P_L(z) = \rho_c g z \qquad 0 \leqslant z \leqslant H_o$$
$$P_L(z) = \rho_c g H_o + \rho_m g(z - H_o) \qquad H_o \leqslant z$$

where ρ_c and ρ_m are the densities of the crust and mantle, assumed constant here, and H_o is the crustal thickness. In a mountainous region, with an elevation of h and a crustal root extending a distance ΔH beneath the normal depth of the Moho, H_o, the lithostatic pressure, P_M, is given by:

$$P_M = \rho_c g(z + h) \qquad -h \leqslant z \leqslant H_o + \Delta H$$
$$P_M = \rho_c g(h + H_o + \Delta H) + \rho_m g(z - H_o - \Delta H) \qquad H_o + \Delta H \leqslant z.$$

The value of the potential energy per unit area depends on the reference level chosen (here, the earth's surface) and the depth to which we integrate. We are interested, however, not in this potential energy, but in the change due to crustal thickening or in the difference between that beneath mountains and that beneath lowlands. Recall that isostatic compensation implies that the lithostatic pressure at a depth $z = H_o + \Delta H$ is constant, $P_M(H_o + \Delta H) = P_L(H_o + \Delta H)$. Therefore:

$$\Delta\rho\Delta H = \rho_c h, \text{ where } \Delta\rho = \rho_m - \rho_c. \qquad (1)$$

Thus the difference in potential energy per unit area beneath mountainous regions from that beneath lowlands is:

$$\int_{-h}^{\infty} [P_M(z) - P_L(z)]dz = \int_{-h}^{H_o + \Delta H} [P_M(z) - P_L(z)]dz.$$

Since $P_M(z) = P_L(z)$ for $z \geqslant H + \Delta H$, there is no need to integrate to greater depths into the mantle, and there is no dependence on the reference level chosen. The resulting difference is:

$$\rho_c g h[h/2 + H_o + \Delta H/2] = 1/2\rho_c g h^2 + \rho_c g h H_o + 1/2\Delta\rho g \Delta H^2. \qquad (2)$$

All this analysis ignores strain energy, which would not be negligible, for instance, if vertical shear stresses supported a substantial fraction of the load of the mountains. Strain energy clearly exists where a strong plate redistributes the support of a range, or where dynamically induced (viscous) stresses contribute to the support of mountain ranges. In such cases, part of the work done creating the mountain ranges is either stored as elastic strain energy or is dissipated viscously in the deforming rocks. Thus, the gravitational potential energy created during crustal thickening is only a lower bound on the work done by the driving forces causing crustal shortening. Probably it most closely approximates the work done in cases where both elastic stresses and creep strengths are small, such as where high, broad plateaus are underlain by weak substrata.

For our discussion here, which ignores elastic plates and viscous stresses, consider only the first and third terms in Equation 2, those proportional to h^2 and ΔH^2 or $h\Delta H$. Two important conclusions may be drawn.

First, the third term is much larger than the first; the ratio is $\Delta H/h = \rho_c/\Delta\rho$. Consequently, if compensation occurs because

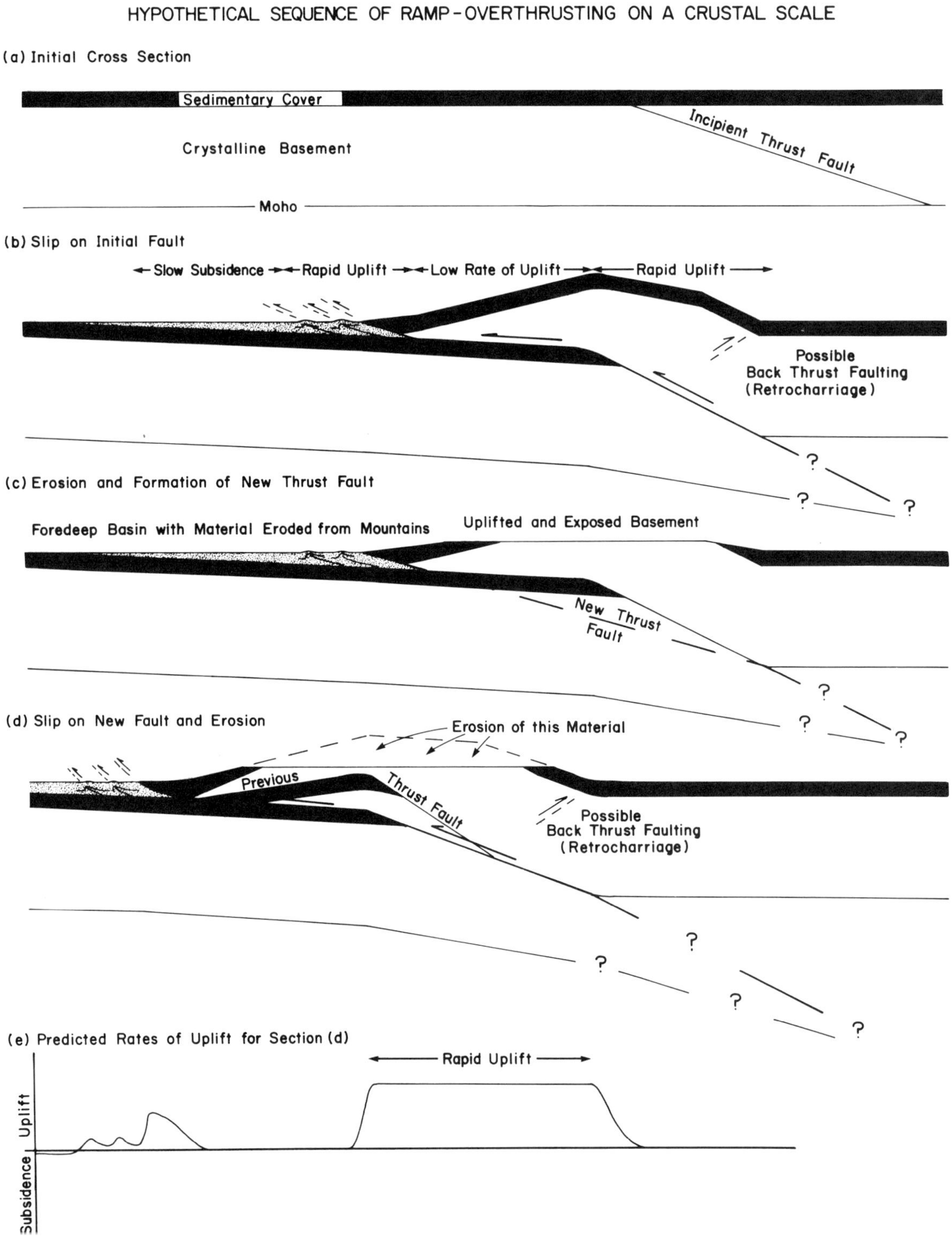

Figure 11. Sketches of a series of migrating large-scale ramp overthrust faults. (a) Simple cross section of basement and overlying sedimentary cover. (b) Balanced cross section after 30 to 50 km of slip. Slip on the initial fault lifts one block up and onto the other. An underthrust block is flexed down to form a foredeep basin. Note regions of varying rates of uplift. No erosion is included. (c) Profile b after erosion reduces elevations to a few kilometers at most. A new fault is shown forming. (d) Balanced cross sections after another 30 to 50 km of slip and erosion. Structures associated with slip on the earlier fault are rotated, and basement from the lower crust is brought near the surface. This process presumably continues with the initiation and slip on later faults. (e) Profile of rates of uplift. Maximum rates occur above steep segments of active faults; no significant uplift may occur where faults dip at gentle angles.

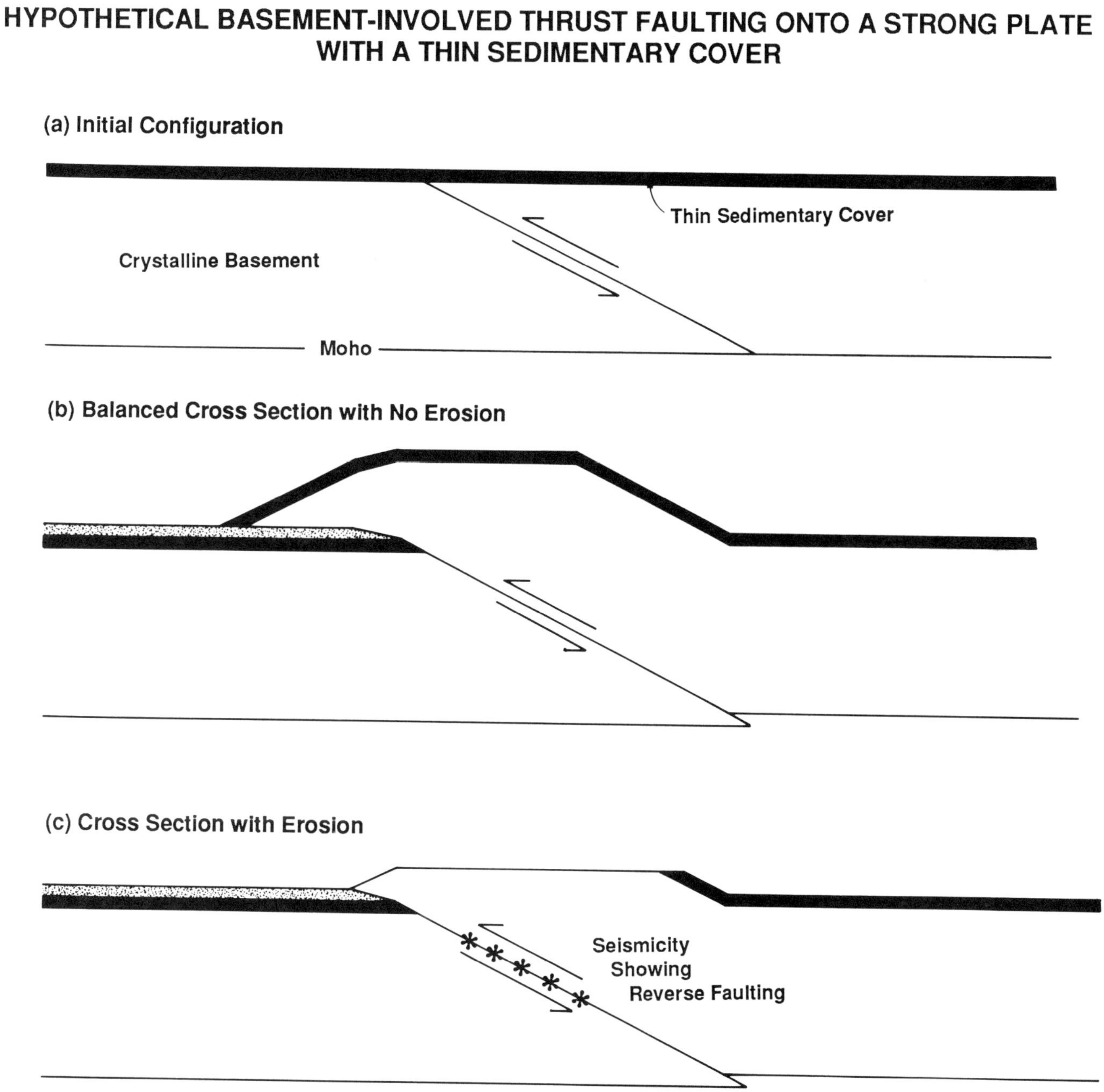

Figure 12. Sequence of cross sections similar to those in Figure 9 but for a strong plate overlain by a thin sedimentary cover. (a) Initial configuration. (b) Balanced cross section ignoring erosion. The strong plate is flexed down slightly so that wide shallow basin forms. Décollement cannot occur on a large scale in the thin sedimentary cover. (c) Section b, but with erosion. Basement is ramped onto the underthrust plate so that the dip of the faults is steep, and the contact between basement and the young deposits in the basin is clear (compare with Fig. 9). Seismicity shows faulting on planes with dips similar to those observed at the surface.

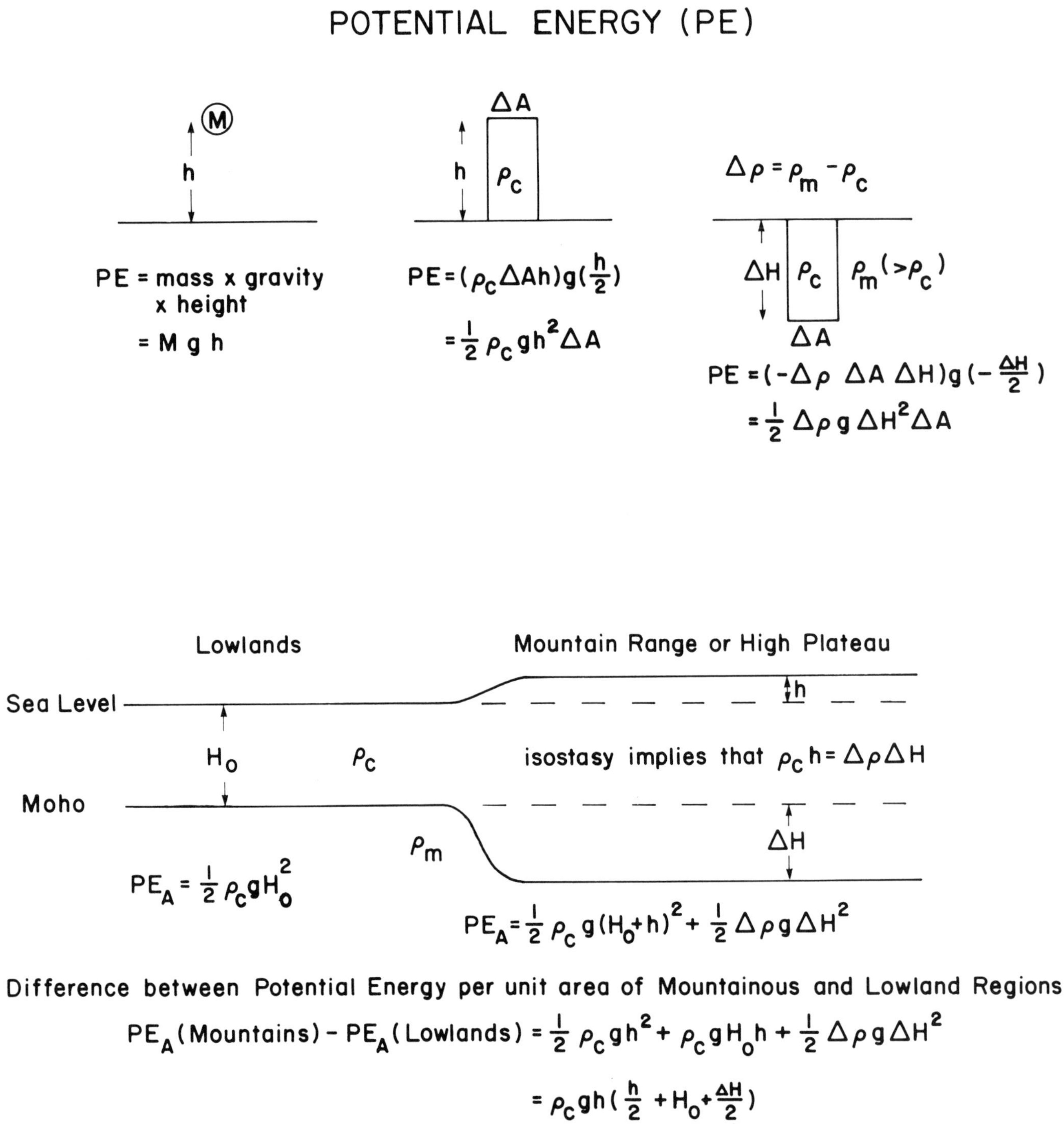

Figure 13. Gravitational potential energy for various situations. Top figure shows potential energy referenced to zero PE at the earth's surface for a mass M (left), for a block of density, ρ_c, cross sectional area ΔA and height h, and for a block of less dense mass ρ_c ($<\rho_m$), and with cross sectional area ΔA and submerged in the more dense material to a depth of ΔH. Bottom shows the difference in potential energy per unit area between areas of high mountains and a deep crustal root and lowland areas. Potential energy for each is calculated taking the depth to the Moho beneath lowlands as a reference and assigning no potential energy to the mantle beneath it. The potential energy in each column depends on which such depth is used, but as long as isostasy prevails, the difference in potential energy per unit area is independent of such choices.

of a very thick (100 km) layer of light mantle beneath the crust, the difference in potential energy beneath mountains and lowlands will be even more than for the case where a crustal root reaches only to a depth of 70 km or so and is underlain by mantle of normal density.

Second, the differences in potential energy per unit area expressed by both the first and third terms in Equation 2 are proportional to the squares of the elevation (h) and the excess crust in the root (ΔH). Thus, as a range or plateau is built, further uplift of it to a higher elevation and further deepening of the root each requires that increasingly more work be done by the driving forces. As the mountains grow higher and the root deepens, more energy is eventually needed to cause a small amount of additional crustal thickening than can be supplied by the energy source responsible for driving the plates together. This will be true regardless of the strengths of the materials or the frictional resistance to deformation. As a result, for a specified driving force, both the mean height of the range and the thickness of the crust should reach limiting values (Molnar and Tapponnier, 1978).

LIMITING ELEVATIONS OF MOUNTAIN RANGES AND PLATEAUS AND MIGRATION OF THRUST FAULTING

If eventually the forces driving the plates together can no longer supply the energy needed to elevate a high range or a plateau higher, the convergence will not necessarily stop. Instead, the mountain range is likely to grow laterally in width instead of increasing in height.

Consider a crustal layer initially of thickness H_o and displaced horizontally an amount Δx so that a mountain range of width w and height h forms, with a root extending a distance of ΔH below H_o (Fig. 14). The potential energy per unit length of range created by this horizontal displacement is:

$$\Delta PE_L = \rho_c\, g\, w\, h(h/2 + H_o + \Delta H/2).$$

Now consider two possible consequences of a further displacement Δx. The elevation could increase to $2h$, and the thickness of the root could increase to $2\Delta H$ (Fig. 14b), or the width of the range could increase to $2w$ (Fig. 14c). In the first case, the potential energy per unit length will be:

$$\Delta PE_L(2h) = 2\rho_c\, g\, w\, h\,[h + H_o + \Delta H].$$

In the second, it will be:

$$\Delta PE_L(2w) = 2\rho_c\, g\, w\, h\,[h/2 + H_2 + \Delta H/2].$$

The difference in potential energy per unit length of a range created by doubling its width from that created by doubling its height is:

$$\delta PE_L = \Delta PE_L(2h) - \Delta PE_L(2w) = \rho_c\, g\, w\, h\,(h + \Delta H) = \frac{\rho_c \rho_m}{\Delta\rho}\, gwh^2. \quad (3)$$

Clearly, more work must be done if the range increases in height than if it broadens. This arises from the linear dependence on w and the square dependence on h.

Thus we expect that, when a range reaches a limiting mean elevation, the zone of active thrust faulting—which presumably dips beneath the range—will migrate away from the edge of the range into the lowlands. Crustal shortening will occur where mountains are low and the thickness of the crust is normal. The uplift of lowland terrains and the creation of a shallow root will not require much work be done, at least until the elevations of such areas reach those of the neighboring highlands. By this mechanism, mountain ranges and high plateaus will tend to grow outward while the forces causing crustal shortening remain essentially constant (e.g., Molnar and others, 1977; Molnar and Tapponnier, 1978; Suárez and others, 1983). In terms of this simple system, only if the driving forces were to increase, would the mean elevation of the range be likely to increase. Note that if deformation occurs by power-law creep, so that strain rates are proportional to the differential stress raised to some power, then a reduction in strength, for instance by heating of the material, could enable the mountains to grow higher, if the stress and the driving force remained constant. Numerical experiments of Housman and England (1986) and England and Houseman (1986) illustrate this.

The migration of active thrust faulting from regions of higher elevation to lower ones is observed at nearly all scales, from snow plowed by snowplows, to accretionary prisms (e.g., Davis and others, 1983; Seeley and others, 1974), to thin-skinned fold and thrust belts, for instance in Canada (e.g., Bally and others, 1966; Price, 1981) or in the western United States (e.g., Armstrong and Oriel, 1956; Burchfiel and Davis, 1972, 1975), and to deep crustal levels such as in the Himalaya (e.g., Gansser, 1964; LeFort, 1975; Mattauer, 1975) or in the Andes. To see this for the Andes, compare the ages of deformation in the Cordillera Occidental (Vicente and others, 1979), the high plateaus (Mégard, 1978, 1984; Mégard and others, 1984; Noble and others, 1979, Wilson and others, 1967), and the Cordillera Oriental and sub-Andes (Mégard, 1978; Dalmayrac and others, 1980). At all of these scales, among the work done by the driving forces, the portion that is transformed into gravitational potential energy will be less when low areas are elevated instead of high ones. Consequently, there will be a tendency for thrust zones to cease being active when they must lift high mountains and for new thrust zones to form beneath regions where the elevations are low.

The strength of materials (or of the crust or the lithosphere as a whole) ignored in the discussion above, plays a key role in the migration of thrust faults. Were the earth liquid, devoid of strength, mountains would not form. Instead, the entire region between converging, snowplow-like blocks would rise uniformly. It follows that the simple migration of thrust faulting can be prevented, or at least modified, if the strength of the material in which new faults must form is large (see below). Recall, however, that one area where the thrust faulting does not show a simple

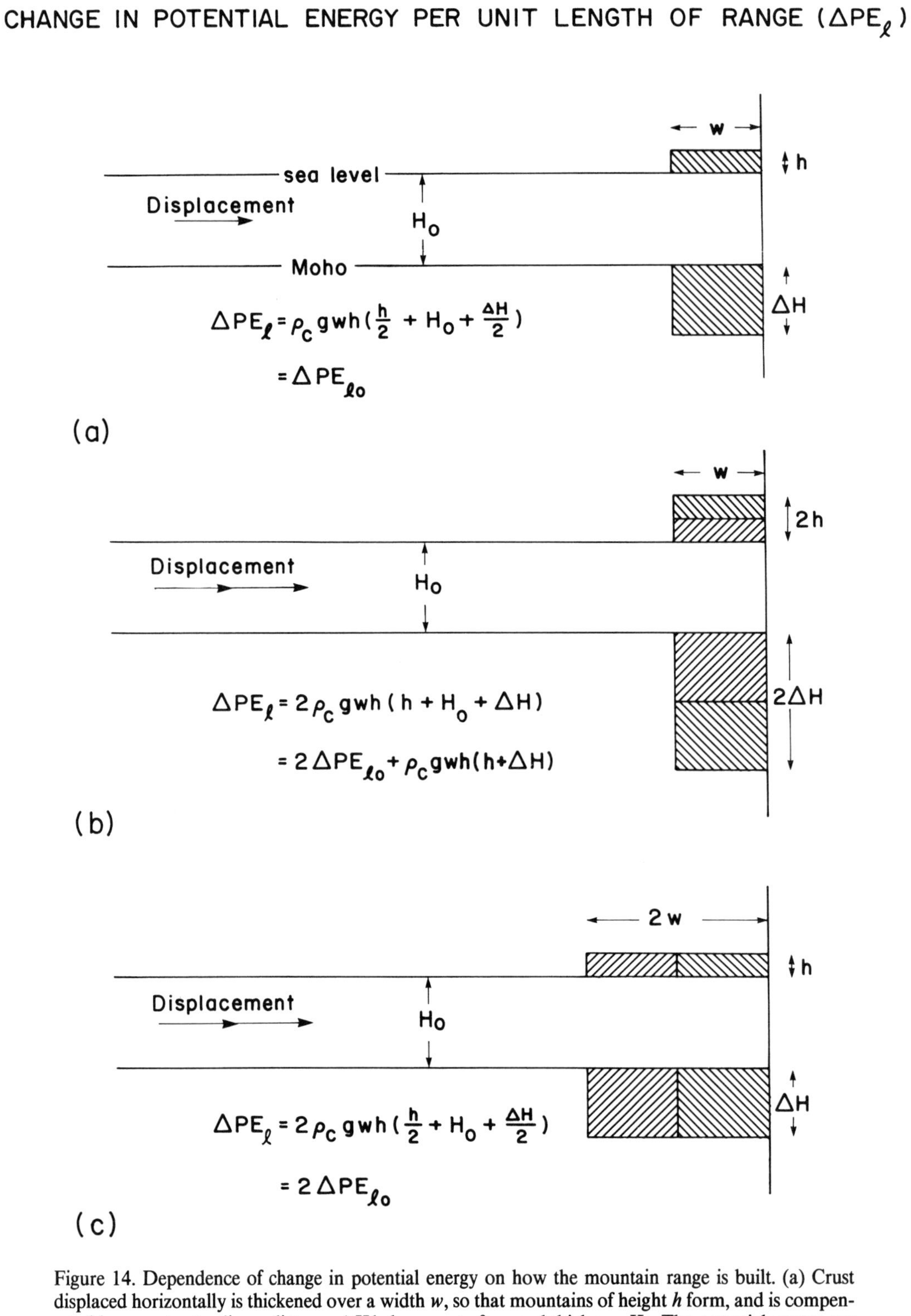

Figure 14. Dependence of change in potential energy on how the mountain range is built. (a) Crust displaced horizontally is thickened over a width w, so that mountains of height h form, and is compensated by a root extending a distance ΔH below crust of normal thickness H_0. The potential energy per unit length of range is the potential energy per unit area given in Figure 13 times the width w of the range. (b) Let the crustal shortening double so that the height and root both double. The change in potential energy per unit length increases to more than twice the value in a. (c) Alternatively, if the width of the range doubles but h and ΔH remain unchanged, then ΔPE_l merely doubles. Thus, mountain ranges cannot increase in height indefinitely, and the crust cannot continue to increase in thickness; eventually, elevating the range and thickening the crust will require more work than is necessary to break through the crust and to form new faults in lowlands.

progression of younger faults breaking undeformed terrain is in the southern part of the Sevier belt, in southern Nevada and California (Burchfiel and Davis, 1975). There, the thick sequence of sedimentary rocks, deposited on the old shelf of western North America and deformed elsewhere during the Sevier orogeny, is virtually absent. Instead, faults associated with the Sevier orogeny have sliced up the basement, which probably was stronger than the sedimentary cover. The work done against gravity in this area may have been much smaller than that done in creating these thrust faults.

EXTRACTION OF GRAVITATIONAL POTENTIAL ENERGY DURING CRUSTAL THINNING AND EXTENSION

The gravitational potential energy created by the formation of the mountain belts, high plateaus, and their roots is not lost and can be recovered; extraction of this energy by crustal thinning, and hence crustal extension, is associated with the divergence of neighboring regions of the belt, a phenomenon sometimes misleadingly called "gravity spreading."

Note that because more potential energy is created by the formation of a deeper than shallower root, a high plateau underlain by hot, light mantle material (asthenosphere) will be associated with more potential energy than one compensated solely by a thickened crust. As a result, if extension is accommodated by continued intrusion of hot asthenospheric material between diverging cold lithospheric plates, extension and crustal thinning can continue until the thickness of the crust is less than that of the initial reference state, H_o. In such a case, more potential energy eventually is extracted than was originally created by crustal thickening. The source of this additional energy is heat; when the mantle beneath the region of thinned crust cools and subsides below sea level, the thin crust and cold upper mantle will store less gravitational potential energy than does crust of normal thickness overlying mantle of normal temperature.

In any case, this section is meant to present in simple physical terms the well-known idea that when erosion is insufficiently rapidly elevated areas should be susceptible to crustal extension and thinning. Moreover, if hot mantle material contributes to the isostatic compensation, then this extension could be sustained longer than otherwise. The notoriously thin crust and hot mantle beneath the Basin and Range Province may exemplify this. As noted elsewhere (e.g., Coney and Harms, 1984; Molnar and Chen, 1983), the crustal thinning might be the response of a thickened crust to a decrease in the horizontal forces that caused this crustal thickening. The continued intrusion of hot asthenosphere, regardless of whether a hot spot (or mantle plume) underlies the region, might then sustain the extension. Note that the proclivity of mountain belts and crustal roots to collapse and extend might explain why ocean basins seem to have closed and reopened in the same localities, viz. the separation of the Atlantic along the Appalachian system (e.g., Houseman and others, 1981; Vink and others, 1984).

This discussion of gravitational potential energy has ignored dynamic considerations altogether, and specifically stress. Let us now address the same topic and also consider stress.

TRANSMISSION OF HORIZONTAL FORCES ACROSS MOUNTAIN RANGES

The existence of crustal extension across a high mountain range or plateau does not imply that the range is being pulled apart or that the neighboring lowlands of the range are moving apart. Crustal shortening occurs on both flanks of the Andes, but normal faulting and crustal extension are clearly active processes within some portions of the high Andes (e.g., Dalmayrac, 1974; Mercier, 1981; Suárez and others, 1983). Crustal shortening occurs on nearly all margins of Tibet, but the active tectonics of the high plateau are dominated by crustal extension in an east-west direction (e.g., Molnar and Tapponnier, 1975, 1978; Ni and York, 1978; Tapponnier and others, 1981). Considerable confusion exists in the literature both because displacement fields and "regional stresses" are often mistakenly interpreted as similar evidence and used interchangeably, and because lateral variations in crustal thickness or density can affect the orientations of principle stresses in a region. This section reviews a simple argument showing that, regardless of the style of deformation within a mountain range or on its flanks, the horizontal force oriented perpendicular to the range, per unit length parallel to the range, is likely to be (or at least can be) constant across the range (e.g., Dalmayrac and Molnar, 1981). Thus, whether underlain by weak or strong material, the range can transmit such a horizontal force across it.

Consider the cross section of a simple plate with variable topography that is isostatically compensated (Fig. 15). For two dimensions the equations of equilibrium reduce to:

$$\frac{\partial \tau_{zz}(x,z)}{\partial z} + \frac{\partial \tau_{xz}(x,z)}{\partial x} = \rho\,(x,z)g \tag{4}$$

$$\frac{\partial \tau_{xz}(x,z)}{\partial z} + \frac{\partial \tau_{xx}(x,z)}{\partial x} = O, \tag{5}$$

where τ_{zz} and τ_{xx} are the vertical and horizontal normal stresses, and τ_{xz} is the shear stress on vertical or horizontal planes. These equations simply state that the divergence of the stress tensor is equal to the body force per unit volume, which in our case is gravity acting on the density. Ignoring the small deviations from horizontality, the top surface, $-h(x)$, is free of stress parallel to the horizontal plane: $\tau_{zz}(x,-h) \approx \tau_{xz}\,(x,-h) \approx O$. Let us also start by assuming that the bottom of the plate, $z = z_o$, is free of shear stress: $\tau_{xz}(x,z_o) = O$. With these conditions, bounds can be placed on the vertically averaged stresses as a function of horizontal position.

Let us first integrate Equation 5 over depth from the top (at a height of $-h$) to the bottom of the plate (z_o):

$$\int_{-h}^{z_o} \frac{\partial \tau_{xx}(x,z)}{\partial x} dz + \int_{-h}^{z_o} \frac{\partial \tau_{xz}(x,z)}{\partial z} dz = O.$$

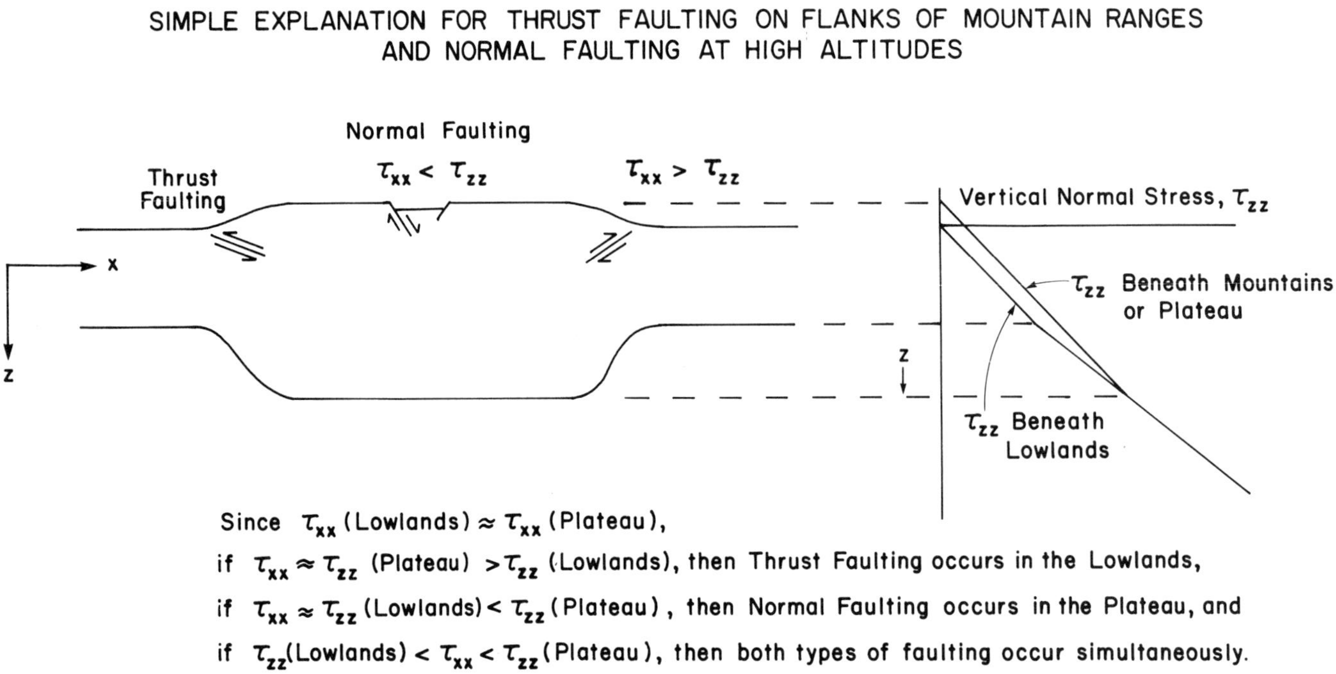

Figure 15. Simple explanation for juxtaposition of thrust and normal faulting on flanks of mountain ranges and beneath their high plateaus. If stresses applied to the base of the lithosphere are negligible, then average horizontal stress, τ_{xx} is constant, independent of horizontal position x. Vertical stress, however, depends markedly on elevation. The drawing on the right shows schematically τ_{zz} as a function of depth beneath mountains or plateaus and beneath lowlands. At all depths, τ_{zz} (plateau) $\geqslant \tau_{zz}$ (lowlands). Since average horizontal stress τ_{xx} does not depend on x, we might expect τ_{xx} to be less than τ_{zz} in the plateau but greater than τ_{zz} in the lowlands and on the flanks of the ranges. In such a case, normal faulting would prevail at high altitudes while thrust faulting occurred on the flanks of the range. Increases or decreases in τ_{xx} might suppress (or prevent) completely either the normal faulting at high altitudes or the thrust faulting on the flanks.

The second term is simply $\tau_{xz}(x,z_o) - \tau_{xz}(x,-h) = 0$, because both terms are zero. Rewriting the first term, assuming that $\tau_{xx}(x,z)$ is a well-behaved function, and ignoring the small variations in the topography, $h(x)$, compared with the thickness of the plate z_o, we have:

$$\frac{\partial}{\partial x} \int_{-h}^{z_o} \tau_{xx}(x,z)dz = 0.$$

Therefore, by the mean value theorem, the product of the average horizontal stress $\bar{\tau}_{xx}(x)$ and the plate thickness, $z_o + h(x)$ is essentially constant, independent of horizontal position, x, (Dalmayrac and Molnar, 1981).

Insofar as the assumption that the shear stress at the bottom of the plate vanishes is correct, the horizontal force per unit length, which is $\bar{\tau}_{xx}(x)[z_o + h(x)]$, is constant across the range, and since $h(x) << z_o$, the average horizontal stress $\bar{\tau}_{xx}(x)$ is essentially constant. Therefore, *high ranges or plateaus not only can transmit forces (or stresses) that are applied to one flank across them to the other flank, but they must do so unless some assumption made above is in error.* The only assumption that could be challenged here is that the shear stress at the base of the plate does not vanish. Indeed distributions of shear stresses at the base of the plate could be concocted that might prevent a horizontal force per unit length from being transmitted across the range. In general, however, we should expect the existence of mountain ranges and compensating roots to have little effect on this horizontal force per unit length, and, thus, on the average horizontal stress. Note that this result is completely independent of rheology, or constitutive relations, and is simply a consequence of mechanical equilibrium.

Now consider the idealized case for which the shear stress on vertical and horizontal planes vanishes at all depths: $\tau_{xz}(x,z) = 0$. Making such an assumption is not wise for regions where the topography or the depth of the crustal root varies with horizontal position, and it is categorically false where an elastic plate is flexed. It probably is reasonable, however, in regions either where the topography is relatively flat, such as beneath lowlands or high plateaus, or beneath crests of symmetric ranges. When $\tau_{xz}(x,z) = 0$, it follows from Equation 4 that:

$$\tau_{zz} = \int_{-h}^{z_o} \rho(x,z)g dz.$$

The vertical stress is simply the lithostatic pressure.

Notice that the vertical stress, τ_{zz}, is not independent of horizontal position x, as the horizontal force per unit length is. At a given depth below sea level, τ_{zz} will be greater beneath high mountains and plateaus than beneath lowlands: τ_{zz} (plateau, z) $\geqslant$ τ_{zz} (lowland, z) (Fig. 15). Of course, complicated profiles of stress with depth can be concocted, but assume that the horizontal stress $\tau_{xx}(x,z)$ increases smoothly with depth, so that not only is $\tau(x)[z_o + h(x)]$ = constant, but also $\tau_{xx}(x,z)$ depends very little on x.

First, consider the case in which the horizontal and vertical stresses are approximately equal beneath a high plateau, $\tau_{xx}(x,z) \approx \tau_{zz}$ (plateau, z). Then, because the vertical stress in the lowlands is less than that beneath the plateau, the horizontal compressive stress in the lowlands will exceed the vertical stress: τ_{zz} (lowlands, z) $\leqslant \tau_{zz}$ (plateau, z) $\approx \tau_{xx}(x,z)$. Thrust faulting and crustal shortening will be likely in regions of low elevation and relatively thin crust on the flanks of the range (Fig. 15).

Alternatively, consider the case in which the horizontal and vertical normal stresses beneath lowlands are approximately equal: $\tau_{xx}(x,z) \approx \tau_{zz}$ (lowlands, z). In the high plateau, the vertical normal stress will exceed the horizontal normal stress: τ_{zz} (plateau, z) $\geqslant \tau_{zz}$ (lowlands, z) $\approx \tau_{xx}(x,z)$. Hence, normal faulting and crustal extension will be likely beneath areas of high elevation.

Finally, imagine an intermediate case in which the horizontal normal stress is less than the vertical stress beneath the plateau but exceeds the vertical stress in the lowlands: τ_{zz} (plateau, z) $>$ $\tau_{xx}(x,z) > \tau_{zz}$ (lowlands, z). For this case, normal faulting beneath the high plateau will occur simultaneously with thrust faulting in, or at the edge of, the lowlands (Dalmayrac and Molnar, 1981).

For these cases, it is *not* lateral variations in the horizontal regional stress field that causes different styles of deformation; instead, it is lateral variations in the vertical stress. Variations in style of deformation do not necessarily reflect variations in a horizontal regional stress.

These conclusions follow from a number of assumptions, all simple and reasonable. Other stress distributions, some of which may be plausible, might allow for normal faulting and thrust faulting to occur near one another, or even above one another. Our point here, however, is not to claim that one and only one explanation can explain, or is required by, varying styles of deformation in adjacent regions. Rather, it is to show that simple, reasonable assumptions for the normal and shear stresses on horizontal and vertical planes in the crust lead logically to such variations. More specifically, although one can imagine possible explanations for juxtaposed normal and thrust faulting, simple physical arguments and simple assumptions about stress, in fact, predict thrust and normal faulting in adjacent regions of different elevations and crustal thicknesses.

The arguments noted above clearly demonstrate the need for caution with the use of the term regional stress. They also are meant to discourage the use of the term gravity spreading. The preceding discussion contends that crustal shortening and mountain building differ from crustal extension (or gravity spreading) not by the role played by gravity, but by the magnitude of the horizontal force per unit length across the range. Gravity resists crustal thickening and encourages crustal thinning (and extension), regardless of what the horizontal compressive stress is. What differ from one range to another are the horizontal forces across them and the mean elevations of mountains created by those forces. Gravity, however, is remarkably constant over the surface of the earth and throughout the crust and mantle.

The discussion above has omitted consideration of displacements and rates of displacement. One can conceive of mountain ranges in which a small amount, or small rate, of extension occurs within the range compared with the amount, or rate, of crustal shortening on the flanks. We presume this to be the case with the Andes, where the rate of underthrusting of the Brazilian shield beneath the range almost surely exceeds the rate of extension in the high Andean plateau (Suárez and others, 1983). Similarly, a high plateau could extend more rapidly than the surrounding lowlands can underthrust the flanks of the plateau. In such a case, although thrust faulting would occur in the surrounding lowlands, or on the flanks of the mountain range, the lowlands would move apart from one another. We also presume this to be the case in eastern Tibet. As Tibet extends in an east-west direction, material on its eastern flank overthrusts southeast China, yet southeast China appears to move away from, not toward, western Tibet at a rate of a centimeter or more per year (Molnar and Deng, 1984). Thus the rate of east-west extension of the Tibetan plateau, which occurs by both normal and strike-slip faulting, apparently exceeds the rate of crustal thickening at the eastern margin of the plateau. It is important to remind the reader that, while the displacement field can be used to infer strains, and then (with greater risk) to infer stresses, displacements cannot necessarily be inferred from the orientations of strain (or stress). The strain must be integrated to obtain displacements, and that is usually very difficult.

A BOUND ON THE AVERAGE STRENGTH OF THE LITHOSPHERE

The differences in elevations and depths of the Moho beneath mountains or high plateaus and lowlands imply that there must be stress differences in the crust (e.g., Artyushkov, 1973; Frank, 1972; Jeffreys, 1962; Kuchai, 1974; McKenzie, 1972). If the mountains or plateaus undergo crustal extension, however, then vertical stress as a function of depth beneath the plateau, τ_{zz} (plateau, z), defines an upper bound for the horizontal normal stress, $\tau_{xx}(x,z)$. The horizontal compressive force per unit length must be less than the product of the average vertical stress and the thickness of the crust beneath the plateau, or the depth of compensation:

$$\bar{\tau}_{xx}\,[z_o + h] < \bar{\tau}_{zz}\,(\text{plateau})\,[z_o + h\,(\text{plateau})].$$

Similarly, if thrust faulting occurs on the flanks of the range, then vertical normal stress there, τ (lowlands, z), defines a lower bound for the horizontal normal stress:

$$\bar{\tau}_{xx}\,[z_o + h] > \bar{\tau}_{zz}\,(\text{lowlands})\,[z_o].$$

Thus the horizontal driving force per unit length, $F_{x,L}$, applied to the mountain range, is less than the difference between the vertically averaged, vertical normal stresses beneath the mountains or plateaus and beneath the lowlands:

$$F_{x,L} < \int_{-h}^{z_o} [\tau_{zz}\,(\text{plateau}, z) - \tau_{zz}\,(\text{lowlands}, z)]dz$$

$$< [\bar{\tau}_{zz}\,(\text{plateau}, z) - \bar{\tau}_{zz}\,(\text{lowlands})]z_o. \quad (5)$$

The integral in Equation 5 is simply the area between the curves in Figure 15 (e.g., Dalmayrac and Molnar, 1981; Frank, 1973; Tapponnier and Molnar, 1976). In this respect, the high plateaus and mountain ranges serve as "pressure gauges" for the average pressures, or the average forces per unit length, applied to the flanks of the ranges (Molnar and Tapponnier, 1978).

The relatively uniform altitudes of high plateaus like Tibet or the Altiplano and the depths to the Moho beneath them can be used to quantify, if still crudely, the horizontal force per unit length, $F_{x,L}$. The integral in expression (6) or the area between the curves in Figure 15 is, in fact, simply the expression (2). Thus,

$$F_{x,L} < \rho_c g h \{\frac{1}{2} h + H_o + \frac{1}{2}\frac{\rho_c}{\Delta\rho} h\}.$$

If $\rho_c = 2.8$ g/cm^3, $\rho_m = 3.35$ g/cm^3, ($\Delta\rho = 0.55$ g/cm^3) and $g =$ 980 cm/s^2, then for h = 5 km for Tibet or h = 4 km for the Altiplano and high Andes, and for H_o = 35 km, we obtain:

$$F_{x,L} = 6.9 \times 10^{15} \text{ dynes/cm} = 690 \text{ bars} \times 100 \text{ km (Tibet)}$$
$$F_{x,L} = 5.2 \times 10^{15} \text{ dynes/cm} = 520 \text{ bars} \times 100 \text{ km (Andes)}.$$

By writing in these units of bars × kilometers, we can visualize these forces per unit length as average horizontal compressive stresses applied to layers of a given thickness. Clearly, changing the value of this thickness requires a proportional change in the average horizontal stress.

If, as we have suggested above (e.g., Dalmayrac and Molnar, 1981; Molnar and others, 1977; Suárez and others, 1983; Tapponnier and Molnar, 1976), mountain ranges reach limiting heights and then grow laterally as zones of thrust faulting migrate outward, we can use these forces per unit length to place bounds on the average strengths of the plates whose convergence causes the ranges. If the plates were infinitely strong except at the mountain range, then the range would increase in height until the forces driving the plates together could no longer provide the energy needed to lift the mountain range higher. At this time, convergence would presumably stop. If the converging plates were weak enough, or if the forces driving them were strong enough, when the mountain range reached its limiting elevation, new faults should form in the lowlands when the shear stress there exceeded the strengths. Thrust faulting would occur near the foot of the range, or in the lowlands, because it is in these areas that stress differences, or shear stresses, are at a maximum. When normal faulting occurs at high altitudes, the horizontal driving forces presumably are not strong enough to elevate the range higher, but if new thrust faults continue to form on the flanks of the ranges, the stresses apparently have exceeded the strengths there. Recall that the maximum shear stress is half the difference between the maximum and minimum normal stress. Thus, if a new fault forms in the lowlands, the maximum, vertically averaged shear stress responsible for it can be estimated by dividing half the horizontal driving forces per unit length by the thickness of the layer involved.

Consider the thrust zones surrounding Tibet, for which we estimate a force per unit length of 690 bars × 100 km. If a layer 100 km thick were to break with the creation of new faults, an upper bound for the average shear stress needed to cause it to break would be about 350 bars. If, however, essentially all of the force per unit length were transmitted through a strong layer only 30 km thick, then the average shear strength of such a layer would be 1 kbar. The differences between values of strength inferred for Tibet and for the Andes are probably not significant, given the simplicity of the calculations. Note, however, that either the strong part of the plate must be confined to a thin layer or one of the assumptions must be totally in error for the average strength of the entire lithosphere (or even the entire crust) at the edge of mountain ranges to be several kilobars.

This analogy with pressure gauges and the deduction of average strengths of plates is valid insofar as isostatic compensation prevails, insofar as vertical shear stresses are negligible, and insofar as shear stresses at the bases of the plates are also small. Surely, inclusion of realistic rheologic distributions and driving forces at the bases of plates will render this analogy crude and ultimately make it only of pedagogical use. Nevertheless, until we understand better the processes taking place deep within the crust and upper mantle beneath mountain belts, so that we can know how to include these "realistic complications," maintaining a simple approach seems justified.

VARIATIONS OF TECTONIC STYLE IN TIME AND SPACE WITHIN MOUNTAIN BELTS

The preceding discussion emphasized differences in the style of deformation associated with differences in elevation and crustal thickness, but it noted that whereas buoyancy of crustal material is important, so is the strength of the lithosphere, or crust, depending on the particular problem. This section uses the simple ideas discussed above to consider how mountain belts might evolve.

Virtually all mountain ranges are built by crustal shortening, which is necessarily driven by forces that cause convergence between two plates. Underthrusting of one plate beneath the range will lead to elevated regions, thickened crust, and an adjacent foredeep basin. As these processes continue, the mountain belt is likely to "ramp" up onto the strong underthrusting plate and override the basin, plowing the foredeep deposits into folds or detached thrust sheets in front of it. The zone of thrust faulting is likely to shift into the lowlands, once the crust is sufficiently thick to resist further thickening. Thus, the rocks in a given part of the fault may undergo deformation for only portions of the time during which the mountain belt is active.

Usually this sequence of events can only be incompletely preserved in the geologic record, so that the continuous convergence and possible constancy of the convergence rate can be obscured and overlooked. Only a finite number of major fault zones will be active during the orogeny. For some such faults, stratigraphic and cross-cutting relationships may allow the timing of slip on faults to be bounded. For many such faults, however, neither the time when slip initiated nor the duration of slip can be constrained tightly. There has been a tendency of workers in some ranges to assume that tectonic activity occurred only during short intervals (defined by cross-cutting relationships on some faults), separated by long intervals of quiescence. For instance, papers on the Andes, beginning with Steinmann (1929) and continuing to the present, are dominated by this opinion. As careful work has documented more events in different localities, however, additional "phases" have been recognized (e.g., Dalmayrac and others, 1980; Mégard, 1978, 1984; Mégard and others, 1984). It seems equally probable to us that convergence was steady, possibly even at a constant rate, but with deformation concentrated in particular regions for only short times. When one region ceased being active, another presumably became actaive or was so already. Thus, the division into separate short phases of activity may be valid only on a very small scale, but not on a regional scale.

It has been noted above that the styles of deformation in separate parts of a belt can simultaneously be different. The ramp overthrusting of crystalline basement onto the surface of one plate will cause rotations and reorientations of the sense of shear formed at earlier stages as packages of already deformed rocks slide up and onto the gentle part of the thrust surface (Fig. 11). A particular region might undergo two or more *very different* styles of deformation during the history of a belt (e.g., Mattauer and Tapponnier, 1978; Molnar and Chen, 1982; Tapponnier and Molnar, 1976). This phenomenon is clear in both the Himalaya and the Andes. Several large, north-trending normal faults slice across the areas of high altitude north of the high peaks of the Great Himalaya (Molnar and Tapponnier, 1978; Ni and York, 1978; Tapponnier and others, 1981). These graben systems affect rocks south of the Indus-Tsangpo suture zone, rocks that were folded and thrust upon one another in mid-Tertiary time. Similarly, most Cretaceous and older sedimentary rocks in the high Andes are folded and show evidence of crustal shortening perpendicular to the range (e.g., Dalmayrac and others, 1980; Megard, 1978). Many of these rocks are now cut by normal faults. Thus in both the Andes and the Himalaya—belts that are clearly still active—the style of deformation has changed dramatically during the evolution of those belts.

We should expect such changes if the simple logic of the preceding sections has any foundation (e.g., Molnar and Chen, 1982; Tapponnier and Molnar, 1976). First, if the forces driving plates together decreased, the style of deformation in lowlands might not change, but in regions of high altitude, the orientations of maximum and minimum compressive stresses could reverse. Likewise, if a change in the horizontal compressive stress parallel to the belt changed, the style of deformation at high altitudes could also change. In Tibet, where normal faulting occurs on north-trending planes, together with conjugate strike-slip faulting (e.g., Molnar and Chen, 1983; Rothery and Drury, 1984; Tapponnier and others, 1982), a decrease in the east-west normal stress may have made that stress smaller than the vertical normal stress, while the north-south and vertical normal stresses remained nearly equal. In both areas, the changes in the driving (or constraining) forces need not be large to cause a reorientation of the local principal stresses.

The creation of a high range by the underthrusting of one plate or block beneath another need not remain asymmetrical with a clear polarity. A terrain originally thrust in one direction over the downflexed underthrusting plate, could, as the range built out over that plate, be thrust back onto lowlands of the hinterland. This "retrocharriage" probably will not occur early in the history of a range; instead, it probably becomes likely only after the elevations are high and the crust is thick and possibly weak, and where the hinterland has remained at low elevations. For example, the Alps are characterized by retrocharriage on the south side. Excluding some minor back folding in Tethys Himalaya (e.g., Bordet and others, 1971; Fuchs, 1977, 1979, 1982; Fuchs and Frank, 1970), however, retrocharriage in the Himalaya occurs as an important phenomenon 1,000 km north and east of the range on the northeastern and eastern margins of the Tibetan Plateau. In a sense the deformation in the sub-Andes is all retrocharriage.

We expect that both the crustal and the mantle parts of the underthrusted lithosphere will weaken as they warm, so that the strength of the crust in high plateaus will become lower than that of the underthrusting plate or the margins of the belt. This weakening will be enhanced if mantle lithosphere delaminates (Bird, 1978) or if the thickened thermal boundary layer at its base detaches and is replaced by hotter asthenosphere (Houseman and others, 1981). Although this weakening might make high plateaus more susceptible to deformation than the stronger, underthrusting plate, the elevations of high plateaus make the vertical normal stress large, which resists thrust faulting. The increased vertical stress, therefore, makes further crustal thickening of high plateaus energetically inefficient, despite the weakening of material there. When the stress system is reoriented, however, the weak plateaus may be susceptible to deformation at stress differences lower than those that exist where one plate underthrusts the range. This weakening will be greatly enhanced if crustal rock is much weaker than the mantle at the same temperature, which is suggested by depths of foci of earthquakes and by measured flow laws of common minerals, but not required by either (e.g., Chen and Molnar, 1983).

Given the role played by variations in elevation and crustal thickness on the style of deformation in an active belt, we should not be surprised to find polyphase deformation, with a full spectrum of styles including thrust, reverse, strike-slip, normal faulting, and corresponding senses of shear. Moreover, these different styles should be coeval: simultaneously, décollement could occur at the edge of the range, thrust faulting on steeper planes (or even reverse faulting) would be occurring at greater depths and farther into the range, normal or strike-slip faulting could occur where altitudes are high, and back-thrusting or retrocharriage could occur on the other flank. We think that the Cenozoic and active tectonics of both the Himalaya-Tibet region (e.g., Molnar and Chen, 1982; Molnar and Tapponnier, 1975, 1978; Tapponnier and Molnar, 1976; Tapponnier and others, 1981, 1982) and the Andes (e.g., Dalmayrac and Molnar, 1981; Suárez and others, 1983) provide evidence for these phenomena.

The geologic evolution of the western United States since the Jurassic period reveals similar sequences of deformation. The huge Jurassic-Cretaceous batholith along the western edge of North America attests to subduction of oceanic lithosphere during Mesozoic time. While this subduction occurred, crustal shortening occurred in the Sevier belt of eastern Nevada and Utah and its extensions in to Canada (e.g., Armstrong, 1968; Burchfiel and Davis, 1972, 1975). Like many others, we view this setting as being very similar to the present Andes (e.g., Burchfiel and Davis, 1976; Suárez and others, 1983). The thrust faulting thickened the crust in what is now western and central Nevada, and the zone of thrust faulting migrated eastward, except in southern Nevada (Burchfiel and Davis, 1975). While this crustal shortening occurred, sediment was shed eastward onto what are now the Colorado Plateau and the Rocky Mountains (e.g., Cook and Bally, 1975) and onto the area east of the Idaho-Wyoming thrust belt (Armstrong and Oriel, 1965). Since the crustal thickness of the Colorado plateau is about 40 km, for the plateau to have been a depocenter, the crustal thickness of the mountains to the west probably was greater than 40 km. Although we are aware of no definitive evidence, we suspect that beneath central Nevada the crust reached a thickness of 50 to 70 km.

In late Cretaceous time, the Rocky Mountains in Colorado, Utah, and Wyoming began to form, where, unlike in Nevada, thick sequences of sedimentary rock had not accumulated. We presume, as Hamilton (1978) did, that the strength of the Precambrian lithosphere beneath the Colorado Plateau prevented deformation of it and that eastward or northeastward displacement with respect to the rest of North America caused crustal shortening where the Rocky Mountains are now. For us, the pressure that maintained the thickened crust of Nevada was transmitted by the strong Colorado Plateau just as the strong Tarim Basin transmits the pressure across Tibet to the Tien Shan (Molnar and Tapponnier, 1975; 1978; 1981).

It may be fortuitous that the present Tien Shan are built largely by steep thrust or reverse faults, as were the Rocky Mountains, with little evidence of décollement or thin-skin deformation. Nevertheless, we believe that the formation of both ranges far from the active plate margin and from antithetic fold and thrust belts may have been prevented deposition of thick sedimentary sequences before these ranges formed. Such thick sedimentary rock can obscure basement involvement and often can deform easily by décollement and large horizontal transport. Accordingly, we think that low angle underthrusting of oceanic lithosphere beneath the whole of the Western United States may not have occurred, and that if it did, it need not have been a cause of the style, location, or even timing of the Laramide deformation in Colorado, Utah, or Wyoming (e.g., Burchfiel and Davis, 1975; Coney and Reynolds, 1977; Cross and Pilger, 1978). In fact, if this underthrust lithosphere was young, it may not have extended as far east as Colorado as a slab of lithosphere (e.g., Molnar and others, 1979).

Finally, when subduction of oceanic lithosphere beneath western North America slowed and finally ceased, the horizontal forces that held together the presumed high plateau in western and central Nevada and that drove the Laramide crustal shortening also decreased. This drop in the horizontal force per unit length culminated in the collapse and extension of this presumed plateau in Nevada to form the Basin and Range Province (e.g., Coney and Harms, 1984; Molnar and Chen, 1983). Note, however, that the extension and collapse of this plateau could have begun well before subduction ceased, much as normal faulting presently occurs in the Andes or Tibet. The prevalence of normal faults cutting volcanic rocks 15 to 20 m.y. old (e.g. McKee, 1971) may indicate a marked drop in the force (per unit length) across western North America at that time, but the older crustal extension revealed by exhumed low-angle normal faults (e.g., Miller and others, 1983; Schultejann, 1984) is consistent with the analogy with the Andes. Thus, we do not share the view of Coney and Harms (1984) that the low-angle normal faulting necessarily occurred after crustal shortening in the western United States ceased.

CONCLUSIONS

The forces driving mountain building must do work against two opposing forces. One might be called mechanical strength, for the stronger the material, the greater the stress needed to deform it. Clearly, more work is done in deforming strong than weak material. For small amounts of deformation, some of the work done can be stored as elastic strain energy, but for the amount of deformation typically observed in mountain belts, most of the work done in deforming the rock must be dissipated by frictional or viscous processes. The forces driving continental plates or blocks together to form mountain ranges must also do work against gravity. The formation of mountain ranges and crustal roots is associated with the creation of gravitational potential energy, which can be increased by crustal thickening and which can be extracted during crustal thinning.

In the foregoing, we discussed the role played by a strong underthrust plate at a mountain belt. A mountain range thrust onto a strong plate will flex it down in front of the range to form a foredeep basin. The width of the basin will be greater, but the depth is likely to be shallower for a strong or thick plate than for a weak one. Analyses of gravity anomalies over mountain ranges and foredeep basins and of profiles of the basement beneath foredeep basins corroborate this inference and show a tendency for older, colder lithosphere to be stronger than younger, warmer, thinner lithosphere (e.g., Karner and others, 1983). A large variation in equivalent elastic thicknesses of flexed plates, however, implies that temperature is either not the only important quantity affecting the strength of the plate or that the dependence on temperature is complex (e.g., Lyon-Caen and Molnar, 1985). Moreover, analyses of gravity anomalies and depths to basement show that the weight of the mountain range thrust on top of continental lithosphere is not the only force deflecting such plates. Either gravity acting on the excess mass in downgoing slabs of oceanic lithosphere, or continental lithosphere stripped of its upper and middle crust, apparently can pull the following, intact continental lithosphere down and exert a bending moment that flexes part of it up. Thus, the simple analysis of mountain ranges supported by flexed elastic plates allows us to quantify some differences among ranges. Nevertheless, a satisfactory understanding of how ranges are supported must await both data that constrain the deep structure of mountain belts and an improved understanding of the dynamic processes occurring in the mantle beneath mountain ranges.

The formation of a thrust fault in continental lithosphere and the subsequent overthrusting onto the intact, strong lithosphere can bring to the earth's surface middle and lower crustal rocks metamorphosed to medium or high grade. Such crustal-scale "ramp-overthrusts" can cause localized belts of rapid uplift, as well as rotations of both fabrics associated with earlier deformation and previously active thrust faults. The existence or absence of broad fold and thrust belts may depend on the strength of the plate, the thickness of weak, detachable sedimentary rock, or the amount of displacement on the thrust fault.

The formation of both mountain ranges and crustal roots creates gravitational potential energy. Moreover, part of this energy increases as the square of the mean elevation, or of the excess thickness of crust in the root. Consequently, as mountain ranges and high plateaus are elevated and the crust is thickened, an increasingly larger amount of work must be done for each subsequent increment of uplift. This result suggests that mountain ranges should reach limiting elevations and limiting crustal thicknesses, regardless of how strong are the materials constituting them.

When a mountain range reaches a limiting elevation, and when the forces driving convergence cannot elevate the range higher or thicken the crust to a greater amount, mountain building does not necessarily cease. Instead, the range is likely to grow laterally and thus to broaden by the appending of material to its flanks. Thrust faulting should migrate outward into the lowlands so that material near sea level is uplifted and crust of normal thickness is thickened.

The potential energy created by the formation of a mountain range is not lost; it can be extracted during crustal thinning and extension. If the cold mantle lithosphere beneath the range can be replaced by hotter, lighter material, additional potential energy can be added to the system. Hence, during crustal thinning and extension, more energy can be extracted than was originally accumulated by crustal thickening. It follows that high plateaus and areas of thick crust are very susceptible to normal faulting and crustal extension.

The occurrence of crustal extension in a high plateau or high range does not imply that the flanks of the range are being pulled or pushed apart. Insofar as tractions on the base of the lithosphere at a mountain range can be neglected, a horizontal force per unit length of range will be transmitted across the range. Depending on the magnitude of this force and on the height of the plateau or range and the thickness of the crust, different styles of deformation will be likely. For large forces, thrust faulting, particularly on the flanks of the range, and crustal shortening will be likely. For small forces, normal faulting, particularly at high altitude, and crustal extension will be likely. For intermediate forces, both can occur simultaneously. It follows that, in some areas, the displacement associated with crustal shortening may greatly exceed those of crustal extension, but in other areas the converse will hold. Moreover, the existence of only a minor amount of crustal extension in a high plateau or mountain range can allow placement of an upper bound on the horizontal force per unit length, or on the average horizontal pressure, that one plate exerts on the other. Thus the mean elevations of high plateaus and mountain ranges can serve as crude "pressure gauges" for the driving forces of mountain building.

In mountain ranges simultaneously undergoing crustal extension at high altitude and growing laterally by the initiation of and subsequent displacement along thrust faults, we can make a crude estimate of the average strength of the lithosphere. From the elevation and crustal thickness, we can place a bound on the force per unit length along the range. Since new thrust faults are forming, this force per unit length is sufficient to rupture the strong plate adjacent to the range. Dividing the force per unit length by the thickness of the strong part of the plate yields a measure of the average strength of the plate.

During the evolution of a mountain range, a given portion of the range is likely to undergo more than one style of deformation. These can include thrust faulting on a steep plane (a crustal ramp-overthrust), thrust faulting on a gentle plane, strike-slip or normal faulting when the material is uplifted to high elevations, and possibly back-thrusting or retrocharriage if the foreland is uplifted higher than the hinterland. As a result, not only ought the styles of deformation to be different in different parts of a mountain belt at any given time, but polyphase deformation is likely to be part of the continuous evolution of a mountain range.

We have presented our conclusions as logical deductions from elementary physics and derivations from simple and reasonable assumptions. In any case, the work of England and McKenzie (1982), England and Houseman (1985, 1986), England and Searle (1986), and Houseman and England (1986)—which accounted for the buoyancy of continental crust and the finite strength of material not only has corroborated many of the conclusions presented herein but also has presented a series of numerical experiments that examine the relative importances of buoyancy and strength. As work of this type and that of others (e.g., Dahlen, 1984; Dahlen and others, 1984; Davis and others, 1983; Fleitout and Froidevaux, 1982, 1983) systematically examines aspects of dynamic and thermal processes of mountain building, many of the simple concepts discussed above will become useful only as pedagogical tools, surely to be supplanted by more rigorously derived concepts. Nevertheless, it is important to emphasize that each of the concepts outlined above evolved from painstaking—and initially puzzling—geologic or geophysical observations about how the earth should behave, and not from casual musings.

ACKNOWLEDGMENTS

The sections on foredeep basins and underthrust plates and thrust zones of this chapter present a brief summary of work that we have been engaged in for the last six years. The remaining sections of the chapter constitute a summary of much of the thinking that Molnar and Paul Tapponnier shared during several years of collaborative work beginning in 1974. We first discussed much of the material in the latter section in 1975 upon reading the *American Journal of Science* volume dedicated to John Rodgers. Unfortunately, time constraints and distance have prevented the authors and Tapponnier from reaching agreement on some geological details presented herein. Although Tapponnier is not listed as an author, we nevertheless acknowledge his substantial influence. We also have benefited from numerous discussions and collaboration with B. C. Burchfiel, who introduced us to many basic geological concepts; with W. P. Chen, whose work on the structure and seismicity of Asia was done in parallel with much of Tapponnier's work; with P. C. England, whose recent work, undertaken independently of ours, has been a strong stimulus in our data analysis and field investigations; and with G. Suárez, who carried out much of the analysis of the Andes on which we base our thinking. Useful discussions with L. Royden are also acknowledged. Finally we thank S. Luria for typing the manuscript and R. Price for providing Figure 10. In addition, the work described here has depended on continued support from the National Science Foundation and from the Geodynamics Program of the National Aeronautics and Space Administration.

REFERENCES CITED

Allmendinger, R. W., Ramos, V. A., Jordan, T. E., Palma, M., and Isacks, B. L., 1983, Paleogeography and Andean structural geometry, northwest Argentina: Tectonics, v. 2, p. 1–16.

Armstrong, F. C., and Oriel, S. S., 1965, Tectonic development of Idaho-Wyoming thrust belt: American Association of Petroleum Geologists Bulletin, v. 49, p. 1847–1866.

Armstrong, R. L., 1968, Sevier orogenic belt in Nevada and Utah: Geological Society of America Bulletin, v. 79, p. 429–458.

Artyushkov, E. V., 1973, Stresses in the lithosphere caused by crustal thickness inhomogeneties: Journal of Geophysical Research, v. 78, p. 7675–7708.

Bally, A. W., Gordy, P. L., and Stewart, G. A., 1966, Structure, seismic data and orogenic evolution of southern Canadian Rockies: Bulletin of Canadian Petroleum Geologists, v. 14, p. 337–381.

Baranowski, J., Armbruster, J., Seeber, L., and Molnar, P., 1984, Focal depths and fault plane solutions of earthquakes and active tectonics of the Himalaya: Journal of Geophysical Research, v. 89, p. 6918–6928.

Barazangi, M., and Isacks, B. L., 1979, Subduction of the Nazca plate beneath Peru: Evidence from spatial distribution of earthquakes: Geophysical Journal of the Royal Astronomical Society, v. 57, p. 537–555.

Bei, M., and Fang, Z., 1981, Relationship between phenomena of surface breaking and geological studies of the Malasi earthquake (M = 8) in 1906 in the southwestern Xinjiang region: Northwest Seismological Journal, v. 3, no. 2, p. 100–102 (in Chinese).

Berg, R. R., 1962, Mountain flank thrusting in Rocky Mountain foreland, Wyoming and Colorado: American Association of Petroleum Geologists Bulletin, v. 46, p. 2019–2032.

Bird, P., 1978, Initiation of intracontinental subduction in the Himalaya: Journal of Geophysical Research, v. 83, p. 4975–4988.

Bodine, J. H., and Watts, A. B., 1979, Lithospheric flexure seaward of the Bonin and Mariana Trenches: Earth and Planetary Science Letters, v. 43, p. 132–148.

Bogdanovich, I. I., Kark, I. M., Korolkov, B. Ya., and Mushketov, D. I., 1914, The earthquake in the northern part of the Tien Shan of 22 December 1910 (4 January 1911): Trans. Geol. Comm., v. 89, p. 1–270.

Bonini, W. E., de Gaete, C. P., and Graterol, V., 1977, Mapa de anomalias gravimetricas de Bouguer de la parte norte de Venezuela y areas vecinas: Ministeria de Energia y Mines, Direction de Geologia, Caracas, scale 1:1,000,000.

Bordet, P., Colchen, M., Krummenacher, D., LeFort, P., Mouterde, R., and Remy, M., 1971, Recherches Géologiques dan L'Himalaya du Népal Region de la Thakola: Editions du Centre National de la Recherche Scientifique, Paris, 279 p.

Bowin, C. O., 1976, The Carribean; Gravity field and plate tectonics: Geological Society of America Special Paper 169, 79 p.

Burchfiel, B. C., and Davis, G. A., 1972, Structural framework and evolution of the southern part of the Cordilleran orogen, western United States: American Journal of Science, v. 272, p. 97–118.

——, 1975, Nature and controls of Cordilleran orogenesis, western United States; Extensions of an earlier synthesis: American Journal of Science, v. 275-A, p. 363–396.

——, 1976, Compression and crustal shortening in Andean-type orogenesis: Nature, v. 260, p. 693–694.

Chapple, W. M., and Forsyth, D. W., 1979, Earthquakes and bending of plates at trenches: Journal of Geophysical Research, v. 84, p. 6729–6749.

Chen, W.-P., and Molnar, P., 1983, Focal depths of intracontinental and intraplate earthquakes and their implications for the thermal and mechanical properties of the lithosphere: Journal of Geophysical Research, v. 88, p. 4183–4214.

Chinn, D. S., and Isacks, B. L., 1983, Accurate source depths and focal mechanisms of shallow earthquakes in western South America and in the New Hebrides island arc: Tectonics, v. 2, p. 529–563.

Coney, P. J., and Harms, T. A., 1984, Cordilleran metamorphic core complexes; Cenozoic extensional relics of Mesozoic compression: Geology, v. 12,

p. 550–554.
Coney, P. J., and Reynolds, S. J., 1977, Cordilleran Benioff Zones: Nature, v. 270, p. 403–406.
Cook, T. D., and Bally, A. W., 1975, Stratigraphic atlas of North and Central America: Princeton, New Jersey, Princeton University Press, 272 p.
Cross, T. A., and Pilger, R. H., 1978, Constraints on absolute motion and plate interaction inferred from Cenozoic igneous activity in western United States: American Journal of Science, v. 278, p. 865–902.
Dahlen, F. A., 1984, Noncohesive critical Coulomb wedges; An exact solution: Journal of Geophysical Research, v. 89, p. 10115–10133.
Dahlen, F. A., Suppe, J., and Davis, D., 1984, Mechanics of fold-and-thrust belts and accretionary wedges; Cohesive Coulomb theory: Journal of Geophysical Research, v. 89, p. 10087–10101.
Dalmayrac, B., 1974, Un exemple de tectonique vivante; Les failles subactuelles du pied de la Cordillère Blanche (Pérou): Cahiers ORSTOM, Serie Geologique v. 6, no. 1, p. 19–27.
Dalmayrac, B., and Molnar, P., 1981, Parallel thrust and normal faulting in Peru and constraints on the state of stress: Earth and Planetary Science Letters, v. 55, p. 473–481.
Dalmayrac, B., Laubacher, G., and Marocco, R., 1980, Géologie des Andes péruviennes; Caractères généreaux de l'evolution, géologiques des Andes péruviennes: Travaux et Documents de l'ORSTOM no. 96.
Das, D., Mehra, G., Rao, K.G.C., Roy, A. L., and Narayana, M. S., 1979, Bouguer, free-air, and magnetic anomalies over northwestern Himalayas, *in* Himalayan Geology Seminar, Section 3, Oil and Natural Gas Resources: Geological Survey of India Miscellaneous Publication 41, p. 141–148.
Davis, D., Suppe, J., and Dahlen, F. A., 1983, Mechanics of fold-and-thrust belts, and accretionary wedges: Journal of Geophysical Research, v. 88, p. 1153–1172.
England, P. C., and Houseman, G. A., 1985, The influence of lithospheric strength heterogeneities on the tectonics of Tibet and surrounding regions: Nature, v. 315, p. 297–301.
—— , 1986, Finite strain calculations of continental deformation; 2, Comparison with the India-Asia collision zone: Journal of Geophysical Research, v. 91, p. 3664–3676.
England, P. C., and McKenzie, D. P., 1982, A thin viscous sheet model for continental deformation: Geophysical Journal of the Royal Astronomical Society, v. 70, p. 295–321.
England, P. C., and Searle, M., 1986, The Cretaceous-Tertiary deformation of the Lhasa block and its implications for crustal thickening in Tibet: Tectonics, v. 5, p. 1–14.
Fleitout, L., and Froidevaux, C., 1982, Tectonics and topography for a lithosphere containing density heterogeneities: Tectonics, v. 1, p. 21–57.
—— , 1983, Tectonic stresses in the lithosphere: Tectonics, v. 2, p. 315–324.
Folinsbee, R. A., 1972, The gravity field and plate boundaries in Venezuela [Ph.D. thesis]: Cambridge, Massachusetts Institute of Technology.
Frank, F. C., 1972, Plate tectonics, the analogy with glacier flow and isostasy, *in* Heard, H. C., Borg, I. Y., Carter, N. L., and Raleigh, C. B., eds., Flow and fracture of rocks: American Geophysical Union Geophysics Monograph 16, The Griggs Volume, p. 285–292.
Fuchs, G., 1977, The geology of the Karnali and Dolpo regions, western Nepal: Beihefte zum Geologische Jahrbuch, Bd. 120, H. 1.
—— , 1979, On the geology of western Ladakh: Beihefte zum Geologische Jahrbuch, Bd. 122, H. 2, p. 513–540.
—— , 1982, Geological map of western Zanskar, Ladakh: Beihefte zum Geologische Jahrbuch, Bd. 125.
Fuchs, G., and Frank, W., 1970, The geology of west Nepal between the rivers Kali Gandaki and Thulo Bheri: Beihefte zum Geologische Jahrbuch, Sonderband 18, 103 p.
Gansser, A., 1964, Geology of the Himalayas: London, Wiley-Interscience, 289 p.
Goetze, C., and Evans, B., 1979, Stress and temperature in the bending lithosphere as constrained by experimental rock mechanics: Geophysical Journal of the Royal Astronomical Society, v. 59, p. 463–478.
Gries, R., 1983, Oil and gas prospecting beneath the Precambrian of foreland thrust plates in Rocky Mountains: American Association of Petroleum Geologists Bulletin, v. 67, p. 1–28.
Gunn, R., 1943, A quantitative evaluation of the influence of the lithosphere on the anomalies of gravity: Journal of the Franklin Institute, v. 236, p. 47–65.
Hamilton, W., 1978, Mesozoic tectonics of the western United States, *in* Howell, D. G., and McDougall, K. A., eds., Mesozoic paleogeography of the western United States: Pacific Section, Society of Economic Paleontologists and Mineralogists, p. 33–70.
Hanks, T. C., 1971, The Kuril trench–Hokkaido rise system; Large shallow earthquakes and simple models for deformation: Geophysical Journal of the Royal Astronomical Society, v. 23, p. 173–189.
Houseman, G. A., and England, P. C., 1986, Finite strain calculations of continental deformation; 1. Method and general results for convergent zones: Journal of Geophysical Research, v. 91, p. 3651–3663.
Houseman, G. A., McKenzie, D. P., and Molnar, P., 1981, Convective instability of a thickened boundary layer and its relevance for the thermal evolution of continental convergent belts: Journal of Geophysical Research, v. 86, p. 6115–6132.
Jackson, J. A., and Fitch, T. J., 1981, Basement faulting and the focal depths of the larger earthquake in the Zagros mountains (Iran): Geophysical Journal of the Royal Astronomical Society, v. 64, p. 561–586.
Jackson, J., and McKenzie, D., 1984, Active tectonics of the Alpine-Himalayan belt between western Turkey and Pakistan: Geophysical Journal of the Royal Astronomical Society, v. 77, p. 185–264.
Jeffreys, H., 1962, The Earth: London, Cambridge University Press, chapter 5.
Jordan, T., Isacks, B., Allmendinger, R., Brewer, J., Ramos, V., and Ando, C., 1983, Andean tectonics related to the geometry of the subducted plate: Geological Society of America Bulletin, v. 94, p. 341–361.
Kadinsky-Cade, C., and Barazangi, M., 1982, Seismotectonics of southern Iran; The Oman line: Tectonics, v. 1, p. 389–412.
Karner, G. D., and Watts, A. B., 1983, Gravity anomalies and flexure of the lithosphere at mountain ranges: Journal of Geophysical Research, v. 88, p. 10449–10477.
Karner, G. D., Steckler, M. S., and Thorne, J. A., 1983, Long-term thermomechanical properties of the continental lithosphere: Nature, v. 304, p. 250–253.
Kono, M., 1974, Gravity anomalies in east Nepal and their implications to the crustal structure of the Himalayas: Geophysical Journal of the Royal Astronomical Society, v. 39, p. 283–300.
Kuchai, V. K., 1974, The level of the free mantle and pecularities of the stress distribution of the earth's crust in central Asia: Doklady Akademii Nauk Tadzhikskoy SSR, v. 17, no. 10, p. 29–33 (in Russian).
Le Fort, P., 1975, Himalayas; The collided range; Present knowledge of the continental arc: American Journal of Science, v. 275-A, p. 1–44.
Lyon-Caen, H., and Molnar, P., 1983, Constraints on the structure of the Himalaya from an analysis of gravity anomalies and a flexural model of the lithosphere: Journal of Geophysical Research, v. 88, p. 8171–8183.
—— , 1984, Gravity anomalies and the structure of western Tibet and the southern Tarim Basin: Geophysical Research Letters, v. 11, p. 1251–1254.
—— , 1985, Gravity anomalies, flexure of the Indian plate, and the structure, support, and evolution of the Himalaya and Ganga Basin: Tectonics, v. 4, p. 513–538.
Lyon-Caen, H., Molnar, P., and Suárez, G., 1985, Gravity anomalies and flexure of the Brazilian shield beneath the Bolivian Andes: Earth and Planetary Science Letters, v. 75, p. 81–92.
Martinez, C., 1980, Structure et évolution de la chaine Hercynienne et de la chaine andine dans le nord de la Cordillère des Andes de Bolivie: Travaux et Documents de l'ORSTOM no. 119.
Mattauer, M., 1975, Sur le mécanisme de formation de la schistosité dans l'Himalaya: Earth and Planetary Science Letters, v. 28, p. 144–154.
Mattauer, M., and Tapponier, P., 1978, Tectonique des plaques et tectonique intracontinentale dans les Alpes franco-itallиennes: Comptes Rendu de l'Academie de Science, Paris, v. 287, Série D, p. 899–902.
Mattauer, M., Collot, B., and Van den Driessche, J., 1983, Alpine model for the

internal metamorphic zones of the North American Cordillera: Geology, v. 11, p. 11–15.

McAdoo, D. C., Caldwell, J. G., and Turcotte, D. L., 1978, On the elastic–perfectly plastic bending of the lithosphere under generalized loading with application to the Kurile Trench: Geophysical Journal of the Royal Astronomical Society, v. 54, p. 11–26.

McKee, E. H., 1971, Tertiary igneous chronology of the Great Basin of the western United States; Implications for tectonic models: Geological Society of America Bulletin, v. 82, p. 3497–3502.

McKenzie, D. P., 1972, Active tectonics of the Mediterranean region: Geophysical Journal of the Royal Astronomical Society, v. 30, p. 109–185.

McNutt, M., and Menard, H. W., 1982, Constraints on the yield strength in the oceanic lithosphere derived from observations of flexure: Geophysical Journal of the Royal Astronomical Society, v. 71, p. 363–394.

Mégard, F., 1978, Etude géologique des Andes de Pérou central: Memoires d'OSTROM, v. 86, 310 p.

——, 1984, The Andean orogenic period and its major structures in central and northern Peru: Journal of the Geological Society of London, v. 141, p. 893–900.

Mégard, F., Noble, D. C., McKee, E. H., and Bellon, H., 1984, Multiple pulses of Neogene compressive deformation in the Ayacucho intermontane basin, Andes of central Peru: Geological Society of America Bulletin, v. 95, p. 1108–1117.

Mercier, J.-L., 1981, Extensional-compressional tectonics associated with the Aegean arc; Comparison with the Andean cordillera of south Peru-north Bolivia: Philisophical Transactions of the Royal Society of London, v. A300, p. 337–355.

Miller, E. L., Gans, P. B., and Garing, J. D., 1983, The Snake Range décollement; An exhumed mid-Tertiary ductile-brittle transition: Tectonics, v. 2, p. 239–263.

Mingramm, A., Russo, A., Pozzo, A., and Cazau, L., 1979, Sierras Subandinas: Segundo Simposio, Geologia Regionale, Argentina Cordoba, Academia Nacionale Ciencias, v. 1, p. 95–138.

Molnar, P., 1984, Structure and tectonics of the Himalaya; Constraints and implications of geophysical data: Annual Reviews of Earth and Planetary Sciences, v. 12, p. 489–518.

Molnar, P., and Chen, W.-P., 1982, Seismicity and mountain building, *in* Hsü, K., Mountain building processes: London, Academic Press, p. 41–57.

——, 1983, Focal depths and fault plane solutions of earthquakes under the Tibetan Plateau: Journal of Geophysical Research, v. 88, p. 1180–1196.

Molnar, P., and Tapponnier, P., 1975, Cenozoic tectonics of Asia; Effects of a continental collision: Science, v. 189, p. 419–426.

——, 1978, Active tectonics of Tibet: Journal of Geophysical Research, v. 83, p. 5361–5374.

——, 1981, A possible dependence of tectonic strength on the age of the crust in Asia: Earth and Planetary Science Letters, v. 52, p. 107–114.

Molnar, P., and Deng Qidong, 1984, Faulting associated with large earthquakes and the average rate of deformation in central and eastern Asia: Journal of Geophysical Research, v. 89, p. 6203–6227.

Molnar, P., Chen, W.-P., Fitch, T. J., Tapponnier, P., Warsi, W.E.K., and Wu, F. T., 1977, Structure and tectonics of the Himalaya; A brief summary of relevant geophysical observations: Paris, Editions du Centre National de la Recherche Scientifique, Sciences de la Terre, p. 267–294.

Molnar, P., Freedman, D., and Shih, J.S.F., 1979, Lengths of intermediate and deep seismic zones and temperatures in downgoing slabs of lithosphere: Geophysical Journal of the Royal Astronomical Society, v. 56, p. 41–54.

Molnar, P., Chen, W.-P., and Padovani, E., 1983, Calculated temperatures in overthrust terrains and possible combinations of heat sources responsible for the Tertiary granites in the greater Himalaya: Journal of Geophysical Research, v. 88, p. 6415–6429.

Ni, J., 1978, Contemporary tectonics of the Tien Shan region: Earth and Planetary Science Letters, v. 41, p. 347–355.

Ni, J., and Barazangi, M., 1984, Seismotectonics of the Himalayan collision zone; Geometry of the underthrusting Indian plate beneath the Himalaya: Journal of Geophysical Research, v. 89, p. 1147–1163.

Ni, J., and York, J. E., 1978, Cenozoic extensional tectonics of the Tibetan plateau: Journal of Geophysical Research, v. 83, p. 5377–5384.

Noble, D. C., McKee, E. H., and Mégard, F., 1979, Early Tertiary "Incaic" tectonism, uplift, and volcanic activity, Andes of central Peru: Geological Society of America Bulletin, v. 90, p. 903–907.

Norin, E., 1941, Geologic reconnaissances in the Chinese Tien Shan; Report from the scientific expeditions to the northwest provinces of China under the direction of Dr. Sven Hedin: Stockholm, Geology 6, publication 16, no. 3, 229 p.

Philip, H., and Mégard, F., 1977, Structural analysis of the superficial deformation of the 1969 Pariahuanca earthquakes (central Peru): Tectonophysics, v. 38, p. 259–278.

Price, R. A., 1981, The Cordilleran foreland thrust and fold belt in the southern Canadian rocky Mountains, *in* Thrust and nappe tectonics: Geological Society of London, p. 427–448.

Rothery, D. A., and Drury, S. A., 1984, The neotectonics of the Tibetan plateau: Tectonics, v. 3, p. 19–26.

Royden, L., and Karner, G. D., 1984, Flexure of lithosphere beneath Appennine and Carpathian foredeep basins; Evidence for an insufficient topographic load: American Association of Petroleum Geologists Bulletin, v. 68, p. 704–712.

Schultejann, P. A., 1984, The Yaqui Ridge antiform and detachment fault; Mid-Cenozoic extensional terrane west of the San Andreas fault: Tectonics, v. 3, p. 677–691.

Seeber, L., Armbruster, J., and Quittmeyer, R., 1981, Seismicity and continental subduction in the Himalayan arc, *in* Gupta, H. K., and Delany, F. M., eds., Zagros, Hindu Kush, Himalaya; Geodynamic evolution: American Geophysical Union Geodynamics Series, v. 3, p. 215–242.

Seely, D. R., Vail, P. R., and Walton, G. G., 1974, Trench slope model, *in* Burk, C. A., and Drake, C. L., The geology of continental margins: New York, Springer-Verlag, p. 249–260.

Sheffels, B., and McNutt, M., 1986, Role of subsurface loads and regional compensation in the isostatic balance of the Transverse Ranges, California; Evidence for intracontinental subduction: Journal of Geophysical Research, v. 91, p. 6419–6431.

Smithson, S. B., Brewer, J. A., Kaufman, S., Oliver, J. E., and Hurich, C. A., 1979, Structure of the Laramide Wind River uplift, Wyoming, from COCORP deep reflection data, and from gravity data: Journal of Geophysical Research, v. 84, p. 5955–5972.

Stauder, W., 1975, Subduction of the Nazca plate under Peru as evidenced by focal mechanisms and by seismicity: Journal of Geophysical Research, v. 80, p. 1053–1064.

Steinmann, G., 1929, Geologie von Peru: Heidelberg, Karl Winter, 448 p.

Suárez, G., Molnar, P., and Burchfiel, B. C., 1983, Seismicity, fault plane solutions, depth of faulting, and active tectonics of the Andes of Peru, Ecuador, and southern Colombia: Journal of Geophysical Research, v. 88, p. 10403–10428.

Tapponnier, P., and Molnar, P., 1976, Slip-line field theory and large scale continental tectonics: Nature, v. 264, p. 319–324.

——, 1977, Active faulting and Cenozoic tectonics of China: Journal of Geophysical Research, v. 82, p. 2905–2930.

——, 1979, Active faulting and Cenozoic tectonics of the Tien Shan, Mongolia, and Baykal Region: Journal of Geophysical Research, v. 84, p. 3425–3459.

Tapponnier, P., Mercier, J. L., Armijo, R., Tonglin, H., and Ji, Z., 1981, Field evidence for active normal faulting in Tibet: Nature, v. 294, p. 410–414.

Tapponnier, P., Peltzer, G., Le Dain, A. Y., Armijo, R., and Cobbold, P., 1982, Propagating extrusion tectonics in Asia; New insights from simple experiments with plasticine: Geology, v. 10, p. 611–616.

Turcotte, D. L., and Schubert, G., 1982, Geodynamics; applications of continuum mechanics to geological problems: New York, John Wiley & Sons, 450 p.

Vicente, J. C., Sequeiros, F., Valdivia, M. A., and Zavala, J. A., 1979, El sobreescurrimiento de Cincha-Luta; Elemento del accidente mayor andino al NW de Arequipa: Boletin de la Sociedad Geologia del Peru, v. 61, p. 67–100.

Vink, G. E., Morgan, W. J., and Zhao, W. L., 1984, Preferential rifting of continents; A source of displaced terranes: Journal of Geophysical Research, v. 89, p. 10072–10076.

Walcott, R. I., 1970, Flexural rigidity, thickness, and viscosity of the lithosphere: Journal of Geophysical Research, v. 75, p. 3941–3954.

Wilson, J., Reyes, L., and Garayar, J., 1967, Geologia de los Cuadrangulos de Mollembamba, Tayambamba, Huaylas, Pomabama, Carhuaz y Huari: Lima, Peru, Servicio de Geologia y Mineria Boletin no. 16.

MANUSCRIPT ACCEPTED BY THE SOCIETY JUNE 1, 1987

Printed in U.S.A.

Index

[Italic page numbers indicate major references]

Typeset by WESType Publishing Services, Inc., Boulder, Colorado
Printed in U.S.A. by Malloy Lithographing, Inc., Ann Arbor, Michigan